AF784820

Springer Series in
MATERIALS SCIENCE 57

Springer
Berlin
Heidelberg
New York
Barcelona
Hong Kong
London
Milan
Paris
Tokyo

Physics and Astronomy ONLINE LIBRARY

http://www.springer.de/phys/

Springer Series in

MATERIALS SCIENCE

Editors: R. Hull R. M. Osgood, Jr. J. Parisi

The Springer Series in Materials Science covers the complete spectrum of materials physics, including fundamental principles, physical properties, materials theory and design. Recognizing the increasing importance of materials science in future device technologies, the book titles in this series reflect the state-of-the-art in understanding and controlling the structure and properties of all important classes of materials.

51 **Microscopic and Electronic Structure of Point Defects in Semiconductors and Insulators**
Determination and Interpretation of Paramagnetic Hyperfine Interaction
Editors: J. M. Spaeth and H. Overhof

52 **Polymer Films with Embedded Metal Nanoparticles**
By A. Heilmann

53 **Nanocrystalline Ceramics**
Synthesis and Structure
By M. Winterer

54 **Electronic Structure and Magnetism of Complex Materials**
Editors: D.J. Singh and A. Dimitrios

55 **Quasicrystals**
An Introduction to Structure, Physical Properties and Applications
Editors: J.-B. Suck, M. Schreiber, P. Häussler

56 **SiO_2 in Si Microdevices**
By M. Itsumi

57 **Radiation Effects in Advanced Semiconductor Materials and Devices**
By C. Claeys and E. Simoen

Series homepage – http://www.springer.de/phys/books/ssms/

Volumes 1–50 are listed at the end of the book.

C. Claeys E. Simoen

Radiation Effects in Advanced Semiconductor Materials and Devices

With 331 Figures

Springer

Prof. Cor Claeys
Dr. Eddy Simoen
IMEC Leuven/Belgium, Kapeldreef 75, 3001 Leuven, Belgium

Series Editors:

Prof. R. M. Osgood, Jr.
Microelectronics Science Laboratory
Department of Electrical Engineering
Columbia University
Seeley W. Mudd Building
New York, NY 10027, USA

Prof. Dr. Jürgen Parisi
Universität Oldenburg
Fachbereich Physik
Abt. Energie- und Halbleiterforschung
Carl-von-Ossietzky-Str. 9-11
26129 Oldenburg, Germany

Prof. Robert Hull
University of Virginia
Dept. of Materials Science and Engineering
Thornton Hall
Charlottesville, VA 22903-2442, USA

ISSN 0933-33X
ISBN 3-540-43393-7 Springer-Verlag Berlin Heidelberg NewYork

Library of Congress Cataloging-in-Publication Data

Claeys, C:
Radiation effects in Advanced Semiconductor Materials and Devices / C. Claeys; E. Simoen. – Berlin; Heidelberg; New York; Barcelona; Hongkong; London; Milan; Paris; Tokyo: Springer, 2002
(Springer series in materials science; v. 57) (Physics and astronomy online library) – Includes biographical references.
ISBN 3540433937 (alk. paper)

This work is subject to copyright. All rights are reserved, whether the whole or part of the material is concerned, specifically the rights of translation, reprinting, reuse of illustrations, recitation, broadcasting, reproduction on microfilm or in other ways, and storage in data banks. Duplication of this publication or parts thereof is permitted only under the provisions of the German Copyright Law of September 9, 1965, in its current version, and permission for use must always be obtained from Springer-Verlag. Violations are liable for prosecution act under German Copyright Law.

Springer-Verlag Berlin Heidelberg New York
a member of BertelsmannSpringer Science + Business Media GmbH

http://www.springer.de

© Springer-Verlag Berlin Heidelberg 2002
Printed in Germany

The use of general descriptive names, registered names, trademarks, etc. in this publication does not imply, even in the absence of a specific statement, that such names are exempt from the relevant protective laws and regulations and therefore free for general use.

Typesetting: Camera-ready copy from the authors
Cover concept: eStudio Calamar Steinen
Printed on acid-free paper SPIN: 10832598 57/3020/kk 5 4 3 2 1 0

Preface

There is a growing tendency for using commercial state-of-the-art microelectronic components for space applications. This is driven, on the one hand, by the so-called custom-off-the-shelf (COTS) approach, where commercial components and circuits are increasingly replacing dedicated expensive radiation hardened electronics. On the other hand, scaling of silicon technologies brings about an inherent hardening against permanent damage, as thin gate dielectrics become less and less prone to it. Furthermore, the use of smart or integrated sensors and MEMS will further stimulate the use of silicon microelectronics in space and other radiation environments, like CERN's future Large Hadron Collider. In fact, in the ultimate limit of scaling, complete systems-on-chip (SOC's) are expected to emerge, combining different technologies and new materials on the same substrate. Furthermore, in order to meet the requirements of the International Technology Roadmap for Semiconductors (ITRS), scaling of the main technology (CMOS) will require the use of novel materials and processing steps. For example, SiGe epitaxy will be implemented more and more for high-speed telecom applications, replacing III-V materials. Novel gate dielectrics (high-k materials) and intermetal layer dielectrics (low-k) will be introduced. Ferroelectrics are becoming of growing interest for memory applications. Device isolation in upcoming technologies will no longer be achieved by LOCOS techniques but requires advanced schemes like Shallow Trench Isolation. For high-speed satellite communication and for on-chip communication, the use of opto-electronics will strongly increase. At the moment, most of the components and systems are based on direct-gap III-V materials, but there is intensive search for silicon-based and silicon-compatible optical interconnects. Powering of satellites is based on solar energy conversion using low-weight high-efficiency tandem solar cells. Currently, the system GaAs on Ge substrates is finding progressive application in satellites. The future use of so-called nano-satellites will trigger the implementation of state-of-the-art microelectronic components and systems.

It is clear from the above that these developments in the semiconductor industry are not driven primarily with space applications or radiation hardness in mind. It is felt, therefore, that there is a need to have a clear view of potential radiation damage problems, even at an early stage of the development of the latest technology generations. This is not only important for the space community itself but can be beneficial during the process/technology development as well. The reason is that during device or circuit fabrication more and more processing steps use an aggressive environment where irreversible radiation damage can occur. So a fundamental understanding of radiation damage mechanisms and degradation is not only of use for the nuclear/space engineer, but may be helpful for the process engineer as well. This monograph is oriented in the first place towards post-graduate researchers who want to enter the field and wish to obtain a good overview of the radiation damage in semiconductor materials and advanced devices. A background in semiconductor and device physics and its interaction with radiation is assumed, although some basic concepts will be briefly summarized. Furthermore, whenever possible, an outlook towards future developments and experimental or modeling needs/shortcomings is provided so that even for the experts in the field, the book could provide significant added value.

The book contains 9 chapters and analyses radiation effects in a variety of semiconductor materials and devices. A kind of justification for the book and a brief discussion of the different radiation environments are addressed in Chap. 1. Information is also given about the component selection strategies for space applications. The basic radiation damage mechanisms in semiconductor materials and devices form the subject of a second chapter. A good fundamental insight into material science and device physics is essential for a proper understanding of the following chapters. Chapter 3 reviews the knowledge related to displacement damage in group IV semiconductor materials such as silicon, germanium and silicon-germanium alloys. Attention is mainly paid to the present understanding of the fundamental mechanisms involved. The potential and drawbacks of several characterisation techniques are outlined whenever appropriate. The device applications of these materials are discussed in later chapters. Due to its importance for both micro- and opto-electronics applications, a fourth chapter is devoted to GaAs. Radiation aspects of silicon bipolar technologies, including vertical bipolar junction transistors (BJTs), lateral transistors and SiGe heterojunction bipolar transistors (HBTs) are critically reviewed in Chap. 5. As already mentioned in the introduction, the key microelectronic technology, which is also driving the activities for scaling down the minimum feature size, is based on CMOS. The corresponding radiation aspects are studied in Chap. 6. Important issues such as ultra-thin gate oxides, alternative gate dielectrics based on nitrided and reoxidised nitrided oxides and device isolation are covered. A special section deals with silicon-on-insulator (SOI) CMOS technologies, as they are no longer limited to niche markets but are also gaining more and more interest for commercial applications. GaAs-based field effect transistors, such as MESFETS and HEMTs and their radiation response are reported in Chap. 7, while the opto-electronic components for space are given attention in Chap. 8. Attention is also paid to different types of components, including light emitting diodes (LEDs), laser diodes, photodetectors and optocouplers. Due to space restrictions not all potential advanced semiconductor materials and devices can be covered in the book. Therefore the last chapter only briefly addresses some hot topics such as non-volatile memories, high-*k* dielectrics for 100 nm and beyond CMOS and SiC and gives an outlook on component requirements for future space applications.

As the advances in the field are appearing so fast, a book can only give the status at a certain moment in time. Therefore the aim was not to look for completeness, but rather to lay a sound physical basis and to give a critical overview of the type of semiconductor materials and devices presently used for microelectronics in a radiation environment and to focus attention on some emerging technologies with a strong potential for use in future space missions.

Over the years a large number of scientists and researchers from all over the world have greatly contributed by their discussions and critical comments to enlarge the knowledge of the authors in the radiation field. The authors are in the first place very grateful to all their past and present IMEC colleagues in the field for stimulating discussions over the years. A special word of thanks has to go to ESTEC who has financially supported the radiation research activities during the past 15 years. A large part of the book is based on numerous discussions with L. Adams, B. Johlander, R. Harboe-Sørensen and A. Mohammadzadeh. The authors also wish to acknowledge Ms. Kathleen Mertens for her support with the scanning of the figures.

Leuven, April 2002

Cor Claeys
Eddy Simoen

Table of Content

List of Symbols

A : device area (cm^2)
A_E : emitter area of a bipolar junction transistor (μm^2)
A_{EX} : extrinsic base area of a bipolar transistor (μm^2)
a_n : carrier removal rate constant (cm^2)
a_0 : lattice constant (Å)
b_c : hopping conductivity parameter ($K^{1/4}$)
b_h : hole trapping yield (cm^{-3})
b_μ : mobility degradation damage constant (cm^2)
c : speed of light ($3x10^8$ m/s)
C : capacitance (F)
C_{CB} : base-collector capacitance (pF)
c_n : capture rate for electrons (cm^3/s)
c_p : capture rate for holes (cm^3/s)
C_{ox} : oxide capacitance per unit area (F/cm^2)
D : dose (rad or Gray)
D^* : responsivity of a photo-detector (A/W)
D_h : hole diffusivity (cm^2/s)
D_H : density of states in the 2-dimensional electron gas ($cm^{-2}V^{-1}$)
D_I : diffusivity of the self-interstitial I (cm^2/s)
D_{it} : density of interface traps ($eV^{-1}cm^{-2}$)
D_n : electron diffusivity (cm^2/s)
d_n : channel depth of a MESFET (nm)
D_{ot} : density of oxide traps ($eV^{-1}cm^{-2}$)
D_p : hole diffusivity (cm^2/s)
d_s : spacer layer thickness of a HEMT (nm)
D_{th} : threshold implantation dose for implant isolation (cm^{-2})
dE/d x: stopping power (MeV cm^2/g)
E : particle energy (keV, MeV)
E_c : minimum of the conduction band (eV)
E_d : threshold energy for atomic displacement (eV)
E_e : electron confinement energy in a QW (meV)
E_{eh} : electron-hole pair ionisation threshold (eV)
E_F : Fermi energy (eV)
E_{fc} : cluster formation energy (eV)
E_G : band gap (eV)
E_h : heavy hole confinement energy in a QW (meV)
E_i : intrinsic level (eV)

E_{ion} : ion energy (keV)
E_m : maximum energy transmitted to the target nuclei (keV)
e_{max} : emission rate window (DLTS) (s^{-1})
E_{ph} : phonon energy (keV, MeV)
E_r : average recoil energy (eV)
E_T : trap energy level (eV)
E_v : maximum of the valence band (eV)
E_x : binding energy of the free exciton (eV)
F : electric field (V/cm)
FF : fill factor of a solar cell
F_{crit} : critical field for single event gate oxide rupture (V/cm)
F_i : electric field at the interface of a heterojunction FET (V/cm)
F_k : critical gate oxide field for radiation-induced leakage current (V/cm)
$f_m(x)$: mismatch parameter of a $Si_{1-x}Ge_x$ alloy
f_{max} : maximum oscillation frequency (GHz)
F_{ox} : oxide field (V/cm)
f_y : hole yield
f_T : cutoff frequency (GHz)
g : degeneracy factor
g_0 : free electron g factor (2.002319)
G_{chan} : channel conductance of a MOSFET (S)
g_d : channel conductance $\partial I_D/\partial V_{DS}$ of a FET (S)
G_{H0} : fitting parameter for the drain current of a MESFET (V^{-1})
g_{ij} : spin second rank tensor
g_m : transconductance $\partial I_D/\partial V_{GS}$ of a FET (S)
GN_{EX} : extrinsic Gummel number of a bipolar transistor (s/cm^4)
GN_{IN} : intrinsic Gummel number of a bipolar transistor (s/cm^4)
h : Planck 's constant ($=6.62617\times10^{-34}$ J s)
$\hbar$: reduced Planck's constant $h/2\pi$ ($=1.05458\times10^{-34}$ J s)
$\boldsymbol{H}$: magnetic field
H : magnitude of magnetic field (G)
h_c : critical thickness of an epitaxial layer (nm)
i_0 : inclination of a satellite orbit with respect to the Earth's axis (degrees)
I_B : base current of a bipolar transistor (A)
I_{bias} : operation current of a LD or a LED (mA)
I_C : collector current of a bipolar transistor (A)
I_D : drain or channel current (A)
I_{Do} : dark current of a photodiode (A)
I_{DSAT} : saturation drain current of a FET (A)
I_e : excess gate current due to radiation-induced leakage current (A)
I_F : forward current (A)
I_G : gate current (A)
I_{Lo} : photocurrent of a photodiode (A)
I_{mg} : midgap current (A)
$\boldsymbol{I_N}$: magnetic moment of the nucleus (vector)
I_{op} : operation current of a laser or light emitting diode (mA)

I_{PL} : photoluminescence intensity (a.u.)
I_R : reverse current (A)
I_{RSB} : excess gate current due to radiation-induced soft breakdown (A)
I_{sc} : short circuit current of a solar cell (A)
I_T : threshold current (A)
I_{th} : threshold current for lasing action (A)
J_e : excess gate current density (A/cm^2)
J_{EX} : extrinsic base current density (A/cm^2)
J_F : forward diode current density (A/cm^2)
J_G : gate current density (A/cm^2)
J_R : reverse diode current density (A/cm^2)
J_+,J_- : hole (+), electron (-) current density (A/cm^2)
k : Boltzmann's constant (1.38066x10^{-23} J/K)
K : damage factor (a.u.)
K_β : damage factor for the static current gain of a bipolar transistor (cm^2)
K_B : boron deactivation damage factor (cm^{-1})
K_{Do} : dark current damage factor of a photodiode (A cm^2/particle)
K_I : leakage current damage factor (A cm^2/particle)
K_{Io} : laser diode threshold current damage factor (cm^2/particle)
K_J : leakage current density damage factor (A/particle)
K_L : diffusion length damage factor
K_{Lo} : photocurrent damage factor of a photodiode (A cm^2/particle)
K_μ : mobility damage factor (Vs)
K_n : carrier removal rate (cm^{-1} per particle)
K_N : carrier removal rate for neutron irradiation (cm^{-1} per neutron)
K_P : carrier removal rate per proton (cm^{-1} per proton)
K_{PL} : photoluminescence damage factor (a.u.)
k_s : technology scaling factor (dimensionless)
K_T : pre factor in the emission rate (K^{-2} s^{-1})
K_τ : lifetime damage factor (cm^2 s/particle)
L : transistor length (μm)
l_c : laser diode cavity length (μm)
L_D : extrinsic Debije length (μm)
L_E : emitter length of a bipolar transistor (μm)
LET : linear energy transfer of ionising radiation (MeV/cm^2/mg)
L_{eff} : effective channel length of a MOSFET (μm)
L_n : electron diffusion length (cm)
L_o : light output of a laser diode or a light emitting diode (a.u.)
L_p : hole diffusion length (cm)
m : particle mass (g)
m_e : electron mass (0.91095x10^{-30} kg)
M_2 : atomic mass (g)
n : free electron or carrier density (cm^{-3})
$n^\uparrow$: number of spin up electrons
$n_\downarrow$: number of spin down electrons

N : Avogadro's number (6.02204×10^{23} mol^{-1})
N_A : acceptor density (cm^{-3})
N_A^* : surface acceptor density in a HEMT (cm^{-2})
N_{AA} : deep acceptor density (cm^{-3})
N_{at} : Atomic density of the target material (at/cm^3)
N_D : donor or doping density (cm^{-3})
N_{DD} : deep donor density (cm^{-3})
N_{eff} : effective doping density (cm^{-3})
NF : high frequency noise figure (dB)
NF_{min} : minimum noise figure (dB)
n_i : intrinsic carrier concentration (cm^{-3})
n_{if} : ideality factor of a junction diode (usually $1 \leq n_{if} \leq 2$)
N_{it} : interface trap surface density (cm^{-2})
$N_{itacceptor}$: acceptor-like interface trap surface density (cm^{-2})
$N_{itdonor}$: donor-like interface trap surface density (cm^{-2})
NM : noise margin of an inverter (V)
NM_{HI} : high noise margin of an inverter (V)
NM_{LO} : low noise margin of an inverter (V)
N_{ot} : oxide trap surface density (cm^{-2})
n_s : electron density in the 2-DEG of a HEMT (cm^{-2})
N_s : doping density at the surface (cm^{-3})
N_T : trap density (cm^{-3})
p : free hole density (cm^{-3})
P : stopping number of the material
P_a : amorphous volume generation rate ($cm^{-3}s^{-1}$)
P_E : perimeter of the emitter of a bipolar transistor (μm)
P_L : light power of a LD or a LED (W/cm^2)
P_{max} : maximum solar cell power (W)
P_p : point defect production parameter (…)
q : electron charge in absolute value (1.60218×10^{-19} C)
R : parameter describing the recombination of point defects from different damage cascades (cm^2)
r_B : base resistance (Ω)
r_{BE} : total base-emitter resistance (Ω)
R_{chan} : channel resistance of a FET (Ω)
R_E : emitter resistance (Ω)
R_{out} : output resistance of a FET (Ω)
R_p : projected range of a particle (μm)
R_s : surface recombination rate (cm^3/s)
R_{speak} : peak surface recombination rate (cm^3/s)
R_{SD} : series resistance of a FET (Ω)
R_T : trap introduction rate (cm^{-1})
R_{tot} : total resistance of a FET (Ω)
s : ion implantation damage ratio
S : magnetic dipole moment of the electron spin (Am^{-2})

S_d : deposited non-ionising energy (keV cm^2/g)
s_g : surface generation velocity (cm/s)
s_r : surface recombination velocity (cm/s)
t : time (s)
T : absolute temperature (K)
T_0 : temperature above which no ion implantation damage is formed (K)
T_A : annealing temperature (K)
t_{BD} : time to breakdown of a gate oxide (s)
T_{IRR} : irradiation temperature (K)
T_{max} : DLTS peak maximum temperature (K)
t_{ox} : oxide thickness (nm)
v : particle velocity (cm/s)
V_{A1} : part of the threshold voltage of a HEMT (V)
V_B : substrate voltage (V)
V_{BE} : base-emitter voltage (V)
V_{bi} : built-in potential of a junction (V)
V_{cl} : damage cluster potential barrier (V)
V_d : full depletion voltage of a radiation detector (V)
V_{DS} : drain-to-source voltage (V)
V_F : forward diode bias (V)
V_{FB} : flat-band voltage (V)
V_{FD} : full depletion voltage of a detector (V)
V_G : gate voltage (V)
V_{GS} : gate-to-source voltage (V)
V_{in} : input voltage of an inverter (V)
V_{it} : voltage related to the interface-trapped charge (V)
V_{mg} : gate voltage corresponding with the mid gap current (V)
V_{oc} : open circuit voltage of a solar cell (V)
V_{off} : cut-off voltage of a HEMT (V)
V_{OT} : voltage related to the positive (or negative) oxide charge (V)
V_{out} : output voltage of an inverter (V)
V_P : bias during filling pulse (DLTS) (V)
V_{Po} : pinch-off voltage (V)
V_R : reverse diode bias (V)
v_{sat} : saturation velocity of carriers (cm/s)
V_{so} : stretch-out voltage (V)
V_T : threshold voltage (V)
v_{th} : thermal velocity of carriers (cm/s)
V_{tr} : transition voltage for an irradiated bipolar transistor (V)
W : depletion width (μm)
W_E : emitter width of a bipolar transistor (μm)
W_{eff} : effective width of a MOSFET (μm)
W_{IB} : intrinsic base width (μm)
x_j : junction depth (μm)
Z, Z_1, Z_2: atomic numbers or charges

Z_W : width of a FET (μm)

List of Greek Symbols

α : absorption coefficient (cm^{-1})
α_0 : model parameter threshold voltage HEMT (=$2.5x10^{-12}$ $eVm^{4/3}$)
α' : model parameter saturation drain current MESFET (V^{-1})
α_{bb} : band-to-band contribution to the absorption coefficient (cm^{-1})
α_c : cavity loss coefficient (cm^{-1})
α_{ex} : excitonic contribution to the absorption coefficient (cm^{-1})
α_{it} : interface trap scattering parameter (cm^2)
α_{ot} : oxide trap scattering parameter (cm^2)
β : current gain of a bipolar transistor
β' : model parameter saturation drain current MESFET ($AV^{-\kappa}$)
β_e : Bohr magnetron ($9.274015x10^{-28}$ J/G)
β_{pk} : peak current gain
ΔE_c : conduction band offset at the heterojunction of a HEMT (eV)
ΔH_T : trap level enthalpy (eV)
$\Delta I_{g.s}$: surface generation current (A)
ΔI_B : change in base current (A)
ΔI_C : change in collector current (A)
$\Delta L, \delta L$: channel length change or reduction (μm)
ΔQ_{ot} : radiation-induced trapped oxide charge (C/cm^2)
ΔS_T : trap level entropy (eV/K)
ΔV_{fb} : change in flat band voltage (V)
ΔV_{GIDL} : change in the gate-induced drain leakage voltage (V)
ΔV_{it} : change of the threshold voltage due to radiation-induced interface traps (V)
ΔV_{ot} : change in the threshold voltage due to radiation-induced oxide-trapped charge (V)
ΔV_T : change in threshold voltage (V)
ε_g : permittivity of GaAs (F/cm)
ε_{ox} : permittivity of silicondioxide (F/cm)
ε_{si} : permittivity of silicon (F/cm)
ϕ : particle flux ($particles/cm^2$ s)
Φ : particle fluence ($particles/cm^2$)
ϕ_b : Schottky barrier height (V)
ϕ_m : metal workfunction (V)

Γ_{bb} : broadening parameter of the band-to-band part of the absorption coeffi cient
Γ_x : broadening parameter of the excitonic part of the absorption coefficient
η : conversion efficiency solar cell (%)
η_o : slope efficiency of a laser or light emitting diode (mW/mA)
λ : wavelength of light/photon (nm)
λ_o : decay constant of plasma damage concentration profile (nm)
λ_H, λ_{H0} : fitting parameter saturation drain current MESFET (V^{-1})
μ, μ_0 : carrier mobility (cm^2/Vs)
μ_m : magnetic dipole moment (vector)
μ_{eff} : effective mobility (cm^2/Vs)
μ_H : magnetic dipole moment (A^{m-2})
μ_{max} : maximum mobility (cm^2/Vs)
μ_n : electron mobility (cm^2/Vs)
ν, ν_0 : frequency, resonance frequency (s^{-1})
ρ : resistivity (Ω cm)
ρ_f : facet reflectance
ρ_m : mass density of target (g/cm^3)
σ : capture cross section (cm^2)
σ_c : conductivity (Ω^{-1} cm^{-1})
σ_n : capture cross section for electrons (cm^2)
σ_p : capture cross section for holes (cm^2)
σ_s : capture cross section of interface traps (cm^2)
τ : lifetime of a LD or a LED (ns)
τ_A : annealing time constant (s)
τ_c : capture time constant (DLTS) (s)
τ_e : emission time constant (DLTS) (s)
τ_g : generation lifetime (s)
τ_{nr} : non-radiative recombination lifetime (s)
τ_o : initial lifetime of a LD or a LED (ns)
τ_r : recombination lifetime (s)
τ_{ra} : radiative recombination lifetime (s)
ω : radial frequency ($=2\pi\nu$) (s^{-1})
Ψ_s : surface potential (V)

1 Radiation Environments and Component Selection Strategy

1.1 Introduction

The amount of radiation that semiconductor devices and materials encounter during their lifecycle strongly depends on the radiation environment and their operating conditions. For space missions and military applications it is obvious that there is a radiation-harsh environment. However, also during their fabrication process and even for standard terrestrial operation the devices may suffer from ionising radiation. Therefore, Sect. 1.2 briefly reviews the different radiation environments and points out the typical particle spectrum that the devices will face. Only some essential features are addressed as whenever needed more details are given in the different chapters. Another important aspect is the trend to use more and more the Custom-Off-The-Shelf (COTS) approach for designing systems to be used in satellite and space programs. This COTS approach, which to some extent was the driving force for writing this book, is treated in Sect. 1.3, dealing with the component selection strategy.

1.2 Radiation Environments

In general, one can differentiate between the following different environments:

- Space
- High-energy physics experiments
- Nuclear
- Natural environments
- Processing-induced radiation

Each of these environments is characterized by its own spectrum of particles and energy distribution. There may be an interaction between the different environments as for instance the inherent built-in damage during the device fabrication may impact the device performance during a space mission. Detailed discussions on radiation environments can be found in [1 - 2]. Within the scope of the present book only a brief overview will be given of some key aspects.

1.2.1 Space Environments

The main sources of energetic particles in a space environment are

- Protons and electrons trapped in the Van Allen belts
- Heavy ions trapped in the magnetosphere
- Cosmic ray protons and heavy ions
- Protons and heavy ions from solar flares.

The radiation level of these sources strongly depends on the activity of the sun. The solar cycle is normally divided in two main activity phases, i.e., the solar minimum and the solar maximum. On the average the cycle lasts for eleven years with approximately four years of solar minimum and seven years of solar maximum. For the cycle itself a minimum of 9 and a maximum of 13 years has been reported [3]. The available models of the radiation environment have to take this into account.

The Van Allen belts, first discovered by *Van Allen and coworkers* on Explorer I, consist mainly of electrons up to a few MeV in energy and protons of up to several hundred MeV trapped in the earth's magnetic field [4]. There is only a small percentage of heavy ions such as O^+. At low altitudes (from 100s of km to 6,000 km) one speaks about the 'inner' belt, while at high altitude (up to 60,000 km) the 'outer' belt with high-energy electrons is observed. Once the charged particles are trapped the Lorentz force controls their motion in the earth magnetic field.

From a radiation viewpoint, the total ionising dose (TID) is a severe problem as it may increase to several hundred of kilorads during the lifetime of a space mission. For a detailed calculation of the dose one has to take into account the altitude and inclination of the satellite, and the shielding used. Protons are especially found in the 'inner' Van Allen belt, which is important for e.g. the MIR station and shuttle flights. Beside TID, another phenomenon is single event upsets (SEU). A good review of the problems with existing models of the radiation belt, such as e.g. the AP-8 and AE-8 whereby 'P' refers to protons and 'E' to electrons and '8' is the version number, has been given by *Daly et al.* [5]. Originally, these models (which are designed in both a solar minimum and a solar maximum version) were developed for simple total dose applications and not for the more complex problems related to single event upsets, charging/discharging, or non-orbit mitigation. In the last decade a huge amount of data has been collected from space missions or dedicated satellites, allowing a refining or an update of the used models [6 - 8]. For low altitude missions, special attention has to be given to the South Atlantic Anomaly (SAA) which is the inner edge of the inner zone [3]. The offset and tilt of the geomagnetic axis with respect to the earth rotation axis cause the anomaly, whereby the particle fluxes are strongly anisotropic depending on the gyro-radius and the particle energy. The standard models like AP-8 fail to predict SAA anomalies [9]. For missions where the spacecraft has random orientations, the effect is averaged out.

The galactic cosmic ray particles are originating outside the solar system and include a large range of elements from atomic number 1 to 92. The flux levels are low, but because they include high energetic particles (10s of MeV up to 100s of GeV) of heavy elements they may produce ionisation effects by passing thought the material. In this case shielding is not obvious and they will mainly cause Single Event Effects (SEE). There is an association with the solar cycle, with a peak flux level during the solar minimum. Also in this case, the altitude and the inclination of the aircraft are important parameters. Commonly used simulators including cosmic ray models are CRÈME and CHIME [10]. The composition of particles consists of about 85% protons, 14% alpha particles and 1% heavy ions [1].

Solar flares cause protons and heavy ions. These flares, which can last for several hours to a few days with proton energies higher than 100 MeV, mainly occur during the solar maximum period and are attenuated by the earth's' magnetosphere. Different models have been developed to predict the impact of the protons and heavy ions [10]. At a system level, one may not forget that during a solar flare or substorm different parts of the spacecraft can be differently charged leading sometimes to potential differences exceeding 10 kV. Especially for solar cell arrays this can lead to a dangerous situation as mostly the array overglasses are the only non-conducting surface on the satellite.

Essential to calculate the radiation environment encountered during a space mission are the altitude of the orbit and the inclination of the spacecraft or satellite [12]. For low earth orbits (LEO) the satellite passes several times a day through the Van Allen belts, i.e., trapping electrons and protons. The highest flux level increase is for inclinations in the range $0° < i_0 < 30°$, whereas it more gradually increases up to 60°. For higher inclination angles the impact on the flux level is marginal. The largest variations with altitude are between 200 and 600 km.
Highly elliptical orbits (HEO) are very similar to LEO orbits. However, as their altitude is more than 30.000 km, they are also suffering long exposures to cosmic and solar flares environments. For geostationary orbits (GEO) only trapped protons with energy levels below the threshold for initiating nuclear reactions are present. In the case of planetary and interplanetary missions the radiation environment is much more complex and depends on the number of times the satellite passes through the earth's' radiation belt and on how close it comes to the sun. In addition, the solar maximum and minimum has to be taken into account. An overview of the radiation sources in a space environment is given in Table 1.1.

A rather special environment compared to the one in orbit is encountered during a Mars mission ,such as for example the Pathfinder and the Surveyor [13-14]. In particular the low atmospheric pressure, the atmospheric dust and the lower operating temperatures can play an important role.

1.2.2 High-Energy Physics Experiments

This radiation environment is characterized by high intensity and high rate particles, such as e.g. in the ATLAS, CMS and LHC-B experiments at the Large Hadron Collider (LHC) at CERN. In the high-energy physics experiments electron or

Table. 1.1. The radiation source in a space environment. (After LaBel et al. [12])

Radiation source	Models	Effect of Solar Cycle	Variations	Type of Orbits affected
Trapped protons	AP8-Min AP8-Max	Solar Min-Higher; Solar Max-Lower	Geomagnetic Field, Solar Flares, Geomagnetic Storms	LEO, HEO, Transfer orbits
Trapped electrons	AES-MIN AES-MAX	Solar Min-Lower, Solar Max-Lower	Geomagnetic Field, Solar Flares, Geomagnetic Storms	LEO, HEO, Transfer orbits
Galactic Cosmic Ray ions	CRÈME, CHIME	Solar-Min-Lower, Solar Max-Lower	Ionisation Level, Orbit Attenuation	LEO, GEO, HEO, Interplanetary
Solar Flare Protons	King, JPL92	During solar max only	Distance from Sun, Outside 1 Au, Orbit attenuation, location of flare of sun	LEO (i_0>45°), GEO, HEO, Interplanetary
Solar Flare Heavy Ions	CRÈME JPL92 CHIME	During solar max only	Distance from Sun, Outside 1 Au, Orbit attenuation, location of flare of sun	LEO, GEO, HEO, Interplanetary

1 Au = 1 Astronomical unit = 1.496×10^{11} m

proton beams having energies as high as 500 GeV are used to study elementary particles with a short lifetime such as e.g. bosons, muons and quarks. The primary hadron-hadron interactions produce a high flux of charged particles, gammas and neutrons, with an expected overall integrated fluence of about 1.6×10^{14} 1 MeV equivalent neutrons cm^{-2}. This imposes severe constraints on the radiation hardness of the instrumental semiconductor electronics. Although some radiation aspects will be discussed in the following chapters, a good overview of the status of the research activities in the field can be found in [15].

1.2.3 Nuclear Environment

In this case reference is made to the nuclear reactors, whereby one has to differentiate between i) the reactor core, ii) the environment of the reactor, and iii) the possibility of accidents. In this case one mostly deals with gamma and neutron irradiation. Compared to the space environment, very high doses and fluxes are involved. In most of the cases, special shielding is used. This special field is falling out of the scope of the book and will therefore not be addressed.

1.2.4 Natural Environment

A significant source of ionising radiation in packaged devices is due to alpha particles coming from the radioactive impurities, such as uranium and thorium and their daughter products, in the used materials. Additional sources may be the lead-based isotopes in solder bumps in case of a flip-chip technology, the gold used for the bond wires and lid plating, the alumina in ceramic packages, lead-frame alloys and interconnect metallizations [16]. These alpha particles lead to soft error rate (SER) in logic circuits [17 - 18]. Especially for scaled down devices even a low flux can already be harmful. Radioactive impurities can also be introduced by the reactants used during the device processing steps [19]. Boron is commonly used as p-type dopant for junction formation and to influence the stress and flow properties of SiO_2 layers, such as in borophosphosilicate glass (BPSG). Boron has two isotopes ^{10}B and ^{11}B, of which ^{10}B is unstable when exposed to thermal neutrons coming from cosmic rays. The reaction process is schematically shown in Fig. 1.1 [20]. In the $^{10}B(n,\alpha)^7Li$ reaction the lithium nucleus is emitted with a kinetic energy of 0.84 MeV 94% of the time and 1.014 MeV 6% of the time. The gamma photon has an energy of 478 keV, while the alpha particle is emitted with an energy of 1.47 MeV.

Depending on the process parameters of the BPSG layer, SRAM devices fabricated in a 0.25 μm technology showed an order of magnitude greater SER than similar non-BPSG devices [16]. This may be especially important for low altitude aircrafts or for avionics in general [21]. Upset mechanisms occur also in DRAM in both the cell-mode and the bitline-mode and are often studied by using nuclear microprobes with focussed high-energy proton or helium ions [22 - 23]. The use of a retrograde p-well with double buried p^+ layers is giving a higher SER immunity than devices processed on conventional p/p^+ epitaxial material. In addition, for future deep submicron DRAM generations, optimal SER suppression is expected for devices made on Silicon-on-Insulator (SOI) technologies [23].

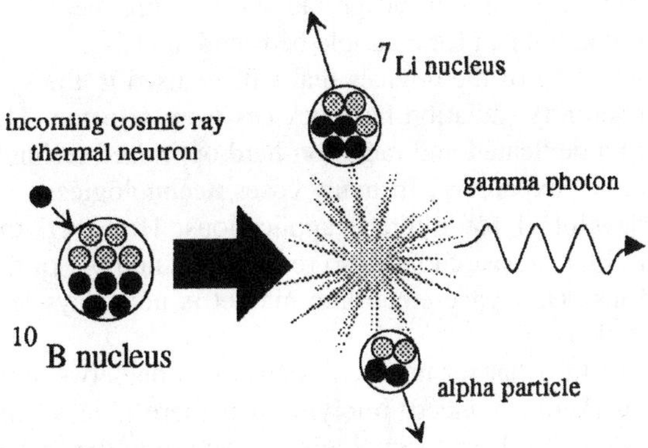

Fig. 1.1. Fission of 10B induced by the capture of a neutron. [After Baumann et al. [20]]

1.2.5 Processing-Induced Radiation

The fabrication of modern semiconductor devices relies on a great number of processing steps which can introduce radiation-damage [24]. Typical examples are ion implantation, dry etching, resist stripping, e-beam or X-ray lithography, plasma enhanced chemical vapor deposition, ion milling, sputtering of barriers and metal layers, etc.. Although the degree of introduced damage is much lower than for the other radiation environments mentioned above, state-of-the-art devices are very vulnerable to it. This is especially the case for ultra-thin gate dielectrics and spacer oxides, a topic that is in depth discussed in Chap. 6. However, also in the other chapters special attention will be given to the impact of processing-induced radiation damage. From a physics viewpoint, in many cases the same modeling approaches can be applied. In some cases, the understanding and control of radiation damage is even crucial for process optimisation. As will be seen in Chap. 3, this is the case for the so-called Transient Enhanced Diffusion (TED) phenomenon, which interferes with the formation of (ultra)-shallow junctions.

1.3 Component Selection Strategy

To design a system for operation in a radiation environment a large variety of parameters have to taken into account. In the case of a space application, one first of all needs details on the mission itself, including the type of mission, the altitude and latitude, when and where the mission will fly, how long it will last in space, which are the worst conditions that may be expected, which are the critical issues for the success of the mission, the amount of shielding that can be used, etc. Based on that one can simulate the radiation environment and predict which type of radiation artifacts the devices will encounter. The more experimental data available the better the simulation will be and the better one can determine the terrestrial type of radiation testing required. A discussion of the radiation testing methodology is out of the scope of this book, but can for example be found in [1].

The following step is the selection of the devices that will be used in the system. In the early days, only inherently radiation-hard devices were selected. The fact that they were fabricated in a dedicated and radiation-hard optimised technology made these components rather expensive. In many cases, technologies were used which were originally developed for military applications. However, the strong drop in investments in military research has also reduced the number of the so-called radiation-hard foundries. The space electronic market is nowadays less than 0.5% of the commercial market.

Another important trend is the fact that for many missions the time between its concept and the launching is markedly reduced, implying that there is less time available to develop dedicated radiation-hard technologies. In addition, the spacecraft should be as light as possible, as small as possible, consuming a low-power level and be more and more powerful in performance. Features such as e.g. imaging, real-time data analysis, and on-board data processing are no longer options but requirements. This leads to the need to use state-of-the-art devices, whereby a natural choice is to switch over to Custom-Off-The-Shelf components, made in mainstream deep submicron CMOS technologies. That is one of the reasons why during the last decade a lot of attention has been given to the testing of commercial devices. An illustration of how successful commercial technologies are utilised in spacecraft applications has been discussed in [12, 25, 26]. NASA has developed a roadmap for radiation effects research in emerging microelectronics and photonics [27]. The basis of their radiation tolerant roadmap leading to radiation-hard, high-performance electronic and photonic systems is illustrated in Fig. 1.2. The interaction between each of the blocs, such as e.g. simulation, on-ground testing and validation with real orbit data, is very crucial. The use of COTS implies that risk analysis has become an important element in the system design cycle. To facilitate the choice of the different devices, work has started to develop an integrated web database for radiation hard components (see e.g. [28]).

The use of COTS has also some drawbacks. State of the art devices have nowadays a rather short lifecycle, as every 2 to 3 years a new submicron technology is being introduced [29]. Therefore attention has to be given to the long-term availability of the components, as otherwise the radiation testing has to be repeated. Even during the period when the devices are commercially available on the market, there is a chance that the manufacturer has slightly modified the fabrication process in order to achieve the electrical specifications with a much higher reproducibility and yield. This may lead to differences in the radiation performance for nominally identical device types but ordered at different times. The only

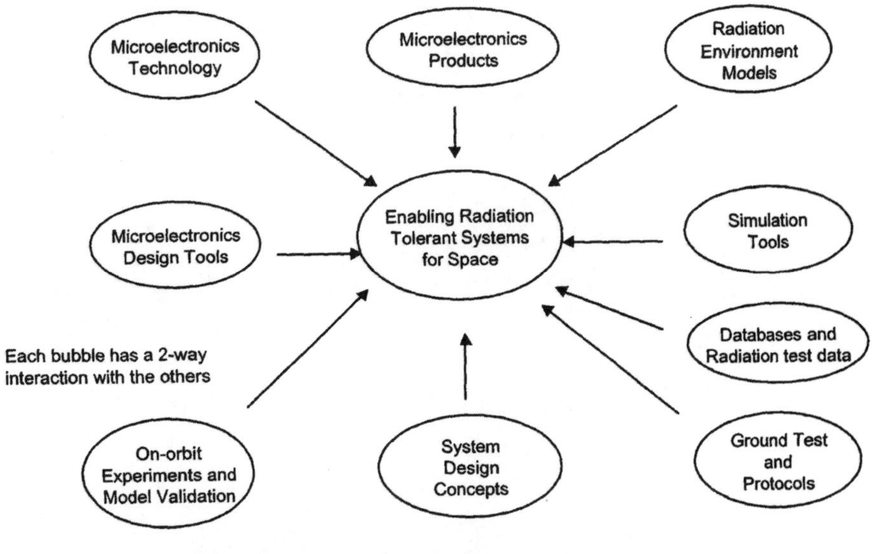

Fig. 1.2. Illustration of a radiation tolerance roadmap. [After Label et al. [27]]

solution to the test problems, which may otherwise result in additional ground-testing, is a very close interaction between the manufacturer and the user

It is important to remark that the COTS approach is referring to standard technologies for commercial devices. However, by using silicon foundries it always remains possible to implement in standard technologies design methodologies to improve the radiation hardness. In this case one could speak about a partial COTS approach as the technology is commercially available but the design has been dedicated for space applications. However, hardness-by-design techniques mostly result into an area penalty [30]. Typical design examples are implementing guard bands to prevent latch-up, the use of edgeless transistors, and built-in redundancy to mitigate SEU [31].

1.4. Conclusions

This chapter has given a brief overview of the different radiation environments and their peculiarities. Once the radiation environment is known, the impact on the device performance can be predicted. As more and more field data is becoming available the simulation tools can be improved leading to a much higher confidence level. The device selection strategy has been addressed and the advantages and disadvantages of the COTS approach discussed. The overall goal remains to achieve more performing systems at a lower cost and to minimize the system design time and radiation testing.

2 Basic Radiation Damage Mechanisms in Semiconductor Materials and Devices

2.1 Introduction

The semiconductor component and circuit operation in a radiation environment, as outlined in Chap. 1, is generally subjected to a mix of particles and photons, with quite a range of energies. Upon its trajectory through the material, a high-energy nuclear particle or photon may lose its kinetic energy in different ways, thereby creating various types of damage. This chapter describes the different interaction and damage mechanisms. Roughly speaking, they can be divided into ionisation and displacement damage effects, which are discussed in the Sect. 2.2. The impact of both damage mechanisms on material and device parameters and characteristics is the subject of Sect. 2.3. Examples are given for the case of silicon and the silicon-silicondioxide (Si-SiO_2) interface, as they are the dominant systems in advanced microelectronics and, therefore, also the best studied. Whenever relevant, results for other semiconductor materials (Ge, GaAs,...) are given. Specific damage mechanisms for III-V and opto-electronic components are addressed in the chapters dealing with these applications. In Sect. 2.4, a brief description is provided of the most powerful spectroscopic techniques, enabling a microscopic study of the radiation-induced defects. Final conclusions are drawn in Sect. 2.5.

2.2 Fundamental Damage Mechanisms

This section defines the fundamental interactions between an energetic particle and a semiconductor device. A distinction can be made between ionisation damage, on the one hand, which creates free electron-hole pairs by disrupting electronic bonds and displacement damage, on the other, giving rise to atoms which are displaced from their usual lattice site, leaving behind a vacancy denoted by V. The former mechanism requires generally far lower energies than is necessary for displacement damage. A further distinction can be made between prompt or fast effects, like Single Event Upsets (SEU) or Latch-up (SEL) and more permanent (hard) damage. In addition, there exists a range of transient effects, which will be described in the following to some extent. A more detailed description of the damage mechanisms can be found in text books [1, 2, 3].

2.2.1 Nomenclature

An energetic photon or particle (ion) travelling through a solid-state system can lose its energy in various ways, whereby the amount of energy deposited not only depends on the projectile mass and energy but also on the atomic number and mass of the target material. The energy deposited per gram of material is called the dose (D) and is expressed in rads or Grays, whereby:

$$1 \text{ rad} = 100 \text{ erg/g} = 6.24\text{x}10^{13} \text{ eV/g} \tag{2.1}$$

and

$$1 \text{ Gray} = 1 \text{ J/kg} = 100 \text{ rad} \tag{2.2}$$

It is clear from the definition that the energy loss also depends on the density of the target so that one usually speaks about rad(Si) or rad(SiO_2) for silicon and its oxide, respectively. As conversion factor one uses 1 rad(Si) = 0.58 rad(SiO_2) = 0.94 rad(GaAs). The particle density impinging on the material per unit of area and time is called the flux (ϕ, in units of particles/cm^{2}s), while the flux integrated over the total exposure time is named the fluence Φ, expressed in particles/cm^2.

2.2.2 Ionisation Damage

Ionisation damage is the dominant mechanism when energetic photons (γ and X-rays) interact with solid-state matter. Depending on the photon energy E_{ph}, different fundamental interactions can take place, as shown in Fig. 2.1 [4], which result in the creation of an energetic free carrier (or pair) in the material. At the lower end of the spectrum (soft X-rays of a few keV) complete absorption of the photon energy by the emitted electron occurs, which is the well-known photoelectric effect. In case of a semiconductor, a bonding valence band electron is emitted into the conduction band across the band gap E_G, leaving behind a free hole. The net result is the creation of a free electron-hole pair. In silicon, an average energy E_{eh} of 3.6 eV is required, while for SiO_2, the threshold is 17 to 18 eV/pair [5], due to the larger band gap energy.

For intermediate photon energies the Compton effect dominates, whereby an incomplete absorption of the photon energy occurs, i.e., a secondary lower energy photon is emitted in addition. If the energy is above the threshold, this photon can again be absorbed to generate a second electron-hole pair, etc. As a result, an average number of E_{ph}/E_{eh} carrier pairs will be generated in the semiconductor material, which is proportional to the original photon energy. This forms the basis of nuclear-radiation spectroscopy using semiconductor detector diodes. Usually, a reversely-biased high-resistivity germanium or silicon p-n junction is used for that purpose, whereby the electric field in the depletion region which extends over most of the detector volume separates the holes from the electrons.

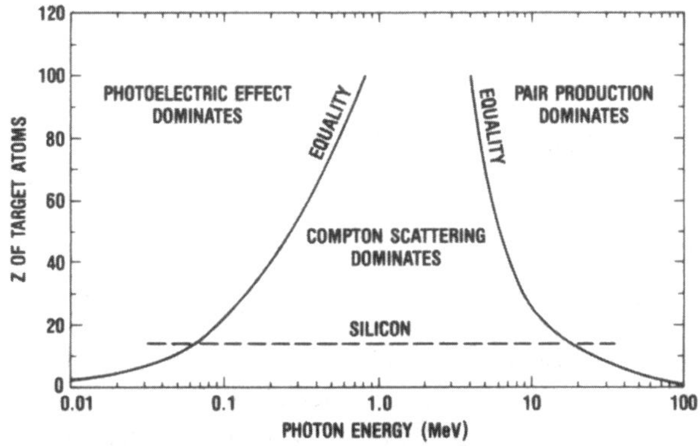

Fig. 2.1. Illustration of the relative importance of the three photon interactions with target materials. The solid lines correspond to equal interaction cross sections for neighboring effects. The dashed line represents the situation for silicon (Z=14). [After McLean and Oldham [4]]

Finally, for energies above the threshold of 1.02 MeV, electron-positron pair production will become important. The limits of equal probability are represented in Fig. 2.1. The probability for the different processes is generally described by the interaction cross section, which can show strong discontinuities (e.g., thresholds) as a function of the photon/particle energy. It should be noted that electron-hole pair formation is also an important energy loss mechanism for high-energy electron and ion irradiation, while it is of secondary importance for neutron exposures.

The amount of energy deposited in the material through ionising interactions is determined by the Linear Energy Transfer (LET) function ρ_m^{-1} dE/dx (in MeV cm^2/g), where ρ_m is the density of the material and E the radiation energy; dx is an elementary trajectory in the material. The stopping power is a function of the target material, the type of particle and the its energy and can be theoretically expressed by the following formula [3]:

$$-\frac{dE}{dx} = 2Pq^4 Z_1^2 Z_2 N_{at} \frac{M_2}{m} \frac{1}{E} \ln\left(\frac{4E}{E_{eh}}\right) \tag{2.3}$$

where Z_1 and Z_2 are the atomic charges of the incident particle and the target material, N_{at} is the atomic density of the target, M_2 the corresponding atomic mass, m and E are the mass and energy of the incident particles and E_{eh} is the mean ionisation energy, defined above. Furthermore, q is the electron charge in absolute value and P the stopping number of the material, which rises slowly with the energy. It has experimentally been demonstrated that the average stopping power in silicon is given by [6]:

$$\frac{dE}{dx} = \frac{80}{(v/c)^2} \tag{2.4}$$

where c is the speed of light and v is the particle velocity.

2.2.3 Displacement Damage

For sufficiently high particle energies, the energy transferred during an elastic or inelastic nuclear collision may be large enough to knock an atom from its lattice site. This creates a vacancy (V) and an atom in an interstitial (I) position. Such one-atom disorder in a crystalline lattice is called a point defect. Group IV elemental semiconductors possess a diamond cubic lattice; for silicon, the lattice parameter is 0.543 nm. Possible structures for the silicon vacancy and interstitial are represented in Fig. 2.2 [7]. It can be deduced that depending on the charge state of the point defect, a slightly different configuration results. As will be seen in more detail below, these primary radiation defects are highly mobile at room temperature and will, therefore, migrate a long distance. They can either disappear from the material, by recombination at the substrate or other sinks or become trapped by impurity atoms, giving rise to more stable secondary defects or to defect complexes.

The probability for such atomic displacements is higher for higher E and m. In principle, high energy photons can not directly produce such events. However, for E_{ph}>170 keV energetic back-scattered Compton electrons may form their own damage cascade in the semiconductor lattice. This explains why also MeV γ-irradiations have been considered for lifetime control in silicon power devices [8]. On the other hand, for energetic ions, only a fraction ($\leq$0.1 %) of the deposited energy goes into displacements (Fig. 2.3), which may result from different types of interactions: Coulombic Rutherford type for ions and elastic nuclear interactions, each having their own probability cross section. For very high energies, also inelastic nuclear collisions occur. In the latter case, the energetic particle is absorbed by a lattice nucleus, which induces the emission of energetic nucleons (alpha particles and photons) and the associated nuclear reactions/decay. The contribution of the different processes is illustrated in Fig. 2.4 for the case of proton irradiation in silicon [9]. The part of the energy that goes into lattice damage is often called Non Ionising Energy Loss (NIEL) [10].

For neutron irradiations, on the other hand, the ionising and Rutherford-type of interactions can be neglected in the first instance as most of the neutron energy loss goes into elastic nuclear displacements. If inelastic neutron absorption takes place, so-called neutron transmutation doping (NTD) can occur. For silicon, the following reaction is used in practice [11]:

$$^{30}\text{Si}(n,\gamma) \rightarrow {}^{31}\text{Si} \rightarrow {}^{31}\text{P}+\beta^- \tag{2.5}$$

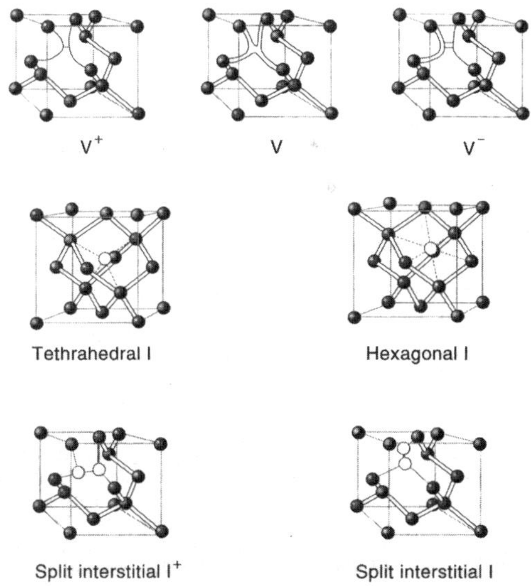

Fig. 2.2. Vacancy, interstitial and split interstitial point defects in three dimensions. Vacancy in the positive, neutral and negative charge state; tetrahedral and hexagonal interstitial; positive and neutral split interstitial defects. [After Privitera et al. [7]]

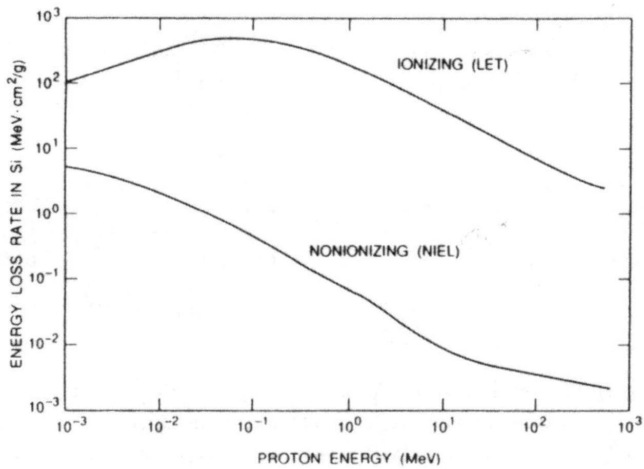

Fig. 2.3. The energy loss rate through ionisation and excitation of the Si lattice (LET) and through atomic displacements (NIEL) versus of proton energies. Only a fraction of one per cent of the energy loss goes into displacement processes. [After Dale et al. [10]]

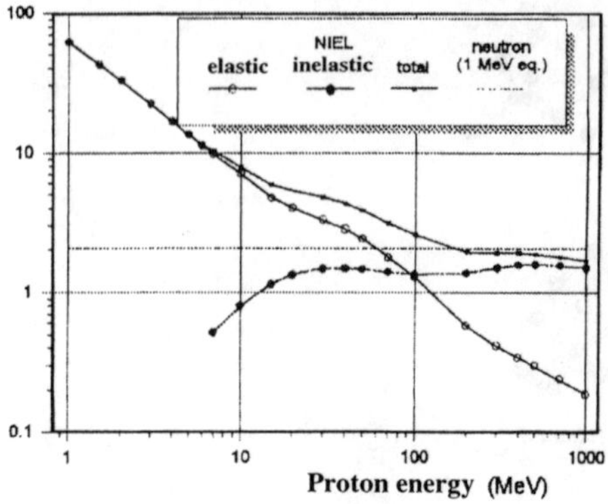

Fig. 2.4. Variation of NIEL as a function of the proton energy in silicon. [After Buisson et al. [9]]

whereby the ^{30}Si nucleus absorbs a thermal neutron (n) and transforms via an unstable silicon isotope into a P donor atom (n-type doping by emission of an electron (β^-)). The usual starting material for NTD is high-resistivity p-type Si, which becomes n-type after a certain exposure time in a nuclear reactor. High-temperature annealing is required to remove the neutron-irradiation damage and to obtain useful material characteristics [11]. In this way, n-type Si with a tightly controlled and radially very uniform resistivity is obtained (spread of a few %), which is necessary for the production of power devices or nuclear-radiation detectors [12, 13].

In the case of Ge, the opposite holds, i.e. n-type Ge can be transmuted into p-type material, whereby the three following reaction chains occur [14]:

$$ {}^{70}_{32}\mathrm{Ge} + \mathrm{n} \rightarrow {}^{71}_{32}\mathrm{Ge} \rightarrow {}^{71}_{31}\mathrm{Ga} + \nu_e (T_{1/2} = 11.2\mathrm{days}) \tag{2.6.a} $$

$$ {}^{74}_{32}\mathrm{Ge} + \mathrm{n} \rightarrow {}^{75}_{32}\mathrm{Ge} \rightarrow {}^{75}_{33}\mathrm{As} + \bar{\nu}_e (T_{1/2} = 82.8\mathrm{min}) \tag{2.6.b} $$

$$ {}^{76}_{32}\mathrm{Ge} + \mathrm{n} \rightarrow {}^{77}_{32}\mathrm{Ge} \rightarrow {}^{77}_{33}\mathrm{As} + \bar{\nu}_e T_{1/2} = 11.3\mathrm{hr}) \tag{2.6.c} $$

$$ \rightarrow {}^{77}_{34}\mathrm{Se} + \bar{\nu}_e (T_{1/2} = 38.8\mathrm{hr}) $$

The first reaction leads to the creation of acceptors; the two others generate a single (As) and a double (Se) donor, respectively. Since the abundance of the ^{70}Ge isotope is more than 95%, the first reaction will dominate, leading to a homogeneous p-type doping of germanium after NTD and annealing.

In silicon, the creation of one V-I pair requires on the average an energy between 15 and 40 eV. From that number follows that a MeV particle will not lose all its energy in one displacement and will create several V-I pairs along its track, i.e., a collision cascade. In fact, most of the energy loss will take place in the final part, in a narrow interval around what is called the projected range R_p. This parameter is a sensitive function of the starting energy E, of the target material and the particle mass and is for example crucial for the application of ion implantation in microelectronics [7]. For simulating the expected doping profile, the use of numerical codes, based on Monte Carlo simulations are very useful. Very popular is the Transport of Radiation In Matter (TRIM) code [15], which enables to calculate the number of displacements (vacancies) along the particle track. An example for 6.1 MeV proton irradiation in silicon is given in Fig. 2.5 [16]. It can be deduced that in the tail region, the damage is quite uniform, while a peak is obtained around R_p, where the particle comes to rest in the material (here, $R_p \approx 300$ µm).

TRIM can also be used for a more complex radiation environment, as in the case of the fission products of a ^{252}Cf source, which is often used for SEU testing, since it offers a mix of particles with different energies and masses [17]. In this case, both alpha's (≈6 MeV) and other heavy fractions (79 MeV Ba and 104 MeV Mo) are emitted, resulting in a double damage peak shown in Fig. 2.6. The first peak corresponds to the measured damage profile derived from the leakage current of silicon a n^+-p junction diode, giving a measure for the macroscopic damage on the device level. In addition, the microscopic damage as revealed by Deep Level Transient Spectroscopy (DLTS, see Sect. 2.4) and probing the concentration of the silicon-interstitial-related radiation-induced point defects show qualitatively the same depth profile [18].

The primary knock-on atoms (PKA's) or recoil atoms displaced by a high-energy particle can vary in kinetic energy from near zero up to some maximum determined by collision kinematics. The resultant PKA spectrum is unique to a given incident energy and target atom and describes the probability of a given recoil energy following a collision. These PKAs in turn can produce additional vacancy-interstitial displacements by further collisions and if they are generated with sufficient energy additional damage cascades will be formed. The resulting damage along an individual particle path becomes quite complex and may be composed of the initial cascade and branching terminal subclusters, represented in Fig. 2.7 [10]. This is also often termed cluster damage. Note that this branching only occurs for high-energy ions and neutrons, while the displacement damage associated with electrons or γ's is more uniform along their trajectory. As a consequence, no clear damage peak is formed, like for the case of protons for example (Fig. 2.5). Regardless of whether an atom is displaced as part of a damage cascade or as an individual lower energy PKA, most of the initial vacancy-interstitial pairs recombine and no permanent disorder results. Only a small fraction of V-I pairs survives and migrates in the material, where they can undergo different types of interactions with impurities in the lattice. In the end (more) stable radiation dama ge is formed, which consists generally of small point defect complexes. In n-type silicon, well-known radiation defects are the A centre (a Vacancy-Oxygen: V-O pair), the divacancy (V-V), the E centre (Phosphorous-Vacancy: P-V pair), etc. These centres will determine the electrical properties of the irradiated material, as will be seen below.

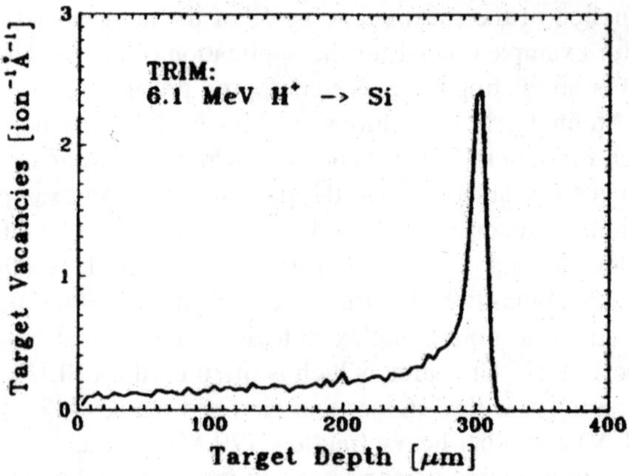

Fig. 2.5. Monte Carlo simulation (TRIM) of the vacancy profile for 6.1-MeV protons in silicon. [After Linnros et al. [16]]

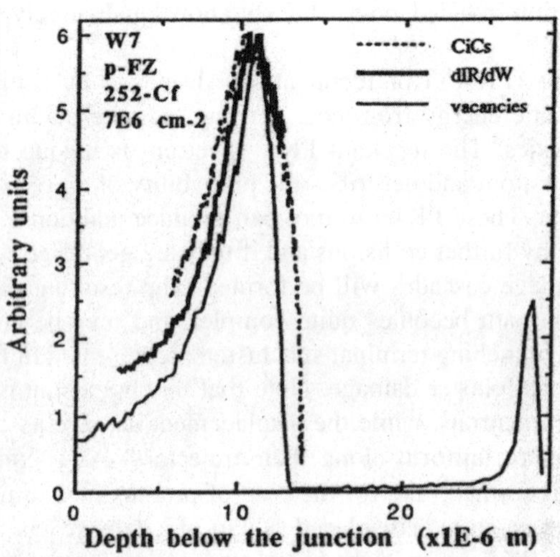

Fig. 2.6. Depth profile of the derivative of the leakage current with respect to depth for an n-type Float-Zone silicon diode after irradiation using a ^{252}Cf source (fluence 3.92x10^6 p/cm^2). Superimposed are the interstitial carbon/substitutional carbon (C_i-C_s)-related deep level concentration profile obtained on a p-type diode exposed to a fluence of 7x10^6 p/cm^2 and the damage depth profile obtained from TRIM assuming perpendicular incidence. [After Vanhellemont et al. [18]]

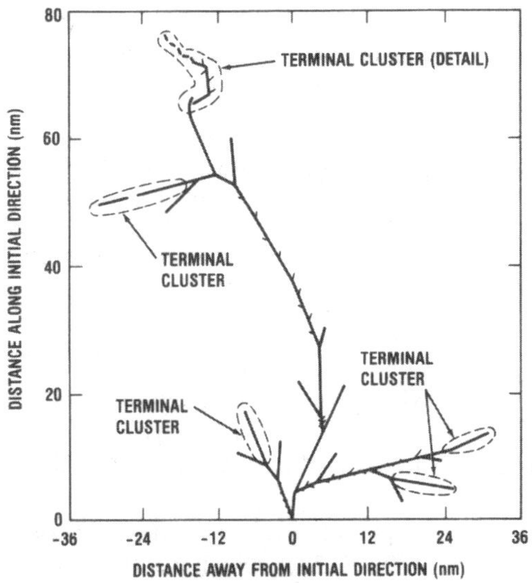

Fig. 2.7. A pictorial representation of the branching of the initial collision cascade into subcascades. According to Monte Carlo simulations, recoil energies above 2 keV result in more but not larger subcascades. [After Dale and Marshall [10]]

It is clear from the above that displacement damage is a complex process, which is not so easy to model. In order to enable the calculation of the fractional energy going into displacements, one needs a parameter which plays the same role as the LET for ionisation damage. This task is fulfilled by the so-called Non-Ionising Energy Loss (NIEL) [10, 19, 20]. The non-ionising energy deposition S_d is calculated by summing the product of the cross section σ_i and the avarage recoil energy $E_r(E)$ of the PKAs for each incident particle energy, E, and for each kind of nuclear interaction, i, of interest, i.e.,:

$$NIEL = S_d = (N/M_2)\ \Sigma_i\, \sigma_i(E)\, E_r(E) \tag{2.7}$$

Here, N is Avogadro' s number andM_2 is the gram atomic mass of the target.

In principle, the NIEL contains Rutherford type of interactions, prevailing at low particle energies and elastic and non-elastic nuclear interactions, becoming progressively more important for higher particle energies. An analytical expression for the displacement cross section of relativistic electrons is given by [20]:

$$\sigma = \pi \left[\frac{Z_2 q^2 \gamma_c}{m_e c^2 (\gamma_c^2 - 1)}\right]^2 \left(\frac{E_m}{E_d} - 1\right) \tag{2.8}$$

with $\gamma_c = \sqrt{1 - v^2/c^2}$, E_m the maximum energy transmitted to the nuclei and E_d the threshold energy for displacement. As mentioned above, several values have been quoted for silicon in the literature, ranging from 15 to 40 eV. Often, 21 eV is used [19, 20], while for Ge $E_d \approx 27.5$ eV [20].

Under a constant energy of irradiation E, the concentration of primary radiation defects (V,I) denoted by N_T is given by:

$$N_T = \sigma N_{at} \Phi \tag{2.9}$$

which links the NIEL ((2.7)) with the created microscopic damage in the lattice of the semiconductor material. The calculated NIEL for silicon and germanium, for different incident particles is represented in Fig. 2.8 [10] and 2.9 [19]. Data for GaAs [21] and other III-V materials can be found in the literature, under tabular form [22].

The NIEL concept has been successfully applied to explain the degradation of various kinds of devices (for silicon, see e.g. [10, 23, 24]), which is rather surprising given the complexity of the problem. The link between the original displacement damage and the final stable radiation defects formed is not straightforward. In addition, as will be outlined in more detail in the next chapters, the different types of radiation defects have different electrical impact. Fortunately, it has been shown empirically that it is not necessary to calculate the final defect inventory from first principles. Experimentally, it has been demonstrated that the radiation response of many devices can be predicted based on calculations of the amount of damage energy imparted to PKAs [10]. In other words, the NIEL can be calculated from first principles based on differential collision cross sections and interaction kinematics. The first order agreement between the measured device degradation and the calculated NIEL shows that to a larger extent, the nature of the stable displacement damage does not depend on the energy spectrum of the PKAs.

An additional advantage of using NIEL is that it provides a means to normalise the energy fraction going into lattice damage for different particles, with different energies, In this way, it becomes feasible to explain the device degradation in a realistic radiation environment, based on a limited amount of radiation testing at some reference energies (usually 10 MeV for protons and 1 MeV for electrons and neutrons). This, combined with the use of damage factors is a convenient tool for displacement damage assessment.

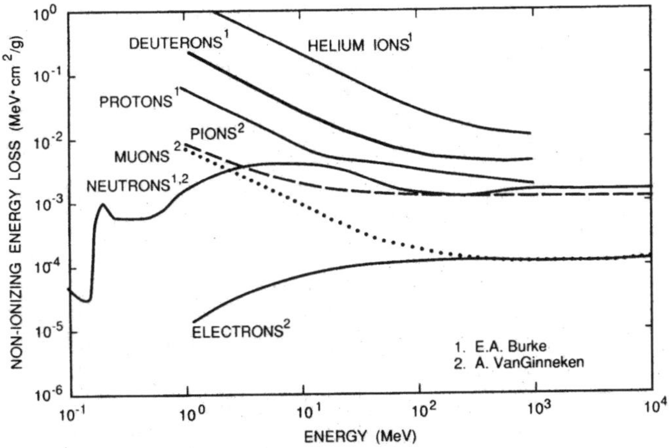

Fig. 2.8. Calculation of the non-ionising energy loss rate in silicon, for various incident particles over a wide energy range. NIEL quantifies displacement damage for a given radiation in exactly the manner that stopping power determines dose for ionising radiation. [After Dale and Marshall [10]]

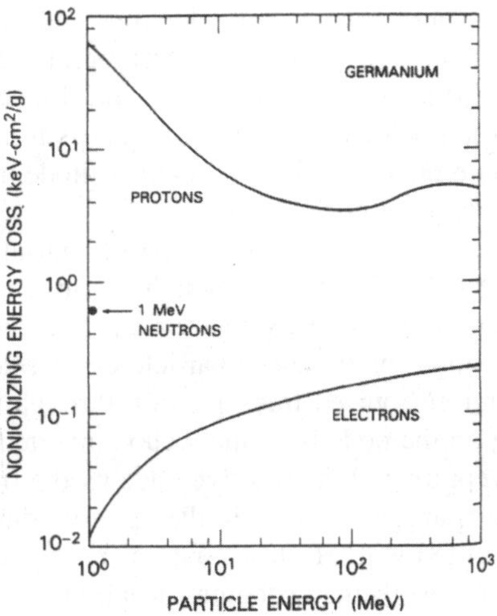

Fig. 2.9. Calculations of the NIEL in Ge due to incident protons and electrons. [After Marshall et al. [19]]

2.3 Impact of Radiation Damage on Device Performance

In the foregoing, two types of basic radiation-damage mechanisms have been described. As will be seen next, they also have a different impact on the device performance, which can be defined as the macroscopic damage or degradation. Besides the damage mechanism, one should also consider the time scale on which the effects play, as the primary damage (i.e. electron-hole or V-I pairs) undergo different kinds of interaction after their creation. In many cases, the degradation of a certain device parameter – for example the reverse current I_R of a diode – can be described by a so-called damage factor or parameter K_{IR}. It is defined as the change of the parameter with total dose (ionisation) or particle fluence (displacement). As will be shown, often there exists a good correlation between these factors and the energy deposited by ionising or non-ionising energy loss. In other words, they are useful in the modeling or even prediction of the device/circuit degradation in a radiation environment, without the need to know the exact details of the microscopic defects formed during the exposure. In the next part, it will be briefly outlined how these damage factors can be used for that purpose.

2.3.1 Ionisation Damage

Ionisation of the material creates free charge which can move in the material, either by drift or diffusion. When an electric field F is present, electrons and holes will be separated and drift to the ohmic contacts. In field-free neutral regions, carrier transport will be dictated by thermal diffusion. The net result is the flow of a current during some time after the irradiation, which is generally harmless for the device operation. In the case of a solid-state nuclear-radiation detector diode, this current forms the signal, used for spectroscopic purposes.

One important degradation mechanism, related to the collection of ionisation induced charge is the Single Event Upset (SEU) [25, 26, 27], which is a major reliability concern for dynamic nMOS and static CMOS memories. The basic mechanism is illustrated in Fig. 2.10. A single high-energy particle can strike a critical node of the device, leaving behind an ionised track passing through the well area or storage capacitor. Depending on the node bias, the well is inverted or depleted. The created electrons will be swept toward the positive node by the field present in the structure, while the positive charge will move in the opposite direction. If the well (storage node) is already filled with electrons, no change of state (bit flip) will occur. However, initially empty wells may become partially filled by a SEU. If enough electrons are collected, a bit flip from the "0" to the "1" state occurs, which is called a soft error. This is opposite to hard errors, associated with permanent damage in the structure. A soft error can in principle be removed by resetting the memory or by software correction for example. A key device parameter is the critical charge or number of electrons associated with the difference between the "0" and the "1". Depending on the technology and the design rules used, typical charge packets are approximately 10^5-10^7 electrons. For example, for a 5-MeV

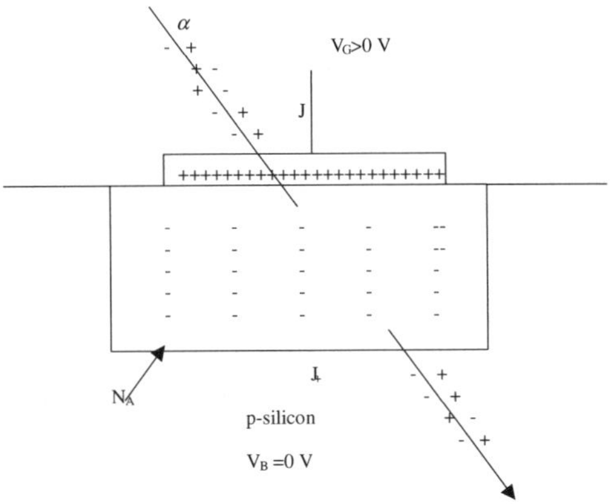

Fig. 2.10. Schematic illustration of impinging particles on the active device area, indicating single event upsets. V_G and V_B are the gate and the substrate (bulk) bias, respectively. J_+ and J_- is the hole and electron current density and N_A the acceptor (doping) concentration. [After McLean and Oldham [26]]

alpha-particle about 1.5×10^6 electron-hole pairs are generated along its track. Alpha-particles are of particular concern, as they are emitted by ceramic materials used for packaging memory chips. Device scaling strongly enhances SEU sensitivity, as the critical charge lowers for reduced storage node dimensions. Similar considerations hold for high-energy ion (or radiation) –induced latch-up (SEL) in CMOS circuits [28], which is a hard or permanent error and can lead to the destruction of the circuit by self-heating. It is outside the scope of this work to give a detailed discussion of the vast amount of literature on SEU and SEL. For a thorough discussion, the interested Reader is referred to [3].

The best known degradation effect of ionisation damage is the occurrence of charge trapping by pre-existing or radiation-induced trap sites. Trapping means that the charge is stored for some time at a defect and is released subsequently. It causes degradation of the resolution of a solid-state nuclear detector and generates an increase in the charge transfer inefficiency (CTI) of Charge-Coupled-Devices (CCDs) used for imaging or spectroscopy purposes [29]. However, for mainstream silicon CMOS, the major concern is hole trapping in the gate, spacer and field dielectrics, which is so far mainly SiO_2. The basic radiation mechanisms in SiO_2 and their time dependence are schematically represented in Fig. 2.11 and 2.12, respectively [5]. The response of SiO_2 on a short radiation pulse is dominated by the creation of free electron-hole (e-h) pairs and their transport through the oxide. It is assumed that the initial creation of e-h pairs only depends on the total ionisation energy deposited and not on the nature of the incident particle/photon. Due to the presence of an electric field in the oxide F_{ox}, the fraction of carriers which escapes direct recombination are separated by it. The highly mobile electrons will be swept out of the oxide in a matter of ps after irradiation towards the anode. However, the low-mobility holes will migrate slowly to the cathode, as depicted in Fig. 2.11.

The fraction of holes surviving the initial recombination [the hole yield $f_y(E,F_{ox})$] strongly depends on the incident energy and particle, and the field F_{ox} [5, 30, 31]. This positive charge causes an initial shift in the flat-band voltage V_{FB} (i.e., the gate voltage corresponding with flat conduction and valence band energy in the substrate) of the MOS structure given by [30]:

$$\Delta V_{FB} = \Delta V_{ot}(t=0) = -3.23\text{x}10^7 f_y(E,F_{ox}) D t_{ox}^2 / E_{eh} \tag{2.10}$$

ΔV_{ot} is the (threshold) voltage shift due to the oxide-trapped hole charge. Note that in (2.10) the flat-band voltage shift is proportional to t_{ox}^2 which has important consequences for scaled submicron CMOS technologies (see Chap. 6), where t_{ox} reduces for every next generation of technology. Equation (2.10) assumes a uniform radiation and hole generation across the thickness of the oxide. In addition, (2.10) predicts a linear increase of ΔV_{FB} with dose D. This is confirmed by experiments [30] and implies that the time-zero damage factor K_{VFB} is just equal to $\Delta V_{FB}/D$. Only at high total dose, the initial flat-band shift will start to saturate [31]. This is mainly related to the change in the oxide field by the high radiation-induced mobile charge density [32].

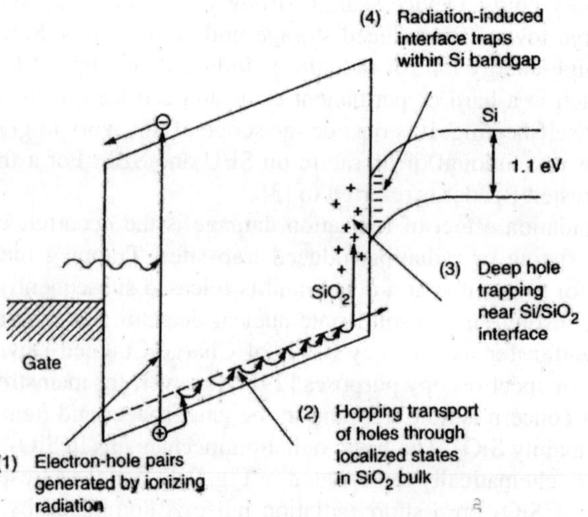

Fig. 2.11. Schematic representation of basic radiation effects in SiO_2. [After Oldham et al. [5]]

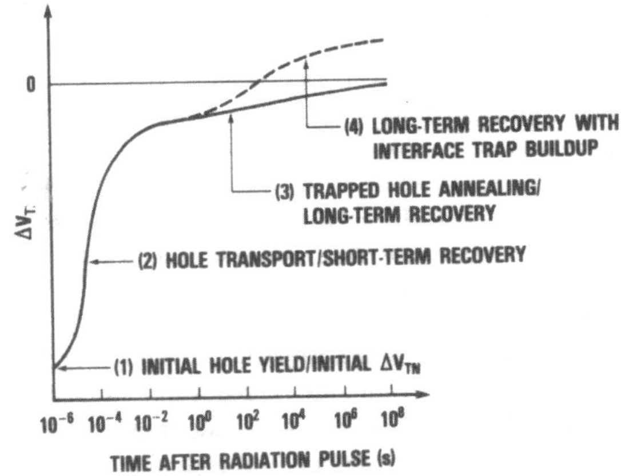

Fig. 2.12. Time-dependent threshold voltage (V_T) recovery of n-channel MOSFETs. The labeled regions correspond to Fig. 2.11. [After Oldham et al. [5]]

After creation and separation by the electric field, the surviving holes undergo a highly time-dispersive field- and temperature-dependent transport toward the negative interface. This leads to the short-term recovery of the threshold voltage shift in Fig. 2.12 (phase (2)) [5]. It has for example been noted that in a 30 nm hard oxide, irradiated with F_{ox}=1 MV/cm at 300 K, 50% of the holes will reach the interface in about 10 μs [31]. In other words, most of the holes leave the dielectric and only a few % will remain trapped near the negative interface. This results in a significant recovery of the threshold voltage shortly after the end of the irradiation (Fig. 2.12). For irradiation of soft oxides, however, a large fraction of holes become trapped in preexisting trapping centres near the negative interface so that the short-term recovery is only partial and significant annealing will follow. The holes are trapped in levels typically 0.3 to 0.4 eV above the oxide valence band. In addition, the transition zone near the silicon interface is such that the highest density of hole traps occurs there – this is related to both the local composition and mechanical stress at the interface region.

Finally, the holes trapped near the negative interface undergo a long-term annealing (phase (3) and (4) in Fig. 2.12) which occurs with an ln t (t=time) dependence. Different mechanisms have been proposed for that, the details of which are outside the scope of this work. One assumes tunneling of electrons from the substrate into the oxide, followed by recombination with the trapped holes. Another model considers trapped-hole tunneling towards the silicon valence band. This trapped-hole removal is accompanied by the creation of interface states with a density N_{it}. Since these interface traps generally behave as negatively charged acceptor states, when located in the upper half of the silicon band-gap (see Fig. 2.11), they compensate the trapped hole charge. As a result, the threshold voltage of an nMOS shows further recovery with time and may even exhibit an overshoot

– often called rebound behaviour – which is a highly undesirable effect for standard MOS transistors. The effect of trapped holes on p-channel devices is opposite. Usually, a more negative threshold voltage results, when radiation induced holes are present near the interface, which may recover during the annealing phase.

It can also easily be explained why the application of a positive bias to the gate electrode during irradiation and subsequent anneal leads to a higher degradation than zero or negative bias. The positive bias repels holes towards the opposite Si-SiO_2 interface, where the effect of the trapped-hole charge on the MOSFET parameters is maximal. For negative gate bias, holes are attracted to the gate, where they have little impact on the flat-band voltage. In addition, the interface-state generation is also suppressed in that case.

Besides the threshold voltage, another MOS parameter which becomes degraded is the inversion-layer mobility μ. The low-field mobility reduces due to Coulombic scattering at the radiation-induced charged interface and oxide traps, with density ΔN_{it} and ΔN_{ot}, respectively [33]. The mobility reduction is given by:

$$\mu = \frac{\mu_0}{1 + \alpha_{it}\Delta N_{it} + \alpha_{ot}\Delta N_{ot}} \tag{2.11}$$

Hereby is μ_0 the pre-radiation mobility and α_{it}, α_{ot} are the respective scattering parameters in (cm^2), for the trap densities in cm^{-2}. They are in the range of 10^{-12} cm^2 [33]. In this case, the inverse mobility μ^{-1} can be used to define a mobility damage coefficient K_μ. The mobility damage factor will then depend on the scattering coefficients and on the relationship between total dose D and interface and oxide charge build-up. Usually, a sublinear relationship with D is observed.

Although the impact of ionisation damage on other MOS parameters has been studied (i.e., series resistance, subthreshold leakage, low-frequency noise, gate delay time,…) the threshold voltage and to a lesser extent the mobility are used as key parameters for the study of the impact of irradiation. They also offer the opportunity to gain better insight in the responsible degradation mechanism, i.e. oxide trapping, interface-state generation, etc. One popular technique in this respect is the so-called charge-separation method introduced by *McWhorter and Winokur* [34] and illustrated by Fig. 2.13. It represents a number of subthreshold characteristics of the drain current (I_D) versus the gate-to-source voltage (V_{GS}) of an n-MOSFET in the ohmic regime, i.e., corresponding with small drain-to-source bias V_{DS}. The curves are before and after four ^{60}Co radiation doses.

The shift in the threshold voltage due to interface traps ΔV_{it} is related to the stretch-out voltage V_{so}, which is given by:

$$V_{so} = V_T - V_{mg} \tag{2.12.a}$$

where V_{mg} is the gate voltage corresponding with the midgap current I_{mg}. ΔV_{it} is then given by [34]:

$$\Delta V_{it} = (V_{so})_{post} - (V_{so})_{pre} \tag{2.12.b}$$

and the corresponding change in interface trap density follows from:

$$\Delta N_{it} = \frac{\Delta V_{it} C_{ox}}{q} \text{ (cm}^{-2}\text{)} \tag{2.12.c}$$

C_{ox} is the gate capacitance per unit of area (pF/cm^2) and given by ε_{ox}/t_{ox} (ε_{ox} the permittivity of SiO_2).

On the other hand, the contribution of oxide-trapped charge is in principle independent of the gate voltage and thus corresponds to the horizontal shift of the linear characteristic of Fig. 2.13; in other words,

$$\Delta V_{ot} = (V_{mg})_{post} - (V_{mg})_{pre} \tag{2.13.a}$$

and:

$$\Delta V_{ot} = \frac{\Delta V_{ot} V_{ox}}{q} \text{ (cm}^{-2}\text{)} \tag{2.13.b}$$

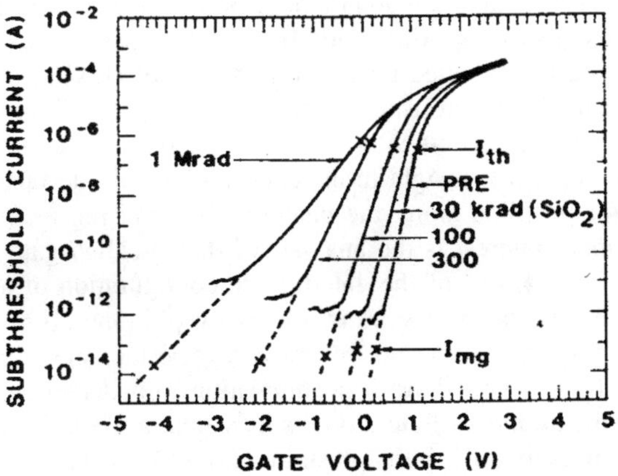

Fig. 2.13. Subthreshold current curves for a n-MOSFET before irradiation and at four different levels of ^{60}Co radiation. Threshold I_{th} and midgap I_{mg} currents are marked on each curve. [After McWhorter and Winokur [34]]

where:

$$\Delta V_T = \Delta V_{ot} + \Delta V_{it} \tag{2.14}$$

In writing (2.13.a) it is tacitly assumed that the interface traps at mid gap are charge neutral, which has been derived from Electron Paramagnetic Resonance (EPR) results of the dangling bond centres at the Si-SiO_2 interface (see Sect. 2.4.1).

The charge-separation method has been widely used for studying radiation effects in the MOS system. It can be applied to transistors in the ohmic regime, using the current-voltage characteristics, as described above or to a MOS capacitor, based on high-frequency Capacitance-Voltage (*C-V*) measurements. The technique has been corrected for the case that passivation of the p-type dopants (boron) takes place by hydrogen released from the oxide by the irradiation [35]. The reduction of the substrate doping density may cause an additional shift in V_T, related to the change in the depletion charge and the surface potential. Often, the thus obtained ΔN_{it} and ΔN_{ot} densities are compared with the data derived from charge-pumping [36] or low-frequency noise [37] measurements. In the latter case, the density of so-called "slow" or border traps is probed, which occur in close vicinity of the interface (the first 2 nm).

Although the charge-separation method has proven very useful, it has also its limitations. Particularly when entering (deep) submicron channel lengths, it becomes doubtful whether the technique still works properly. There are several arguments for that: for short-channel devices, the subthreshold slope becomes length dependent, due to the impact of the lateral electric field. In addition, it is strongly believed that for submicron transistors, the radiation damage is no longer uniform along the channel [38 - 40] (see also Chap. 6). In addition, particularly for low temperature irradiations (77 or 4.2 K) so-called local non uniformities (LNUs) in the damage degrade the subthreshold characteristics [32]. The same applies if significant field-oxide related edge leakage occurs after exposure. Therefore, alternatives have been looked for, which provide insight in the contributions of interface and oxide charges and allow for profiling along the channel. A promising techniques which gains more and more interest is the analysis of the so-called gate-induced drain-leakage (GIDL) [39 - 42] or of the gated-diode configuration of a MOSFET [43 - 46]. In both cases, the gate of the transistors is swept in the off region, well below the threshold voltage, so that no inversion layer establishes. In the GIDL technique, the gate bias is varied from accumulation into depletion, whereby the reverse current of the drain-substrate diode is measured at fixed V_{DS}. The gated-diode configuration measures the drain-substrate current as a function of V_{GS} at a slightly forward fixed V_{DS}. The principle of the latter type of measurement is represented in Fig. 2.14, for a 10 MeV proton irradiated gated diode. One can clearly see that different regimes can be distinguished. For V_{GS} in accumulation (< -4 V), the diode leakage current generated in the bulk depletion region is measured. Changing V_{GS} towards depletion, a step in the collected current is observed which is proportional to the surface generation (s_g – reverse biased diode)

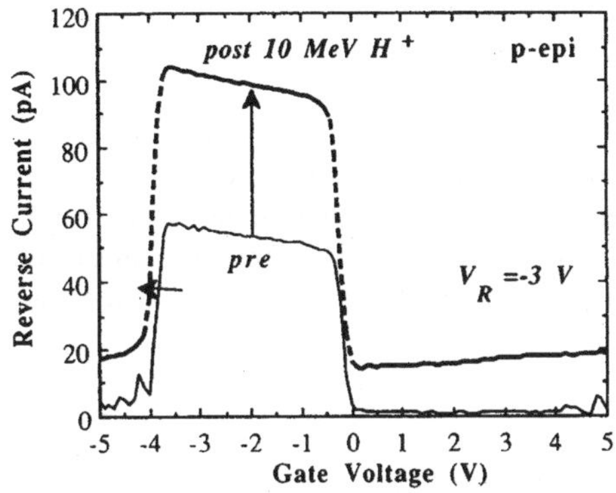

Fig. 2.14. The total gated diode characteristic before and after a 10 MeV proton exposure to a fluence of 4×10^{10} cm^{-2}. The reverse diode bias is –3 V. The n^+-p diode has been fabricated in an epitaxial (epi) substrate

or recombination (s_r – forward biased diode) velocity. For reverse operation, the step height of the surface generation current $\Delta I_{g,s}$ of a gated diode is given by [47]:

$$\Delta I_{g,s} = qs_g n_i A \tag{2.15}$$

with:

$$s_g = \frac{\pi}{2}\sigma_s v_{th} n_i (kTD_{it}) \tag{2.16}$$

from which the density of interface traps D_{it} (per cm^2 and per eV) can be derived. In (2.15), A is the gate area and n_i the intrinsic carrier concentration (for silicon ~10^{10} cm^{-3} at 300 K). Furthermore, k is the Boltzmann constant, T the absolute temperature, v_{th} the thermal velocity (~10^7 cm/s) and σ_s the effective carrier capture cross section, in the range 10^{-15} – 10^{-16} cm^2 for silicon interface traps. Similar formulas can be found for the MOSFET gated-diode [46]. The difference in the step height in Fig. 2.14 yields the radiation induced interface state generation, suing (2.15) and (2.6) and provided σ_s is not affected by the exposure. Furthermore, the shift to negative gate bias indicated by the horizontal arrow in Fig. 2.14 points to the creation of radiation-induced positive oxide charge, given by:

$$q\Delta N_{ot} = \Delta V_{GS} C_{ox} \tag{2.17}$$

for the midgap gate bias. It is clear that this method provides a rather straightforward way to analyse the radiation damage in the MOS system in terms of interface and oxide charge. In principle, one can also obtain profiling information along the channel [46], although this requires a numerical simulation of the surface potential. In addition, for sufficiently large lateral electric fields (i.e. large V_{DS} or V_{GS})

field-assisted carrier generation mechanisms (i.e., trap-assisted tunneling and Poole-Frenkel effect) should be taken into account as well [42], making the analysis quite complex.

It has also been proposed to use the gated-diode technique for the Si-SiO_2 interface characterisation of radiation detectors on high-resistivity silicon material, before and after irradiation [48]. It is clear from this that although the original concept is already more than 30 years old [47], there is still a bright future ahead for the gated-diode (MOS) technique in the field of reliability and radiation damage assessment.

2.3.2 Displacement Damage

Displacement lattice damage in semiconductor materials can have a significant impact on their electrical properties, through the creation of stable radiation defects, which have one or more levels in the bandgap. Following *Srour and McGarrity* [29] and as depicted schematically in Fig. 2.15 the following basic electrical degradation effects can be distinguished:

- the generation of e-h pairs by radiation-induced defects
- the recombination of e-h pairs by radiation-induced defects
- trapping of carriers by radiation-induced defects
- compensation of donors and acceptors
- tunneling of carriers

Consequently, important material parameters such as free carrier mobility and density, resistivity ρ and generation (τ_g) and recombination (τ_r) lifetimes will be affected by displacement damage. After irradiation, an increase in the resistivity is mostly observed, which results from a combination of two effects [49-54]. First, the removal or reduction of the free carriers occurs due to either direct removal of dopants from active (substitutional) lattice sites by interaction with the created V' s and I' s giving rise to stable point defect complexes or to dopants in a "neutral" (i.e. interstitial) lattice site. Well-known examples for silicon are the creation of group V donor – vacancy pairs (E centres), or the creation of interstitial B (B_i). More details about the dominant radiation defects in Group IV and III-V semiconductors will be provided in Chap. 3 and 4, respectively. Furthermore, the presence of radiation-induced deep levels changes the charge balance in the material and – through Fermi-Dirac statistics – the position of the Fermi level, which impacts on the net free carrier density. In many cases, the charge of the shallow dopants is compensated by the radiation levels, resulting in a lowering of the free carrier density and an increase in ρ. This is illustrated in Fig. 2.16 for the case of high-energy proton irradiated n-type silicon, where for each energy a resistivity peak is obtained which nicely tracks the damage peak [54].

In extreme cases, like for heavily irradiated high-resistivity silicon radiation detectors [55, 56] type inversion occurs, which is detrimental for their operation. Similar carrier removal and type inversion effects have recently been observed for high-fluence irradiated p-type silicon solar cells [52, 53]. On the other hand, the

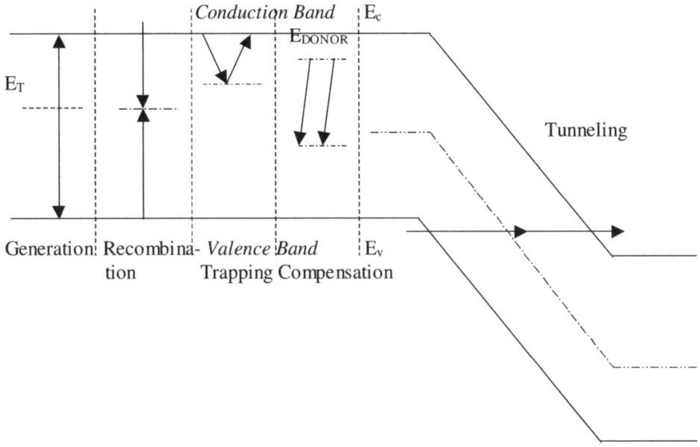

Fig. 2.15. The five basic effects of a radiation-induced defect level (E_T) on the electrical device performance. [After Srour and McGarrity [29]]

application of a high-fluence proton-irradiation of silicon [57 - 59] or GaAs [59] (see also Chap. 4) creates very high resistive material. In the case of silicon, MΩcm layers can be achieved which is of potential interest as an alternative for silicon-on-insulator (SOI) or as starting material for microwave applications. In the latter case, a clear improvement of the rf performance is expected, by the reduction of transmission-line and inductor losses or the suppression of cross coupling [59].

Another application of the radiation-induced carrier removal is to use spreading resistance profiling (SRP) measurements for the study of the migration of radiation or ion-beam induced point defects [7, 60]. In this way, some idea about the diffusivity of V and I at room temperature can be gained, which is of great importance for the modeling of radiation-defect formation processes, during ion implantation for example. The lower limit of the diffusivity of Si self-interstitials (10^{-10} cm^2/s) obtained by this technique is much higher than that predicted by theory [7] and could point to the occurrence of ionisation-enhanced diffusion at 300 K.

The carrier removal rate (K_n) in n- and p-type silicon has been more systematically studied by *Pease et al.* [61]. K_n is defined by:

$$n(\Phi) = n(0) - K_n\Phi \tag{2.18}$$

with $n(\Phi)$ and $n(0)$ the free carrier concentration after and before exposure to a fluence Φ, respectively. K_n turns out to be a constant for low Φ, in the first approximation, so that n reduces linearly with the fluence. In addition, it is about two times larger in p- compared with n-type silicon, for 1 MeV equivalent neutron or high-energy proton irradiation [61]. Values are reported in the range 2-3 cm^{-1}(n-type) and 4-5 cm^{-1} (p-type). Moreover, it was found that under a reverse junction

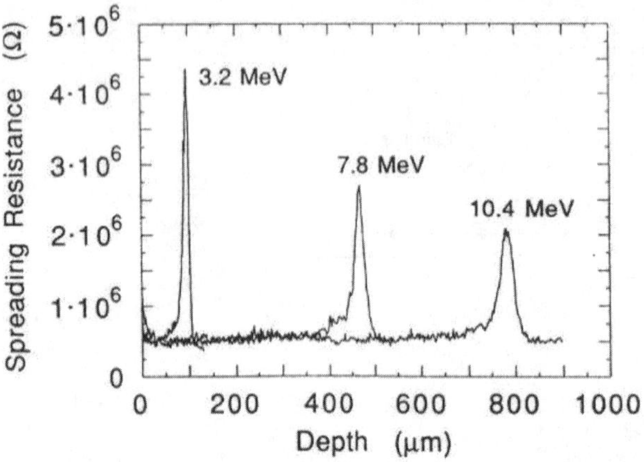

Fig. 2.16. Spreading resistance measurements of proton irradiated n-type silicon. Proton energies are 3.2, 7.8 and 10.4 MeV and the fluence is $1 x 10^{11}$ cm^{-2}. [After Keskitalo and Hallén [54]]

bias, the carrier removal rate for both neutrons and protons becomes smaller for p-type silicon, while little effect was seen for n-type material. It was observed that the experimentally determined ratio of proton (K_P) to 1 MeV equivalent neutron carrier removal rates (K_N) versus proton energy match quite accurately the calculated NIEL, as shown in Fig. 2.17 [61].

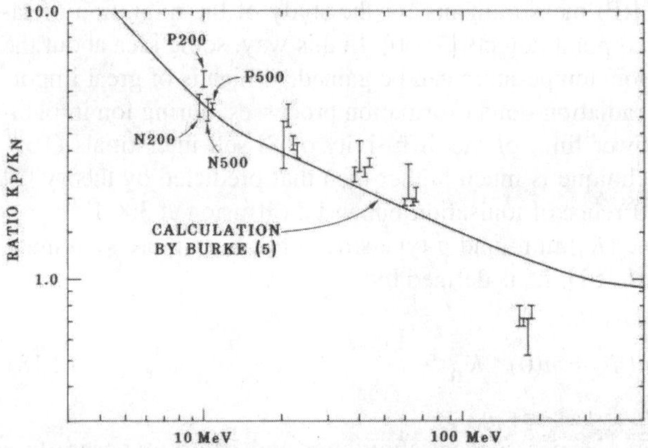

Fig. 2.17. Comparison of K_P/K_N versus proton energy to calculation based on non-ionisation energy loss in silicon. N200 and N500 refers to n-type Si with a breakdown voltage of 200 and 500 V, respectively. P200 and 500 refer to p-type Si. [After Pease et al [61]]

It should be remarked that under certain conditions, the shallow doping density increases. This is for example the case after high-energy proton implantation plus thermal anneal in the range 300 to 400°C [50, 51]. In the damage peak region additional, hydrogen-related shallow donors are formed, which increase the doping density in n-type silicon.

As mentioned in Sect. 2.2.2, a fast ion or neutron creates both point defects (low energy PKA) or damage clusters (a high-energy PKA subcascade, see Fig. 2.7) along its track. This will be particularly the case in the region of maximum energy deposition, i.e., in the vicinity of the projected range R_p of the ions. A damage cluster in this context can be considered as an aggregate of several different lattice defects such as vacancies, dopant atoms, interstitial and site impurities[1]. In *Gossick's* classical model for a damage cluster [62], it is surrounded by a potential barrier qV_{cl} that screens out free charge carriers (Fig. 2.18) [63]. As a result, a local high resistive region exists, which impacts on the material and device properties. Such cluster damage has been studied experimentally in GaAs [63], Ge [64] and Si [65]. One of the important findings is that due to the high density of defects in the cluster, inter-trap transitions become possible, whereby carriers are exchanged between neighbouring deep levels [66, 67]. This deviation of the standard Shockley-Read-Hall (SRH) generation-recombination mechanism may explain the anomalously high leakage current in neutron irradiated silicon radiation detectors. For more details about the physics, the Reader is referred to [68 – 70].

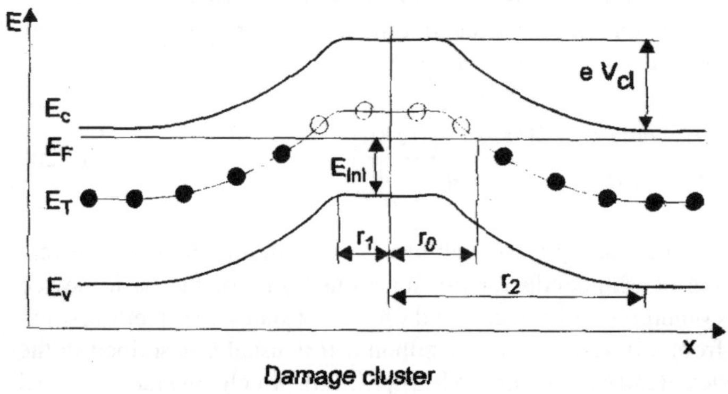

Fig. 2.18. The energy band diagram of a damage cluster in fast neutron and high energy proton irradiated n-GaAs. [After Brudnyi et al. [63]]

[1] It should not be confused with an interstitial or vacancy cluster (I_n or V_n with n=4,5,...) which also forms after post-radiation annealing; the higher the annealing temperature, the higher the size of the cluster and its stability upon annealing.

It has also been suggested that the lattice strain present in such a cluster impacts on the deep level behaviour. This is supposed to be the case for the two divacancy (often represented by V-V or V_2) acceptor levels in silicon, occurring in the damage peak region after ion implantation [71].

For very high ion fluences, individual damage cascades overlap, resulting in a heavily damaged region. In the case of high-dose ion implantation, amorphisation of the lattice occurs at the projected range, above a certain damage threshold, which depends on the ion mass and energy. A high temperature thermal anneal is required to remove the damage and activate the dopants. More detail will be provided in Chap. 3.

To a far lesser extent, the impact of displacement damage on the carrier mobility (scattering) has been studied. For silicon, it was observed that 17 MeV proton irradiation in the range 10^{12} to 10^{15} p/cm^2 mainly affects the carrier density and has only a minor effect on the hole mobility [72]. This is in agreement with earlier reports [49, 61]. Defining the majority carrier mobility damage factor (or coefficient) K_μ as [61]:

$$\frac{1}{\mu(\Phi)} - \frac{1}{\mu(0)} = K_\mu \Phi \tag{2.19}$$

a value for n-type silicon in the range -4.5×10^{-19} Vs has been reported for neutron-irradiated material [61].

From a bulk semiconductor device operation viewpoint perhaps the most important effect is the degradation of carrier lifetime τ. The corresponding damage coefficient K_τ is usually defined as [18, 66, 73, 74]:

$$K_\tau = \frac{\partial(1/\tau)}{\partial \Phi} \approx \frac{[1/\tau(\Phi) - 1/\tau(0)]}{\Phi} \tag{2.20}$$

with $\tau(\Phi)$ and $\tau(0)$ the post and pre-rad lifetime. Carrier lifetime to a large extent governs the operation of semiconductor p-n junctions – one of the basic device structures, which is omnipresent in micro- and opto-electronics and, therefore, extremely important from a device physics standpoint. It is usually described in the frame of the Shockley-Read-Hall theory (SRH) [75, 76], which considers thermal (i.e. phonon-assisted) carrier transitions to and from the bands.

One can distinguish two important lifetimes, governing the recombination (τ_r) and generation (τ_g) of carriers [75, 76]. The first one is important for the diffusive transport of minority carriers in neutral semiconductor regions. It sets the diffusion length L_n or L_p for holes and electrons, respectively, given in low injection by $\sqrt{D_n \tau_r}$ or $\sqrt{D_p \tau_r}$, with D_n and D_p the diffusivity of electrons or holes. For low minority carrier injection conditions, the recombination lifetime corresponds in first order to the minority carrier lifetime [75, 76]. In the case *j* independent recombination levels are active, with respective concentrations N_{Ti} and minority carrier capture cross section σ_i, τ_r can be described by [73, 76, 77, 78]:

$$\frac{1}{\tau_r} = \sum_{i=1}^{j} N_{Ti}(\Phi)\,\sigma_i\,v_{th} \tag{2.21}$$

From (2.21) follows that the recombination lifetime damage coefficient is a weighed sum of the trap introduction rates R_{Ti}, given by $\partial N_{Ti}/\partial\Phi$. For the measurement of the recombination lifetime, a rich palette of techniques has been developed, an overview of which can be found in Chap. 8 of [75]. It can be directly measured as a material (substrate) parameter or derived from device characteristics, using for example the saturation diffusion current of a forward biased diode [73, 79] or of the base-emitter junction in a bipolar transistor [10]. Note in addition that the actually measured recombination lifetime is usually affected by surface recombination [75, 76] and is, therefore, to be considered an effective τ_r. As shown above, surface recombination is determined by the density of interface traps and the surface potential. In contrast to the bulk recombination lifetime, it is more sensitive to ionisation damage, and only marginally to displacements. Using a gated diode as a test vehicle may enable to separate surface from bulk effects. Correction or separation of surface effects is also possible in some of the other lifetime techniques, but renders the analysis more complex.

The generation lifetime defines the leakage current I_R in a reverse biased junction and is often approximated by [75, 76]:

$$I_R = \frac{qn_iWA}{\tau_g} \tag{2.22}$$

with W the depletion width for a given reverse bias V_R. The corresponding leakage current density J_R is given by I_R/A. For fully depleted radiation detectors one normally describes the leakage per unit of depletion width, i.e., $I_R/100\ \mu$m for example. It is clear from (2.22) that the generation lifetime can be derived from the measured diode leakage [73, 75, 76, 79]. Alternative techniques use MOS capacitors, which are pulsed into depletion. The time evolution of the capacitance after the voltage pulse is analysed to extract τ_g [75, 76]. Recently, also non-destructive – i.e., contactless – techniques have become available to measure the generation lifetime in silicon wafers, without the need for device fabrication [75]. One important aspect is that while τ_g usually probes only a small part of the substrate[2] which typically amounts to a μm or so (depending on the doping density and reverse bias), the recombination lifetime can be determined by a much larger material thickness (several hundreds of μm up to 1 mm even), depending on its quality.

[2] Except for the case of a high-resistivity nuclear-radiation detector, where the depletion region may extend over the full wafer thickness.

Therefore, one usually finds different values for both lifetimes, in the same material or device. In the first approximation, the following relationship holds [76]:

$$\frac{\tau_g}{\tau_r} = \exp\left(\frac{|E_T - E_i|}{kT}\right) \tag{2.23}$$

with E_i the intrinsic Fermi level position (~ mid gap). From (2.23) it follows that the generation lifetime is significantly larger than τ_r if the energy level is more than a few kT from mid gap. For high-quality silicon, one finds in practice a value of the lifetime ratio of 50 to 100 [76]. After irradiation by high-energy protons, a similar value has been found in n- and p-type Float-Zone (FZ) silicon [73]. Note that also the leakage current of a diode may suffer from peripheral, i.e., surface generation. It is necessary to separate bulk from surface leakage effects in order to assess the true τ_g and, hence, the non-ionising energy loss in the semiconductor material. This can be achieved either by combining a number of diodes with different area to perimeter ratio, or by using a gated diode [79].

The recombination lifetime is of direct importance for the operation of e.g. nuclear-radiation detectors and opto-electronic devices (solar cells) where a long lifetime (carrier diffusion length) needs to be preserved during its operation. Therefore, its degradation by the creation of displacement damage is of great concern. This is particularly true for indirect band gap semiconductors like silicon and germanium, where direct optical recombination is unlikely, while deep level assisted thermal recombination is the dominant mechanism. For direct gap materials like GaAs, optical recombination is dominant, explaining their suitability for optical applications (see Chap. 8). In contrast, high-energy particle irradiation for local lifetime control is a common practice for fabricating power devices [80 - 82]. An example of the lifetime profile for 6.1 MeV proton irradiated n-type FZ silicon is given in Fig. 2.19 [16]. The leakage current is important for the resolution and noise in a nuclear or photo-detector; it also degrades the CTI of a CCD. In case of Dynamic Random Access Memories (DRAM' s) increasing node leakage reduces the refresh time and, hence enhances the power consumption. Some of these effects will be discussed in more detail in the following chapters.

The correlation between the lifetime damage factor and the NIEL has been studied extensively for the case of silicon. Figure 2.20 shows the good agreement between the recombination lifetime factor, normalised to 1 MeV (Si) neutrons and the NIEL [10]. In Fig. 2.21, the leakage current density damage factor K_{JR} is represented for n- and p-type FZ silicon diodes versus proton energy, measured at 296°C and a V_R=6 V [83]. According to (2.22), K_{JR} is proportional to $K_{\tau g}$. A fair agreement with NIEL is again observed. From this, one can conclude that the electrical parameter degrades linearly with particle fluence in a restricted range and for a broad range of energies. This implies that no detailed knowledge of the microscopic damage is required to model the electrical device degradation. One only needs to determine experimentally the proportionality constant between the particular damage factor $K(E)$ and the NIEL at a few or even only one particle energy. The damage at another particle energy (or even for another particle with the same or different energy) can then be determined from the theoretical NIEL. If one takes for example 1 MeV as the reference value, the average degradation at an energy E is given by [10, 22]:

$$\Phi(E)\,K(E) = \Phi(E)\,K(1\ \mathrm{MeV})\,\frac{\mathrm{NIEL}(E)}{\mathrm{NIEL}(1\,\mathrm{MeV})} \tag{2.24}$$

If the particle spectrum for a given space mission or environment is known, one can integrate (2.24) over the energy distribution in the following way [10]:

$$\text{Mission damage} = \int_0^\infty K(1\mathrm{MeV})\frac{\mathrm{NIEL}(E)}{\mathrm{NIEL}(1\mathrm{MeV})}\frac{\mathrm{d}\Phi(E)}{\mathrm{d}E}\mathrm{d}E \tag{2.25}$$

where $\mathrm{d}\Phi(E)/\mathrm{d}E$ is the differential proton fluence. Combining damage factors of electrical parameters, experimentally determined at one particle energy (or a few) and the theoretical NIEL, one has a powerful tool to predict the radiation response, without the need for extensive (and expensive) radiation testing nor a detailed knowledge about the microscopic damage mechanisms. However, for a fundamental understanding of radiation damage, careful spectroscopic studies of radiation-induced defects are mandatory, as will be outlined next..

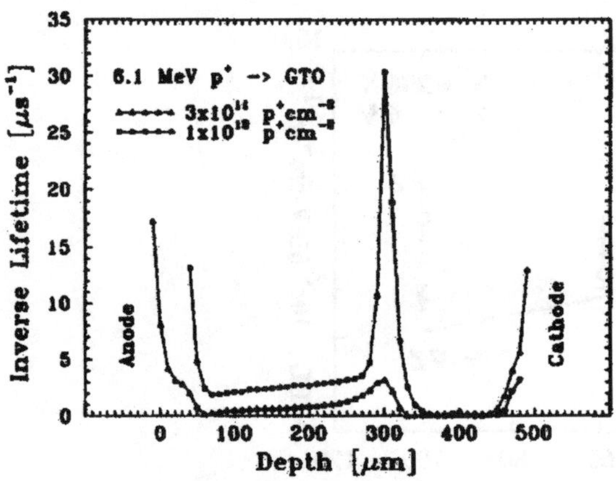

Fig. 2.19. Extracted depth profiles of the inverse recombination lifetime for two 6.1 MeV proton fluences. [After Linnros et al. [16]]

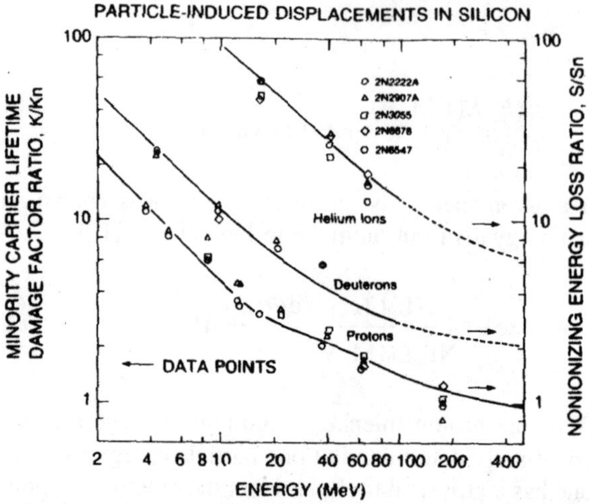

Fig. 2.20. Damage factors for bipolar transistors for protons, deuterons and helium ions normalised to 1 MeV equivalent (Si) neutron damage factors, as a function of energy. The solid lines (right-hand ordinate) are calculations of the corresponding ratios of the non-ionising energy loss. [After Dale and Marshall [10]]

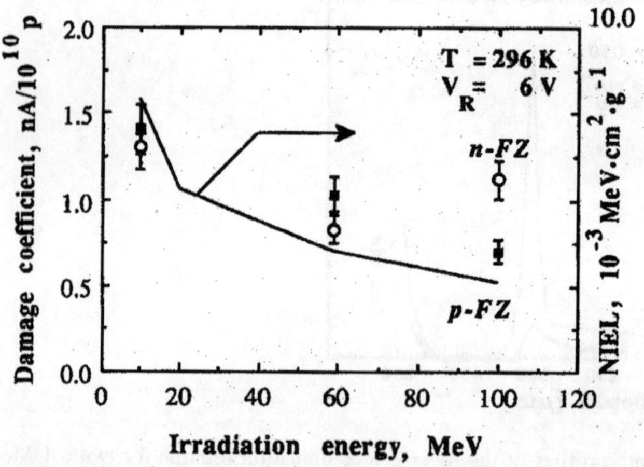

Fig. 2.21. Leakage current density damage coefficient in function of the proton energy for p- (■) and n-type (o) FZ-Si junctions at 296°C and a reverse bias of 6 V. The calculated NIEL corresponds with the full line

2.4 Spectroscopic Study of Microscopic Radiation Damage

Although the use of damage factors is of great practical interest, they are in general only relevant for a given type of device or technology. This implies that for each new technology generation or device type, a minimum radiation testing has to be performed in order to model the radiation response. On the other hand, the development of truly predictive tools is based on a thorough understanding of the fundamental radiation damage mechanisms. This requires the application of a combination of spectroscopic techniques which enables the investigation of the radiation-induced electrically active defects. In this paragraph, three techniques will be described, which have been extensively used in the context of radiation defect studies and have contributed significantly to the basic understanding of their formation, structure, composition and stability. Most importantly, one should be able to detect the electrically active centres and to determine their relevant parameters, with sufficient sensitivity and resolution. As will be pointed out, the results of these spectroscopic studies can be used as input for physics-based simulation tools. Such simulators have become available for certain applications and are reasonably successful in predicting the device performance.

2.4.1 Electron Paramagnetic Resonance (EPR)

Electron paramagnetic resonance (EPR) is a tool which provides a highly detailed picture of defects on an atomic level [84, 85]. One obtains a spectrum – i.e. a coded picture of the defect – which can be translated in structural and chemical information. In order to be detectable by EPR, a defect has to be paramagnetic, i.e. it must possess a magnetic dipole moment of an unpaired spin. Defects with even numbers of electrons are often non-magnetic and unobservable with EPR, since the bonding electrons usually occur into bonds between atoms, with their spin paired off and hence their magnetic dipoles cancelling. However, what generally matters are electrically active defects, which have a level in the band gap. Depending on the position of the Fermi level, there will always be a charge state, corresponding to an unpaired electron which will be EPR sensitive [84].

In EPR, the sample is placed in a slowly varying magnetic field ($\boldsymbol{H}$) and exposed to a microwave frequency magnetic field oriented perpendicularly to the applied field. Usually, the measurements are made at an X-band (9.5 GHz), Ka-band (23 GHz) or Ku-band (35 GHz) [85, 86]. Due to the applied field, the spin degeneracy of the level is lifted, since there are two possible orientations (parallel and anti parallel) and the electrons and holes are forced to polarise along the static field. The level splits in its Zeeman components. From classical electricity and magnetism, the energy of a magnetic moment μ_{H} in a magnetic field $\boldsymbol{H}$ is $-\mu._{\mathrm{H}}\boldsymbol{H}$. Magnetic resonance occurs when the energy difference between the two electron orientations is equal to Planck' s constant, h, times the microwave frequency. For the very simple case of an isolated electron, the resonance requirement may be expressed as [84, 85]:

$$h\nu = g_0 \beta_e H \tag{2.26}$$

where g_0 = 2.002319 (for a free electron) and β_e is the Bohr magneton, $qh/4\pi m_e$ (m_e the electron mass). The resonance condition for an electron which does not otherwise interact with its surroundings is described by (2.26). At resonance, both spin up and spin down populations will be equalised (spin flip), while otherwise, the net polarisation at temperature T will be [87]:

$$P = (n^{\uparrow} - n_{\downarrow})/(n^{\uparrow} + n_{\downarrow}) \approx g_0 \beta_e H / 2kT \tag{2.27}$$

with $n^{\uparrow}$ and $n_{\downarrow}$ the number of up and down spins.

The structural and chemical information provided by EPR is due to deviations from this simple expression. Generally speaking, these deviations are due to spin-orbit coupling and electron-nuclear hyperfine interactions. The spin-orbit coupling may be included in the EPR resonance condition by replacing the constant g_0 in expression (2.26) with a second rank tensor g_{ij} [84 - 86]. The symmetry of this tensor, reflects the symmetry of the paramagnetic centre. In other words, studying the resonance for different orientations of the magnetic field with respect to the sample axis may reveal structural information of the underlying defect. The second important source of deviation from (2.26) is the hyperfine interaction of the unpaired electron with the nearby nuclei. Certain nuclei have magnetic moments I_N, which may take $2I_N$+1 different orientations in the applied magnetic field. The impact of these hyperfine interactions are schematically depicted in Fig. 2.22 for an unpaired electron interacting with a single nucleus, I_N=1/2. The significant magnetic nuclei in the Si-SiO_2 system are summarised in Table 2.1 [84, 85]. Together with the natural abundance of the isotopes, they provide a unique fingerprint of the nearby nucleus in a paramagnetic centre. In other words, the number of EPR lines and the ratio of their amplitudes points to the presence of a certain element in the paramagnetic defect, thus yielding chemical information.

Table 2.1. Nuclear spin I_N and natural abundance of some elements

Isotope	I_N	Abundance [%]
^{27}Al	5/2	100
^{1}H	1/2	99.9
^{14}N	1	99.6
^{18}O	0	99
^{31}P	1/2	
^{28}Si	0	95
^{29}Si	1/2	4.7

The fact that EPR provides both structural and chemical information of defects makes it a very powerful tool. A minor aspect of the technique is its rather low sensitivity. Generally, quite large volumes in the range 1 cm^3 (or 1 cm^2 for a MOS system) are required to generate a detectable EPR signal, i.e., a sufficient density of spins should be available (about 10^{15} cm^{-3}). Its accuracy with respect to defect concentration is quoted to be better than a factor 2 in absolute precision and better than ±10 % for relative densities [85]. In order to enhance the sensitivity, different modifications have been introduced. Perhaps the most powerful technique is based on the so-called spin-dependent recombination (SDR) method, originally proposed by *Lepine* [88]. The basic idea is that there is an effect of magnetic alignment of the electron spin on the electron-hole recombination current associated with a paramagnetic defect. Out of resonance, both the free electron and the defect spin will be aligned predominantly parallel with ***H***. This means that recombination is less probable. Simply speaking, the applied magnetic field has the effect of reducing the deep level' s capture cross section. At the resonance condition, the spins flip to equal up and down orientation, so that the transition becomes more probable, resulting in a detectable increase of the recombination current.

The advantage of SDR is that it can be applied to small devices (MOSFETs, diodes) which are relevant for practical applications. An example is shown in Fig. 2.23, for the case of a silicon MOSFET, which is measured in the gated diode configuration, at slight forward bias and with the gate in depletion [89]. It is shown there that after the application of hot-carrier stress, SDR is observed. The corresponding *g*-factor is ascribed to the P_{bo} silicon dangling bond centre at the (100) Si-SiO_2 interface [80], the structure of which will be described below. SDR has also been applied to the study of recombination active defects in γ-irradiated oxides [90] and has led to the identification of a neutral oxide trap mediating stress-induced leakage current in thin gate oxides [87]. In the latter case, trap densities as low as $1x10^9$ cm^{-2} can be detected. It has furthermore been noticed that SDR not necessarily observes the same defects (densities) as EPR [91]. It is mainly sensitive to defects producing maximum recombination current – they are at the same time the most relevant for device operation. A kind of SDR measurements has also been performed on proton-implanted silicon, for the detection of radiation defects (A-centres and carbon-related defects) [92]. In low-temperature electron-irradiated silicon solar cells, an SDR spectrum ascribed to the divacancy has been observed in the p-n junction depletion region [93].

A more detailed discussion on the most important radiation defects in semiconductor materials will be given in Chap. 3 and 4. Here, only a few examples of basic defects in the Si-SiO_2 system are reported. The spectra and structure of the lattice vacancy (V) in silicon are shown in Fig. 2.24 [84], for the single positive (V^+) and negative (V^-) charge state. The lattice vacancy can occur in four different charge states, only two of which are paramagnetic. In order to be able to detect the silicon vacancy, irradiations have to be performed at temperatures well below 100 K (see Chap. 3). At higher temperatures the vacancy is unstable and highly mobile, so that it will be captured by impurities or other defects to create stable point defect complexes. Therefore, the lattice vacancy cannot be observed at room temperature irradiations. The silicon interstitial in p-type Si is even mobile at liquid

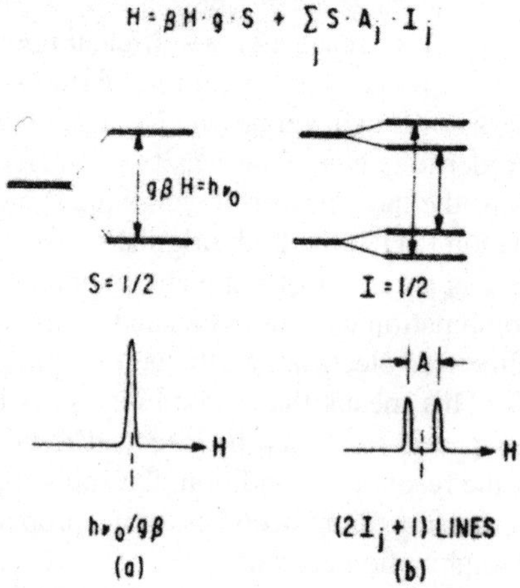

Fig. 2.22. Hamiltonian and spectrum (**a**) for an unpaired electron and (**b**) interacting with a single nucleus, I_N=1/2. [After Watkins [84]]

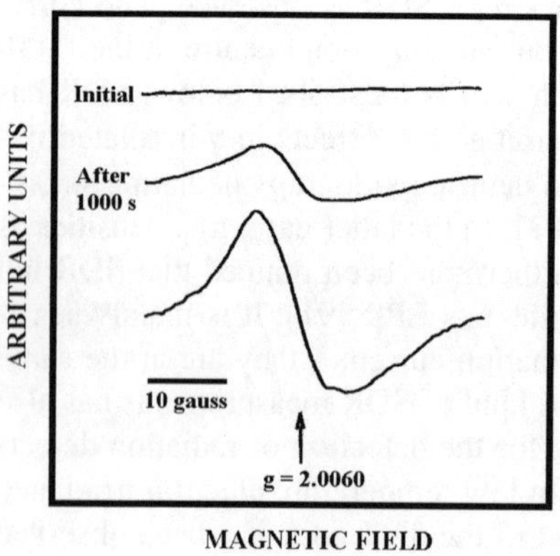

Fig. 2.23. SDR spectra for a transistor which has been subjected to channel hot hole stressing. The applied magnetic field is oriented perpendicular to the (100) Si-SiO_2 plane. The SDR measurements were performed while biasing the transistor at gate voltages corresponding to the peaks in the recombination current, in the forward gate diode configuration. V_F=0.25 V. [After Krick et al. [89]]

helium temperatures, explaining that there is so far, no unambiguous experimental observation of its EPR signal [94, 95]. This is derived from the fact that even at 4.2 K in p-type Si, interstitial reaction products – i.e., interstitial Al or B – are detected in equal concentrations as the created vacancies. In n-type silicon, defect production is seen to be much lower at cryogenic temperatures, but the available evidence indicates that interstitial migration may not occur until annealing to ~ 140-175 K [94, 95], where the trapped interstitial configurations are observed to emerge. A spectrum labeled Si-G25 was found to anneal in this temperature range and it has been tentatively ascribed the isolated interstitial silicon atom [94]. However, to date, no generally accepted experimental evidence of the structure or the electrical properties of I is available [95].

In the Si-SiO_2 system, one can distinguish two important families of radiation defects [85], one associated with the interface and another with the bulk oxide. The structure of the P_b centre is given in Fig. 2.25a; it has been observed for (111) silicon interfaces. Such a P_b centre corresponds to a so-called dangling bond, where one silicon atom at the interface is back-bonded to three other silicon atoms, whereby one unpaired electron is oriented normal to the surface. For the technologically more important (100) interface two EPR centres have been identified, which are very similar to the P_b defect, namely, P_{b0} and P_{b1}. P_{b0} is schematically illustrated in Fig. 2.25b [85]. The structure of the P_{b1} defect is as yet under discussion, although it is clearly a silicon "dangling bond" at the interface. It has also been demonstrated extensively that the P_b family of defects is responsible for the electrically active interface traps, which largely dominate the surface recombination at the Si-SiO_2 interface [85].

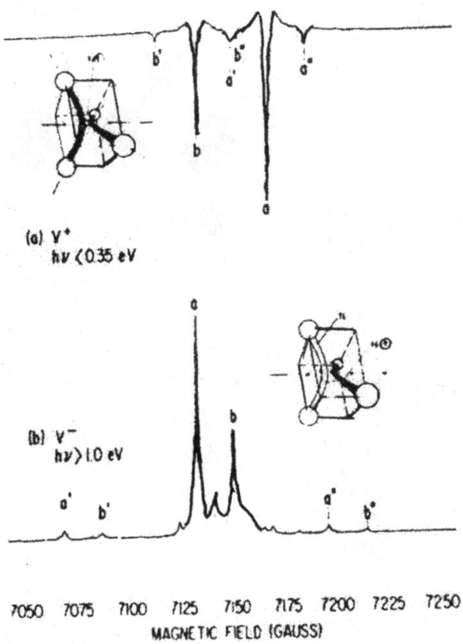

Fig. 2.24. The vacancy spectra, $\boldsymbol{H} \parallel$ <100>, ν_0=20.5 kMc. (a) V^+ at 4.2 K, illumination with light <0.35 eV; (b) V^- at 20.4 K, illumination with light > 1.0 eV. [After Watkins [84]]

A second important class of radiation induced EPR centres are the E' centres, which involve an unpaired electron localised on a silicon back-bonded to three oxygens (oxygen vacancy), as shown in Fig. 2.26. Usually, though not always, the paramagnetic site is coupled to a positively charged diamagnetic silicon [85]. There exists a general consensus that the E' centres are dominating trapped hole centres in technologically relevant grown oxide films. So far, five different but related E' variants have been identified. Another important finding is that both trapped holes and E' centres preferentially occur at the Si-SiO_2 boundary, after irradiation under a positive gate bias [85], again emphasising their close connection. For more details about the EPR defects in the Si-SiO_2 system the Reader is referred to the overview paper by *Lenahan and Conley, Jr* [85]. The nitrogen-related oxide defects will be discussed in the context of Chap. 6.

It should finally be remarked that based on the fundamental insight provided by EPR in correlation with electrical investigations, one can conceive a physics-based
model for radiation-induced hole trapping [96] and interface-trap formation [85, 97], which yields a good first-order quantitative agreement and has demonstrated promising predictive capability.

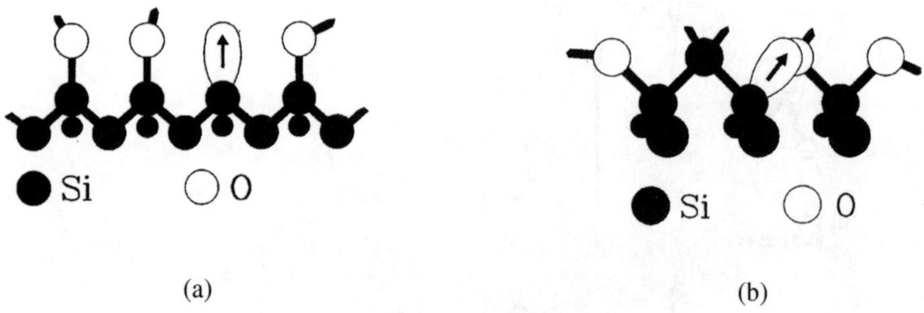

Fig. 2.25. Schematic illustration of a P_b centre at the (111) Si-SiO_2 interface (a) and (b) of a P_{b0} centre at the (100) Si-SiO_2 interface. [After Lenahan and Conley, Jr [85]]

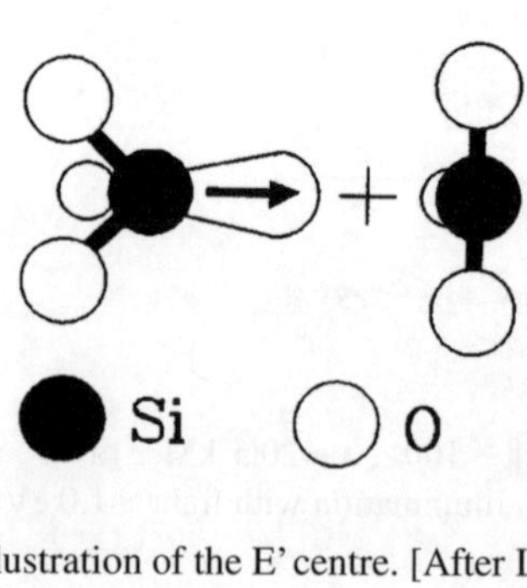

Fig. 2.26. Schematic illustration of the E' centre. [After Lenahan and Conley, Jr [85]]

2.4.2 Deep Level Transient Spectroscopy (DLTS)

While EPR enables to identify structurally and compositionally the important radiation defects, from a practical viewpoint, their electrical properties are even more important. In addition, EPR requires generally quite large samples and rather high spin densities (trap concentrations), which may not always be representative for real applications. Important electrical trap parameters are its energy position E_T in the band gap, the corresponding charge states, the concentration N_T (or for radiation defects: their introduction rates $R_T = N_T/\Phi$) and the capture cross section for electrons (σ_n) and for holes (σ_p). When these parameters are known, one can substitute the numbers in the SRH model to calculate the generation or recombination rate and, hence, the electrical activity of a trap.

The invention of Deep Level Transient Spectroscopy (DLTS) by *Lang* [98] has meant an enormous breakthrough in the study of electrically active radiation defects in semiconductors. The technique is nowadays widely used, even in production environments for quality control and defect/contamination monitoring [99]. The basis of DLTS (or capacitance transient methods) consists of the repetitive filling and emptying of deep levels in the depletion region of a diode by a bias pulse. The resulting diode capacitance transient is filtered electronically and monitored as a function of the sample temperature. Frequently, a p-n junction or a Schottky barrier, deposited on the semiconductor material under investigation, is employed, although the technique can also be adapted to MOS devices, for interface-trap studies. The diode is operated under reverse bias V_R, which is during a bias pulse reduced to V_P (i.e. 0 V). Empty traps, residing in the former depletion region, will be able to capture free carriers (electrons in n-type and holes in p-type material) and become occupied (filled). After restoration of the original bias V_R, the charge in the depletion region will be lower than before in the case of a majority carrier pulse, due to the carriers stored in the deep levels. These carriers will be released again through thermal emission, which proceeds exponentially in time, with a time constant which is dictated by detailed balance and given by:

$$\tau_g = \frac{1}{K_T T^2} \exp\left(\frac{E_T}{kT}\right) \tag{2.28}$$

As can be seen from (2.28), τ_e is thermally activated. The set of parameters (K_T,E_T) is called the trap signature. K_T is proportional with the majority carrier capture cross section.

This thermal discharging of the occupied trap is monitored generally by measuring the capacitance of the reverse biased diode in function of time after the filling pulse [98]. Alternative methods follow for example the current, or operate under constant-capacitance by varying the bias V_R. In the first instance, for not too large point defect concentrations, the resulting capacitance transient proceeds also exponentially with time, having the same time constant, i.e.,

$$C(t) = C(\infty) - \Delta C \exp\left(-t/\tau_g\right) \tag{2.29}$$

with ΔC the absolute value of the capacitance transient amplitude and $C(\infty)$ is the steady-state capacitance for t going to ∞ . From (2.29) it follows that for a majority carrier pulse, i.e. with $V_P \leq 0$ V, the sign of the transient is negative. Given the thermally activated nature of τ_e, the shape of the transient will also vary strongly with temperature, as illustrated in Fig. 2.27 [98]. At low T and long τ_e, the capacitance will vary very slowly with time, while the signal has died out soon after the end of the filling pulse for sufficiently high T.

The salient feature of DLTS is how these transient signals are converted into a quasi-spectrum, as a function of temperature. In the original technique by *Lang*, the capacitance transient is measured at two fixed times t_1 and t_2 after the pulse and the signal $C(t_1)$-$C(t_2)$ is measured. From Fig. 2.27, one can easily see that when the temperature is varied, a peak-shaped signal will result. In addition, it can be demonstrated that the peak maximum corresponds with a time constant τ_{max}, only defined by the selected instrumental times t_1 and t_2, namely [98]:

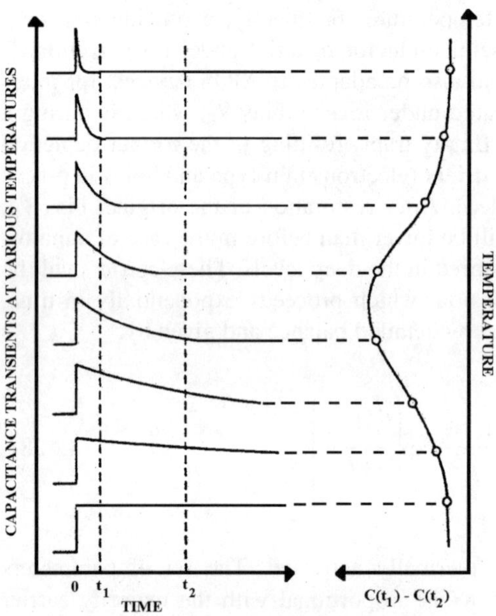

Fig. 2.27. Illustration of how a double boxcar is used to define the rate window in DLTS. The left-hand side shows capacitance transients at various temperatures, while the right-hand side shows the corresponding DLTS signal resulting from using the double boxcar to display the difference between the capacitance at time t_1 and time t_2 after the end of the filling pulse, as a function of temperature. [After Lang [98]]

$$\tau_{max} = (t_1 - t_2)[\ln(t_1/t_2)]^{-1} \qquad (2.30)$$

The corresponding emission rate $e_{max}=\tau_{max}^{-1}$ is often called the emission rate window. Changing t_1 and/or t_2 will change e_{max} and hence the peak position T_{max}, corresponding with a certain deep level E_T. If one repeats the temperature scan for different well-chosen rate windows, a set of peak maxima T_{max} can be obtained. According to (2.28) plotting τ_{max} versus $1/T_{max}$ in a so-called Arrhenius diagram should result in a straight line with slope E_T/k and intercept K_T [3]. Beside the double boxcar technique, many other analogue filtering operations have been proposed for imposing the rate window, each of which have their advantages and disadvantages. More recently, a digital (numerical) filtering operation is applied directly on the measured transient data. This has the big advantage that in principle, only one temperature scan suffices to determine the activation energy E_T.

For each deep level present in the material above a minimum concentration limit, one will obtain a DLTS peak. A spectrum generally consists of a series of peaks, in the temperature interval studied. For silicon, one normally measures from the freeze-out region (T=20-30 K) up to room temperature. For Ge, one has to scan between 4.2 – 300 K, while in GaAs a suitable range is 4.2 – 450 K. In principle, one can probe the complete band gap of the material by employing an injection pulse, i.e., going slightly in forward bias during the pulse, when using a p-n junction. In the case of a Schottky barrier, minority carriers can be supplied by applying an optical pulse. If only a majority carrier pulse is used, the upper or lower band half is probed, respectively, for n- and p-type material. It is also fairly easy to determine the deep-level concentration from a DLTS peak, with transient amplitude ΔC. It can be demonstrated that the relative trap concentration is given by [98]:

$$\frac{N_T}{N_D} \approx 2\frac{\Delta C}{C(\infty)} \qquad (2.31)$$

with N_D the back-ground doping concentration, derived from a high-frequency C-V measurement on the diode for example. The measured DLTS peak amplitude is proportional to ΔC, from which N_T can be calculated. From (2.31) one derives that the concentration detectivity is mainly determined by the doping concentration of the material. A detection limit better than 10^{-4} is often quoted. This implies that for a starting concentration of 10^{12} cm^{-3}, typical for the substrate of silicon radiation detectors, one can measure deep levels at the 10^8 cm^{-3} range. There are few analy-

[3] Often, $\tau_{max}T_{max}^2$ is plotted versus $1/T_{max}$ in order to account for the T^2 temperature dependence of the emission rate.

sis techniques – if any – which can handle such a low density of impurities/ point defects, explaining the success of DLTS. However, for microelectronic device operation, such concentrations of metallic contaminants are of technological relevance.

Another useful feature of DLTS is that it is possible to record profiles as a function of the depletion depth *W*. In this way, the peak damage after particle irradiation can be recorded and compared with TRIM modeling, like in Fig. 2.6. To do this, one has to measure the peak height in function of the reverse bias V_R, keeping the pulse height constant, for example. The extrinsic Debije length of the material [100] limits the depth resolution.

A final parameter, which can be determined by DLTS, is the majority carrier capture cross section and its temperature dependence. A first way is to extract $\sigma_{n,p}$ from the K_T pre-factor of an Arrhenius diagram, although this generally gives only a rough order of magnitude estimate. A better way is to monitor the DLTS peak height in function of the filling-pulse duration. In principle, the trap filling will also proceed exponentially, with a capture time constant given by:

$$\tau_c = \frac{1}{n\sigma_n v_{th}} \tag{2.32}$$

for the case of an electron trap (n-type material). Plotting the peak amplitude versus the pulse duration in a semilog plot should yield a straight line with slope given by (2.32). Based on the order of magnitude of the capture cross section, one can derive the nature of the trap, i.e., whether it is a Coulombic attractive trap (a deep donor in n-type or a deep acceptor in p-type material) for σ_n or $\sigma_p > 10^{-14}$ cm^2. Neutral traps have a cross section in the range 10^{-15} cm^2. Deep levels with cross sections below 10^{-16} cm^2 can be considered repulsive for the majority carrier, i.e., they correspond with deep donors in p-type and deep acceptors in n-type material[4]. An alternative way to distinguish the charge state of the defect is by studying the electric field dependence of the activation energy. An important remark is that it is far more difficult to study the minority carrier cross section in DLTS [101], although from a recombination lifetime view this is the important one. It has only been possible in a few well-defined cases. In general, other, more sophisticated modifications of the DLTS technique are required for that purpose [78, 81, 93, 102].

According to Fig. 2.27 and to the spectra shown in later chapters, the DLTS peaks are quite broad, limiting the achievable energy resolution to a few *kT* typically. This is one of the limitations of the technique. However, applying the so-called Laplace-DLTS method [103] enables to drastically enhance the energy resolution to a few meV so that formerly overlapping peaks can now be resolved. In this case, not the temperature is varied, but the signal frequency is swept. This

[4] Often a deep donor is confused with an electron trap and a deep acceptor with a hole trap. All peaks in a majority carrier n-type spectrum are electron traps. Whether they act as deep donors or acceptors depends on σ_n. The same goes for p-type material.

requires a very stable temperature control to within 0.1 K or better and quite some computing power. The numerical and experimental difficulties are such that presently, one can not expect it to become as popular as standard DLTS.

An important drawback of DLTS is that it generally lacks chemical information. One only measures a spectrum of peaks – one for each deep level present – without any additional information on the chemical identity of the impurities involved in the defect complex. The energy resolution is generally not good enough to resolve isotope effects like in many optical techniques (e.g. photoluminescence) or structural information by the application of an external stress [104, 105]. In only a limited number of cases, one has undoubtedly identified the nature of a deep level found by DLTS. This can for example be achieved by the implantation of radioactive transition metals [106]. The concurrent growth of one (or more) DLTS level at the expense of another according to the half life of the isotope is a good fingerprint of its identity. Given the high sensitivity of DLTS for low deep level concentrations, it is otherwise not an easy task to identify an unknown peak. Firstly, one can compare with literature data obtained on deliberately contaminated, irradiated or doped samples. If the concentration is sufficiently high, other element specific analysis techniques like EPR, Secondary Ion Mass Spectroscopy (SIMS) can be applied on the same sample or material in order to identify the responsible impurity or constituents of a complex defect. In addition, the study of the thermal stability, i.e., the annealing behaviour, may help to specify unknown deep levels. It has also been demonstrated that Laplace-DLTS offers sufficient energy resolution to study compositional changes in the defect complexes [103, 107].

As mentioned before, DLTS has been and still is extensively used to investigate radiation defects in semiconductors. It has for example been applied to the study of the vacancy in silicon. A spectrum obtained after a 2 MeV electron irradiation at 55 K of p-type silicon is shown in Fig. 2.28 [108]. Besides the double donor level of the single vacancy (V^{++}), one can notice two donor levels ascribed to the Sn-V complex, in the double and the single positive charge state. A total of five charge states have been identified for the vacancy in silicon by EPR [94, 95], represented by V^{++}, V^{+}, V^{0}, V^{-} and V^{--} and schematically shown in Fig. 2.29. One can notice the 'negative U' property of the two donor levels [109], i.e. they occur in reverse order where the second electron is more strongly bound to the V^{++} core than the first one. In other words, the capture of the first hole is immediately followed by a second, so that in DLTS, one can only detect the E_v+0.13 eV level of Fig. 2.28 and the transition from the neutral charge state (indicated by the subscript 0) to the double positive state, i.e. $V^{++/0}$. In addition, due to the double charge emission, the corresponding DLTS peak has double the normal amplitude.

Similar as for EPR, there is not a generally accepted DLTS level assigned to the self-interstitial. One potential candidate, observed in 77 K proton or α-particle irradiated p-type silicon is found at E_c-0.39 eV [110]. Associated with it is an EPR spectrum labeled Si-AA12, which anneals in the range 200-300 K or under carrier injection at 77 K. The level position is in close agreement with theoretical expectations [95]. However, theory also predicts that it may be very difficult, if not impossible to detect the isolated interstitial in n-type silicon [95].

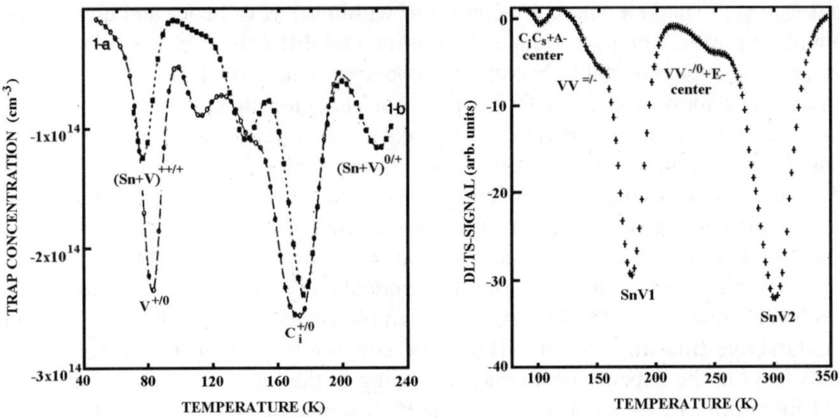

Fig. 2.28. Typical DLTS spectra measured on a p-type sample irradiated at 55 K by a 2 MeV electron beam (a) before and (b) after a 225 K 30 min anneal. [After Goubet et al. [108]]

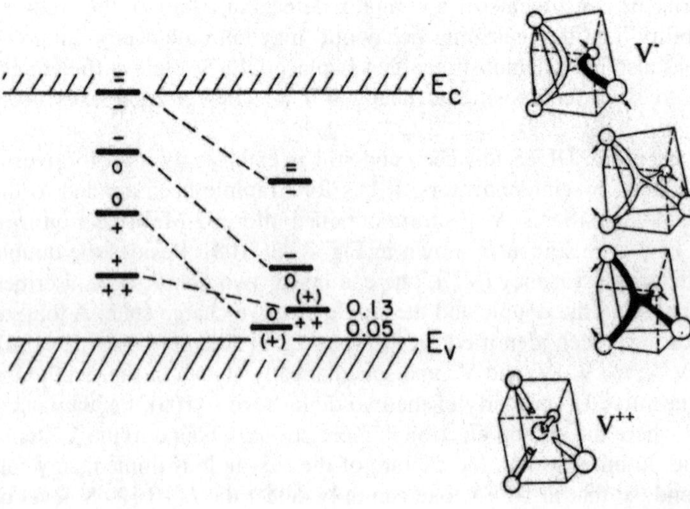

Fig. 2.29. Level position for the vacancy before, and after lattice relaxation, as shown in the insets. The darkened bonds indicate the location of the electronic spin seen in EPR. [After Watkins [95]]

As mentioned above, the microscopic information, derived from DLTS on irradiated material can be substituted in the SRH model for the relevant electrical parameter (recombination lifetime, diffusion length, leakage current...) in order to predict the device response to radiation. This has been applied with reasonable success to the case of high-resistivity silicon radiation detectors [66, 67, 101] for the prediction of the leakage current degradation and the carrier removal rate.

2.4.3 Photoluminescence (PL) Spectroscopy

Besides electronically active defects, other centres give rise to a level falling in the valence or conduction band and are not active. However, such "neutral" radiation defects can affect the optical properties of the semiconductor, which is of strong interest for opto-electronic applications. Well-known examples of such centres are substitutional carbon (C_s) and interstitial oxygen (O_i) in silicon, which are technologically extremely important impurities and as will be seen in Chap. 3, to a large extent interact with radiation-induced V' s and I' s. While these impurities have no electrical impact, they give rise to clear vibrational absorption bands. In addition, some of the deep-level defects also give rise to absorption bands [111 - 113]. An overview of the application of absorption spectroscopy to III-V componds can be found in [114]. Therefore, absorption spectroscopy may provide useful chemical and structural information, as the energy resolution is high, particularly when applied at cryogenic temperatures. One drawback of the technique is that one needs a rather large total number of absorption centres, which implies rather thick, technological less relevant samples, or a high density of defects. This can be achieved by applying high fluence irradiations, which may be far from a realistic operation condition. One optical technique which largely overcomes these difficulties is photoluminescence [115]: it can be applied to low defect concentrations and to standard samples, provided that competitive non-radiative carrier recombination is not overwhelming the optical transitions. This means in practice that the surface quality of the samples needs to be acceptable, in order to reduce the surface recombination.

In PL spectroscopy, free electron-hole pairs are created by the absorption of photons with energy larger than the band gap. Usually, visible-laser excitation operating at ~2 eV is applied. Excess carriers are typically created in the top few μm of the sample, depending on the absorption coefficient of the material at the excitation wavelength. The PL signal, on the other hand, originates from a larger region, as the free carriers can diffuse some distance from their zone of creation. The measurement configuration for PL is schematically depicted in Fig. 2.30.

The created excess carriers recombine by different mechanisms. At low temperatures ($T<30$ K in silicon) excitons[5] may recombine by emission of a photon with lower than band gap energy. For indirect gap materials, this transition involves assistance of a phonon to conserve the momentum. The corresponding energies in silicon have been determined to be [115]:

$$\begin{aligned} \hbar\omega &= 18.4\pm0.2\text{ meV}, \quad \text{TA mode} \\ &= 56.2\pm1\text{ meV}, \quad \text{LO mode} \\ &= 58.0\pm1\text{ meV}, \quad \text{TO mode} \end{aligned} \tag{2.33}$$

[5] An exciton is an electron bound to a hole, with energy below the conduction band. An exciton in silicon has a hydrogenic series of excited states with ground energy 14.3 meV [115].

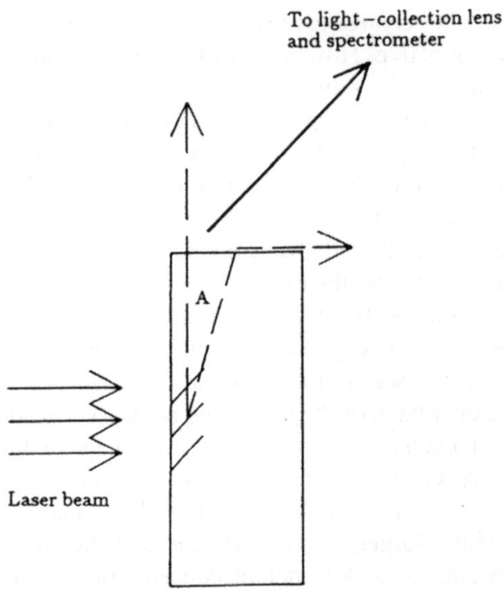

Fig. 2.30. Refraction of the luminescence originating (in the hatched area) near the surface of a cuboid of silicon is efficiently collected with "135° geometry"; total internal reflection occurs for A>17°. [After Davies [115]]

with TA the transverse acoustic phonon mode, TO the transverse optical and LO the longitudinal optical mode, respectively. In principle, also no(zero) phonon (NP) transitions can be observed, close to the band gap energy.

Besides free exciton luminescence, one can also observe the recombination of excitons bound to a defect centre, which may for example be the dopant atoms (B, P,...). The binding energy of the exciton depends on the potential of the defect. As a result, a series of PL lines is obtained for each type of optical recombination defect, with a spacing corresponding to (2.33), for near band gap excitonic type of transitions. Centres with a stronger binding energy and local mode phonons produce a more complex spectrum. For silicon, one usually investigates the energy region 0.6-1.2 eV. An example of a basic radiation defect, namely interstitial carbon (C_i) is given in Fig. 2.31 [116].

One important advantage of PL is its high spectral resolution, which renders the technique sensitive to isotope or mechanical stress effects. In this way, detailed structural and chemical information about the underlying defects can be obtained [115]. A large number of PL lines have been identified in silicon, which were summarised by *Davies* [115]. PL has been extensively applied to irradiated silicon, providing additional insight in the defect formation mechanisms. Moreover, PL is somehow complementary with DLTS, since in the first case, one probes optical and in the second case thermal recombination centres. Generally, PL is quenched in samples containing large DLTS trap concentrations and vice versa.

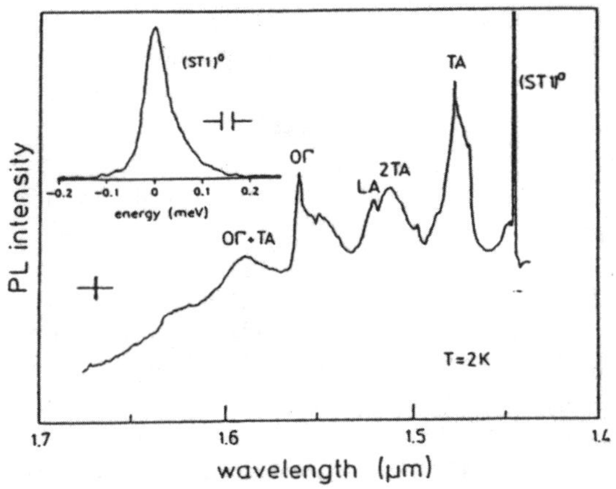

Fig. 2.31. PL from the interstitial carbon atom, and (inset) its zero phonon line. The band is superimposed on a broad background. [After Thonke et al. [116]]

While the spectral resolution is superior, it is more difficult to quantitatively determine the concentration of the PL centres. However, based on PL of electron irradiated material, it is possible to lower the detection limit for substitutional carbon in silicon to the 10^{14} cm^{-3} range [117]. It is believed that a similar approach can improve the detection limit for interstitial oxygen as well [118].

2.5 Conclusions

It is clear from the foregoing that the interaction of high-energy irradiation with semiconductor devices is a complex matter, which requires different levels of understanding. A first level deals with the interactions governing the energy transfer from the particle/photon to the material. The energy deposited will depend on the particle energy, mass and charge and on the target mass. In addition, most of the energy transfer will go into ionising events for most particles/photons, while only a fraction causes atomic displacements. The first mechanism is well described by the LET function, while it has been demonstrated that the NIEL can play a similar role with respect to displacement damage. Both LET and NIEL are very useful tools for the modeling of the expected device degradation. Furthermore, a number of well-established simulation codes, based on Monte Carlo calculations, are available to model the lattice damage profiles. They are not only useful for irradiation damage, but are essential for the modeling of ion-implantation related damage in semiconductor process simulation.

In order to model the macroscopic device degradation, a methodology has been developed, which describes the damage in terms of empirical damage factors. In combination with the LET or NIEL, one can fairly easy model the expected degradation of certain electrical parameters, provided that their degradation proceeds approximately linearly with dose or fluence. The main advantage of this strategy is that it largely reduces the amount of radiation testing for a certain device or technology generation. This is particularly important for the case of displacement damage, where testing for different energies, particles etc. is time consuming and costly. The disadvantage of this method is its largely empirical nature. This implies that for every new technology generation or novel component it is absolutely required to perform in-depth experimental testing before they can be recommended for use in space applications.

The highest level of understanding is forming the link between the primary energy deposition mechanisms and the final stable and electrically active radiation defects formed. In the course of the last four decades, a number of spectroscopic tools have been developed – with EPR, DLTS and PL as the most important ones, in our opinion – which are very suitable for that purpose. Based on these spectroscopic studies, a thorough understanding of the major radiation-induced point defects in the Si-SiO_2 system has been gained. The level of understanding for other semiconductor materials is lower, although progress has been made there as well, as will be described in Chap. 3 and 4. The defect parameters obtained from these studies are valuable input for physics-based damage simulators, which are currently under development. It is expected that this type of modeling will gain more importance in the coming years. This is related to the fact that the smaller the device dimensions become, the more difficult it is to apply a purely macroscopic picture. In the end, one will have to deal with quantum mechanical effects, which require further specialised analytical tools.

3 Displacement Damage in Group IV Semiconductor Materials

3.1 Introduction

This chapter gives an overview of the main radiation defects in the group IV elemental semiconductors silicon and germanium and their $Si_{1-x}Ge_x$ alloys (x the percentage fraction). As shown in Chap. 2, a high energetic particle or ion may loose part of its energy by interaction with the nuclei of the target material. This initially results in the creation of vacancy-interstitial pairs, of which only a small fraction escapes direct recombination. These intrinsic point defects are generally highly mobile and may, therefore, interact with other point defects and impurities to form more stable radiation defects. The nature of these defects depends strongly on the irradiation and/or annealing temperature, so that different damage regimes can be identified. Usually, increasing the annealing temperature leads to a further clustering or aggregation of the point defects into larger, more stable defects. Small clusters may, however, also directly form in the 'cluster damage' region induced by high energy neutrons or ions, where a high density of primary V-I pairs is created.

For each of the materials, an overview will be given of the main radiation defects and their electrical parameters, usually derived from DLTS studies. Whenever relevant, information about radiation damage produced during certain semiconductor processing (ion implantation, dry etching, …) steps are included.

Displacement damage has a major impact on "bulk" devices, which rely on electrical material properties like minority carrier lifetime or carrier density. These include amongst others diodes, bipolar transistors, radiation detectors, power devices and solar cells. In certain cases, it has been possible to identify the main defect level responsible for the operational degradation amongst the various created radiation centres. This is valuable information, as it may lead to a possible improvement of the radiation performance, by applying substrate or device hardening. While for silicon, several hardening schemes have been proposed and implemented for specific applications, this is not yet the case for Ge or SiGe compounds. However, as will become clear, the use of SiGe layers may inherently enhance the resistance against displacement damage.

3.2 Displacement Damage in Silicon

Vacancies and interstitials in silicon are highly mobile, even at low temperatures. In addition, they may occur with different charge states, depending on the position of the Fermi level (see e.g. the V scheme of Fig. 2.29). During irradiation most of the energy is disposed by ionisations, so that a high density of free carriers is present, which can modify the charge state and the diffusion and reaction properties of the point defects. In addition, capturing of carriers at the defect site can serve to dump vibrational energy into the system, as the atoms move toward their configurations in the new charge state. This, in turn, may assist the migration, a phenomenon known as "recombination-enhanced migration" (REM) [1, 2]. It can lead to a high mobility, particularly for the self-interstitial in p-type silicon, which diffuses even at 4.2 K. It is believed that a combination of a low diffusion barrier for the I^+ state and REM induced by carriers created by the irradiation results in an athermal diffusion of interstitials [1, 2].

Both I and V have their own chain of reactions. While the vacancy tends to pair with impurities, leading to P-V, V-O, and/or V-V centres, an interstial may kick out impurities from their substitutional lattice site, which in turn become mobile and interact with other species. The properties of the main reaction products will be described to some extent, with particular emphasis on their electrical parameters, which have been usually determined by DLTS. One important remark, which is generally valid for all studies trying to link the device performance to certain defect centres, needs to be made. The DLTS parameters of a deep level are derived in a certain temperature range, which may not be identical to the device operating temperature (mostly 300 K). It is for example often found that the capture cross section for holes or electrons is thermally activated (or has a power law dependence on temperature), so that its measured value has to be extrapolated to room temperature. Substituting these extrapolated data in the SRH model may lead to significant discrepancies between theory and practice, as demonstrated on a number of occasions [3, 4]. In spite of these difficulties, successful microscopic damage modeling has been achieved for specific cases, some of which will be highlighted further. These physics-based insights also indicate possible routes to silicon material and device hardening, as briefly outlined in the last section.

3.2.1 Radiation Defects in Silicon

The vacancy EPR spectrum in low-temperature irradiated silicon disappears after a 15 min annealing sequence at ~70 K in n-type, ~ 150 K in p-type and ~200 K in high resistivity (near intrinsic) material [1, 2]. At these temperatures, the Fermi level is such that the V transforms in a different, more mobile charge state. The kinetics of the processes have been determined, yielding for the vacancy migration activation energy 0.18±0.02 eV in n-type, 0.45±0.04 eV in high resistivity and 0.32±0.02 eV in p-type material. This presumably reflects the different properties of the V^{--}, V^0 and V^{++} charge states, which are dominant for each of the Fermi level positions in the respective cases [1, 2]. It should also be remarked that because of the low stability of vacancies in n-type material, it has so far not been possible to study the electrical acceptor levels shown in the DLTS spectra of Fig. 2.29. Their position has only indirectly been estimated from EPR investigations.

Once the vacancy starts to diffuse, it can be trapped by immobile point defects, leading to a whole series of complexes. The structure of the most prominent ones, identified by EPR is represented in Fig. 3.1, together with the associated energy level. Their formation and stability range is schematically represented in Fig. 3.2 [1, 2], from which one can derive that at room temperature, the V-P (or vacancy-group V pairs, also called E-centres), V-O (or A-centre) and V-V (divacancy or V_2) are stable radiation defects, in n-type silicon. In p-type silicon, V-O and V-V are the stable vacancy related centres, while V-B is unstable.

In the case of heavy particle damage, as occurs in reactive ion etching and ion implantation processes, the recoiling atoms have substantial kinetic energy which, in turn, can produce many displacements in a highly localised region. As a result, vacancy aggregation is expected to dominate over interactions with the more distant isolated dopant impurities. Based on EPR results of neutron-irradiated material, structures for several small vacancy aggregates (di-, tri-, tetra-, penta-vacancy,…) have been proposed [1]. In addition, complexes of the di- and tri-vacancy with interstial oxygen, resulting in V_2O and V_3O centres have been observed [5]. It is inferred that oxygen plays a stabilising role by tying up the various

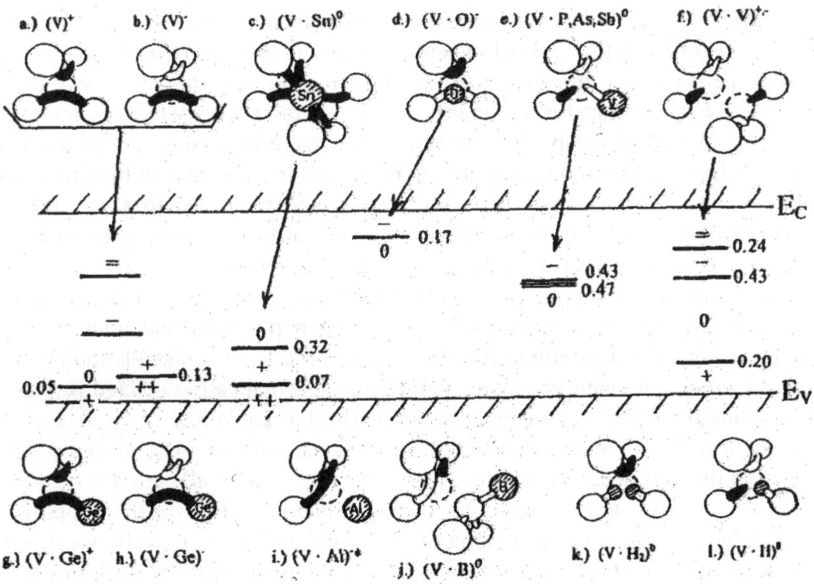

Fig. 3.1. Structures of trapped vacancies and their electrical levels (eV from nearest band edge). [After Watkins [1]]

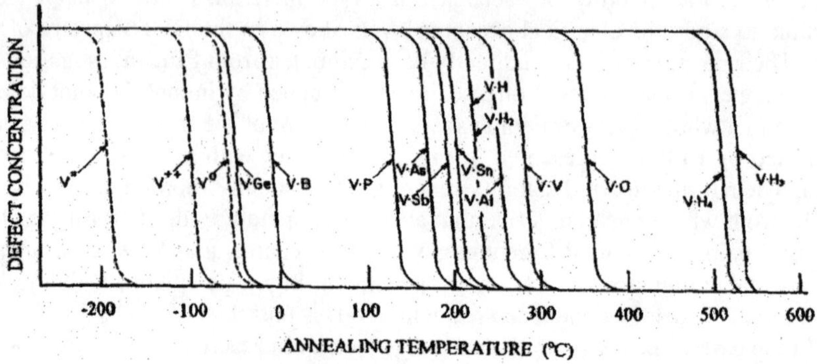

Fig. 3.2. Schematic of vacancy and vacancy-defect pair annealing stages (~15 min isochronal). [After Watkins [1]]

reconstructing bonds, like in the A-centre. The occurrence of these centres will be particularly important in Czochralski (Cz) silicon, which is known to contain a high concentration of interstital oxygen (O_i). For state-of-the-art Cz material, typical values are in the range 6-7x10^{17} cm^{-3}, which is several orders of magnitude higher than in Float-Zone (FZ) silicon[1]. This most likely explains the transformation of the V-V into V_xO_y upon annealing of irradiated p-type Cz silicon at 200°C. This is illustrated in Fig. 3.3 [6], where can be seen that the H(0.19) single donor level of the divacancy transforms in the H(0.24) hole trap, 0.24 eV above the valence band E_V. This transformation proceeds much more slowly in FZ silicon [6].

As mentioned in Chap. 2, there is no clear experimental evidence about the isolated silicon interstitial during low temperature-irradiation. On the other hand, values for the self-interstitial diffusivity D_I have been derived from the diffusion behaviour of dopants, e.g. B, or transition metals at high temperatures. However, the resulting values differ by orders of magnitude (for an overview, see e.g. [7]), so that extrapolation to room or even lower temperature leads to a large spread in D_I. Indirectly, however, ample evidence has been provided that interstitials are mobile in a broad temperature range [7] and will react to form stable radiation defects. The reaction chain versus temperature is schematically depicted in Fig. 3.4 [1, 2], while the defect parameters are summarised in Table 3.1.

It is clear from Fig. 3.4 that the impurity content of the material has a strong impact on the 300 K defect stability, as has been systematically studied by *Kimerling et al.* [8]. The high oxygen content in Cz material will favor the trapping of I by O_i at the expense of the formation of interstitial carbon (C_i) and boron (B_i). Important room temperature defects are C_iO_i, shown as the H(0.36) peak in Fig. 3.3 [5], C_iC_s, B_iO_i, B_iB_s and B_iC_s. In addition to the typical as-irradiated DLTS spec-

[1] High-resistivity material, used for nuclear-radiation detector fabrication is also grown by the FZ method and normally contains O_i in the 10^{15} cm^{-3} range.

trum for p-type silicon of Fig. 3.3, C_iO_i and C_iC_s also produce a characteristic PL spectrum, given in Fig. 3.5, with a no phonon line at 789.4 and 969.5 meV [9, 10].

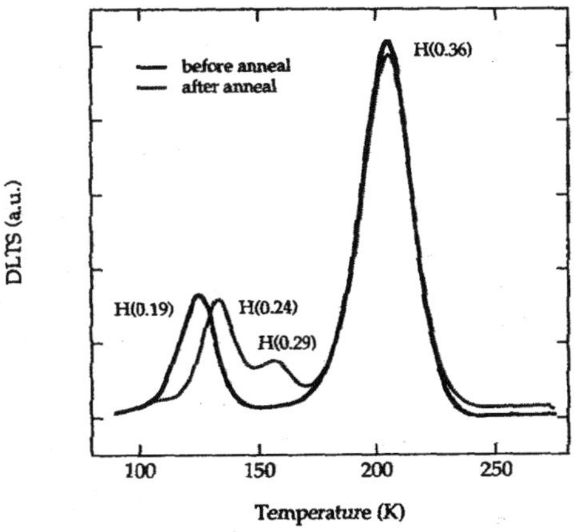

Fig. 3.3. Typical DLTS spectra obtained on 2 MeV irradiated p-type silicon before (full line) and after (dotted lines) a 10 day anneal at 200°C. H refers to a hole trap, with the hole emission activation energy with respect to the valence band given in eV between brackets. [After Trauwaert et al. [6]]

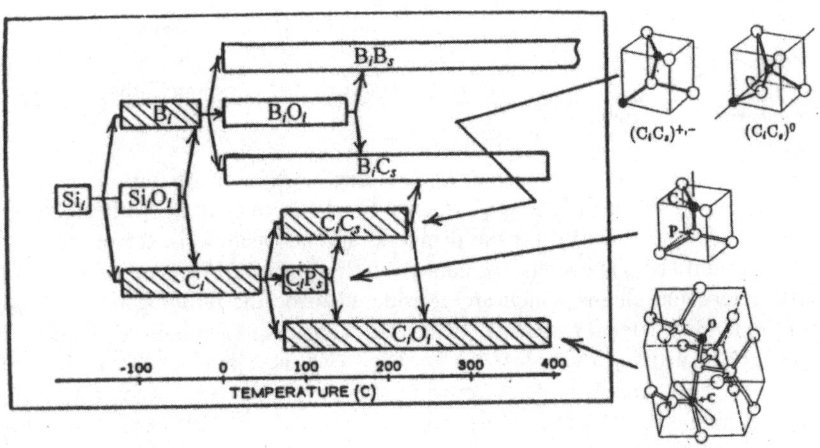

Fig. 3.4. Annealing reactions for interstitial-related defects (~15 min isochronal). The cross-hatched blocks have been followed by EPR, with the identified structures for the carbon pairs indicated in the insets. The un-hatched reactions have been deduced from DLTS measurements. [After Watkins [1]]

Table. 3.1. Commonly observed deep levels in irradiated silicon at room temperature

Energy level Position [eV]	Capture Cross Section[a] [cm^2]	Identity	Charge states	Remarks
E_c-0.17	~10^{-14}	V-O	(0/-)	Stable up to ≥300°C
E_c-0.17		C_iC_s	(0/-)	Bistable, anneals out
E_v+0.09		C_iC_s	(+/0)	at ~200°C
E_c-0.23	10^{-15} – 10^{-16}	V_2	(-/--)	Anneals out between
E_c-0.42	~10^{-15}	V_2	(0/-)	200 and 300°C
E_v+0.20	~10^{-15}	V_2	(+/0)	with a lower rate in
				high-purity samples
E_c-0.10		C_i	(0/-)	Long range migration
E_v+0.27	~10^{-17}	C_i	(+/0)	at RT
E_v+0.35	~10^{-16}	C_iO_i	(+/0)	Stable up to ≥300°C
E_c-0.13		B_i	(0/+)	Negative U; long
E_c-0.45		B_i	(0/-)	range migration at RT
E_c-0.26	10^{-13} – 10^{-14}	B_iO_i		Anneals at ~200°C
E_v+0.30		B_iB_s		Anneals at ≥400°C
E_v+0.29		B_iC_s		Anneals at ~400°C
E_c-0.38		C_iP_s	(0/-)	Metastable, overlaps
E_v+0.48		C_iP_s	(0/+)	with P-V level,
				anneals at 125°C
E_c-0.44	10^{-14} – 10^{-15}	VP	(0/-)	E-centre, overlaps
		(VAs, VSB)		with V_2 level,
				anneals at ~150°C
E_c-0.32	~10^{-15}	Hydrogen-		Stable in excess of
E_v+0.27	~10^{-15}	related centre		300 °C in FZ-Si
E_c-0.45	~10^{-17}	Hydrogen-		Stable up to ≤300°C
		related centre		in FZ-Si

[a]For majority carrier capture.

During heavy particle irradiations, one may expect clustering of interstitials into aggregates. From EPR, some evidence for the di-interstitial has been reported [1, 5, 11]. A new structure for the di-interstitial has been derived from the Si-P6 EPR spectrum [12]. The centre is claimed to disappear at 170°C and can form the <110> interstitial chains which are considered to be the building blocks for the {311} extended defects frequently observed in ion-implanted silicon. Concerning its electrical activity even less is known. Two acceptor levels: E_c-0.07 eV, σ_n=4.6x10^{-16} cm^2 and E_c-0.49 eV, σ_n=6.6x10^{-16} cm^2 have been tentatively ascribed to the di-interstitial [13, 14]. They were observed by DLTS in as-grown high-resistivity n-type FZ silicon and did anneal out at 170°C. A similar level, disappearing at 170°C and with 0.39 eV activation energy has been observed in high energy neutron- or proton- irradiated high-resistivity n-type FZ silicon [15, 16]. This has been ascribed, however, to higher order vacancy clusters [15, 16].

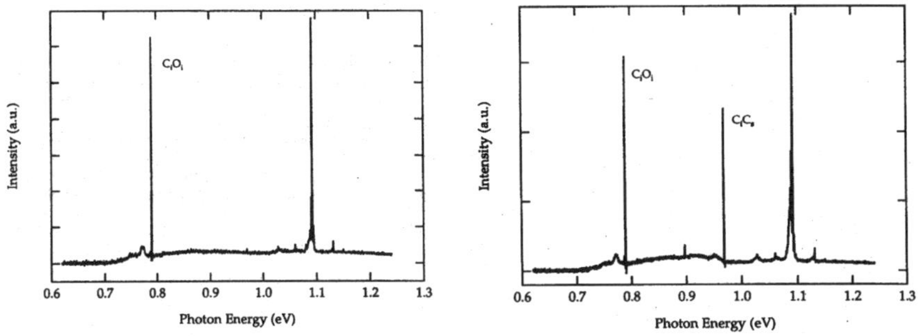

(a) (b)

Fig. 3.5. Photoluminescence spectrum from a Cz (a) and FZ (b) 2 MeV electron-irradiated p-type silicon substrate. [After Trauwaert et al. [9]]

With respect to silicon interstitial clusters, much was learnt from ion implantation damage studies. Considerable effort has been spent in trying to understand the mechanisms involved in the so-called Transient Enhanced Diffusion (TED) phenomenon [17, 18]. TED consists of a point-defect mediated anomalous high diffusion of dopants, during short time activation annealing after a non-amorphising ion implantation. This is of extreme importance for the development of shallow and ultra-shallow junctions in deep submicron CMOS technologies. Its control requires a fundamental understanding of the point defect clustering mechanisms during ion implantation and subsequent annealing. A profound insight of the phenomenon has been achieved, whereby the role of interstitial clusters as the source of the diffusion-assisting point defects is now well-established. A popular, yet too simple model – the so-called "+1" model [19], assumes that only the interstitials generated by the implanted atoms becoming substitutional (i.e. at the end of range) need to be considered, while neglecting the vacancy-interstitial pairs generated in the damage cascade along the track. More accurate models have become available, explaining the deviations from the simple +1 approximation (see e.g. [20]).

One of the interesting aspects of these TED studies are the insights gained in the formation mechanisms of interstitial clusters after ion implantation, upon thermal annealing. A whole range of cluster sizes has been revealed, starting from simple point-defect like aggregates observed by DLTS after low-temperature annealing, up to extended defects, like the {311} defects, which can be revealed by Transmission Electron Microscopy (TEM) for example [18, 21, 22]. The resulting picture is schematically represented in Fig. 3.6 [22], showing the evolution from simple point defect complexes to large-size clusters. The larger clusters grow at the expense of the smaller ones. Ultimately, even the {311} defects disappear and are unfaulted to perfect dislocation loops [23].

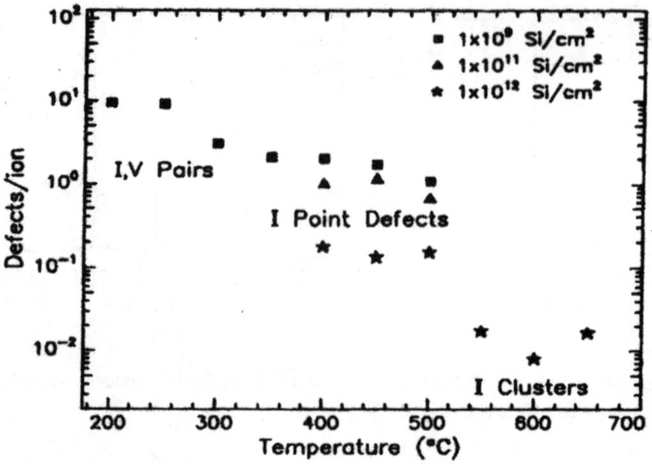

Fig. 3.6. The total defects/implanted ion measured by DLTS for p-type epitaxial silicon implanted with 1.2 MeV Si at three doses. At 1x10^9 Si/cm^2, the point defects anneal until only interstitial point defects remain. At higher implant fluences and at higher anneal temperatures, the defects/ion are reduced, indicating the formation of interstitial clusters. [After Benton et al. [22]]

From the study of TED in the 600 to 800°C range, *Cowern et al.* succeeded in an inverse modeling of the underlying size distribution of the interstitial clusters [24, 25], the result of which is shown in Fig. 3.7. Some 'magic numbers' turn up, suggesting very stable clusters at sizes n=4 and 8 and corresponding with a minimum in formation energy E_{fc}. There exists some experimental evidence to support the idea for an 8-interstitial-controlled nucleation of extended defects in silicon [26].

Irradiation or ion implantation introduces a whole series of PL bands, which have been reviewed by *Davies* [10]. Heavy radiation damage is characterised by the occurrence of the so-called W (or I_1) vibronic band at 1018 meV [27 - 29], after annealing between 400 and 600 K. An example of its spectrum is shown in Fig. 3.8 [28]. The growth of the W-line is characterised by an activation energy of ~0.85 eV [28], while it disappears above 700 K. It is strongly believed that the line is related to an intrinsic defect formed by the diffusion of point defects out of the primary damage cascade of the radiation. It has been ascribed to a vacancy [30] or an interstitial cluster [27 - 29], although more and more evidence points towards the interstitial nature of the defect. It has in fact been suggested that it could be related to a tri-interstitial [31]. Moreover, annealing studies of MeV self-ion implanted n-type silicon in a broad temperature range between 525 and 750°C reveals a whole series of related PL-lines, suggesting a family of interstitial complexes [32]. For higher implantation doses {113} self-interstitial aggregates have been observed, along with the "903 meV" PL signal, associated with the {113} defects, and a DLTS trap at E_c-0.33 eV [32].

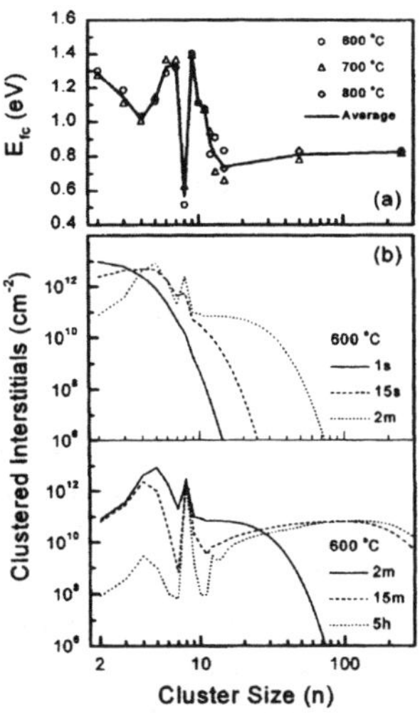

Fig. 3.7. Formation energies of interstitial clusters, estimated from the Ostwald ripening (OR) analysis (a). Symbols show the values extracted from the interstitial supersaturation data at each temperature, and the line represents the average results. (b) Time evolution of the cluster-size distribution at 600°C. [After Cowern et al. [24]]

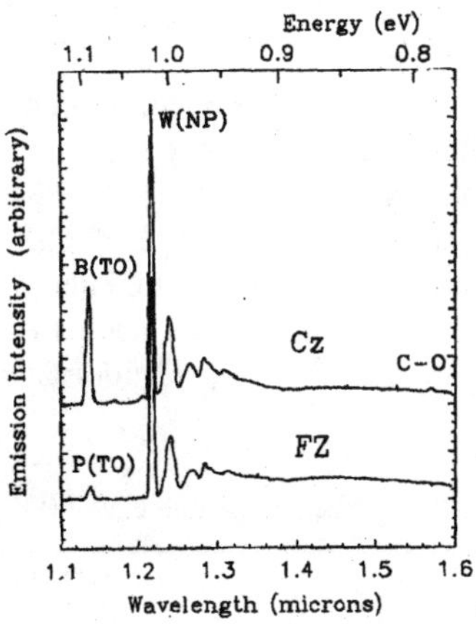

Fig. 3.8. PL emission spectra for B-doped Cz-Si{100} and P-doped FZ-Si {111} implantedat 500 K. The transverse optical phonon replicas of the boron and phosphorous exitonic peaks (at 1093 and 1092 meV, respectively), the 789 meV C-O band and the 1018 meV exitonic (no-phonon) W center are labeled. Broad-band radiative damage from 800 to 1000 meV is of comparable magnitude, but is suppressed on this scale. [After Schultz et al. [28]]

3.2.2 Impact of Radiation Defects on Silicon Devices

It is clear from the previous paragraph that in silicon, a whole range of stable radiation defects are formed, which not only depend on the radiation parameters (energy, dose, particle type, target temperature) but also on the starting material parameters (doping type and density, impurity content, preheat treatment,…). The dominant radiation defects stable at 300 K for p-type Si have been shown in Fig. 3.3, the ones in n-type Si are represented in the top spectrum of Fig. 3.9. This manifold of radiation levels explains the difficulties encountered in the modeling of the radiation damage starting from first principles. Usually, a semi-empirical description is used, based on damage factors and NIEL. However, not all radiation defects stable at room temperature are equally harmful for the device operation. Their electrical activity not only depends on the trap concentration N_T (or the introduction rate R_T) but more importantly on the energy position in the band gap E_T and on the capture cross section for electrons (σ_n) and for holes (σ_p). As explained in Sect. 2.4.2, it is very difficult to achieve a complete electrical characterisation of a deep-level trap, particularly with respect to the minority capture cross section. However, this has been achieved by *Hallén et al.* [33] for the main radiation defects of Fig. 3.3 and 3.9. The data are summarised in Table 3.2. Note that the capture rates c_n and c_p, defined as $v_{th}\sigma_n$ and $v_{th}\sigma_p$, respectively, show a strong temperature dependence and have been derived from DLTS in a narrow temperature range, where a corresponding peak can be measured. This implies that modeling of the electrical device parameter at stake at the operating temperature implies extrapolation of these data – usually to 300 K. This may add already considerable error to the calculations [3, 4].

The data of Table 3.2 have been used to calculate the recombination lifetime in electron and proton-irradiated n- and p-type silicon [33 - 35]. A good agreement with measurements has been found, from which it was concluded that the E_c-0.421 eV level, related with the single-negatively charged acceptor state of the divacancy is the dominant recombination centre in proton or electron-irradiated n- and p-type Si for low injection levels. At high injection, the E_c-0.164 eV level associated with the A-centre becomes the most important one [33 - 35]. The dominance

Table 3.2. Electrical parameters of the dominant radiation defects in Si stable at 300 K

Defect centre	**Band gap enthalpies[a] [eV]**	c_n **[cm³/s]**	c_p **[cm³/s]**	***T* Interval [K]**
$V\text{-}O^{0/-}$	0.164	$1.4\text{x}10^{-8}\text{xT}^{0.5}$	$8\text{x}10^{-8}\text{xT}^{0.7}$	80-108
$V\text{-}V^{-/--}$	0.225	$1.6\text{x}10^{-12}\text{xT}^{1.4}$	$7\text{x}10^{-7}$	105-155
$V\text{-}V^{0/-}$	0.421	$5.4\text{x}10^{-9}\text{xT}^{0.4}$	$2\text{x}10^{-6}\text{xT}^{-0.3}$	182-266
$V\text{-}V^{+/0}$	0.194	$>>c_p$	$2.1\text{x}10^{-9}\text{xT}^{0.2}$	104-146
C_iO_i	0.339	$5.1\text{x}10^{-23}\text{xT}^{5.2}$	$1.2\text{x}10^{-10}\text{xT}^{0.61}$	160-238

c_n electron capture rate, c_p hole capture rate, T temperature

[a]The trap energy E_T is related to the enthalpy ΔH and entropy ΔS through $E_c\text{-}E_T=\Delta H_T\text{-}T\Delta S_T$

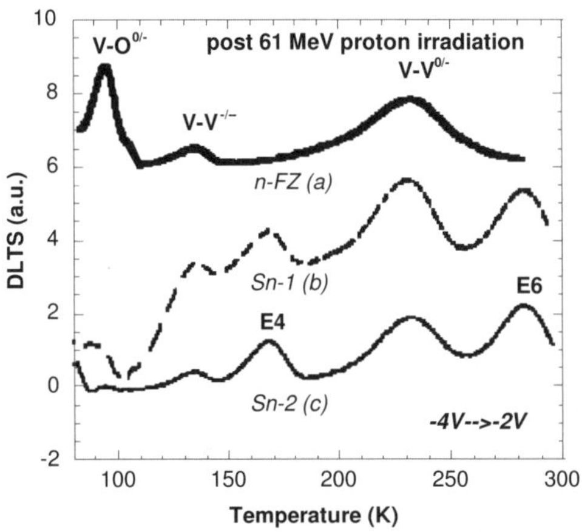

Fig. 3.9. DLTS spectra after a 5x10^{11} cm^{-2} 61 MeV proton irradiation of (a) an n-type FZ-Si p-n junction; (b) a 1.7x10^{18} cm^{-3} Sn-doped n-type Cz-Si Au-Schottky barrier and (c) a 6.5x10^{18} cm^{-3} n-type Cz-Si Au-Schottky barrier. The bias pulse was from –4 to –2 V; the time constant window was 6 ms. The peaks E4 and E6 are believed to correspond to the double- and single-negatively charged acceptor state of the Sn-V complex.

of the E_c-0.42 eV V-V level was already concluded from previous studies [36]. On the other hand, Fig. 3.10 shows that in proton-irradiated n-FZ silicon the temperature dependence of the high-injection (excess carriers ~ 10^{16} cm^{-3}) bulk lifetime is governed by a level at E_c-0.17 eV [37], which coincides with the A-centre. In the same figure, an activation energy of about 0.23 eV was found for p-type FZ material, which indicates the role of the divacancy donor level at high injection. This is in agreement with data for 1.5 MeV electron-irradiated p-type Cz silicon [38]. Other studies pointed out the importance of even other radiation defects for the recombination lifetime τ_r [39, 40], which demonstrates the complexity of the problem. However, in most cases, the divacancy, when present, plays a crucial role in the recombination lifetime of irradiated silicon.

In Sect. 2.3.2 it was demonstrated that the generation lifetime – important for the leakage current of diodes – is not equal to the recombination lifetime. In addition, according to the SRH theory, levels close to mid gap are the preferential leakage current generation centres. A good candidate for this in irradiated silicon is the single acceptor of V-V at E_c-0.42 eV. This correlation has been firmly established in a number of studies [3, 4, 41, 42]. In fact, it has been demonstrated that the derivative of the leakage current can be used to investigate the concentration profile of the divacancy. An example is given in Fig. 3.11 for helium ion-irradiated n-type silicon. In both n- and p-type Si the same effective generation level at $|E_T-E_i|$ was observed [3, 42], showing that also in p-type material, the same deep level governs carrier generation in the depletion region. At the same time, it was found that the leakage current damage factor was 25 % larger in n-compared with similarly doped p-type silicon [3]. The recombination lifetime damage factor $K_{\tau r}$ even differed a factor two, so that for space solar cells for example, a p-type base is preferred owing to the higher radiation tolerance [43].

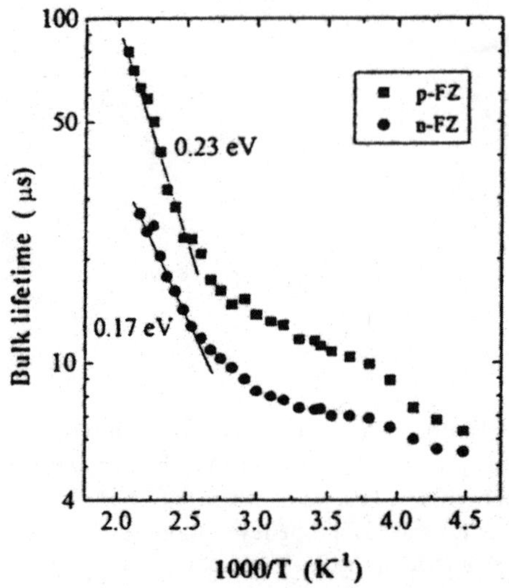

Fig. 3.10. Temperature dependence of the bulk carrier recombination lifetime at an injection level of 10^{16} carriers/cm^3 in p- and n-type FZ silicon, irradiated by 10 MeV protons at a fluence of $4x10^{10}$ cm^{-2}. [After Kaniava et al. [37]]

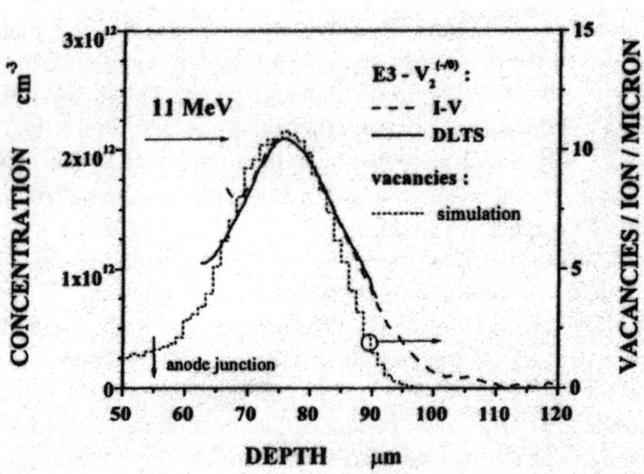

Fig. 3.11. Comparison of V-V$^{-/0}$ profiles obtained from DLTS and *I-V* measurements including the simulated distribution of primary vacancies (11 MeV $2.5x10^9$ cm^{-2} helium irradiation through the anode). [After Hazdra et al. [3]]

Although the understanding of the lifetime and the leakage current evolution in irradiated silicon p-n junctions is quite complete, this is not the case for certain other applications. For example, large efforts have been spent to investigate the radiation phenomena in heavily damaged high-resistivity (HR) silicon detectors, intended for use in CERN's Large Hadron Collider (LHC) experiment [15, 16, 44, 45]. Two phenomena have to be explained: the increase of the reverse dark current with particle fluence, on the one hand and, secondly, the carrier removal and type inversion of the detector material from lowly n- to lowly p-type doping. Associated with the latter effect is the increase of the detector depletion voltage, which is an important operational parameter. Some of the problems related with the modeling are connected with the special nature of the HR-FZ silicon, which contains low concentrations of oxygen (O_i) and dopants (P) so that the standard deep levels like the A and E-centre are formed with low introduction rates, while other unknown peaks may occur in DLTS.

Another difficulty is met when cluster damage is formed, as for instance during high energy proton or neutron irradiation. In that case, some new defects are introduced, which are hardly known from standard resistivity silicon [15, 16, 44, 46]. In addition, some of these peaks overlap with the V-V single acceptor level at E_c-0.42 eV. As can be seen in Fig. 3.9, the corresponding DLTS peak is rather broad and in addition, much higher than the accompanying double acceptor V-V state at E_c-0.23 eV. Normally, equal amplitudes are expected for peaks belonging to the same defect. This difference in appearance, typical for irradiations producing cluster damage has been explained in terms of strain-induced reorientation of the divacancy [47], whereby lattice strain exists in the heavily damaged cluster regions. On the other hand, for γ- or electron-irradiations a perfect one to one amplitude ratio is found for the two acceptor levels [48].

Beside the stress argument, one should also consider the contribution of other defects to the broad divacancy peak. This can be demonstrated by investigating the isochronal thermal annealing behaviour. It has been found that there exist at least two annealing stages for the dissociation of a defect – one at 70°C and a second at 170°C [15, 16]. Moreover, one has to take into account deviations from the standard SRH theory in heavily damaged regions, where locally very high trap densities may exist [44, 45, 49, 50]. Different pictures have been proposed for this case: one can either interpret the carrier transitions based on the space charge regions induced by the compensated damage clusters [49, 50]. Another approach is to rely on intertrap transitions, for example between the single and double acceptor level of two neighbouring divacancies [44, 45]. According to the second model, this leads to an enhanced carrier generation and, therefore, to a markedly higher damage factor K_I, which could explain the observed values.

In spite of these problems, attempts have been made to implement radiation induced deep levels in a numerical device simulator [51, 52]. In order to reduce the computational complexity, an empirical approach is followed, whereby one, two or three deep levels can be introduced to model the leakage current increase and the donor removal rate. It turns out that the simulations rather well reproduce the

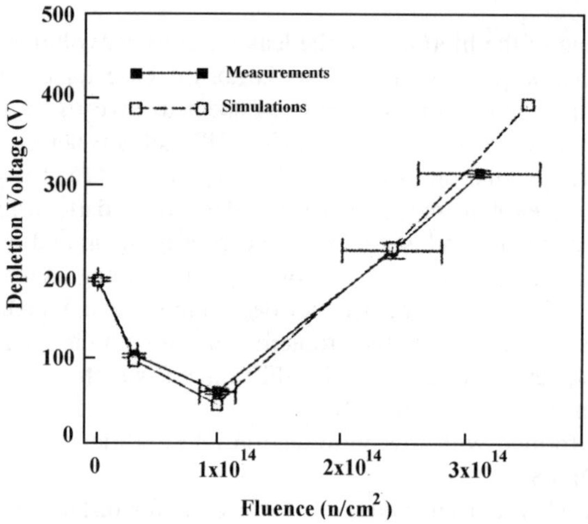

Fig. 3.12. Depletion voltage as a function of neutron fluence: comparison between measurements and simulations for a resistivity of 1.6 kΩ.cm. [After Passeri et al. [52]]

experiments for the three-level case [52]. An example is given in Fig. 3.12, showing the evolution of the detector depletion voltage with neutron fluence [52]. Ideally, the depletion voltage should be kept at its minimum value during the operational life of the detector. Comparable numerical modeling has also been performed for neutron radiation effects in silicon avalanche photodiodes [53].

3.2.3 Substrate and Device Hardening

It is clear from Sect. 3.2.2 that the stable radiation damage formed at a certain operation temperature not only relies on the radiation conditions but also on the starting material and subsequent processing details. This implies that the radiation performance can be optimised or improved by a smart choice of starting material. It has been shown in the previous paragraph that p-type silicon is more resistant than n-type, which explains the preference for n^+p silicon solar cells for space applications [43]. Not only the doping type, but also the resistivity or even the doping atom can play a role. For example, experience has learnt that p-type solar cells with lower resistivity (higher B content) degraded faster. This can be explained by the fact that the creation of B_i by a radiation-induced silicon interstitial is proportional with the concentration of B ([B]) [8]. The formation of the B_iB_s pair proceeds proportional with $[B]^2$ etc. It has also been observed that Ga-doped silicon (Si:Ga) is more radiation tolerant than B-doped material (Si:B) [43]. At high irradiation doses, the carrier removal and type inversion of solar cells is offset to higher fluences for Si:Ga compared with Si:B [54]. DLTS studies have pointed out that in Si (Si:Ga), the dominant donor-like centre at E_c-0.18 eV, believed to correspond to B_iO_i is not formed. The same applies for the recombination centre at E_v+0.36 eV.

For HR-Si radiation detectors on n-type Si on the other hand, a better performance is achieved for material with a lower resistivity [55]. This is illustrated by Fig. 3.13, representing the neutron fluence at which the space charge in an irradiated detector changes its sign, i.e. indicative of type inversion. In order to explain the observations, one has to consider not only the creation of a deep acceptor, compensating for the positive donor charge but also a removal (reduction) of the donor concentration.

Beside the type of dopant and its concentration, non-doping impurities, like oxygen and carbon play a decisive role in the radiation performance of silicon components. In Cz material, the A-centre will have a pronounced introduction rate, compared with FZ-Si. When carbon is present, different interactions between C_i and other impurities may create stable radiation defects. In the early days of space solar cells, it was for example observed that FZ-Si gave a better radiation performance than Cz. $[O_i]$ and $[C_s]$ have also been considered as a hardening parameter for HR-Si detectors [16, 56, 57]. Figure 3.14 illustrates the impact of oxygen and carbon doping on the variation of the effective doping concentration N_{eff} versus proton fluence. On the right-hand axis, the corresponding full depletion voltage V_{FD} for a 300 μm thick detector is shown [56]. One can clearly see that while oxygen-diffused material exhibits a better radiation performance, the opposite holds for carbon-doped material. Furthermore, oxygen-doping by diffusion in the range of a few times 10^{17} cm^{-3} clearly improves the V_{FD} compared with standard undoped material. It is speculated that this behaviour is closely related to the reducd introduction of V_2O centres upon irradiation, which are thought responsible for the carrier removal by introducing a compensation acceptor level near mid gap [44, 45, 56].

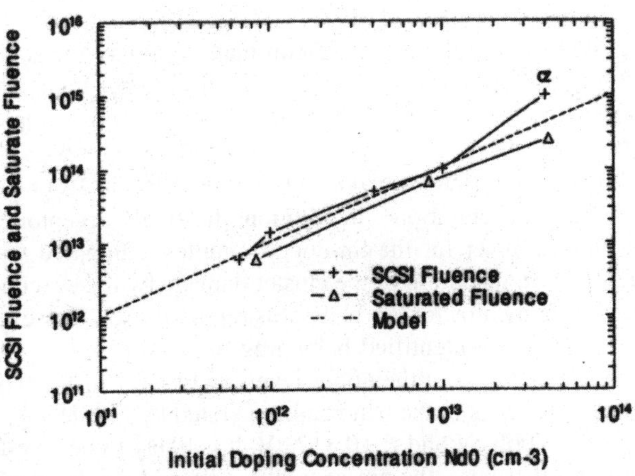

Fig. 3.13. Neutron fluence for type inversion (SCSI: Space Charge Sign Inversion) in a detector diode or for resistivity saturation in a silicon resistor, as a function of the starting doping concentration of the n-type silicon substrate. [After Dezillie et al. [55]]

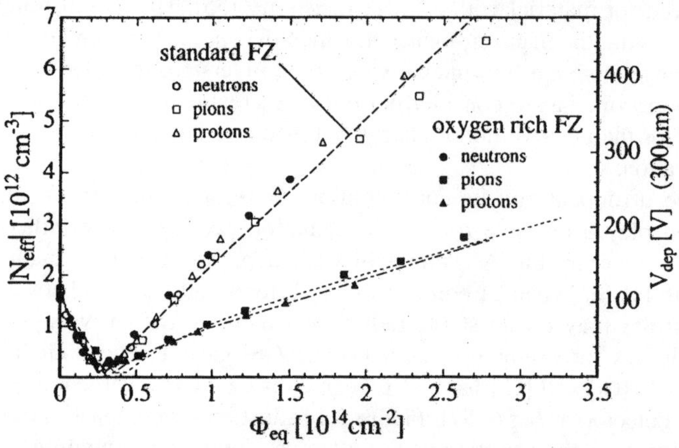

Fig. 3.14. Effective space charge density and full depletion voltage versus proton fluence for standard, carbon-enriched and three types of oxygen diffused samples: 24, 48 and 72 hours diffusion at 1150°C. [After Lindström et al. [56]]

In Cz-silicon, the $[O_i]$ is by definition quite high – in the 5 to $10x10^{17}$ cm^{-3} range typically, although the concentration for state-of-the-art material is at the lower end of the range. Systematic investigations of the impact of the oxygen concentration and the thermal preheat treatment of the p-type wafers, whereby an oxygen-lean denuded zone of ~10 µm was deliberately formed, showed rather subtle effects in the order of 20 to 25% improvement for lower-oxygen doped material [58].

It has also been suggested that Sn-doping of silicon may be a way for achieving harder substrates [59, 60]. The basic reason is that Sn has a rather high solubility in silicon and is known to trap vacancies, giving rise to the Sn-V complex (see Fig. 3.1 and 3.2) [59 - 61]. In this way, the Sn-V formation suppresses the creation of other harmful radiation defects. This can also be inferred from Fig. 3.9, showing a lower A-centre concentration[2]. In addition, the single acceptor divacancy peak is smaller and narrower for the Sn-doped samples, compared with the Sn-free substrate, suggesting a reduction of the cluster damage by tin. A reduction of the trap introduction rate by a factor 2.5 to 10 was reported [62]. However, at the same time, 4 levels have been identified belonging to Sn-V [62, 63]. The two acceptor levels are shown in Fig. 3.9; the donor levels in Fig. 2.28 [64]. This means that Sn-V has five charge states in the band gap, separated by four levels, at E_c-0.214 eV, E_c-0.501 eV, E_v+0.066 eV and E_v+0.317 eV. It is likely to be an efficient recombination centre. Furthermore, the presence of a near mid gap level makes it a rather efficient generation centre as well. Similarly, it is not expected that Ge doping will provide a promising pathway to hardening, as the Ge-V centres are

[2] Note that spectra b and c in Fig. 3.9 correspond with Cz material, which has a much higher $[O_i]$ than the FZ material of spectrum a.

unstable already at 200 K (Fig. 3.1 and 3.2). However, the case is different if we consider SiGe alloys (Sect. 3.4).

In fact, from Fig. 3.2 or 3.4 another way of increasing the radiation resistance of silicon devices can be deduced. Indeed, if the device is operated at sufficiently low temperature – say < 100 K – vacancies and basic interstitial defects (C_i, B_i, Al_i,...) will be frozen, so that only few harmful defects are formed, if any. Of course, the device has to be kept cool all the time, if hardness is to be preserved, because heating to higher temperatures will inevitably lead to mobile point defects and subsequent formation of more stable radiation complexes. Another way to improve the operation of irradiated detectors is to cool it after room-temperature irradiation to sufficient low temperature [65]. The reason behind the improvement of post-radiation cold detector operation is that traps may become saturated by charge carriers, which are permanently frozen on the deep levels. This is because the thermal re-emission is activated and becomes very slow for sufficiently low temperature. The improvement of the detector performance relies strongly on the operation conditions, the cooling bias, etc. These cryo-techniques are at the moment still in their infancy, but it is expected that in the future interest will increase for applications where a cryogenic environment is affordable and the anticipated performance gain is higher than the additional operational cost and complexity.

Finally, one could also rely on a passivation strategy of the radiation defects, for example by Li [43] or hydrogen doping/incorporation at a certain stage. However, given the high mobility of both H and Li, one can seriously question the reliability and stability of such an approach. Furthermore, the presence of these impurities may cause other undesired effects – eventually, they may also diffuse out of the material. Clearly more research is necessary before passivation becomes a mature tool for radiation defect control.

3.2.4 Summary Silicon Radiation Defects

In summary, the insight in the radiation-defect formation mechanisms in silicon has reached a level of maturity, which allows for radiation-defect engineering. As shown in the previous paragraph, a whole arsenal of techniques are available for implementing radiation-hardening. It starts already with a careful selection of the silicon starting material. Next, the device processing can be optimised in order to have some relevant impurities, like oxygen or carbon, in the desired state. This will not prevent radiation-defect formation, but leads to a lower introduction rate of certain harmful ones, or may alternatively introduce other defects, which are less harmful for the considered application. Finally, the operation temperature can be optimised to reduce the electrical effect of the radiation-induced levels. For silicon devices, lowering the device temperature may be an option. In principle, high(er) temperature could also be an alternative from a radiation-damage viewpoint, as this will lead to in situ annealing of smaller point defects and to a higher mobility of vacancies and interstitials, so that they can more easily escape. However, above 45 °C or so, the device leakage becomes such that normal device operation becomes problematic. Alternatively, one could consider applying periodic high temperature annealing to repair the device performance. This could be performed during maintenance periods, or during standby times as frequently happens in space orbits, when crossing the radiation belts.

3.3 Displacement Damage in Germanium

Although the first bipolar transistors have been developed on Ge, during the last three decades silicon has become the dominant semiconductor material, surely for Ultra and Very Large Scale Integration (ULSI-VLSI) technologies. This is due to the favorable oxidation properties of Si, which enabled the development of the planar technology, while it is very difficult to form a stable oxide on the germanium (or $Si_{1-x}Ge_x$) surface. Nevertheless, for some niche markets, Ge is an important material and finds its way to space or nuclear applications as well. Therefore, in a first part, a brief summary of possible applications of Ge in a radiation environment is given.

Due to the limited technological interest in Ge, a large deficit has built up with respect to the knowledge and understanding of defects in the material. This has changed since the early 90ties because of the technological interest in the use of SiGe alloys for the fabrication of high-performance deep submicron devices. In order to understand the impact of alloying on the material parameters, it is essential to have a profound knowledge of both extremes, i.e., Si and Ge. However, there is also a fundamental reason behind the lack of insight, namely the fact that the EPR technique, successful in unraveling the identity and structure of many silicon defects can hardly be applied to Ge. Instead, techniques like DLTS and infrared absorption have been used to fill this gap, though they do not provide generally the same microscopic and chemical information. This explains the uncertainty which still exists concerning the main radiation defects in Ge, i.e., the A, the E-centre and the divacancy [66]. In the next two paragraphs, the defect formation under cryogenic or room temperature irradiation will be discussed. In the latter case, one strongly relies on the work by *Fage-Pedersen et al.* [66], which seems to provide the most reliable deep-level identification for radiation defects in Ge. Finally, some results about the impact of radiation on Ge material and device performance will be described.

3.3.1 Potential Applications of Ge

A first class of applications is infrared and far-infrared spectroscopy using a cooled Ge detector. For example, stressed Ge:Ga detectors will be employed for the FIRST (Far InfraRed Space Telescope) mission [67], for the spectral region 85 to 210 μm. They operate in the liquid helium temperature (LHT) range, i.e., between 1.5 and 2 K. For the wavelength range beyond 200 μm, stressed Ge:Ga detectors provide a unique combination of sensitivity and long-wavelength response. The sensitivity is quoted to be two to three orders of magnitude better than that of a bolometer operated in the same temperature range [68]. Unstressed Ge:Ga and Ge:Sb detectors for astrophysical investigations are also under development [69].

A second related group of devices are the Ge-based thermistors for bolometer applications [67]. Operational temperatures lie in the sub 1 K range. A basic requirement for a semiconductor thermistor is a uniform resistivity and hence doping concentration. This can be achieved by the Neutron Transmutation Doping (NTD) process, described by (2.6) of Chap. 2 [70].

High-purity and ultra-high purity germanium is a material suitable for the fabrication of nuclear-radiation detector diodes, which operate typically at 77 K under full depletion conditions [71, 72]. They are frequently used in γ-ray spectroscopy, because of the high target mass, resulting in a high LET (see (2.3)). Although originally mostly p-type material was used [71], interest focuses more and more on n-type detectors, which have been shown to be more radiation hard. The reason is that irradiation introduces predominantly hole traps in the high-purity material. In a p-type detector, holes have to cross most part of its volume, increasing the possibility of capture and loss of energy resolution and enhancing the noise. In n-type detectors, on the other hand, the electrons have to travel the largest distance before being collected at the positively biased electrode, which should yield lower charge trapping and a higher collection efficiency.

The interest in epitaxial GaAs/Ge solar cells for space applications has drastically increased in recent decades [43, 73, 74]. In that case, a III-V active solar cell is deposited epitaxially by, for example, Metal Organic Chemical Vapor Deposition (MOCVD) on highly doped n^+ Ge wafers (concentration range 10^{16} to 10^{17} cm^{-3}). The advantages are that it is easier to grow III-V compounds pseudomorphically on Ge than on Si substrates, because of the lower lattice mismatch. Another benefit of Ge is its larger mechanical strength compared with GaAs. The wafers can, therefore, be made thinner, e.g. 200-300 μm for a 100 mm diameter. This offers a weight advantage, which is of primary importance for space. Next, if a highly doped p^+ wafer is used, an active tandem cell can be fabricated, whereby the Ge absorbs in the longer wavelength region of the solar spectrum. On top of that, there are a number of economical reasons to prefer Ge wafers instead of GaAs substrates (price, availability, larger diameter, better crystal perfection). Finally, it has been observed that GaAs/Ge solar cells have a larger radiation tolerance than their GaAs/GaAs counterparts [43, 73, 74].

Given this wide range of potential space applications, it is important to study and understand radiation damage in the Ge lattice.

3.3.2 Cryogenic Irradiation of Ge

Since many of the envisaged applications of Ge and Ge-devices are at low temperatures, it is important to have an idea about the cryogenic radiation damage. In addition, as for Si, it enables to study the electrical properties of the primary radiation defects, i.e., the Ge vacancy and interstitial. Heating up after cryogenic irradiation should reveal possible annealing stages and simple point defect reactions.

Originally, 4.2 K [75 - 79] and 77 K [80, 81] particle irradiations were performed to investigate the stability of the primary displacement damage (vacancies and interstitials). Early studies pointed out that independent of the initial doping Ge becomes highly ohmic and p-type after high irradiation fluences. Typical introduction rates for n-type Ge were in the range of 1 cm^{-1}, while for p-type or intrinsic Ge introduction rates are at least two orders of magnitude lower [79]. Similar as for silicon, this could be explained by a high mobility of the Ge interstitial, even at 4.2 K. There are indeed indications of interstitial atoms reacting at low temperatures with group-III (Ga) as well as group-V (As, Sb) dopants [79].

Different annealing stages have been identified: one at ≈35 K and another at 65 K are ascribed to the mobility of the interstitial [78], while the vacancy is believed to become unstable at about 90 K in n-type Ge and around 170 K in p-type Ge [78]. The mobility of vacancies and interstitials thus depends on the position of the Fermi level (charge state of the vacancy) and will reduce going from heavily doped n-type, over high-purity intrinsic material to p-type Ge. In the literature, estimates have been derived from electrical measurements of the activation energy associated with the diffusion of charged and neutral Ge vacancies [82]. A value of 0.42 eV was found for V^- (n-type Ge) and for p-type or intrinsic Ge a migration energy of 0.52 eV was suggested for V^0. These values are higher than in silicon (see Sect. 3.2.1) and furthermore suggest annealing stages at temperatures much higher than 100 K, i.e., in the range of 200 K. There seems to be a general agreement of the mobility of divacancies in Ge around 400 K [79].

More recent experiments at low temperature are based on DLTS measurements and have been performed on 4.2 K electron-irradiated Ge [83] and on neutron-irradiated high-purity Ge [84, 85]. It has been found that the induced defects are totally different from the room temperature ones [83, 85]. The introduction rate strongly depends on the doping level: while values in the range 10^{-3} to 10^{-2} cm^{-1} are found for ≈10^{15} cm^{-3} doped n-type material, lower values (10^{-5} to 10^{-4} cm^{-1}) have been obtained for 10^{13} cm^{-3} n-Ge [83]. In other words, the lower the doping density, the smaller the introduction rate. This could point to a higher mobility of the primary defects in the lowly doped material. Additionally, the chance of getting trapped by a doping atom also reduces in high resistivity Ge. Furthermore, two levels occurring at E_c-0.12 eV are associated with the Ge interstitial or complexes involving an interstitial. Another vacancy related level in the range E_c-0.1 to -0.2 eV anneals at ≈100 K [83]. It has also been noted that 90 % of the created damage, which is derived from the deactivated dopant concentration, is not found by DLTS. It is believed that this missing damage could be primary vacancy-interstitial pairs, which recombine at 65 K [83]. This is confirmed by X-ray diffraction measurements of high-fluence (1-3x10^{19} cm^{-2}) 2.5 MeV electron-irradiated Ge crystals at 4.2 K [79]. It is shown there that a high density of close Frenkel pairs (V-I pairs) are created with an introduction rate of ~3 cm^{-1}. These Frenkel pairs are believed to be electrically inactive and show annealing stages between 8 and 55 K and 100 and 300 K. The first stage is ascribed to the recombination of close Frenkel pairs, due to the mobility of the Ge interstitial. The second stage involves long range migration, whereby the interstitial mobility governs below ~200 K, while vacancy migration contributes in the 200-300 K interval [79].

After a 10 K fast neutron irradiation of n- and p-type high-purity Ge, different annealing stages have been observed [85], namely at ≈40 K and 100 K in n-type Ge in agreement with the earlier results by Swanson [86]. They correspond to the interstitial mobility (35 K) and the vacancy mobility (95 K) region. In p-type material, on the other hand, no significant annealing was found below ≈175 K. Therefore, it is concluded that the primary defects in n-type Ge are stable up to ≈35 K and up to ≈160 K in p-type material [85]. Furthermore, it has been observed that the total introduction rate of primary defects upon 10 K neutron exposure is a factor 2 lower than the theoretical estimates (50 cm^{-1}). This is explained by the fact that even at low temperatures, the interstitials are mobile and part of them escape and recombine. For room temperature irradiations, the measured total introduction rate by high energy neutrons is in the 1.2 cm^{-1} (p-Ge) to 1.9 cm^{-1} (n-Ge) range, suggesting p-type material to be more radiation tolerant, similar as for silicon. It should also be remarked that for p-Ge, the introduction rate after low temperature irradiation and anneal up to 320 K is slightly larger than the room temperature rate. This means that warming up a Ge detector operating in a radiation environment at cryogenic temperatures can have adverse effects on the performance. In many cases, it is better to keep the system cold.

The primary radiation defects in Ge have also been studied using the so-called Perturbed Angular Correlation Spectroscopy (PACS) [87]. In this case ^{111}In doped Ge has been irradiated by 1.2 MeV electrons at 77 K. This is followed by an annealing at room temperature, which allows the In atoms to catch vacancies and interstitials. The In is used as a nuclear probe in PACS. The resulting picture which emerges from these studies is summarised in Fig. 3.15.

The vacancy has an acceptor state at E_v+0.2 eV and the Ge interstitial has a donor state at E_c-0.04 eV. Long range migration of the neutral vacancy and the positive interstitial takes place at 200 and 220 K, respectively. These scattered and contradictory results ilustrate that the understanding of the electrical and structural properties of the primary radiation defects in Ge is far less well developed than in Si. More work is required to come to a consistent picture of the experiments.

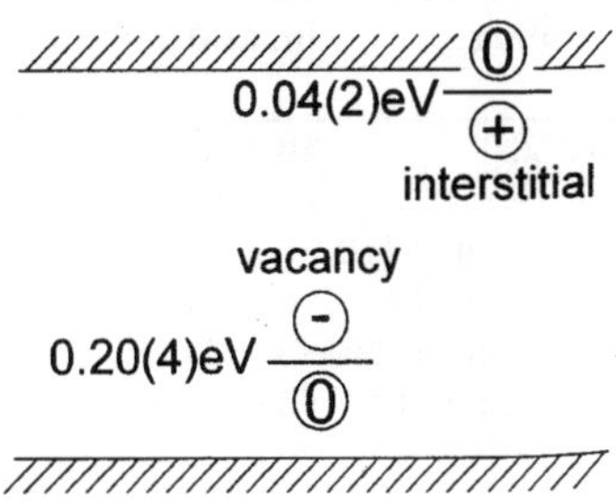

Fig. 3.15. Electrical levels for the vacancy and self-interstitial Ge as determined from PACS. The interstitial level is tentative. [After Haesslein et al. [87]]

3.3.3 Room Temperature Irradiation of Ge

It is only with the advent of DLTS that a more systematic study of Ge radiation defects became feasible. In the meantime, investigations have been performed on high-energy electron [66, 83, 88 - 92], low energy proton [66, 89, 92], gamma and neutron [93] and high-energy ion [94 - 96] irradiations. In addition, the impact of neutron irradiation on high purity germanium (doping density in the range 10^{10} cm^{-3}) has been studied for NTD purposes [84, 85, 92].

Similar as for silicon, a whole set of stable radiation defects has been found at room temperature, which depend a.o. on the starting material properties (doping density and type, dopant, grown-in impurities like hydrogen, oxygen,...). However, there is far from an agreement as to the assignment of these levels. This matter has been critically discussed by *Fage-Pedersen et al.* [66] and has led to a possibly correct identification of the most prominent room temperature radiation defects in Ge. The results are summarised in Table 3.3 and may still be controversial. However, the conclusions are based on an extensive study of irradiated Ge and SiGe alloys over a wide composition range *x*, so that the impact of the Ge content on the same radiation levels can be followed. A typical DLTS spectrum of irradiated n- and p-type Ge is given in Fig. 3.16 [93] and 3.17 [90], respectively. The situation for the hole traps in irradiated p-Ge is less clear and requires further detailed studies.

Comparing with the radiation defects in Si, some striking differences can be noted. With respect to the E-centre, it has been observed that the level has a pronounced dopant dependence in Ge [66]: while the Sb and P related E-centres overlap, a markedly difference position at E_c-0.27 eV has been found for the As counterpart [66, 97]. In addition, the capture cross section for electrons of Table 3.3 is much smaller than for the A-centre, which could point to a double negative charge

Table 3.3. Properties of prominent radiation-induced electron and hole traps in n-type germanium

Label	E_T [eV]	σ [cm^2]	Annealing [°C]	Identification
$H_{0.30}$	0.30	1.6×10^{-13}	in at 150	Sb related
$E_{0.37}$	0.37	1.1×10^{-14}	out at 150	E-centre
$E_{0.23}$	0.23	2.0×10^{-15}	in at RT out at 110	Sb and I related
$E_{0.19}$	0.19	1.5×10^{-14}	in at RT out at RT	Sb and I related
$E_{0.13}$	0.13	3.2×10^{-15}	in at RT out at RT	Sb and I related
$E_{0.21}$	0.21	7.1×10^{-14}	in at 90 out at 180	Sb related ?
$E_{0.29}$	0.29	2.1×10^{-15}	out at 180	divacancy
$E_{0.30}$	0.30	2.9×10^{-14}	in at RT out at 110	I and impurity related?
$E_{0.28}$	0.28	6.2×10^{-15}	in at 190 out at 270	O-impurity complex?
$E_{0.27}$	0.27	2.6×10^{-15}	out at 150	A centre
$E'_{0.19}$	0.19	2.2×10^{-15}	in at 130 out at 190	O related
$E'_{0.14}$	0.14	1.3×10^{-16}	in at 130 out at 190	O related

E_T apparent enthalpy of the deep level, σ majority capture cross section

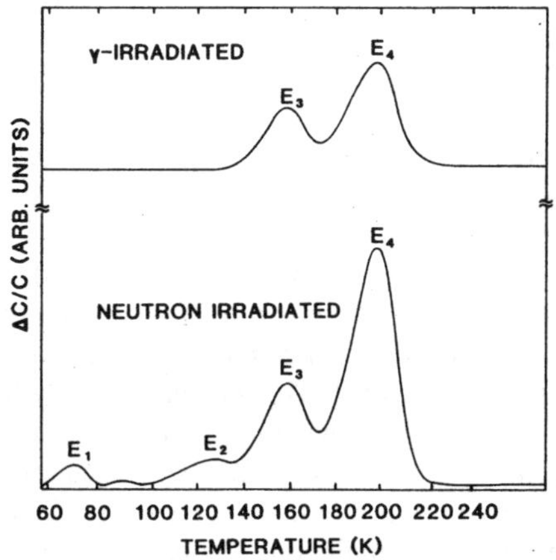

Fig. 3.16. DLTS spectra for γ- and neutron-irradiated n-type Ge. The γ dose was 30 Mrad and the neutron fluence $2x10^{14}$ n(1 MeV equivalent)/cm^2. E_4 corresponds to the E-centre, E_3 to the A-centre. E_1 at E_c-0.09 eV is tentatively assigned to the planar tetravacancy. E_2 could be level $E_{0.23}$ of Table 3.3. [After Nagesh and Farmer [93]]

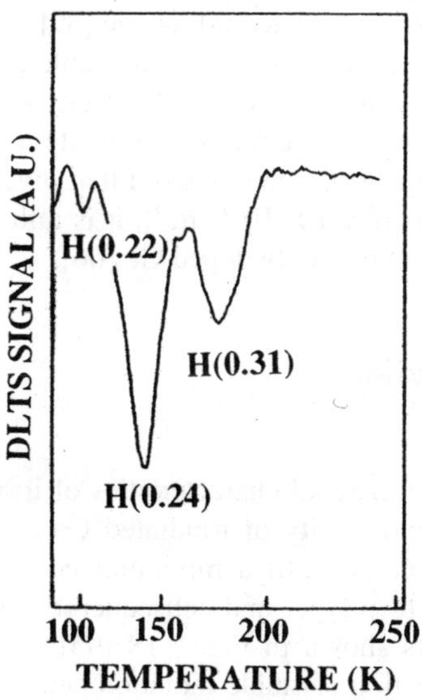

Fig. 3.17. DLTS spectrum after 10 MeV electron irradiation of In doped p-type Ge. Hole trap H(0.24) is possibly In-related. Level H(0.31), which is a majority trap in p-Ge could be the analogon of the $H_{0.30}$ minority trap in n-type Ge. [After Fukuoka and Saito [90]]

instead of a single acceptor state. This is further supported by the annealing behaviour of the level under reverse bias: the annealing is impeded for the Ge E-centre, while there is an enhancement for Si [66]. Moreover, as will be explained in more detail in the next part, the E-centre level moves down from the valence band in $Si_{1-x}Ge_x$ for increasing x. It is no longer observable for x=0.35 or 0.5. Extrapolating this to the case of x=1 would mean that the single acceptor level is shifted in the lower half of the band gap, while the double negative charge state may be split off the conduction band into the upper half.

While the thermal stability of the E-centres in Ge is similar as for Si, the divacancies and A-centres anneal at significantly lower temperatures in Ge. The minority trap $H_{0.30}$ shows a large cross section in Table 3.3 and is, therefore, assumed to be attractive for holes, i.e., it is a deep acceptor level. This is confirmed by the Poole-Frenkel shift towards a lower activation energy, when the trap is measured at a higher reverse bias[3]. The assignment of the $E_{0.29}$ level to the divacancy is based on the fact that it is only clearly observed in proton-irradiated n-type Ge after annealing at 110 °C [66] and not after electron irradiation. The peak appears to be broader than expected for a single defect, so it is the sum of two close-lying levels. A similar complex situation is found for the single acceptor level of V-V in Si (see previous paragraph and Fig. 3.9). From Fig. 3.16 one can also derive that heavy radiation damage can lead to the formation of higher order vacancy/interstitial clusters [93]. A similar conclusion was reached for neutron-irradiated high-purity Ge [85, 92, 96, 98]. Two V_2-H related defects at E_v+0.3 and E_v+0.42 eV were also observed in low-energy proton-irradiated p-type Ge [89].

The identity of the A-centre and of the O-related radiation defects in Table 3.3 has been derived from material which was deliberately doped in the melt with oxygen. The A-centre in fact strongly dominates the spectrum in Ge:O, while it is less pronounced in the undoped material. It should finally be remarked that given the low solubility of substitutional carbon in Ge (order 10^8-10^{10} cm^{-3}), it is rather unlikely to expect C_i-related defects and reactions in normally doped Ge [66].

3.3.4 Impact Radiation Damage on Ge Material and Device Properties

A number of studies have been devoted to the material characteristics of irradiated Ge [84, 85, 95]. When monitoring the conductivity of irradiated Ge as a function of ion fluence, first a reduction is observed until a minimum occurs, which corresponds with type inversion from n- to p-type, and subsequently the conductivity increases again [95]. An example is shown in Fig. 3.18 [95]. The change in resistivity is definitively due to the creation of stable radiation centres, on the one hand, and to the deactivation of dopants, incorporated in V-Donor

[3] Increasing the reverse bias of the diode increases the maximum electric field at the junction. For traps which are attractive for the emitted carrier, this will result in a lowering of the barrier for thermal emission proportional with $\sqrt{F}$.

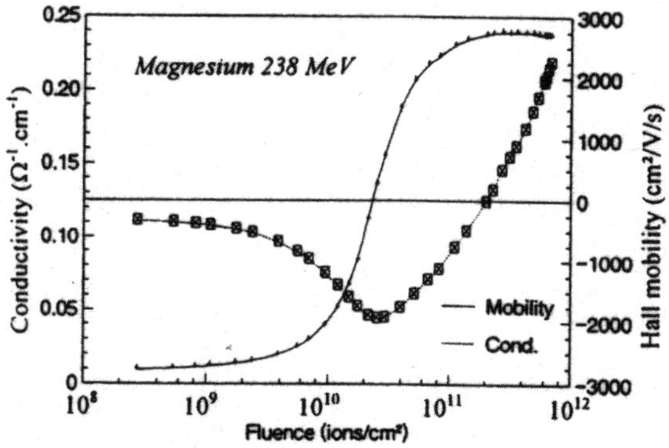

Fig. 3.18. Conductivity-fluence and Hall mobility-fluence experimental curves for n-type Ge (starting concentration 2.4x10^{14} cm^{-3}). [After Marie et al. [95]]

complexes, on the other. These measurements are useful to extract defect introduction rates

So far, only one type of Ge-based devices has shown evidence to suffer from ionisation damage [68, 99]. It has namely been observed that stressed Ge:Ga photoconductors degrade even for a very low back-ground dose of high-energy γ-quanta (< 4 rad). This results in an increase in the conductivity and a reduction of the breakdown voltage. Hits by a few high-energy photons or particles can change the responsivity of the detector, operating in the 1 - 4.2 K regime, by decades. It is believed that the induced changes are related to the creation of electron-hole pairs in the material, which can not escape the detector and, therefore, get trapped in the donor and acceptor sites, respectively. This changes the compensation ratio and the conductivity. An alternative contact configuration has been proposed, to eliminate the barriers at the p^+ contacts, in order to remove the created minority carriers from the volumc of the detector [99].

3.3.5 Summary Germanium Radiation Defects

It is clear from the above that the nature of the radiation-induced defects in Ge is far less well established than in silicon. The reasons are the lower technological relevance and interest and the difficulty to apply analytical techniques which have been proven successful for silicon (EPR,...). From a fundamental viewpoint, many questions remain open. Furthermore, although there are some indications of its role, no real systematic study of the impact of the substrate parameters (crystal growth, doping density and type, O, C and H content,...) on the radiation damage has been performed so far. Hardening of Ge components is at the moment more a matter of experience than of understanding. However, it is expected that the interest in SiGe alloys will also stimulate the investigation of Ge and its applications.

3.4 Displacement Damage in SiGe Alloys

The SiGe alloys form a complete series of solid solution and, therefore, allow performing band-gap engineering between 1.12 eV (Si) and 0.74 eV (Ge). This unique feature, together with the compatibility with silicon processing offers very promising perspectives for a wide-scale application of $Si_{1-x}Ge_x$ hetero-epitaxy in high-speed ULSI technologies. Therefore, in a first section, the relevant material properties and the structure of SiGe-based devices will be briefly outlined. Next, a summary is given on the radiation damage studies in bulk SiGe and in relaxed and strained SiGe-layers. Additional information can be retrieved from the investigation of processing-induced radiation damage, for example caused by plasma etching, metal deposition or ion implantation. Finally, the radiation tolerance of some recent devices and technologies will be discussed.

3.4.1 SiGe Material Properties and Applications

As already mentioned, SiGe alloys can be formed continuously between x=0 and x=1. However, due to the larger atomic radius of Ge (0.123 nm versus 0.117 nm for Si), some strain is introduced in the lattice when more and more Ge atoms are incorporated. The lattice mismatch can amount to 4.2%. This results in a reduction of the alloy band gap E_G, as shown in Fig. 3.19 [100]. Most of the band-gap reduction occurs for the valence band maximum, while the conduction band minimum shows only a minor shift (Fig. 3.20a) [101]. However, above 15 atomic % Si (x=0.85), the position of the band-gap minimum shifts from the Ge-like L points (<111>), to the Si-like X points (bottom along <100> directions), which has consequences for the bulk material properties [102, 103]. The addition of a fraction y of C to form ternary $Si_{1-x-y}Ge_xC_y$ alloys yields an additional degree of freedom for band-gap engineering in this sense that the band-gap change is now divided between the valence and conduction band [104]. This might be of interest for the fabrication of n-channel heterostructure Field Effect Transistors or pnp heterojunction bipolar transistors (HBT).

Recently, interest has developed for the crystal growth of bulk SiGe by the Czochralski (Cz) technique [102, 105], whereby potential applications could be solar cells [106, 107], or substrates for the deposition of epitaxial layers. However, most efforts are devoted to the incorporation of strained or relaxed SiGe layers in silicon processing. An epitaxial layer is obtained either by Molecular Beam Epitaxy (MBE) or by Ultra High Vacuum (UHV) or Low-Pressure Chemical Vapor Deposition (LPCVD). For most ULSI applications, thin (≤100 nm range) strained layers on silicon are necessary. The strain comes from the lattice mismatch between the SiGe alloy and the silicon substrate, and increases for increasing x. For a thickness below a certain critical value h_c, the layer grows pseudomorphically. Above h_c, the strain relaxes, giving rise to the formation of dislocations, which may cross the layer (threading dislocations). Figure 3.21 shows the critical thickness versus the mismatch. The presence of these dislocations degrades not only the structural but also the electrical and optical integrity of the layer. Furthermore, as can be seen

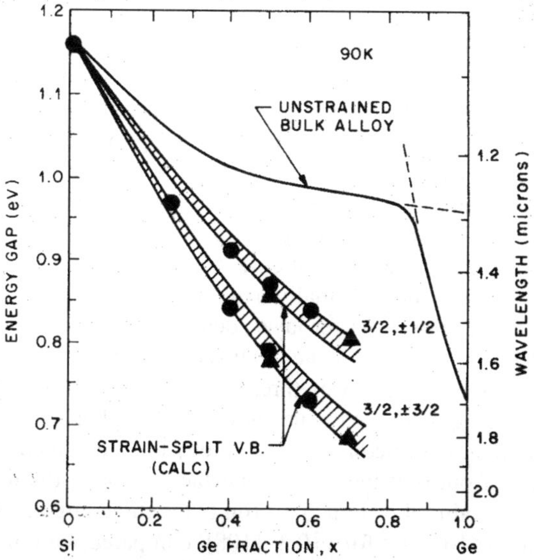

Fig. 3.19. Summary of energy gap values at 90 K. The double points at the same value of x correspond to a splitting of the valence band. The calculations showing the strain-split valence band are indicated by the cross-hatched bands. [After Lang et al. [100]]

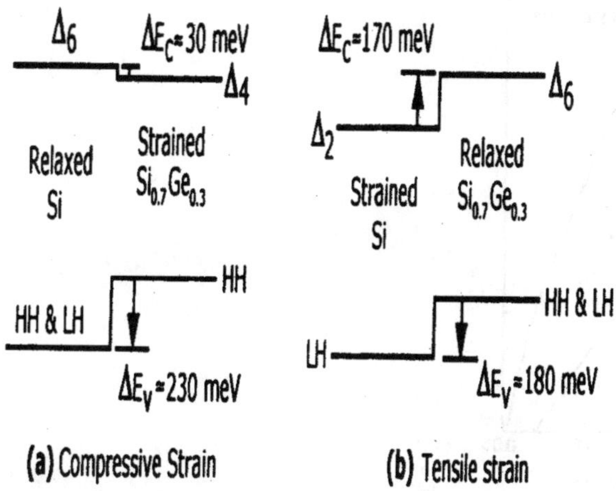

Fig. 3.20. Band alignments between Si and $Si_{0.70}Ge_{0.30}$ on two substrates: (a) Si and (b) $Si_{0.70}Ge_{0.30}$. [After Maiti et al. [101]]

from Fig. 3.21, h_c reduces with increasing temperature, implying that a strained layer deposited at a certain temperature will relax when heated up sufficiently long at a higher temperature. This is one of the important limitations for the use of SiGe in ULSI technologies, requiring a reduced thermal budget after epitaxy. Fortunately, this goes along with the on-going effort for technology scaling to deep submicron dimensions. Strained layers can also be deposited on a relaxed SiGe buffer, confining the threading dislocations to the latter and the silicon substrate. Another handicap with SiGe is the difficulty in forming a good stable oxide on its surface, which is highly recombination active. In most cases, this is solved by the deposition of a thin Si cap layer, which is then thermally oxidized.

For other applications (solar cells, relaxed buffer layer, detectors, sensors,...) or for materials' characterization related purposes thicker, relaxed epi-layers may be useful. However, in order to limit the formation of threading dislocations and to preserve a good quality film, adapted growth techniques have to be applied. In that case, the final layer is deposited on an intermediate one, whereby the Ge content is gradually (i.e., step-wise) changed from 0 at the silicon interface, towards the desired value, allowing for a gradual strain relaxation. This limits the threading dislocation density to reasonable values ($\leq 10^5$ cm^{-2} for x=0.3) [108], compared with material grown without graded buffer (range 10^8 to 10^{12} cm^{-2}). The latter densities are too high to make the material appropriate for ULSI applications. The combination of the graded (stepped x) and relaxed layer (constant x) is often called a virtual substrate

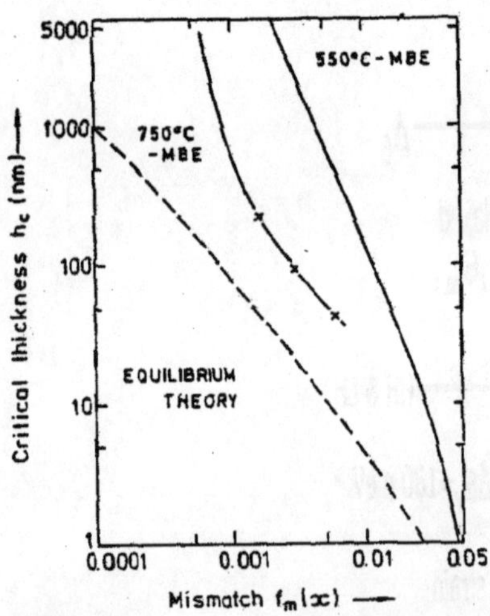

Fig. 3.21. Critical layer thickness h_C for misfit dislocation generation against mismatch parameter $f_m(x)$. [After Jain and Hayes [103]]

Due to the continuous improvement in the SiGe epitaxial layer quality, one should be careful in comparing early results with later reports, with respect to irradiation-induced defects and radiation hardness. Some of the inconsistencies and contradictions could be explained by the presence of grown-in extended defects.

The most successful application, where SiGe heterostructures realized a breakthrough to commercial products is in the field of HBTs (for an overview see e.g. [109]). Replacing the epitaxial silicon base by a SiGe heterojunction, as schematically illustrated in Fig. 3.22, offers several important advantages. First of all, due to the band offset, higher current gain $\beta = I_C/I_B$ values can be reached, with I_C the collector and I_B the base current. This is because the flow of holes from base to emitter is reduced owing to the barrier in the valence band. Second, the hole mobility in the SiGe layer is significantly higher than in a p-Si base, reducing thus the base transit time and hence, increasing the maximum oscillation frequency f_{max} and the cutoff frequency f_T of the devices. This renders HBTs particularly attractive for high-frequency (microwave) applications, where they become a serious competitor for the III-V based technologies, pushing continuously the limits upwards. This could be important for future satellite communications systems. Implementing SiGe in BiCMOS opens up the market for mixed mode analog-digital applications, like portable telecommunication systems. Finally, the HBT can be optimized for cryogenic operation down to 77 K or even 4.2 K, a region where the classical silicon BJT is no longer operational [109]. It has in fact been shown that the performance

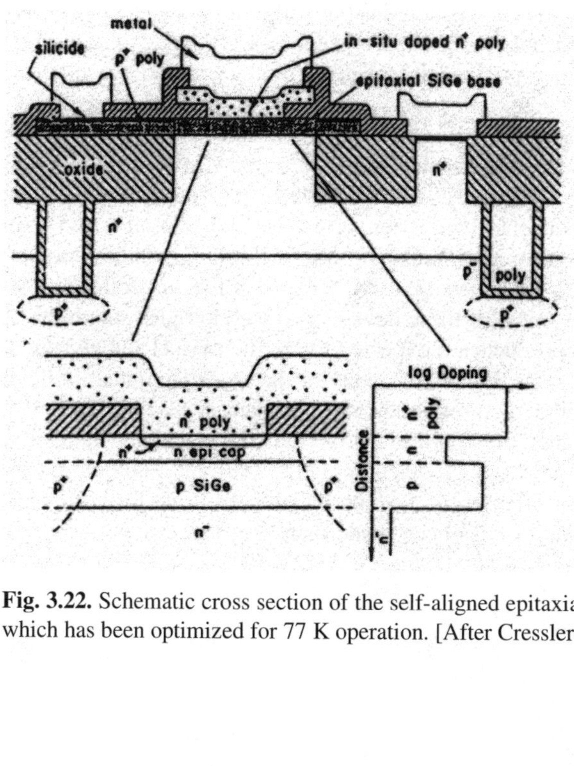

Fig. 3.22. Schematic cross section of the self-aligned epitaxial emitter-cap SiGe technology which has been optimized for 77 K operation. [After Cressler [109]]

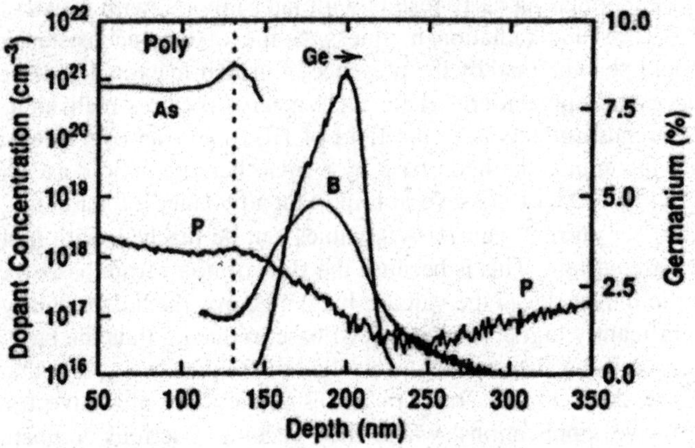

Fig. 3.23. Measured SIMS impurity profile for the epitaxial emitter-cap SiGe technology. [After Cressler [109]]

at 77 K for a cryogenic SiGe HBT exceeds the one at room temperature in many aspects, similar as for a cooled MOSFET. In order to benefit fully from its potential, it is mandatory to use a graded Ge strained epitaxial layer, as shown in Fig. 3.23, for example, whereby the Ge fraction lies typically in the range 8 to 16%.

The enhancement of the hole mobility in p-SiGe can also be exploited for the development of high-mobility p-channel heterostructure (or quantum well) MOSFETs [110, 111], rendering them attractive for high-speed applications and making the deep submicron p-channel device more compatible with its n-channel counterpart. A next step could be the development of SiGe based CMOS, using the layer scheme of for instance Fig. 3.24a [112] or Fig. 3.24b [111], respectively. In recent years even more exotic concepts have been proposed and demonstrated, which could lead to future deep sub micron (sub 100 nm) technologies, like modulation doped FETs or MODFETs [101], vertical MOSFETs [113] or strained p-Si p-MOSFETs [101]. However, most of these devices and technologies should be considered as exploratory and products are expected to see the market somewhere during the coming years. The same applies for many of the so-called quantum devices [114] like resonant tunneling structures and hot-carrier transistors

SiGe has also large potential as photosensitive material, enabling the development of infrared Schottky barrier photodetectors [115, 116] and photodiodes [117, 118]. Less successful are the attempts to develop SiGe based Light Emitting Diodes (LEDs), although some progress has been made using Er-doped material (see e.g. [119]).

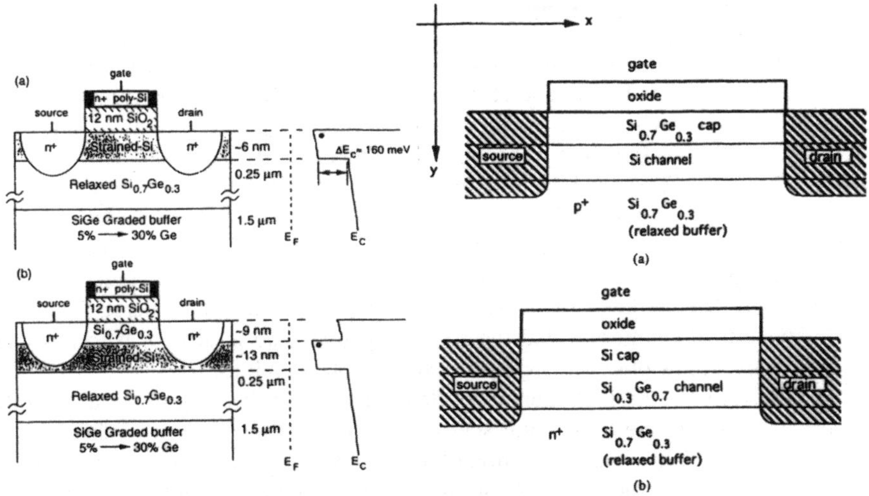

Fig. 3.24. Schematic representation of: Left - a strained-Si n-channel at the surface (a) or-buried below the SiGe cap (b) [After Welser et al. [112]]. Right - p-FET with a strained Si-channel and a SiGe cap (a), and n-FET with a strained SiGe channel and a Si cap (b). [After O'Neill et al. [111]]

3.4.2 Radiation Damage in SiGe

In this paragraph, the radiation defects in SiGe as observed by DLTS will be described. The properties of the deep levels will first of all depend on whether one deals with a SiGe bulk or epilayer material, since this determines the band gap according to Fig. 3.19. For the moment, epitaxial SiGe is technologically far more relevant than bulk crystals or wafers. The thickness of the layers is another important parameter, which determines whether the layer is in a strained metastable or in a relaxed state. As will be shown, this may impact on the radiation response and the deep-level parameters. For comparison the radiation damage during certain ULSI processing steps, like ion implantation, dry etching etc., will be discussed in the next paragraph Sect. 3.4.3). It should finally be remarked that in order to assess the radiation damage in thin epilayers, DLTS and PL have proven very useful, while volume techniques like EPR and infrared absorption can only be applied to bulk samples. This complicates the structural and chemical identification of the observed deep levels or luminescence bands. However, by studying the defects in function of the composition parameter x, one usually observes some clear trends, which help to identify the DLTS peaks.

3.4.2.1 Radiation Damage in Bulk SiGe

The literature on the radiation damage in bulk SiGe material is rather limited. There is one report on the low-temperature irradiation (~30 K; 2-MeV protons) behaviour of bulk n-type SiGe grown by the Cz technique [120]. The resulting DLTS spectrum is shown in Fig. 3.25. The figure illustrates the transformation of the Ge-V level E4 to the A-centre E2 in the material doped with 2.5% Ge. The E4 level has an activation energy of 0.290 eV with respect to the conduction band and an electron capture cross section of $5x10^{-16}$ cm^2. Based on the fact that the annealing at room temperature is retarded under the application of a negative diode bias, it is concluded that the level corresponds with the double-negative charge state of the Ge-V [120]. It is inferred that the single-negative state is near or below mid gap and, hence, below observation for the experimental conditions.

The study of the GeV centre in Ge-doped silicon bears also some important consequences for the V acceptor states in Si. As shown in Fig. 3.1, it has been derived from EPR that the association of a vacancy with a Ge atom has minor effects on the electronic structure [1, 2], so that the levels are expected to be similar as for the isolated (single) vacancy. The main role of the pairing with Ge is to stabilise (immobilise) the isolated vacancy. This is confirmed by the results obtained on the 2.5% doped n-type SiGe, where the GeV level remains stable close to room temperature. This is higher than the ~250 K annealing stage in lowly Ge-doped silicon. Using this enhanced stability, it is possible to obtain indirect information of the V acceptor levels in n-type silicon [120]. From Fig. 3.25, it is also derived that the formation of the A-centre strongly depends on the amount of Ge and is suppressed for the low-temperature irradiation

Other studies have focussed on the formation and thermal stability of the vacancy and vacancy-related defects in bulk Cz SiGe by the positron annihilation

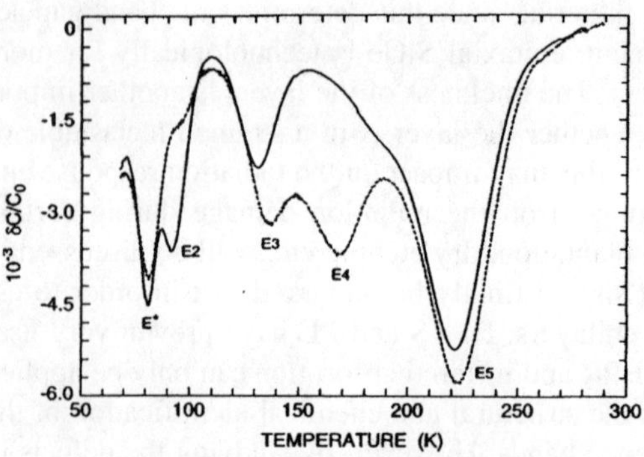

Fig. 3.25. DLTS spectra before (broken curve) and after (full curve) annealing at 295 K for 15 min at zero bias of a proton-irradiated, Cz grown, p^+n diode containing 2.5% Ge. The *y* axis shows the DLTS signal normalised to the diode capacitance. The spectra were recorded using a repetition rate of 25 Hz. [After Budtz-Jørgensen et al. [120]]

technique [121, 122]. It has been observed that an abrupt change in the bulk positron lifetime occurs for a Ge fraction of 0.8 to 0.83, in agreement with the shift in the conduction band minimum from the L (Si) to the X point (Ge). After 10^{18} cm^{-2} 3 MeV electron exposure no stable vacancy clusters were observed for Ge-rich alloys. The absence of vacancy clusters for $x>0.8$ can be explained by assuming that both the monovacancy and maybe the divacancy are unstable in "Ge" and disappear from the crystal before agglomeration to tetravacancies. In other words, the mobility of the vacancy depends on the electronic structure of the crystal and is higher in Ge-rich crystals compared with Si-rich alloys

For Si-rich alloys, on the other hand ($x\leq 0.37$), it is expected that the V-Impurity and divacancies will be the major defect species. For the intermediate Ge range ($0.6\leq x\leq 0.8$), the higher positron lifetimes suggest the presence of larger vacancy clusters. Furthermore, from thermal annealing experiments, it is concluded that the formation of larger vacancy clusters due to the agglomeration of single vacancies is inhibited due to the presence of Ge. This effect is similar as for heavily doped Si compared with undoped material, where vacancy-dopant complexes retard the vacancy clustering. This role is overtaken here by the Ge, which is assumed to pair with the vacancy for reasons of strain compensation. This is also in line with the retarded A-centre formation found in Ge doped silicon [120, 123], whereby the created vacancies are trapped at specific sites, so that they are not available for pairing with interstitial oxygen. In addition, these trapped vacancies are not necessarily electrically active, so that it may not be possible to detect them by DLTS.

3.4.2.2 Radiation Damage in Relaxed SiGe Layers

The isolated vacancy has been studied in p-type relaxed or unstrained $Si_{1-x}Ge_x$ by DLTS, after 30 K 2-MeV proton irradiations [124]. Fig. 3.26 represents the peak corresponding with the $V^{++}\rightarrow V^0$ transition, by the simultaneous emission of two holes for samples with different x (in %). This corresponds with the negative U behaviour described in Sect. 2.4.2. It is clear that the signal shifts to lower temperatures, for higher x and disappears in the freeze-out region for $x>0.15$. When analysing the Arrhenius plot, it turns out that the hole ionisation enthalpy is independent of x, while the emission rate factor enhances drastically for higher x. This is attributed to the entropy term, caused by a lattice relaxation specific to the statistical alloy disorder and responsible for a large carrier-phonon coupling around the vacancy [124].

The vacancy behaviour has also been studied in unstrained n-type SiGe alloys, after 30 K 2-MeV proton exposures [120]. Similar results as for the bulk n-type SiGe samples have been obtained: a level at E_c-0.29 was identified as the $GeV^{-/--}$, showing higher stability for a higher Ge concentration. The results are summarised in Table 3.4.

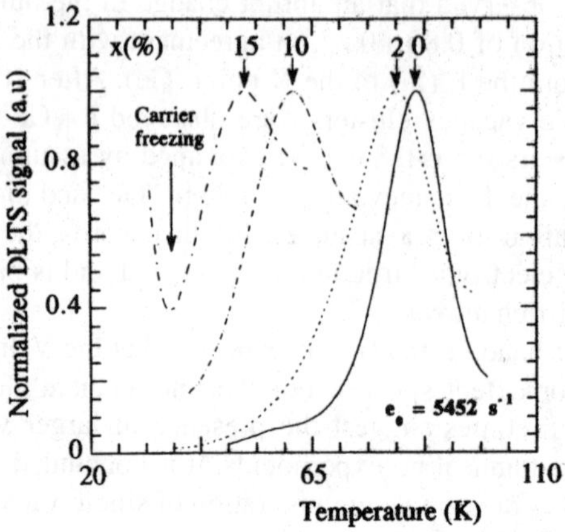

Fig. 3.26. The DLTS spectra recorded after proton irradiation of releaxed $Si_{1-x}Ge_x$ alloy layers. The arrow points out the temperature below which the analysis cannot be continued due to carrier freezing-out. [After Mesli and Nylandsted Larsen [124]]

Room temperature radiation-induced defects in relaxed SiGe layers have been studied by a number of groups, after both high-energy electron [125, 126] and proton irradiation [127- 129]. Early results obtained on low-quality n-type material are somewhat inconsistent, allowing no clear identification of the observed trap levels [125, 126]. More recent experiments on better quality material deposited on a graded buffer layer have enabled a more firm identification of the most prominent radiation-induced defects [128 - 130]. For the case of n-type material spectra for different Ge fractions are summarized in Fig. 3.27 [127, 128], showing some clear trends. First, the concentration of defects for the same irradiation conditions (2 MeV protons at a fluence of $1.5x10^{12}$ cm^{-2}) reduces strongly for increasing x. The introduction rate of trap T4 reduces by a factor 4 in going from a Ge content from 5% (R_T=90 cm^{-1}) to 25% (R_T=20 cm^{-1}). This strongly suggests a hardening effect due to the presence of Ge, in line with the above observations.

A second conclusion to be drawn from Fig. 3.27 is that the same basic radiation defects are present, whereby the level position is composition dependent. This is derived from the shift of the DLTS peak maximum with x to higher temperatures for T1 to T4. However, when analyzing in more detail the Gibbs free energy of the defects, it is concluded that the energy position is not changing with respect to the conduction band (Fig. 3.28) [127]. Also the electron capture cross section σ_n for T4 is independent of alloy composition, yielding an average value of $6\text{-}8x10^{-16}$ cm^2. This order of magnitude suggests an acceptor nature of the trap T4, which is, therefore pinned to the conduction band. Finally, from the constant full width at half maximum of the DLTS peaks, it is concluded that no alloy broadening occurs in the case of T4. Based on the above evidence, peak T4 has been identified as the Sb-V pair (Sb doped n-type material). It is speculated that T3 may be the V-V$^{-/--}$ charge state of the divacancy, while the origin of the other trap levels is unclear.

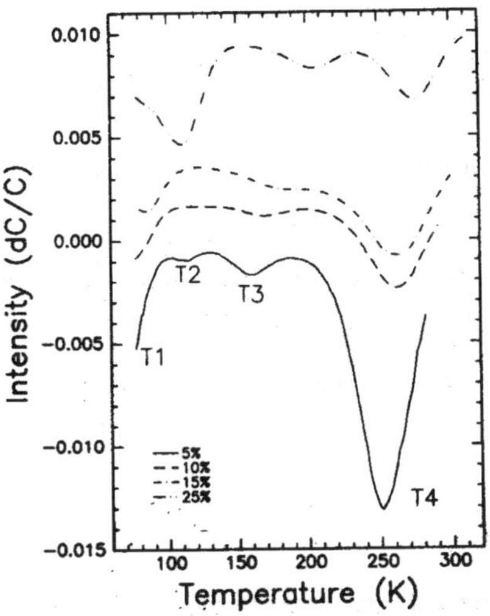

Fig. 3.27. DLTS spectra of $Si_{1-x}Ge_x$ epitaxial layers with different compositions after irradiation with identical fluences of 2 MeV protons. The spectra were recorded using a repetition rate of 1 kHz. [After Kringhøj and Nylandsted Larsen [127]]

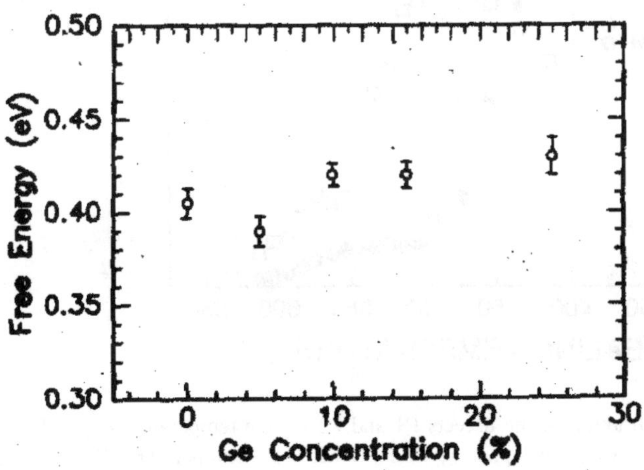

Fig. 3.28. Gibbs free energy of the T4 defect as a function of composition. The uncertainties are statistical uncertainties only; the systematic uncertainties are estimated to be of the order of 5%. [After Kringhøj and Nylandsted Larsen [127]]

The situation in proton-irradiated relaxed p-type SiGe layers is somewhat more complex [129], as illustrated by Fig. 3.30. Beside two hole traps H1 and H2 found for all x, an additional near mid-gap centre is found for $x \geq 5\%$. In this case, the level positions tend to shift towards lower temperature for increasing x, which is opposite to the shift observed in Fig. 3.27. It should be remarked, however, that in Fig. 3.30 hole emission towards the valence band is monitored, while in Fig. 3.27 electron emission towards the conduction band is observed. From the shift of the activation enthalpy with compositions, shown in Fig. 3.11 for the levels H1, H2 and H3, it is concluded that the trap levels are also pinned to the conduction band, i.e, their relative energy position with respect to the conduction band minimum remains fixed. The corresponding defects parameters can be found in Table 3.4.

A new level H3 develops for all x, after annealing at 150°C, whereby there is a clear link between the annihilation of level H2 and the creation of H3. Another interesting observation is that the stability of the levels H1, H2 and H3 increases for increasing x, as evidenced e.g. by Fig. 3.32 [129]. A similar conclusion is reached for the stability of vacancy-related radiation defects in bulk SiGe, derived from positron annihilation lifetime [121, 122].

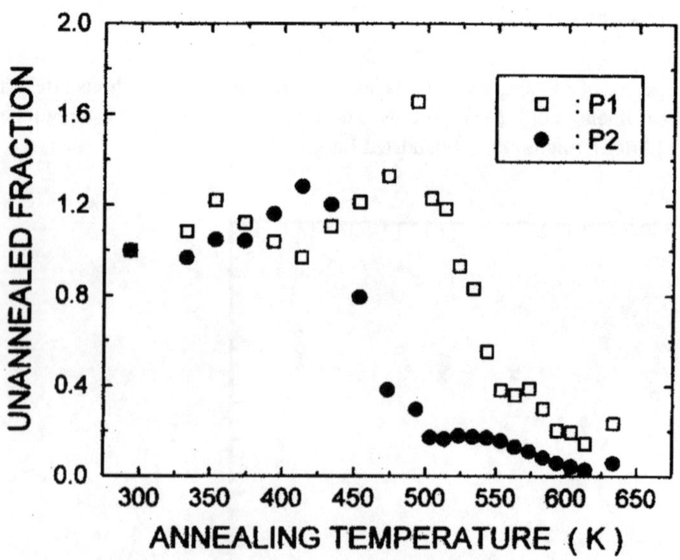

Fig. 3.29. Unannealed fraction of the defects P1 and P2 vs. the temperature of isochronal annealing for 10 min under He atmosphere. [After Goubet and Stievenard [126]]

Table 3.4. Defect parameters for the main radiation defects in relaxed $Si_{1-x}Ge_x$ epilayers

Ge fraction & type [%]	Apparent Activation Energy [eV]	σ_∞ [cm^2]	Annealing T_A [K]	Identity
5 (n-Sb)	E_c-0.101	$9x10^{-16}$	--	C_i
5 (n-Sb)	E_c-0.158	$2.8x10^{-16}$	--	OV (A-centre)
5 (n-Sb)	E_c-0.264	$5x10^{-15}$	--	$VV^{-/--}$
5 (n-Sb)	E_c-0.290	$5x10^{-16}$	--	$GeV^{-/--}$
5 (n-Sb)	E_c-0.434	$1.9x10^{-15}$	295-315	SbV + $VV^{0/-}$
30 (n-P)	E_c-0.32	10^{-15}	560	$VV^{-/--}$
30 (n-P)	E_c-0.49	$2x10^{-15}$	470	PV
25 (n-Sb)	E_c-0.51	$6\text{-}8x10^{-16}$	--	SbV
15 (p-B)	E_v+0.14[a]	$4x10^{-16}$ [b]	550	$VV^{+/0}$
25 (p-B)	E_v+0.13[a]	$5x10^{-15}$ [b]	650-700	B_iC_s

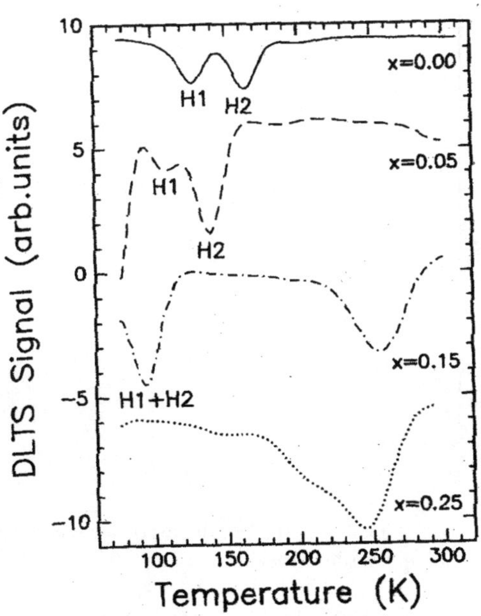

Fig. 3.30. DLTS spectra from 2 MeV proton-irradiated p-$Si_{1-x}Ge_x$ diodes of the mentioned compositions. [After Monakhov et al. [129]]

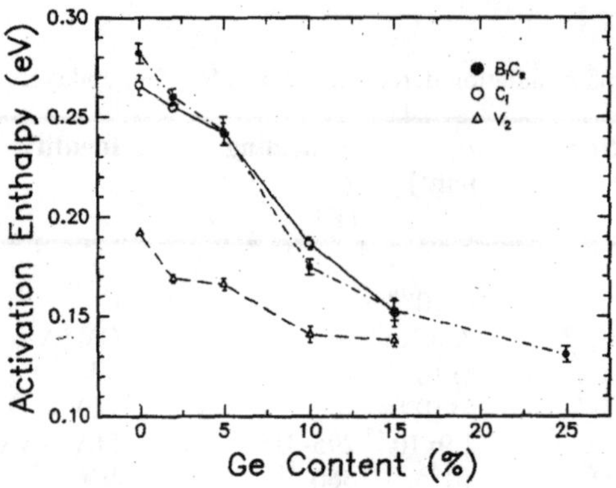

Fig. 3.31. The extracted activation enthalpies relative to the valence band for the three identified levels as a function of Ge content. The error bars represent one sigma values from the Arrhenius fit. [After Monakhov et al. [129]]

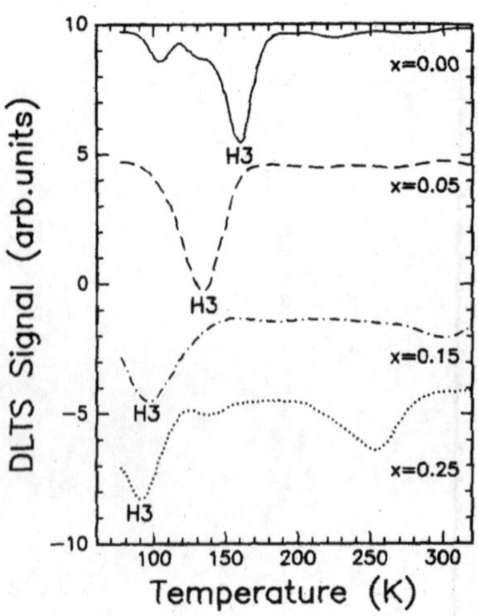

Fig. 3.32. DLTS spectra from 2 MeV proton-irradiated p-$Si_{1-x}Ge_x$ diodes of the mentioned compositions after annealing at 300°C for 30 min. [After Monakhov et al. [129]]

In analogy with the case of pure silicon [130], the levels H1 to H3 have been tentatively assigned to the V-$V^{+/0}$ donor, the interstitial carbon C_i and the B_iC_s complex. The increased stability of the H2 peak is likely due to a reduced mobility of C_i in SiGe compared with pure Si. Here, a similar 'retarded' formation is observed as for the A-centre in Ge-doped Si, whereby C_i is believed to be trapped at a site, which is not electrically active.

3.4.2.3 Radiation Damage in Strained SiGe Layers

Only few reports are available on the radiation damage in thin strained layer n-SiGe [131, 132]. Results of 5.4 MeV alpha particle irradiations on 4% n-type SiGe are in line with the results obtained on relaxed n-SiGe: compare for example the spectra of Fig. 3.33 with the ones displayed in Fig. 3.27. The same trend is observed: a reduction of the trap concentration for the same fluence ($3x10^{11}$ cm^{-2}) going from x=0 to 0.04, while the apparent trap position (DLTS peak maximum temperature) shifts to higher values for the same emission rate window. Level EA2 (P-V) has an apparent activation energy of 0.44 eV and a capture cross section (from the intercept of the Arrhenius plot) of $7.6x10^{-16}$ cm^2. The $V\text{-}V^{-/--}$ EA1 level in α-irradiated 4 % n-SiGe has a signature of E_c-0.24 eV and σ_n=$2.3x10^{-16}$ cm^2. Both observations are in fair agreement with the data in Table 3.4. Defect level EA3 is only detected in material containing Ge, after the alpha particle exposure.
The radiation defects formed in strained-layer p-SiGe under different particle irradiations have been studied in detail [133-144]. It has been shown that 1 and 2 MeV electron irradiations cause severe device and material degradation above a certain threshold fluence [133-136]. At the same time, the creation of radiation induced deep levels is evidenced by DLTS spectroscopy on exposed heterojunction diodes. However, the amount and nature of the stable defects depend markedly on the Ge content. This follows for example from a comparison of Fig. 3.34a and 3.34b [135] and is even more clearly illustrated by the electron trap concentrations versus fluence shown in Fig. 3.35. In the spectra of the x=0.12 layers, both majority (hole) and minority (electron) traps are found after an injection pulse (Fig. 3.34a). The activation energy for the hole traps derived from an Arrhenius plot is

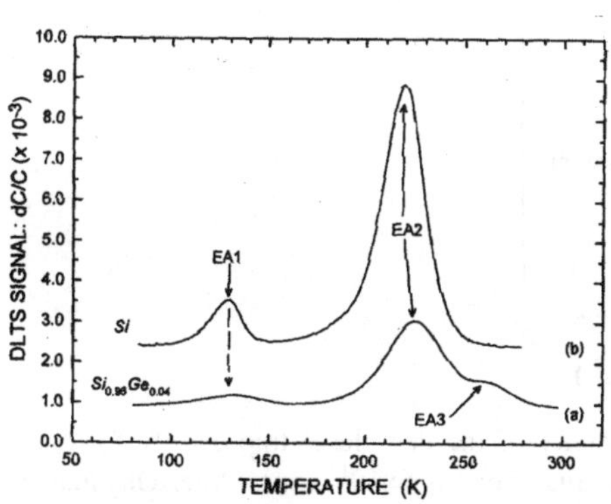

Fig. 3.33. DLTS spectra of 5.4 MeV alpha particle irradiated $Si_{1-x}Ge_x$. Curve (a) is for x=0.04, curve (b) is for x=0. Spectra were recorded at a quiescent reverse bias of 0.5 V, and at a frequency of 46 Hz. [After Goodman et al. [131]]

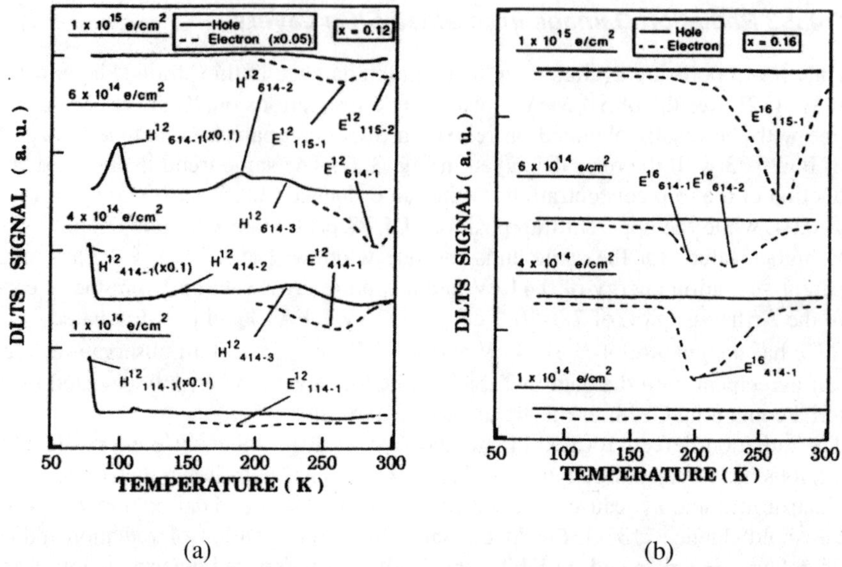

Fig. 3.34. DLTS spectra of a 1 MeV electron-irradiated p-type $Si_{0.88}Ge_{0.12}$ (a) and $Si_{0.84}Ge_{0.16}$ (b) epitaxial layer revealing both hole and electron traps, for different fluences. [After Ohyama et al. [135]]

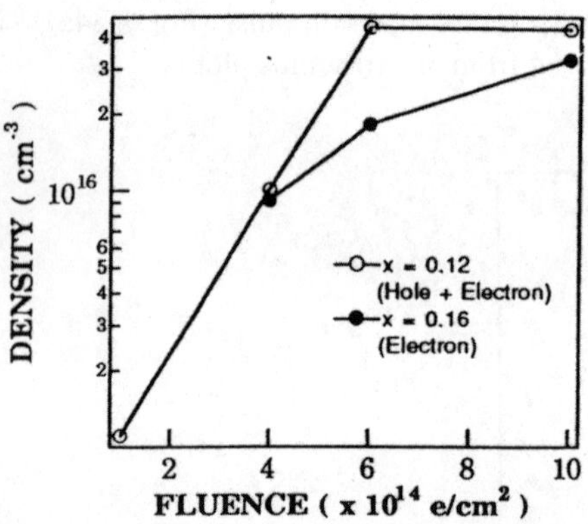

Fig. 3.35. 1 MeV electron fluence dependence of the deep level trap density for two Ge compositions. The B doping density in the SiGe layer is $6x10^{17}$ cm^{-3}. [After Ohyama et al. [135]]

[135]: E_v+0.11 eV; E_v+0.28 eV and E_v+0.46 eV and this for a fluence of $4x10^{14}$ cm^{-2}. This could tentatively correspond to the V-V donor level and either C_i or C_iB_s (compare with Table 3.4). The deepest hole trap could be similar as the one observed in proton-irradiated relaxed p-type layers, occurring for x>0.05 [129].

In addition to the hole traps, also electron traps are observed in Fig. 3.34. For x=0.16 these are the only radiation-induced levels observed above the DLTS detection limit. Activation energies are in the range E_c-0.56 eV (x=0.12) and E_c-0.66 eV (x=0.16). The trap level profile has also been studied for different reverse biases. It is found that the hole traps show a profile which reduces towards the surface. From studies of irradiated silicon it is known that vacancy-related levels show a lower concentration towards the surface [145], which is a perfect sink for vacancies. It is, therefore, speculated that some of the unknown hole traps may be V-related, e.g. V-B complexes [8]. The electron trap, which is close to mid gap and hence an effective generation centre shows a similar profile as the active B, with a concentration which reduces with depth. In analogy with the case of silicon, it is assumed that these levels are related to interstitial B_i [8], possibly a complex with B_s: B_i-B_s. It is furthermore well-known that B is the dominant radiation defect in highly B-doped silicon, irradiated at low temperatures [1, 8], giving rise to an electron trap at E_c-0.45 eV. However, this level anneals at 240 K and is not expected to be stable at room temperature, even for SiGe, with rather high x. Due to its mid gap position this electron trap(s) will have a strong impact on the electrical characteristics of irradiated heterojunction diodes [135].

With respect to the introduction rates $R_T=\partial N_T/\partial \Phi$, the following can be remarked. From the linear part of Fig. 3.35, a slope of 25 (x=12%) and 22 (x=16%) 1/cm is derived [135]. This is, however, significantly lower than the B de-activation rate K_B in the range 100 to 200 cm^{-1} [133 - 135], obtained from C-V measurements. From a comparison with calculated NIEL values for the alloys studied, a good agreement with the damage factors was found [137, 138]. In this respect, it should be remarked that in first order, the NIEL can be calculated by a linear combination of the NIEL for Si and Ge, weighed by their respective fractions (1-x) and x.

Another important issue with respect to radiation damage is its stability upon thermal annealing and possible recovery of the degraded device performance. This has been studied extensively for the 1 MeV electron-irradiated diodes [135]. For example, the electron traps discernable in Fig. 3.34b for x=0.16 are stable up to ≈300°C (Figs. 3.36a and 3.36b), if the stability limit is defined at 10% unannealed fraction (Fig. 3.36b). From Fig. 3.36b also follows that the defect annealing is in the first order composition independent. The high annealing temperature in fact rules out an assignment to B_i, since this level disappears at much lower temperatures [1, 8]. It is more likely a higher order complex involving B and some other impurity or point defect.

Roughly the same qualitative results have been obtained for irradiations with fast neutrons or high-energy ions. Similar electron traps and freeze-out behavior (B de-activation) have been found for 1 MeV neutron and 20 MeV proton irradiations [138, 140]. Only for the α-exposures, a different type of spectrum is observed in Fig. 3.37 [144]. However, as found systematically for all irradiations, the trap concentrations and B de-activation reduce with Ge fraction x, confirming the alloy hardening effect. Furthermore, a similar defect recovery is found, for example after a 20 MeV proton exposure [140]. A slightly higher activation energy for annealing of the electron traps is found in that case (0.27 eV).

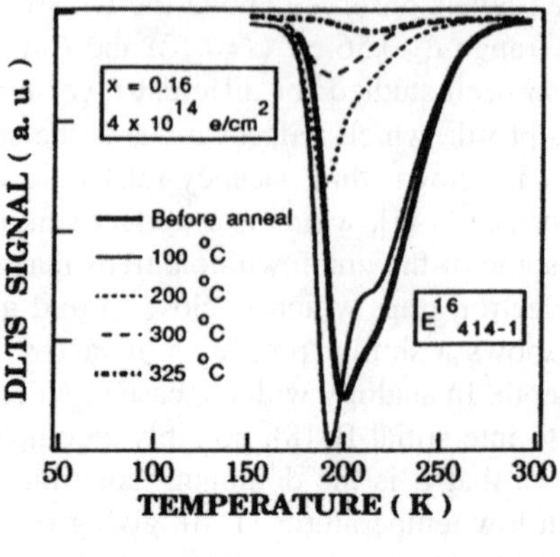

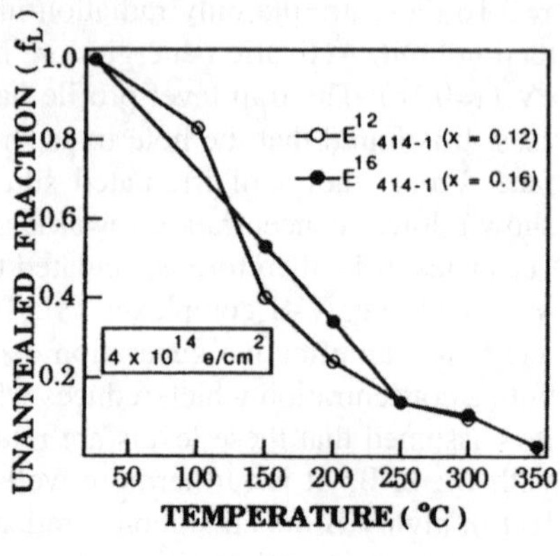

Fig. 3.36. Isochronal annealing of the electron trap level for diodes with x=0.16, after 4x10^{14} cm^{-2} 1 MeV electron irradiation (a) and unannealed fraction of the electron traps for x=0.12 and 0.16 as a function of annealing temperature (b). The anneals were for 15 min. [After Ohyama et al. [135]]

Finally, deep levels have also been studied in the emitter-base and base-collector junctions of irradiated heterojunction npn BJTs [136, 137, 140 - 142]. After a 1 MeV neutron exposure, the same type of electron traps was observed, in addition to a deep hole trap [137]. However, in most cases, it was only possible to detect electron traps in the n-type silicon collector region, showing the "classical" silicon radiation defects (V-O, V-V, P-V).

Summarizing the results obtained for the p-type SiGe strained layers, the following trends have become obvious. First, the microscopic device degradation, i.e., the boron deactivation and the trap level introduction rate, reduce significantly for higher x. In some cases, a factor 3 improvement can be observed, for x going from 0.08 to 0.16. This can be partially explained by the lower number of knock-on atoms created for higher x, although there is not a linear correlation between the calculated damage parameters and the measured damage coefficients. One of the most harmful radiation induced defects in strained p-type SiGe layers is the near mid gap electron trap(s), which should be an efficient leakage current generation centre. The fact that the same annealing activation energy is found for both the contration of the trap level and the reverse current supports this idea. The activation energy for annealing lies typically in the range 0.2 to 0.3 eV. It is a rather stable defect, showing an annealing stage around 500 to 600 K, which is too high to correspond to B_i. A higher order complex involving B is therefore put forward for its origin.

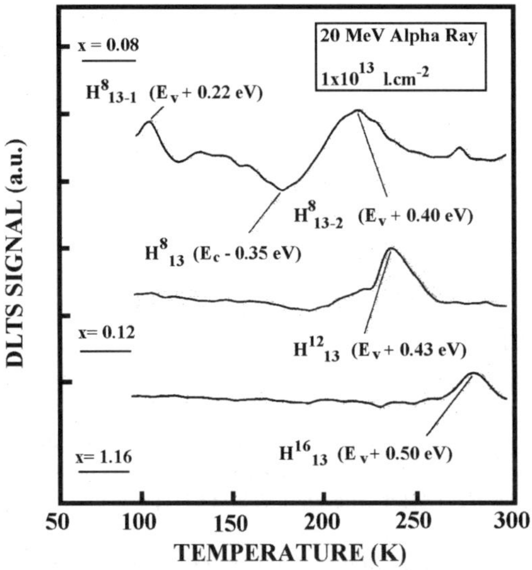

Fig. 3.37. DLTS spectra for $Si_{1-x}Ge_x$ diodes with different germanium content, subjected to 20 MeV alpha particles with a fluence of 10^{13} cm^{-2}. [After Ohyama et al. [144]]

3.4.3 Processing-Induced Radiation Damage in SiGe

Modern processing uses fabrication steps whereby often radiation damage by low energy electrons (metal sputtering), ions (ion implantation, dry etching,...) or photons (X-rays mainly) can occur. Since this degrades the electrical and structural material properties, it is important to understand and control their formation. Quite some information has been gathered in recent decades, showing some overlap with the high energy radiation behavior. These investigations have been extended to the case of SiGe layers, in view of their application in ULSI technologies.

3.4.3.1 Dry Etching and Electron Beam Sputtering Damage

The impact of low-energy 1 keV He ion etching on n-type SiGe layers with different compositions and degrees of relaxation has been studied by *Goodman et al.* [131, 132] and compared with high-energy α-irradiation and radio-frequency sputter-deposition of Au Schottky barrier diodes. In total 6 electron traps have been observed after He-ion etching, four of them are similar to the case of silicon (Fig. 3.38). Most of the deep levels have a uniform profile for the depth range studied. An important remark is that the trap concentration for the 4 %Ge containing material is one decade *larger* than for the pure silicon. This is in strong contrast to what has been reported above for high-energy particle irradiations. Two traps EHe5 and

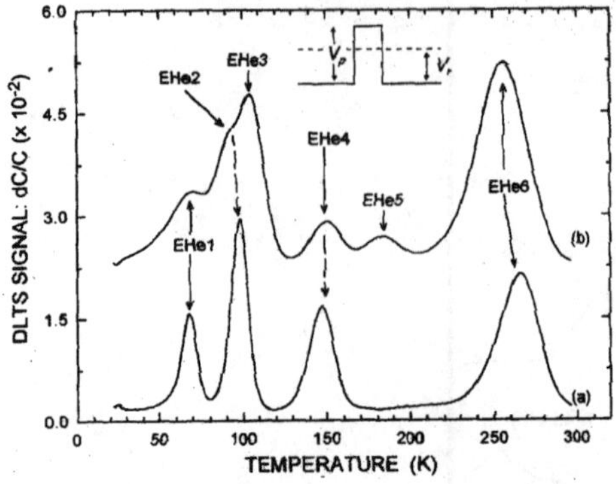

Fig. 3.38. DLTS spectra of 1 keV He-ion sputter-etched $Si_{1-x}Ge_x$ at a fluence of 10^{12} cm^{-2}. Curve (a) is for x=0, curve (b) is for x=0.04. Spectra are recorded at a quiescent bias of 0.5 V. The measurement frequency was 46 Hz. [After Goodman et al. [131]]

EHe3 have only been observed for the Ge containing layers and increase in concentration with x, suggesting that the damage in the SiGe layers *is more severe than in silicon*. Similar spectra and identical defect levels were obtained for the rf deposited Au Schottky barrier diodes without He ion etching [131]. The role of the Ge content on the He-etching induced damage is shown in Fig. 3.39. Both the trap concentration and position behave strangely as a function of x. Opposite shifts have been found for EHe3 and EHe6 as a function of x. The activation energy of EHe6 for x=0.04 is 0.52 eV and for EHe3 0.19 eV. Finally, it has been demonstrated by comparison with the results for 5.4 MeV alpha particles that a totally different defect spectrum is found (compare with Fig. 3.33). This may partly explain the contrasting observation of increasing trap densities with increasing Ge fraction x.

Similar studies have been undertaken for p-SiGe [132, 146, 147], whereby Ar-ion etching was compared with 5.4 MeV α–exposures and e-beam damage. A whole range of deep levels has been found (Fig. 3.40a). However, the main defect He2 is the same for all types of irradiations and has a deep level at E_v+0.529 eV (x=0) and a hole capture cross section of 9.4×10^{-14} cm^2. From the shift of the energy position (activation energy for hole emission) it is derived that the trap is pinned to the conduction band (Fig. 3.40b). The activation energy can be described by the empirical relationship [146]:

$$E_T(x) = 1.3\,E_G(x) - 0.94\ \text{(eV)}\quad 0<x<0.15 \tag{3.1}$$

with $E_G(x)$ the band gap energy corresponding to a Ge fraction x. Trap He2 furthermore shows a pronounced concentration profile [132], which is shown to be exponentially increase towards the surface, with a characteristic length λ_0=0.23 nm. From the observation that the height of the Sc Schottky barrier increases for SiGe exposed to electron-beam damage, it is concluded that most of the created

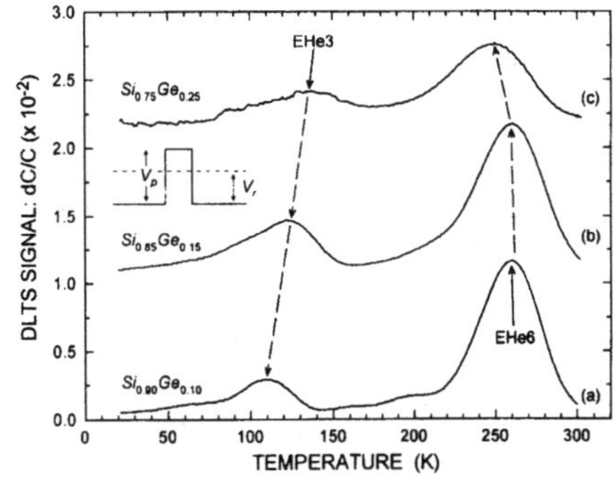

Fig. 3.39. DLTS spectra of 1 keV He-ion sputter-etched $Si_{1-x}Ge_x$ at a fluence of 10^{12} cm^{-2}. Curves (a), (b) and (c) are for x=0.10, 0.15 and 0.25, respectively. Spectra are recorded at a quiescent bias of 0.5 V. The measurement frequency was 46 Hz. [After Goodman et al. [131]]

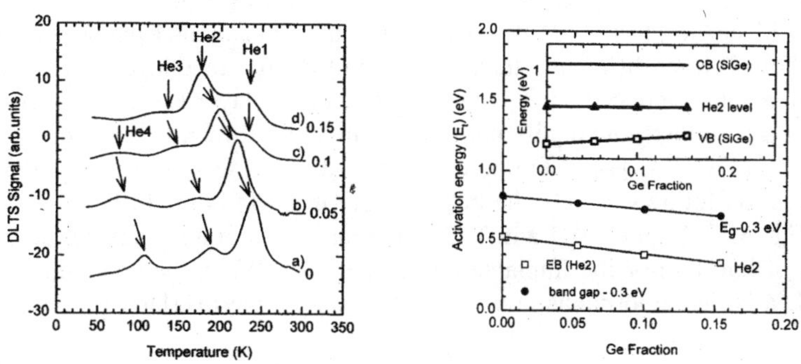

(a) (b)

Fig. 3.40. DLTS spectra of electron beam deposited Sc Schottky barrier diodes on p-$Si_{1-x}Ge_x$ with different Ge content. (a) Variation of $Si_{1-x}Ge_x$ band gap (•) and the activation energy of the most prominent defect (□) as a function of Ge content. (b) Correlation of the change in activation energy and valence band relative to silicon with Ge content is shown in the inset. [After Mamor et al. [147]]

Note finally the peculiar behaviour of level He1, which does not shift with the Ge fraction [131, 146, 147], indicating that it is pinned to the valence band. This means that the shift of a trap centre with x depends on whether it is pinned to the valence or the conduction band and hence on the nature of the trap: donor or acceptor character and hole or electron trap.

3.4.3.2 Ion Implantation Damage in SiGe

Ion implantation has been and still is the technique of choice for the selective doping of silicon in ULSI technologies. However, for deep submicon devices, the doping profiles should be tightly restricted to narrow sub 0.1 μm junction regions. This not only requires a good thermal budget management but at the same time a severe ion-implantation defect control. In spite of decades of research, the understanding of defect formation during ion implantation is not sufficiently complete for that purpose. This is one of the reasons for the interest in MeV self- and ion implantation in silicon, producing a damage layer which is sufficiently far from the surface and thus enabling a full characterization of the damage peak by techniques like Transmission Electron Microscopy (TEM) or Rutherford Back-Scattering (RBS). This can then be compared with a Monte Carlo simulation of the damage peak using for example numerical codes like TRIM. Such studies have been undertaken for epitaxial SiGe as well, since the early 90ties [148 - 152]. Generally, rather thick relaxed layers have been used for that purpose. It has been noted, however, that ion implantation not only gives rise to defects, but can also cause a strain relaxation in the SiGe layers [148].

Extensive studies have first of all pointed out that the ion-implantation damage behavior can not simply be extrapolated from the behavior of the elemental semiconductors Si or Ge [148, 149]. It appears that the fundamental damage mechanisms in alloys behave differently than in Si, even for unstrained layers. In practice, the ion-implantation induced damage produced in a SiGe layer is much higher than in silicon and is furthermore highly non-linear with the Ge fraction (and dose). This is illustrated in Fig. 3.41 [148]. Furthermore, it has been observed that for $x \geq 0.5$ the damage creation follows closely the behavior of Ge ($x=1$) [148, 149]; this is also shown in Fig. 3.42. Figures 3.41 and 3.42 represent the maximum damage (peak) as a function of ion dose or ion implantation temperature. When the relative damage becomes 1, a buried amorphous layer is formed, due to the overlap of the damage cascades.

Comparison with TRIM simulations leads to the conclusion that while the damage peak position is well described, this is not the case for the peak height (Fig. 3.43) [149]. The discrepancy increases even for increasing x. Such an effect could be explained by assuming an increase in the deposited energy or a reduction of the mobility of the primary defects produced or both [148, 149].

It has been shown extensively that the ion-implantation created damage is a strong function of the substrate temperature (Fig. 3.42) [148, 149]. Above a certain temperature T_0 no damage is formed due to a more efficient defect annealing, caused by a larger intrinsic point defect mobility. This threshold temperature increases monotonically with the Ge fraction for $0<x<0.5$, while for $x \geq 0.5$, T_0 is approximately constant. For the same Si ion dose (10^{14} cm^{-2}) a larger fractional peak damage is reached in Fig. 3.42, for higher x. The critical dose for amorphisation reduces with x, as evidenced by Fig. 3.44 [151], for 2 MeV Si-implanted $Si_{1-x}Ge_x$ at room temperature, and can be empirically described by [151]:

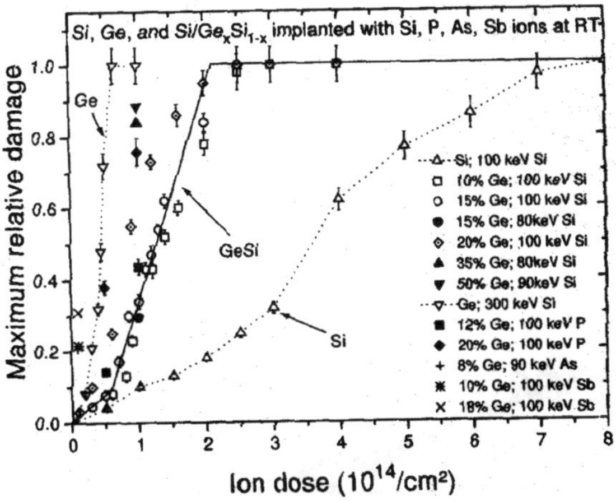

Fig. 3.41. Maximum relative damage values obtained from ^{4}He ion channeling spectra for Si, P, As and Sn implanted Si(100), Ge(100) and pseudomorphic Si(100)Ge_xSi_{1-x} ($0.08 \leq x \leq 0.5$) plotted versus the ion dose. A unity value in the maximum relative damage means that a continuous amorphous layer is formed in the ion implanted sample. A solid line is added as a guide to the eye to indicate the highly non-linear nature of damage buildup as a function of incrementing ion dose in the Ge-Si system. [After Lie [148]]

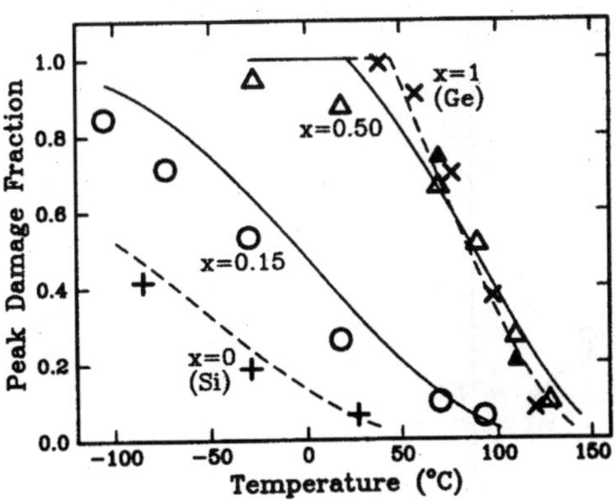

Fig. 3.42. Temperature dependence of the Ge peak damage fraction (the atomic fraction of displaced Ge at the peak of the damage profile) for a dose of 10^{14} ^{30}Si/cm^2 (J=0.05 μA/cm^2) in $Si_{1-x}Ge_x$ layers with x=0 (+), 0.15 (o), 0.8 (▲) and 1 (x) at energies of 70, 80, 90, 90 and 100 keV, respectively, in order to have the same depth of the peak. [After Haynes and Holland [149]]

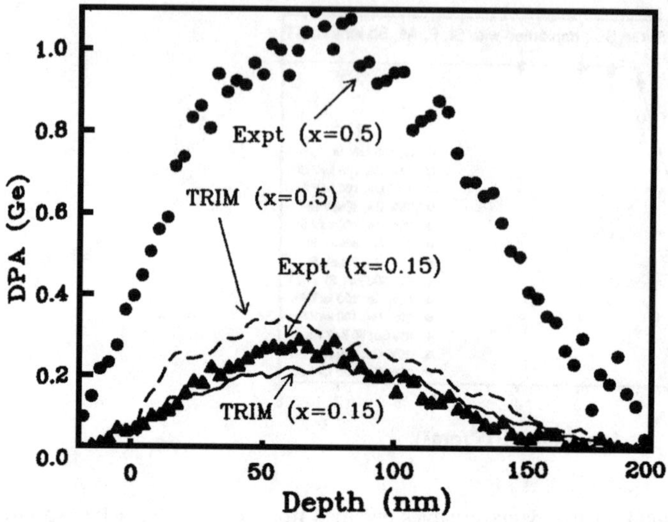

Fig. 3.43. Measured damage profiles extracted from the Ge portion of the 19° channeling spectra are compared to the Ge recoil distributions calculated by TRIM using a displacement energy of 15 eV. [After Haynes and Holland [149]]

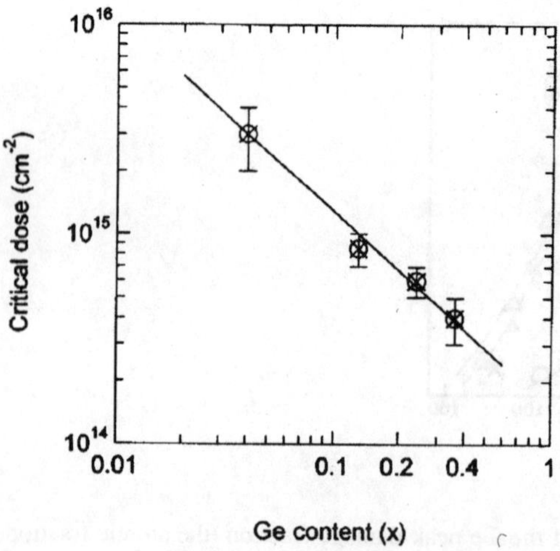

Fig. 3.44. Critical dose as a function of Ge content for the formation of a buried amorphous layer in 2 MeV Si-implanted $Si_{1-x}Ge_x$ alloys estimated by TEM (o) and ORDP (x). [After Nylandsted-Larsen et al. [151]]

$$D^* = a^* x^{-0.93} \tag{3.2}$$

with $a^*=1.5\text{x}10^{14}$ cm^{-2} and for $0.04 \leq x \leq 0.36$.

A final relevant observation is that the resulting damage shows a dose rate effect, as illustrated by Fig. 3.45 [149]. For each alloy, the damage produced increases strongly with the dose rate near T_0, while it is much less sensitive around room temperature. The onset of such dose rate effects indicates that a fundamental change in the mode of damage growth occurs near T_0. From the data presented above, it becomes clear that the addition of Ge in the alloy inhibits the dynamic damage annealing during ion implantation and, therefore, Ge retards the short-range mobility of the primary point defects, i.e. within the damaged volume. For implantations well below T_0, it is assumed that all the damage is formed within the cascades, while near T_0, other damage mechanisms may become competitive [149]. The usual cascade mechanism is considered as prompt, with a homogeneous nucleation of the damage, while a second dose rate dependent mechanism near T_0 has a heterogeneous nucleation and occurs retarded (delayed) because it requires diffusion of primary defect species [150]. This second mechanism is, therefore, strongly temperature and dose rate dependent. Finally, DLTS studies have revealed the presence of three electron traps in ion implanted n-type SiGe, which have been assigned to the A-centre (T_1), the V-V (T_2) and the Sb-V complex (T_3) (Fig. 3.46) [151].

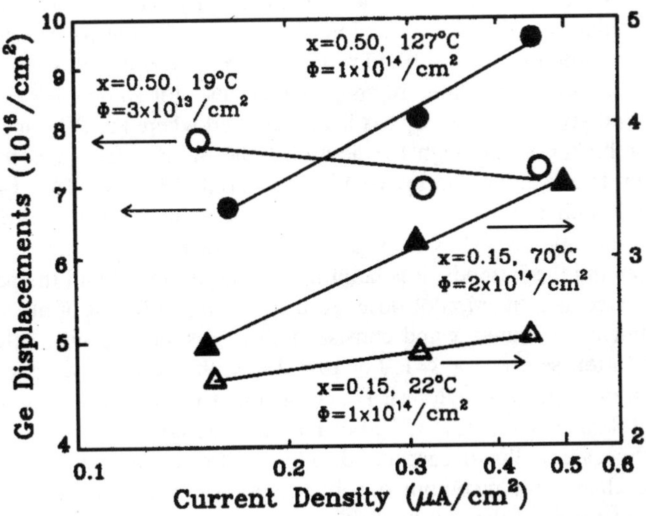

Fig. 3.45. Dose-rate dependence of the depth-integrated Ge damage yield near room temperature and at elevated temperatures. The fluence was selected for each alloy and temperature to give a damage fraction ≈30%. [After Haynes and Holland [149]]

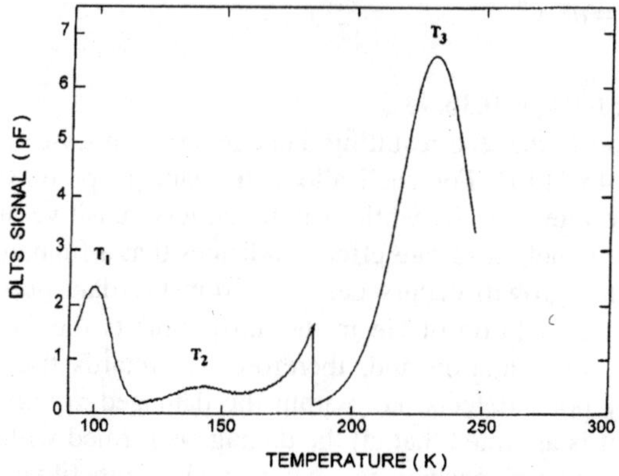

Fig. 3.46. DLTS spectrum of a Schottky diode made on a $Si_{0.96}Ge_{0.04}$ alloy. The gain was multiplied by 10 for data obtained at temperatures below 190 K. Implantation was for 10^{10} Si^{+} cm^{-2} at a dose rate of 10^{9} $cm^{-2}s^{-1}$. [After Nylandsted Larsen et al. [151]]

3.4.3.3 Ion Implantation Damage Model

Comparing the defect behavior in irradiated with implanted SiGe alloys, a strongly contrasting behavior is found. While the stable (and electrically active) point defect (PD) concentration reduces with the Ge fraction, the opposite is true for the peak damage in ion implantations, where besides PDs more extended clusters are formed, resulting in a local amorphous region. In fact, also for ion-etched or e-beam exposed epitaxial layers, a tendency has been observed where the PD concentration is higher for higher x, although the nature of the point defects has been shown to be different than in high energy irradiated material [131, 132, 146, 147]. In summary, the stable radiation damage in the peak region seems to behave differently than the one in the tail, where low damage levels are reached.

In order to understand these trends, it is helpful to study in more detail the peak ion implantation damage as a function of dose. Figures 3.47a and b first of all confirm the higher damage for higher x and consist of three distinct regions, before continuous amorphisation sets in for s=1. For low doses, the damage yield s increases linearly with dose. In this region, it is assumed that no overlap occurs between the individual damage cascades [152]. For intermediate doses a sublinear increase is found. In this case, collision cascades develop in already damaged regions, which enhances the chance for recombination of the point defects. However, for even higher doses, clustering of the point defects in the highly damaged region will compete with recombination and stabilizes the PDs. At the same time the formation and stimulated increase of the amorphous regions is responsible for the superlinear damage increase, until complete amorphisation occurs.

Based on the results of Fig. 3.47 for example, a model has been proposed, which includes a number of important material parameters, represented in Fig. 3.48 [152]. First, the point defect production parameter P_p appears to be a factor 2.8 larger for

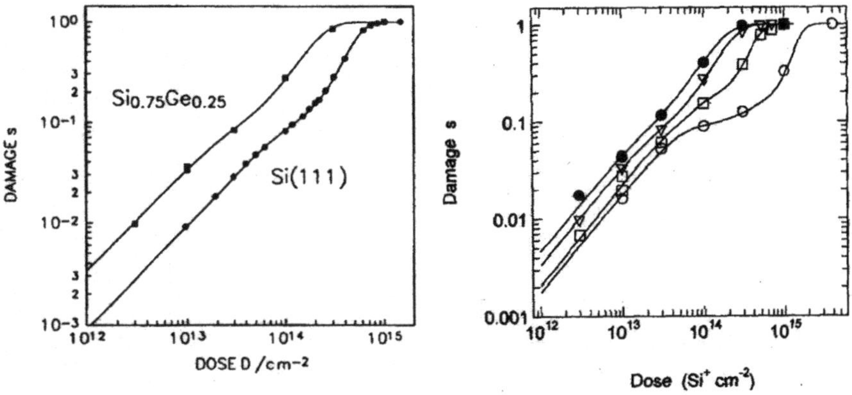

(a) (b)

Fig. 3.47. (a) Dose dependence of the optically determined damage at the depth of the profile maximum for 2 MeV Si^+ implantations into $Si_{0.75}Ge_{0.25}$ and pure Si at room temperature [After Lindner [152]]. (b) Dose dependence of the optically determined damage at profile maximum for 2 MeV Si^+ implantation of the SiGe alloys with x=0.04 (o), 0.13 (□), 0.24 (Δ) and 0.36 (•). [After Nylandsted-Larsen et al. [151]]

$Si_{0.75}Ge_{0.25}$, while the amorphous volume generation rate P_a is even a factor 8.3 larger than in Si. Since the stopping power in both materials is almost the same, these differences have to be attributed to a reduced stability of the chemical bonds and/or to a less effective intra-cascade annealing process. On the other hand, the recombination of PDs from different cascades described by the parameter *R* is 4 times stronger in the SiGe alloy than in Si (Fig. 3.48). The inter-cascade recombination increases with target temperature, which is due to the elevated PD mobility and increase with ion mass, which is attributed to the larger collision cascades of heavier ions. The increased inter-cascade PD recombination could be partially explained by the tendency of Ge to capture a V, which is immobilised for recombination with an interstitial.

3.4.4 Radiation Damage in SiGe Devices

Most radiation studies so far have been focussing on SiGe heterojunction devices (diodes or BJTs); little or no results are related to SiGe based field effect transistors. For the bipolar type of devices, the device degradation is in first instance related to displacement damage. However, as will be seen in Chap. 5, also ionisation damage plays a role for high speed BiCMOS technologies.

Initial studies on heterojunction [153, 154] and homojunction diodes [155] concentrated on the degradation of the forward and reverse current-voltage (*I-V*) characteristics. It was observed that qualitatively, the degradation of the forward *I-V*

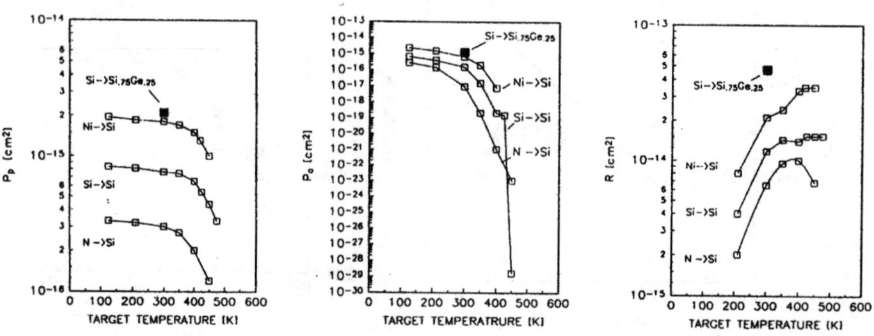

Fig. 3.48. Temperature dependence of defect production and interaction parameters, used to describe the dose dependence of damage generated by Si→$Si_{0.75}Ge_{0.25}$ and Si→Si. *R* is the recombination of point defects from different cascades. [After Lindner [152]]

was similar as for a silicon diode [153]. In homojunctions irradiated by 1 Mrad or 8.5 MeV protons, a reduction of the reverse current was observed, which could be ascribed to an increase of the barrier height for thermionic emission. At the same time, the corresponding Richardson constant of the Schottky barrier increased after the irradiation, which indicates the creation of traps at the p^+-$Si_{1-x}Ge_x$/p^--Si interface [155].

A systematic study of strained layer diodes and BJTs has been undertaken by *Vanhellemont, Ohyama et al.* The results of these studies on diodes can be summarized as follows. Exposure to high-energy particles causes a marked degradation of the *I-V* characteristics, like in Fig. 3.49a [134]. Both the forward and the reverse current increase after irradiation, whereby the increase of the reverse current is linear with fluence, for not too high fluences (Fig. 3.49b). This allows to define a damage coefficient for the reverse current I_R at V_R=0.7 V for example [134]. Similarly, the forward current also increases for low V_F, while in the series resistance dominated region, the current drops gradually with fluence (Fig. 3.49a). Both the increase of the forward and the reverse current indicate the creation of traps: recombination centres in the neutral part of the SiGe layer and generation centres in the depletion region, respectively. The reduction of the forward current for V_F>0.5 V is in line with the reported B de-activation.

Annealing studies combined with DLTS have revealed a close correlation between the reverse current and the near mid gap electron traps around E_c-0.5 eV. The annealing behavior of the diode *I-V* is depicted in Fig. 3.50a for an x=0.16 device; the temperature dependence is shown in more detail in Fig. 3.50b for V_F=0.4 V and V_R=-0.8 V. An activation energy of ≈0.2 eV was derived for the recovery of the reverse diode current. A similar value was found for the annealing of the near mid gap trap levels (see Sect. 3.4.2.3)

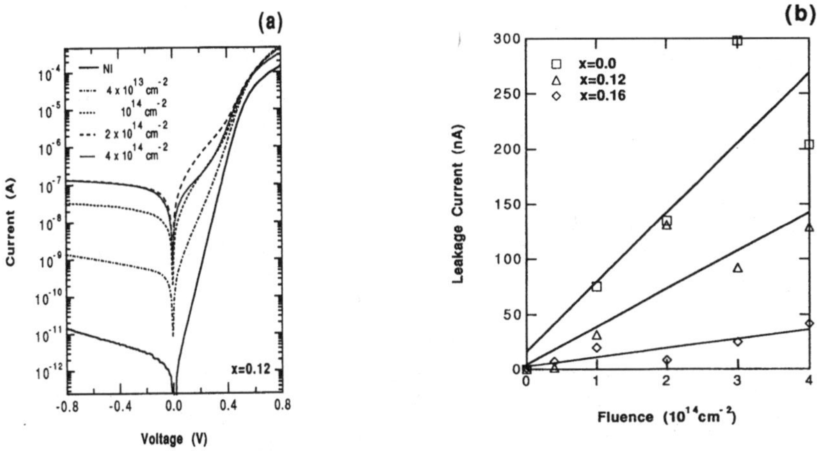

Fig. 3.49. (a) *I-V* characteristics recorded before (NI: not irradiated) and after irradiation with different fluences of 1 MeV electrons for *x*=0.12. (b) Variation of the leakage current at -0.8 V reverse bias for the three types of diodes with 1 MeV electron fluence. [After Vanhellemont et al. [134]]

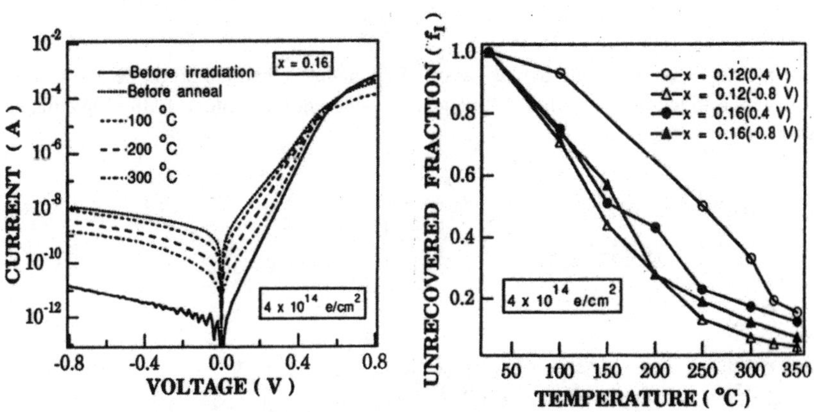

(a) (b)

Fig. 3.50. (a) The recovery of the *I-V* dependence for the $4x10^{14}$ e/cm^2 irradiated *x*=0.16 diode, (b) The temperature dependence of the unrecovered fraction for -0.8 V and 0.4 V. [After Ohyama et al. [135]]

Finally, a clear correlation has been observed between the damage coefficient for the reverse current at -0.8 V (Fig. 3.51a) and for the B de-activation (Fig. 3.51b), for the case of 20 MeV α-irradiations [143, 144]. This correlation is more explicitly shown in Fig. 3.52 and can be fitted by the following empirical model:

$$K_I = 1.58 \text{x} 10^{-24} K_B^{0.24} \tag{3.3}$$

This again emphasizes the relationship between the displacement of B atoms into an interstitial position and the leakage current generation centres in the depletion region.

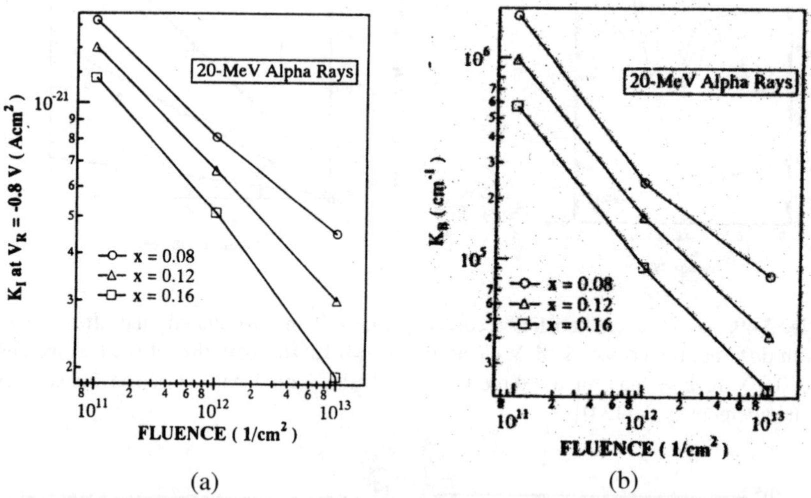

Fig. 3.51. (a) Reverse current damage coefficient at -0.8 V and (b) boron de-activation coefficient as a function of 20 MeV α-fluence, for different Ge content. [After Ohyama et al. [144]]

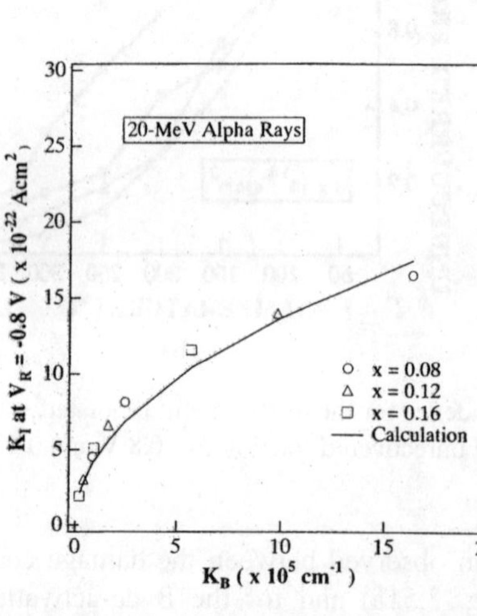

Fig. 3.52. Effective reverse current damage coefficient versus the effective B de-activation coefficient for the 20 MeV alpha fluence and Ge contents studied. [After Ohyama et al. [144]]

3.4.5 Conclusions Radiation Damage in SiGe Alloys

It is clear from the above that the impact of radiation on SiGe alloys first of all depends on whether one deals with bulk material or a strained or unstrained epitaxial layer, as this determines the band gap and the valence band offset. This implies that the electrical parameters (level position, capture cross sections, etc…) not only rely on the alloy composition (on x) but also on the thickness of the layer (band gap). In addition, there are some specific effects related to the statistical disorder in an alloy, which are not observed in the pure group-IV semiconductor materials. This gives rise to a large increase in the vibrational entropy factor for the vacancy donor level in p-type $Si_{1-x}Ge_x$. Evidence exists that for example Laplace DLTS provides the energy resolution to discriminate local Ge environments, i.e., whether there are 1, 2, … nearest neighbour Ge atom close to a deep level.

As to the introduction rate, an alloy hardening effect has been reported by different groups. In other words, the created trap density for the same particle energy and fluence reduces for higher x. However, an opposite trend has been observed for processing-induced damage, by plasma etching or ion implantation. Key parameter in understanding this effect is the dose rate.

From a practical device operation standpoint, the use of epitaxial SiGe heterojunction devices in a radiation environment does not pose particular problems and in some cases may even offer a "harder" alternative.

3.5 General Conclusions Group-IV Semiconductors

For obvious reasons, the comprehension of radiation effects in Si is far better developed than in Ge or in SiGe alloys. Depending on the application, several schemes can be considered to optimise the device behaviour in a radiation environment. Device simulators, taking account of basic radiation damage mechanisms are emerging and may be helpful in predicting the response, without the need for extensive and expensive radiation testing. It should be remarked, however, that certain basic information still lacks for a complete modeling of all processes involved. This is certainly true for the diffusivity of the silicon interstitial, where only theoretical estimates or extrapolated values are available. This holds to a lesser extent for the vacancy, where precise information is still lacking on the two acceptor levels. Given the high mobility of the point defects in silicon, it is probably very difficult to experimentally assess these parameters.

The case is different for Ge, where technological interest in the material was limited until the early nineties. The large potential of SiGe alloys in microelectronics has also triggered research in germanium. New applications are being considered, whereby even pure Ge epitaxial layers can be used in future MOSFETs. Ge island growth also receives considerable attention both from a fundamental and a practical perspective. However, more systematic studies are needed for a full understanding of even the basic radiation defects in Ge. Historically, this can be attributed to the lack of EPR data. One can expect, however, that the study of the SiGe alloys may reveal some useful trends, which can be extrapolated to the 0 or 100% extreme cases.

Particularly for low-temperature irradiations in Ge (and Si!) only a fraction of the created damage may be electrically active. This means that the 'inactive' damage like close Frenkel pairs can not be revealed by DLTS or EPR and should be studied by other, more sophisticated techniques, like X-ray scattering. Although from a device viewpoint only electrically activE-centres are of interest in many cases, a better understanding of this inactive damage may lead to a better overall insight and can shed new light on the interstitial and vacancy mobility at low temperatures.

4 Radiation Damage in GaAs

4.1 Introduction

GaAs is a semiconductor with superior radiation hardness compared with silicon and, therefore, lends itself for applications where this is of concern. In addition, the high electron mobility in GaAs (order 7000 cm^2/Vs [1]) makes the material perfectly suitable for the development of high-speed microwave circuits. However, there is more and more competition of SiGe-based and scaled deep submicron Si microelectronics, which dominate the low-power portable telecommunications market (1-2 GHz frequency range typically). Another main advantage of III-V compounds is the direct band gap, which favors optical recombination of carriers, with the emission of a photon and no involvement of a phonon. This opens up the way for optoelectronics, especially through the use of band gap engineering by modifying the III-V alloy or its composition. In this way, the complete range from the near infrared (InGaP) to the blue wavelength range (GaN) can be covered.

This chapter provides a description regarding displacement damage in GaAs. It is the basic material for most of the III-V based electronics, which cover the field from high-speed microwave applications (Metal-Semiconductor Field-Effect Transistor – MESFET or High-Electron Mobility Transistor – HEMT), nuclear-radiation detectors on Semi-Insulating (SI) n-GaAs, optoelectronic components like Light-Emitting Diodes (LEDs) and Laser Diodes (LDs) and space solar cells. The radiation damage in microwave electronics (HEMTs and MESFETs) will be highlighted in Chap. 7, while the impact of radiation on optoelectronics is given in Chap. 8.

After a brief introduction (Sect. 4.2), summarising some notions, a few words will be spent on native or grown-in point defects in GaAs (Sect. 4.3.1). The basic radiation defects in electron and neutron/ion irradiated GaAs will be reviewed in Sect. 4.3.2. It is shown that there is a close connection between grown-in and radiation-induced defects. In Sect. 4.3.3, we will see what can be learnt from processing-induced radiation defects, created for example by neutron transmutation doping, ion implantation and plasma etching. The impact of radiation on material parameters like the resistivity, the mobility, the recombination lifetime and diffusion length will also be described and the corresponding damage factors presented. Moreover, the correlation with the NIEL, defined in Chap. 2, is investigated in Sect. 4.4. In Sect. 4.5, the effect on the device operation and possible ways of hardening will be provided. Finally, the chapter is wrapped up in the general conclusions section.

4.2 Basic Notions and Definitions

In order to displace an atom in the GaAs lattice, an average threshold energy E_d of ~9.8 eV is required [2], whereby similar values are found for Ga and As. In fact, it has been observed that E_d shows an orientation dependence in the range 7 – 11 eV [3]. This is related to the fact that the GaAs zinc-blende lattice has a polar character, as shown in Fig. 4.1 [3]. Consequently, in the [111]As direction, it is relatively easy to displace an As atom, as the lattice is widely open in front of it, while for the Ga atom, As is in the way of the forward beam direction. The opposite holds of course for the [111]Ga direction of Fig. 4.1. Another important difference with Si and Ge is that there are more simple point defects to be considered beside the As vacancy and interstitial usually denoted by V_{As} and As_i and the Ga vacancy and interstitial (V_{Ga} and Ga_i). In addition, one can consider so-called antisite defects, i.e., an As atom on a Ga site and vice versa; they are abbreviated as As_{Ga} and Ga_{As}, respectively. Antisite defects are characterised by 4 nearest neighbours of the same element and are, therefore, also noted as $AsAs_4$, for example. The simple divacancy is now composed of two different adjacent vacancies, i.e., V_{Ga}-V_{As}. Also V_{Ga}-V_{Ga} or V_{As}-V_{As} could be considered. The presence of the As and Ga sublattices makes radiation defect identification in GaAs more difficult than in the elemental semiconductors, even for the point defects.

There is another factor, which complicates defect identification in irradiated GaAs, namely, the fact that certain popular techniques , like DLTS, EPR and optical absorption spectroscopy are more difficult to apply [4]. In case of EPR, both As and Ga atoms have a nuclear spin I_N=3/2 so that 4 lines are expected if a Ga or

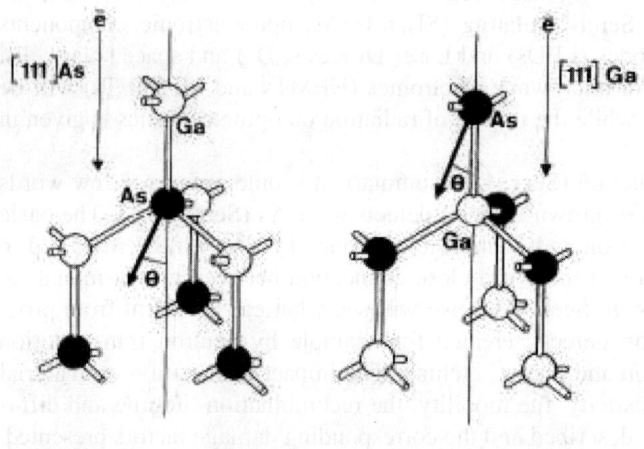

Fig. 4.1. Zinc-blende structure of GaAs. The [111]Ga and [111]As directions are, respectively, "hard" and "easy" directions for displacement of As atoms and conversely for Ga atoms. [After Pons and Bourgoin [3]]

As atom forms the core of the defect. This renders the assignment of a radiation defect to the As or Ga sublattice already more difficult. The superhyperfine structure, amongst others related to the natural abundance of the elements, is generally not resolved since the lines are broad (~300 G). Therefore, the EPR spectrum is insensitive to the presence of weakly interacting defects. One good example is the spectrum of the isolated antisite As_{Ga}, which is nearly identical to the spectrum of the EL2 grown-in defect or to other complexes of the As_{Ga}+X family, having an As_{Ga} core, but another simple defect associated with it. This can for example be an As_i, which moreover can occur at different distances of its core defect [4, 5]. Another problem is that if the free carrier density n is too high, the material is not transparent for microwaves [6]. Irradiation of the sample lowers n [1, 6], so that EPR becomes feasible.

Optical absorption spectra in GaAs are broadened severely by the large electron-phonon interaction, characterising defects in the material. Therefore, the technique is only of limited use for the study of the electronic properties of deep levels. However, room-temperature wafer mapping of the native EL2 density is routinely done, using the 1 μm optical absorption band [7]. In addition, as mentioned in Sect. 2.4, the detection limit of absorption spectroscopy is rather high, so that samples with high defect concentrations, i.e., exposed to large radiation fluences, need to be prepared. The same drawback goes for EPR. DLTS has the advantage that it is a relative technique: its detection limit is a fraction of the background doping density (see (2.31)) so that it is a favorable method to study radiation-induced deep levels in GaAs as well. However, irradiation generally increases the resistivity ρ of the material [1]. Much higher values (>10^7 Ωcm) and lower carrier densities (~10^7 cm^{-3}) than in silicon can be achieved. This is for example already so for SI GaAs which is the base or substrate material for many applications. In such a case, it is difficult if not impossible to apply DLTS. Instead, other, more sophisticated techniques, suitable for insulating materials, have to be used. Thermally stimulated current (TSC) measurements have been applied to fast neutron-irradiated SI GaAs [8, 9]. In such experiments, samples are first cooled to low temperature and then illuminated for a certain time to populate the traps with charge carriers. Next, they are warmed up at a constant rate in the dark, while the current is measured at a fixed bias voltage. What is basically followed is the thermal emission of trapped carriers as a function of temperature. For each deep level present in the material, a peak-shaped feature, similar as in DLTS is obtained.

4.3 Native and Radiation-Induced Point Defects in GaAs

In order to understand radiation-induced point defects in GaAs, one should consider first simple point defects, as they are the basic constituents of more complex deep levels. While in elemental semiconductors one only has to deal with vacancies and interstitials, the situation is more difficult in III-V compounds, as explained in the previous section. Furthermore, the stability of point defects is much higher in GaAs, so that a significant concentration of them can be frozen in during crystal growth (native defects). It is, therefore, important first to discuss these native defects, as they are inevitably present in the starting material. In addition, there exists a close link with the radiation-induced defects. Similar deep levels can also be created during certain processing steps like ion implantation or plasma etching, which will be the subject of the last subsection.

4.3.1 Native Point Defects in GaAs

Depending on the growth conditions, GaAs contains a certain density of native point defects. This is controlled in the first instance by the stoichiometry, i.e., the ratio Ga/As, as schematically represented in Fig. 4.2 [4]. Most of the native defects have corresponding levels in the band gap, which are summarised for the most important ones in Table 4.1. From the Table, one can derive that Ga_{As} is a double acceptor, so that a Ga rich melt will produce p-type material (Fig. 4.2). The opposite holds for the As_{Ga} related grown-in defects, of which the best known is EL2, quoted to be a double donor with energy levels at E_c-0.76 eV (0/+)[1] and probably the double positive charge state at E_v+0.5 eV (+/++) in p-type material [4]. As can be derived from Fig. 4.2, the growth conditions can be selected such that a very high resistivity, SI material can be obtained, which is slightly n-type. In this case, the EL2 deep double donors compensate the density of grown-in acceptors (usually carbon - C), so that $N_{DD} > N_A$-$N_D > 0$, with N_{DD} the concentration of deep donors and N_A and N_D the shallow acceptor and donor concentration [10].

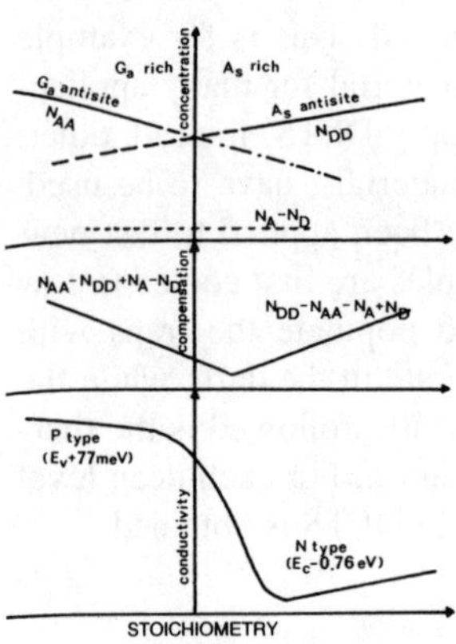

Fig. 4.2. Schematic representation of the variation of the antisite concentrations, of the residual free-carrier concentration (compensation) and the conductivity versus the melt stoichiometry. N_{AA} and N_{DD} stand for the concentration of possible deep acceptors or donors, respectively. [After Bourgoin et al. [4]]

[1] It has been suggested that the true activation energy of EL2 is closer to 0.6 eV [11].

Table 4.1. Deep level parameters of the most important native (EL) and radiation-induced defects (E and H traps) in GaAs

Label	E_T [a] [eV]	$\sigma_{n,p}$ [b] [10^{-14} cm^2]	Thermal stability T_A [K or °C	Suggested Identity	Reference
EL2	0.82	10 (n) 2×10^{-4} (p)	>850 °C	As_{Ga}+As_i	[4]
EL3	0.57	10			[4]
EL4	0.51	100			[4]
EL5	0.42	20			[4]
EL6	0.35	150			[4]
EL12	0.78	500			[4]
EL14	0.21	0.05			[4]
E1	0.045		220 °C (stage III)	$V_{As}^{-/--}$	[4]
E2	0.14	12	220 °C (stage III)	$V_{As}^{0/-}$	[4]
E3	0.30	2×10^{-4}	220 °C (stage III)	V_{As}-As_i	[4]
E5	0.96		220 °C (stage III)	V_{As}-As_i	[4]
H1	0.25		220 °C (stage III)	V_{As}-As_i	[4]
H0	0.06		220 °C (stage III)	V_{As}-As_i	[4]
			>950 °C	As_{Ga}	[4]
	E_v+0.077			$Ga_{As}^{0/-}$	[4]
	E_v+0.230			$Ga_{As}^{-/--}$	[4]
E4	0.76			As_{Ga}+$V_{As}^{+/++}$	[4]
			~200 °C	As_i	[4]
E(0.23)	0.23	3	~235 K (stage I)	V_{Ga}-V_{As}	[12]
HNI(0.42)	0.42	0.54	~280 K (stage II)	V_{Ga}-related	[13]
HNI(0.25')	0.25	0.25	~280 K (stage II)	V_{Ga}-related	[13]
	E_V+0.042			V_{Ga}	[14]

σ_n: apparent electron capture cross section (electron – E or EL – traps), σ_p: apparent hole capture cross section (hole – H – traps)

[a]Apparent activation energy derived from the slope of an Arrhenius plot. For the electron (E, EL) traps it is with respect to the conduction band (E_c-); for the hole traps (H) it is with respect to the valence band (E_v+).

[b]Apparent capture cross section, derived from the intercept of the Arrhenius plot

Bulk GaAs is grown by the Liquid Encapsulated Czochralski (LEC) or the horizontal Bridgman (HB) technique. LEC or HB n-type material always contains EL2 and EL6 centres with a concentration in the range 10^{15}-10^{16} cm^{-3}. In epitaxial materials, defect concentrations are much lower, in the range 10^{13}-10^{14} cm^{-3} and these defects can be varied in a controlled manner through growth condition(stoichiometry). In vapor-phase epitaxy (VPE) materials, EL2 is always present with another defect, EL5 when the growth rate is large enough. In normal molecular beam epitaxy (MBE), i.e., As-rich GaAs, several levels are consistently seen [4]. Liquid-phase epitxay (LPE) appears to produce the purest n-type material since, apparently, it does not contain any trap. However, LPE and VPE layers contain a number of hole traps whose concentration is strongly influenced by the growth conditions [4]. Most of these hole traps have been observed as minority carrier levels, i.e., in n-type material. Several of them are undoubtedly associated with defects involving transition-metal impurities like Cr, Fe, Cu or Zn. However, it has been observed that in Ni and Cu-diffused bulk HB and LEC n-GaAs the minority carrier (hole) lifetime is determined by the E_c-0.42 eV EL5 trap [15]. In HB n-type material, EL6 (E_c-0.35 eV) has been shown to be the dominant recombination centre [16]. For a more complete overview on native defects in GaAs, the Reader is referred to [4].

Before addressing the radiation defects in n-GaAs, a few more words should be spent on the most important grown-in defect, namely EL2. One of its most striking features – and fingerprint – is the persistent quenching of the photoconductivity at low temperature [4]. The defect disappears from EPR observation for example by photoexcitation at 1.1 eV at low temperature and is regenerated by a thermal treatment at 140 K. This quenching corresponds to the transformation of the neutral stable state of the defect into another one which is metastable since at low enough temperature the defect remains in this new state after the optical excitation has been removed. Both the stable and metastable state have the same charge state, but a different configuration. EL2 gives rise to two PL bands in the range 0.63 and 0.68 eV [7, 14, 17, 18] and to extrinsic absorption in SI or n-type GaAs extending from 0.8 eV to the conduction band. It should be remarked further that the EL2 electrical parameters are highly sensitive to the magnitude of the electric field in the depletion region [10]. One should therefore apply DLTS in a low field configuration for an accurate determination of E_T and σ [4].

The microscopic structure of EL2 is still a matter of lively debate. In Table 4.1, the As_{Ga}+As_i model is mentioned [19], which was amongst others supported by optically detected electron-nuclear double-resonance (ODENDOR) experiments [20]. This modified EPR technique resolves not only the hyperfine interaction with the central nucleus but also the ones with the ligands, so that the defect symmetry can be studied. In this way, a distinction can be made between the different members of the As_{Ga}+X family of defects, which are believed to exist in irradiated GaAs [5]. Some researchers claim that EL2 is equal to the isolated As antisite [18], although there exist substantial arguments against this interpretation of facts. It has for example been demonstrated that As_{Ga} does not show the persistent photoquenching behaviour in EPR [21]. Other experimental evidence has been produced which is in strong disagreement with X being an As interstitial in the second nearest neighbour position [22, 23]. It is rather believed that a vacancy is involved. Clearly this matter deserves more careful experiments and/or first-principles calculations before it will be settled definitively.

4.3.2 Basic Radiation Defects in GaAs

Similar as for Si or Ge, it is expected that low-temperature irradiations result in stable, single point defects in GaAs. In addition, simple defects will be formed only for electron or γ-irradiation. In an early assessment, using resistivity measurements after 5 and 77 K electron irradiation, *Thommen* [24] distinguished 3 annealing stages in GaAs. Stage I is centered around 235 K, stage II around 280 K, while stage III occurs at 520 K. It was also noted that the production threshold energies of around 25 eV derived for the stage I and II defects are twice the value as for single point defects annealing at stage III, implying that double displacements are required for the formation of these deep levels.

A more detailed identification of the deep levels has only become possible by a systematic investigation using DLTS [25 - 28]. An example of spectra obtained on 1 MeV electron-irradiated n-type GaAs is given in Fig. 4.3 [26]. The same labeling as in Table 4.1 is used. Note the shift of the peak positions to lower temperature with increasing doping density from spectrum B to C in Fig. 4.3. This indicates the sensitivity of the deep level parameters of E3 to the electric field in the depletion region of the diode. A typical p-type spectrum is represented in Fig. 4.4 [4, 29]. From Table 4.1, it is derived that the deep levels E1 to E5 and H0 and H1 belong to the As sublattice, while DLTS of room temperature irradiated material provides no direct evidence of Ga-related defects.

The assignment of E1-E5 and H0 and H1 to basic radiation defects is based on the following observations [29]. First, the introduction rate of the levels E1-E5 is independent of the irradiation temperature in the range 4 to 300 K. It is also unaffected by the doping concentration or type. The obtained sum of their introduction rates is in agreement with theoretical expectations, assuming a threshold energy for displacement of ~10 eV. Furthermore, the H0 and H1 levels are observed as minority traps in electron-irradiated n-type GaAs [4], while the same holds for the E-levels in p-type material [29]. Third, the annealing kinetics which is first order and the amount of defects (>90 %) which occurs around 200-220°C is consistent with the recombination of a vacancy-interstitial pair [29, 30]. The orientation dependence pointed out that these primary radiation defects occur preferentially in the As lattice, i.e., they are V_{As}-As_i pairs, while it is expected that V_{Ga}-Ga_i pairs do form in irradiated n-GaAs, but immediately recombine by Coulombic attraction, if the V_{Ga} is negatively charged.

The E1 and E2 levels have been attributed to the isolated As vacancy, i.e., the pair where the As_i is at several interatomic distances from V_{As}. Traps E3 and E5 correspond to pairs in which the As_i is near, probably at the third- and second-neighbour interstitial positions [4, 30]. The H0 and H1 traps correspond perhaps to other levels of the above defects. There is some evidence that the E4 trap is associated with the As_{Ga}-V_{As} complex.

Related to the thermal stability of V_{As} in high fluence electron-irradiated n-GaAs, it should be remarked that only a fraction of it, incorporated in V_{As}-As_i pairs anneals at 220°C, due to the As_i mobility. The largest unpaired part disappears around 450°C [4]. Arsenic interstitials can not be directly detected by EPR, electrical or optical techniques. However, their creation during electron irradiation is evidenced by the partial recombination of the As vacancies in n-GaAs at 220°C with mobile As_i. A corresponding migration energy of ~0.51 eV has been derived fro the motion of As_i [4]. It has furthermore been observed that the As_i mobility is charge state (Fermi level) dependent: it is much higher in p-type than in n-type material, explaining the formation of the H2, H3 and H4 complexes [29, 30].

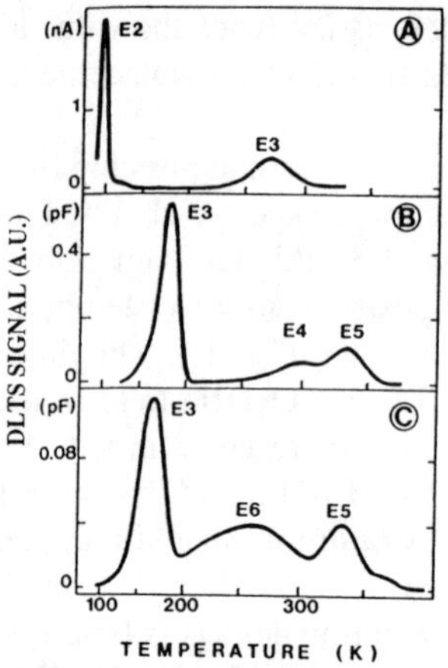

Fig. 4.3. Typical DLTS spectra of irradiated n-type VPE layers. (A) Spectrum for 2x10^{16} cm^{-3} material, irradiated at 300 K with 8x10^{15} cm^{-2} 1-MeV electrons (B) Spectrum for the same sample as (A) with a lower rate window. (C) Spectrum obtained in 5x10^{16} cm^{-3} n-GaAs irradiated with 1.4x10^{16} cm^{-2} 0.4-MeV electrons. The shift in the position of E3 between spectra B and C can be explained by a field effect. [After Pons et al. [26]]

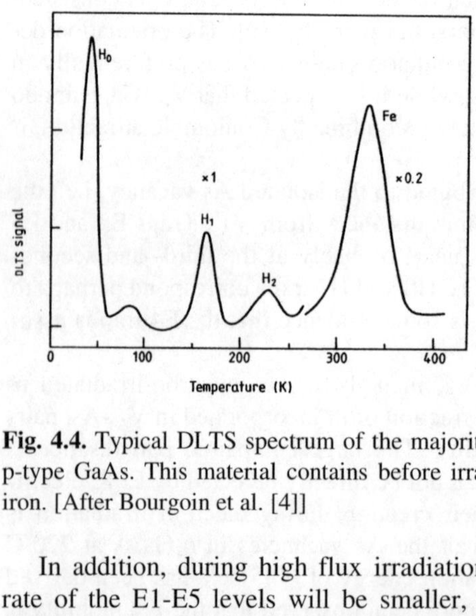

Fig. 4.4. Typical DLTS spectrum of the majority-carrier (hole) traps in electron-irradiated p-type GaAs. This material contains before irradiation a native trap (Fe) associated with iron. [After Bourgoin et al. [4]]

In addition, during high flux irradiations of n-type material, the introduction rate of the E1-E5 levels will be smaller, due to the high free carrier injection, which renders As_i more mobile (recombination enhanced migration – REM - mechanism [25, 31]). This partially anneals the created V_{As}-As_i centres. It is believed that the existence of the various levels of the V_{As}-As_i pair in Table 4.1 corresponds to a distribution in distance between V_{As} and As_i [30]. It should also be remarked that an isolated V_{As} is a double acceptor, according to Table 4.1.

Although DLTS did not provide evidence for radiation defects in the Ga sublattice, other techniques suggested the existence of such deep levels: this was inferred indirectly from carrier density measurements [32 – 34], suggesting the presence of acceptor levels well below E_c-0.15 eV. Later, more direct proof for the existence of negative V_{Ga} centres came from positron-lifetime measurements [35, 36], while an EPR spectrum was assigned to V_{Ga} by *Jia et al.* [37]. The centre was shown to have trigonal symmetry and a –2 charge state of a double acceptor. It corresponds to spectrum (3) of Fig. 4.5, also showing the predominant As_{Ga} spectrum in n-type GaAs.

The annealing stages I and II have been re-investigated by *Palmer and co-workers* using DLTS after low-temperature electron [12] or proton irradiation [13]. Similar results were later obtained for 15 K alpha-irradiated n-GaAs as well [38]. It was concluded that a broad level at E_c-0.23 eV (E(0.23) in Table 4.1) annealed near 235 K (stage I) and could be associated with a set of closely-related divacancy type of defects [12]. Using optical-excitation minority-carrier DLTS of n-GaAs irradiated by 1.0 MeV protons at 120 K, the presence of three new hole traps at E_v+0.16 eV, E_v+0.25 eV and E_v+0.42 eV was demonstrated [13]. They were shown to anneal around 280 K (stage II). The E_v+0.25 eV peak was too broad to be a single defect and was, therefore, assumed to be composed of different similar complexes, having various spacings of the constituents. The E_v+0.42 eV level was reported before in irradiated p-GaAs and appeared to be stable at 300 K in such material [39]. The reduction of the annealing temperature indicates that the trap is less stable in its more negative charge state in n-GaAs [13]. One should account for the fact that the stage II defects are formed by double displacements. Given the low stability, this leaves V_{Ga}-V_{Ga} or the mixed divacancy as candidates for these new levels [13]. Another suggested identity was V_{Ga}, which according to theory should have a level close to E_v+0.44 eV. However, this has been ruled out by a PL identification of the isolated V_{Ga} [14].

The gallium vacancy has been shown to be optically active optical activity: a PL line at 1.475 eV was assigned to a donor-V_{Ga} pair transition in irradiated GaAs [14, 40]. The corresponding spectrum is given in Fig. 4.6. From this, it has been concluded that the gallium vacancy is a shallow acceptor at E_v+0.042 eV. Moreover, it was found that post-irradiation annealing at 550°C largely restored the original, pre-rad PL intensity, as judged by the C_{As} line in Fig. 4.6, corresponding with the transition of a free electron to a carbon acceptor on an As site. Another line at 1.483 eV attributed to a donor (D_o)-Si_{As} acceptor pair recombination emerges in the spectrum after the anneal and is believed to be linked to the V_{Ga} through the reaction [14, 41]:

$$V^{+}_{As} + Si^{+}_{Ga} + 2e^{-} <-> Si^{-}_{As} + V^{-}_{Ga} + 2h \quad (4.1)$$

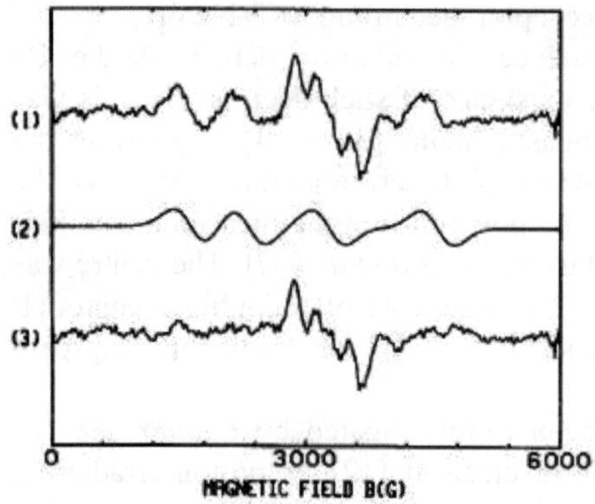

Fig. 4.5. (1) EPR signal at 4-K thermal equilibrium of a Zn-doped p-GaAs sample after a dose of 10^{18} cm^{-2} 1-MeV electrons (irradiation at 300 K). **B**$\parallel$[100]; $h\nu$=9.37 GHz, and the microwave power is 2 mW. (2) Simulation spectrum for the As antisite. (3) After subtraction the Ga vacancy spectrum becomes the dominant signal. [After Jia et al. [37]]

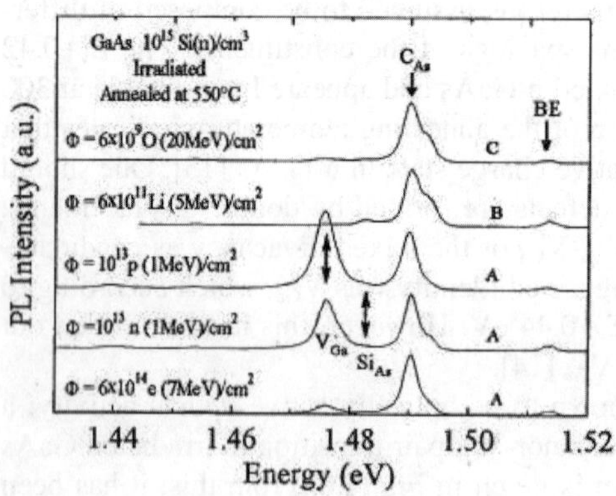

Fig. 4.6. The effect of various irradiations and anneals on the PL spectrum of n-type GaAs doped with silicon to 10^{15} cm^{-3}. The wafer used is indicated above each spectrum. The weak peak at 1.483 eV is attributed to (D_o,Si_{As}). The transition (e,C_{As}) at 1.492 eV is more evident in annealed samples. BE is the bound exciton transition. [After Khanna et al. [40]]

This equation shows the amphoteric character of the silicon dopants, i.e., they can act both as a donor (Si_{Ga}) or as an acceptor (Si_{As}). The donor in the 1.475 eV D_o-V_{Ga} pair transition is most likely Si_{Ga}^+, which is formed during the irradiation and is found to dissociate from complex defects above 450°C. This explains why after post-rad 550°C annealing, the V_{Ga} PL intensity increases. As discussed in the next paragraph, the gallium vacancy PL line can be used as a monitor of the nuclear displacements formed during irradiation and correlates well with the macroscopic damage factors and the NIEL. No V_{Ga} PL has been observed in irradiated and annealed p-type GaAs [42]. This implies that the V_{Ga} is more stable in its more negative charge state, as would be expected for an acceptor.

There has been a long-standing debate whether EL2 is formed during electron irradiation or not, in other words, whether it is identical to level E4 (Table 4.1). According to some researchers (see e.g. [5] and [21]) no EL2 is formed during electron irradiation. The main argument for this conclusion is the absence of the photoquenching behaviour in the antisite EPR spectra or in the photocapacitance [21]. The same was observed for γ-irradiated n-type GaAs [43, 44]. In addition, it was discovered that the post-rad EL2 peak consisted of two closely spaced deep levels both for electron- and γ-irradiated material [45 – 47]. This led to the conclusion that EL2 is a complex defect, belonging to a family of closely related deep levels [5, 48]. This picture has been confirmed by a detailed analysis of optically detected magnetic resonance, where at least four different members of the As_{Ga}+X family have been identified [49].

Another important point to remark is that the radiation-induced defects will depend strongly on the growth conditions of the material. In the case of epitaxial n-GaAs, a systematic study was performed by *Yousefi et al.* [50]. It was concluded that metalorganic vapor phase epitaxy (MOVPE) or MBE grown material behaves differently compared with VPE or bulk GaAs. In addition, MBE material showed a significant reduction in trap densities (introduction rates). In highly doped MOVPE n-GaAs an additional trapping level with high concentration was observed, while the carrier removal rate in this material is not a simple function of the electron fluence.

The data of Table 4.1 allows to conclude that the main electron-irradiation induced defects are composed of point defect complexes. There is little evidence for the Watkins replacement or kick-out mechanism, whereby for example a mobile As_i kicks out a substitutional dopant from its lattice site, to create an interstitial dopant defect. Only in p-type material, where the mobility of the interstitial arsenic is higher, complexes with impurities have been observed (the H2 to H5 peaks in Fig. 4.4) [4, 29]. The same applies for high-flux exposures in n-GaAs, where complexes of As_i with B or C have been reported [51]. Likewise, inpurity complexes are created if n-GaAs is irradiated at elevated temperatures in the range 240-330 °C [51]. It could also explain the EL2 controversy: low-flux irradiation mainly introduces simple point defects, while high-flux irradiations could produce more complex traps, and, hence, EL2 for example.

4.3.3 Neutron and Ion Radiation-Induced Defects in GaAs

The study of neutron-irradiation damage is not only important in its own right, but may also serve as a replacement for ion implantation damage, as it produces high-energy recoil atoms [5]. The difference is that the damage is produced deeper in the bulk of the material, which makes it often more amenable to detection by spectroscopic techniques. In addition to the electron irradiation-induced levels E1 to E5, the EL12, EL14 and the U-band, shown in Fig. 4.7 [52] are observed in neutron- or ion-irradiated n-GaAs. The corresponding level scheme is represented in

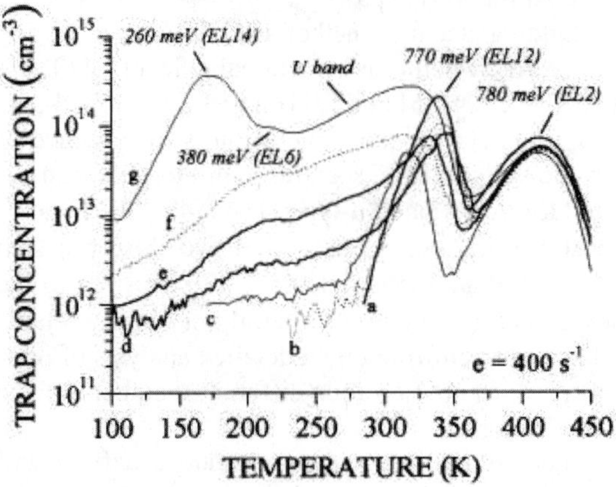

Fig. 4.7. The DLTS spectra of an n-GaAs sample exposed to various neutron fluences: a) 0, b) 10^{12}, c) $3x10^{12}$, d) 10^{13}, e) $3x10^{13}$, f) 10^{14}, g) $3x10^{14}$ neutrons/cm^2. [After Khanna et al. [52]]

Fig. 4.8 [53]. The U-band and EL6 anneal typically between 400 and 500°C, which is sometimes considered as the stage IV [28]. The EL12 trap has been shown to be a surface-related deep level [14]. In p-GaAs, an additional band is observed after neutron or ion irradiation with near mid gap position. It is called the L-band [28] and anneals around 550°C. A connection of both the U- and L-band with the As antisite has been proposed [26]. These extra features in the DLTS specta of massive particle-irradiated GaAs are thought to originate from the high defect concentration produced by the energetic recoiled lattice atoms, at their end of range.

Several models have been proposed for the broad U-band of deep levels [5, 54 - 56]. A first idea considers that for large displacement damage, giving rise to defect clusters, interactions between shallower traps, like EL6, and EL2 start to occur, which alter the electrical properties of the EL2 center [5, 52]. If neighboring point defects occur within 10 nm of each other (typical size of the damage cluster along the track of an ion), the emission path of captured electrons is changed. Phonon-assisted tunneling to the shallower trap may become possible, from which the carrier is subsequently emitted to the conduction band. As a result, the true EL2 concentration can only be found after annealing the other radiation damage at 600°C.

Alternatively, it has been suggested that the U-band is a specific cluster defect not directly related to EL2 and EL6 and caused by the large number of displaced atoms [57 - 59]. In that picture, the U-band defects are complexes between radiation-defects and (dopant) impurities. In another study, the U-band was resolved in 4 single level peaks, suggesting that different defect arrangements in the cluster damage region contribute to it [60].

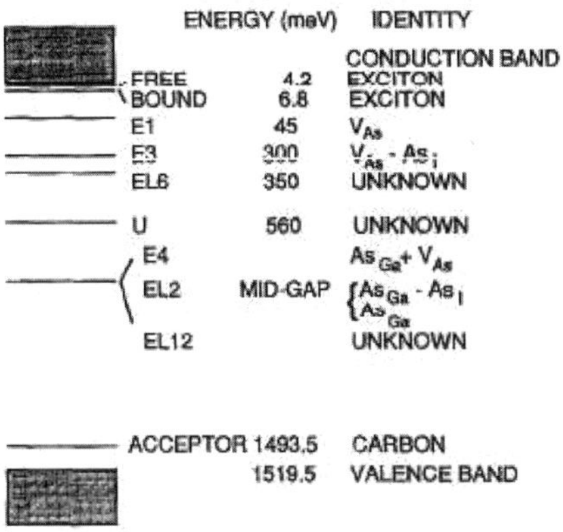

Fig. 4.8. The energy level diagram of some mid gap states in GaAs. Some of the levels have been observed in as grown samples, others in electron- or neutron-irradiated GaAs. [After Jorio et al. [53]]

A final interpretation is based on the cluster-damage model, represented by Fig. 2.18 [55]. Assuming a core of filled traps, surrounded by a space charge region, the broad and asymmetric shape of the U-band peak could be reconstructed reasonably well [55, 56]. The asymmetric low-temperature tail can be explained by electric field enhanced thermal emission. According to the analysis by *Brudnyi et al.* [55], the U-band is in fact composed of two subbands. Although the exact nature of the U-band defects is far from resolved, one can conclude that it is unlikely to be associated with EL2.

Related to the neutron-induced EL2 centres, it has been demonstrated that at least two types of "antisite" defects are created. The majority of them is not photoquenchable in EPR, as observed in Fig. 4.9 [61] and anneals around 600°C. The light-sensitive ones - mostly grown-in centres - remain stable up to at least 800°C. Additional EL2 in excess of a thermodynamic grown-in equilibrium concentration tends to disappear in the 500-600°C range [57]. It has also been observed that the concentration of ion-irradiation or –implantation induced EL2 centres strongly depends on the starting material [28]. It is normally very low in LPE type of material. As will be shown in the next paragraph, this two-stage annealing behaviour is also reflected in the electrical characteristics of neutron-transmutation doped (NTD) GaAs.

4.3.4 Processing-Induced Radiation Defects in GaAs

Similar as for Ge and Si, the processing of GaAs employs steps which can create radiation damage in the material. An obvious candidate is of course ion implantation, which is often used to selectively dope conductive regions. In the case of GaAs there is another unique application of ion implantation, i.e., the formation of

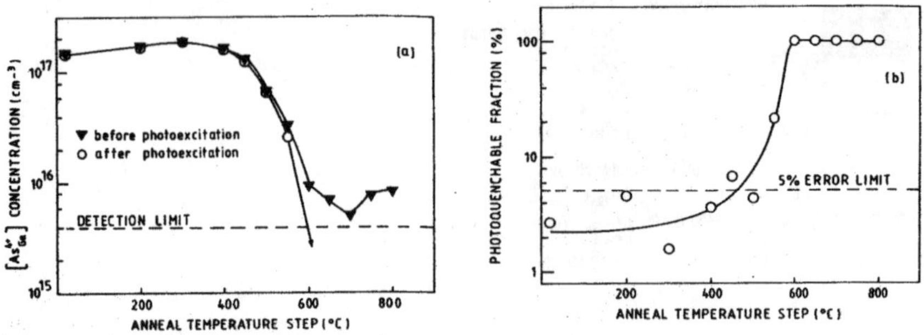

(a) (b)

Fig. 4.9. (a) Thermal decay of neutron-generated As_{Ga}^{4+} centres before and after photoexcitation. (b) Corresponding photoquenchable As_{Ga}^{4+} fraction versus annealing temperature. [After Goltzene et al. [61]]

GaAs there is another unique application of ion implantation, i.e., the formation of high-ohmic isolation regions without changing the wafer topography. Plasma etching is suitable for the removal or etching of layers, whereby the surface is bombarded by low energy ions (Ar^+, He^+,...). Radiation damage can also be introduced during the sputter deposition of a metal gate or a Schottky barrier. This section, however, starts with a short description of NTD of GaAs and the behaviour of the associated radiation damage.

4.3.4.1 Neutron-Transmutation Doping of GaAs

NTD of GaAs can proceed along three possible channels, whereby a ^{69}Ga or ^{71}Ga isotope is transmuted into ^{70}Ge or ^{72}Ge and ^{75}As is transformed into ^{76}Se after capturing a thermal neutron [62, 63]. Both act as donors, so that, starting from SI GaAs, n-type material could be obtained. This requires rather high fluences of thermal or fast neutrons, in the range 10^{17} to 10^{19} cm^{-2}. Consequently, a large amount of radiation damage will be created. In order to remove it and to activate the dopants, so that they take the required substitutional lattice sites, a thermal annealing after neutron exposure is applied. Generally, the recoil energy of the radioactive decay is large enough to displace the transmuted atom from its site [62]. As a consequence, after neutron irradiation, highly disordered and resistive material is obtained (Fig. 4.10 [64]). The high resistivity is related to the high radiation damage, which causes a reduction in carrier mobility and a trapping of the free carriers by the radiation-induced centres. If the density of deep levels is large enough, the charge transport occurs through hopping, whereby a carrier tunnels from one defect to a neighbouring one. This results in a conductivity given by $\sigma_c \sim \exp(-b_c/T^{1/4})$, as represented by Fig. 4.11 [65]. Such a behaviour was originally observed in ion-implantation damaged GaAs crystals [66].

From Fig. 4.10 follows that full dopant activation and damage removal are achieved in the temperature range 650-700°C, which is believed to correspond to a removal of both the Ga and As antisite defects. The Ga_{As} defects annihilate at 700°C and play an important role in the activation process of the transmuted impurities at 600°C. The presence of the Ga_{As} is studied by the PL emissions at 860 and 935 nm at 77 K, corresponding with the E_V+0.077 and E_V+0.2 eV acceptor states [64].

The role of the As antisite defects is highlighted in Fig. 4.11, showing that a photoquenching behaviour is observed after annealing above 250°C [65]. This suggests that in addition to isolated As_{Ga} also EL2 or EL2-like defects can be created after neutron irradiation and anneal [67]. The activation energy for the annealing of the As_{Ga} centres is estimated to be 0.9 eV [68], while the NTD created EL2 centres can be stable at much higher temperatures (up to 850°C [68]). In addition, the presence of Ge_{As} shallow acceptor defects at E_v+0.03 eV has been demonstrated by the PL line at 838 nm (1.48 eV) [63, 69 – 71] and results in a partial activation of the Ge NTD centers, which may range from >98 % (fast neutron fluence below 7x10^{17} cm^{-2}) down to 75 % for high fluence irradiation [70].

It must be remarked that transmutation doping of GaAs can also be achieved by high-energy proton or alpha-particle irradiation [72, 73].

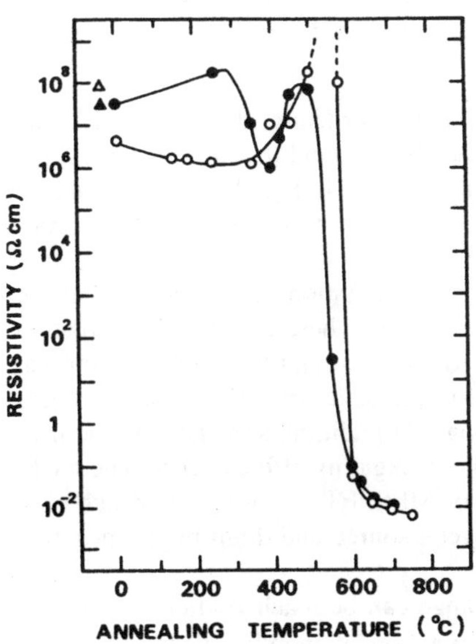

Fig. 4.10. Room-temperature resistivity as a function of annealing temperature for undoped and In-doped semi-insulating GaAs irradiated with thermal and fast neutrons of 1.5x10^{18} and 7.0x10^{17} cm^{-2}. Closed and open symbols represent undoped and In-doped samples, respectively. Triangles represent the unirradiated samples. [After Satoh et al. [64]]

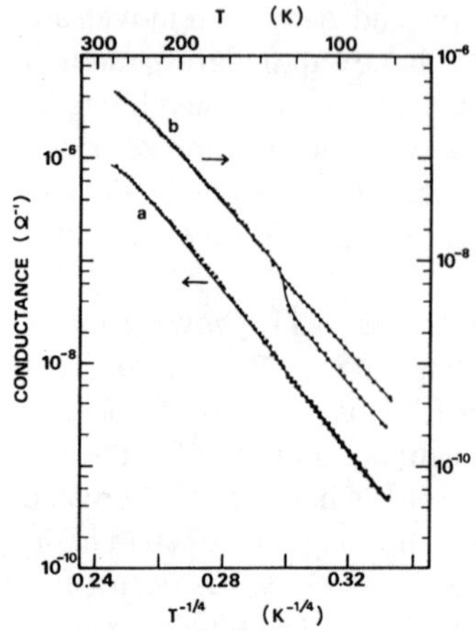

Fig. 4.11. Temperature dependence of conductance for (curve a) as-irradiation and (curve b) annealing at 450°C for 30 min before and after illumination with light of λ=940 nm for 20 min. The annealed sample shows a decrease in conductance below 125 K, indicating the photoquenching. [After Satoh et al. [65]]

4.3.4.2 Ion Implantation Damage

Similar as in Si, ion implantation in GaAs creates structural disorder, which can give rise to deep levels in the band gap. It has been found that in most cases, the same DLTS spectrum as for neutron irradiation shown in Fig. 4.7 is observed [28, 74 – 76]. There is some evidence for element-specific radiation defects, incorporating the implanted dopant impurity [28, 76, 77].

Beside changing the doping density, ion implantation can also be used for achieving local regions with a very high resistivity. These can serve as isolation regions (implant isolation) [78 - 80] or as current (wave) guides for optoelectronic components. The implant isolation is useful in the development of a planar GaAs technology, in contrast with a mesa technology. In addition, associated to the irradiation of the field regions a reduction of the backgating effect occurs. The backgating is a detrimental effect to the HEMT or MESFET performance, which consists in a reduction of the conductivity between source and drain by the negative polarization of a nearby contact [81].

For GaAs, resistivities in the 10^9 Ωcm range can be reached after an isolation implant, whereby ρ is enhanced by a combination of mobility degradation and free-carrier concentration reduction. The latter can be achieved by two mechanisms: one is the trapping of free carriers by the implantation-induced damage levels. The second is the compensation of the shallow dopants by near mid-gap levels of the implanted impurities, which results in s0o-called chemical isolation. The latter effect normally occurs only for impurities on substitutional lattice sites and requires annealing and diffusion. The implantation damage mechanism is characterised by the fact that a high temperature anneal completely restores the original resistivity, while this is generally not the case for the chemical isolation. Usually, light-ion implantation is used (H^+, He^+, B^+, O^+, N^+) so that the damage extends over a region which is much deeper than the conductive regions, using the (low) energies of commercial implanters. However, for photonic applications, isolation over much larger depths is needed, which requires MeV light-ion implantations [82 - 84]. As a rule of thumb, the depth of the implant isolation regions is chosen twice the thickness of the conductive layers. For a homogeneous damage profile, multiple implants are usually performed. The doses required for the isolation are relatively low (<10^{14} cm^{-2}), since tens or hundreds of carriers are removed per incident ion [85]. This implies also that the threshold dose for isolation D_{th}, corresponding to the highest resistivity increases proportionally to the initial doping density, as shown in Fig. 4.12 [81]. Beyond D_{th}, the conductivity remains constant, implying that all free carriers are trapped by the deep levels. The plateau ends when the damage concentration becomes high enough so that intradamage-hopping conduction starts to manifest itself [66, 84]. Usually, an implant dose of a few times D_{th} is selected. This is followed by annealing at temperatures equal to or greater than 300°C [81] to reduce or suppress the hopping conduction.

An important parameter of the implant isolation is its thermal stability [71, 84, 86]. It was found that the main parameter governing the thermal stability is the damage concentration in the doped region. For a chosen ion, the thermal stability increases with increasing dose. The mass of the ions plays only a minor role in the isolation stability. The same applies for the implantation temperature. The upper annealing temperature limit for which the isolation persists is ~650°C in irradiated n- or p-type layers [84, 86]. This rather low stability implies that the isolation has to be performed in the back-end, low-temperature part of the GaAs processing. It has also been concluded that antisite defects play a major role in the implant isolation [87].

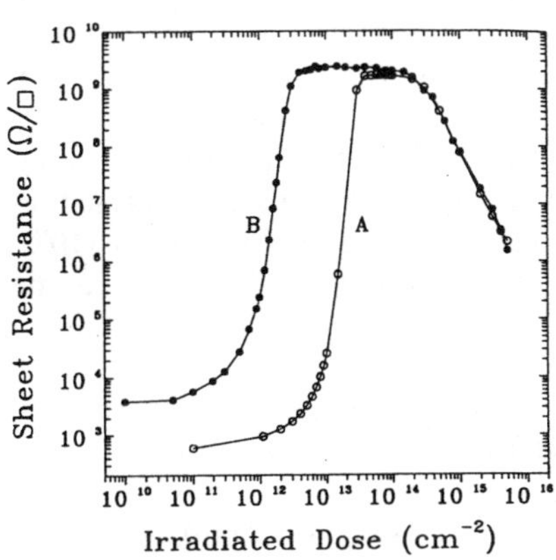

Fig. 4.12. Evolution of the sheet resistance in a n-type GaAs layer with an original sheet carrier concentration of 7.0×10^{12} cm^{-2} (A) , or 6.8×10^{11} cm^{-2} (B). [After de Souza et al. [81]]

This implant isolation not only works for GaAs, but has also been developed for $Al_xGa_{1-x}As$ [88], GaP, InGaP and InGaAlP [89 – 92] and GaN [84, 93 – 96]. In the latter case, extremely high resistivities, up to $\sim 10^{11}$ Ωcm can be achieved. The technique is less successful for InP [89, 97, 98], where for n-type InP the resistivity can only be increased up to the 10^3-10^4 Ωcm range, while for p-InP this may amount to 10^7 Ωcm. This behaviour is explained by considering that the created defects tend to pin the Fermi level in the upper half of the band gap [89]. However, one way to improve the implant isolation technique for n-InP is to perform the implantations at low temperatures, e.g. ~77 K. It is observed that in this case, highly polycrystalline or amorphous regions are formed, resulting in a significantly higher sheet resistivity [89].

4.3.4.3 Plasma Damage

Plasma or ion-beam etching of GaAs is accompanied by damage creation in a near surface layer [99 – 106]. The damage extends far beyond the ion range and shows steep profiles towards the surface [99]. Globally speaking, the same deep levels are found for low-energy Ar or He-ion bombardment compared with high-energy α-particle irradiation [106], albeit in different relative concentrations. This is evidenced by Fig. 4.13 [106]. In order to properly interpret the defect parameters, one should account for the high electric field existing close to the surface [106]. This implies that DLTS measurements should be performed at fairly low reverse bias and pulse height. In addition, a band of defects (EHe5 in Fig 4.13) close to the GaAs surface with continuous energy distribution has also been observed for low-energy He exposures [106].

The sputter deposition or cleaning of metal contacts on GaAs using Ar ions increases the barrier height ϕ_b on p-GaAs and reduces it on n-GaAs [105, 106]. The damage created during the formation of metal gates by using sputtering or e-beam evaporation is of great technological relevance for GaAs circuits as it may form a reliability hazard [107]. In addition, the transistor parameters can be degraded by the damage. A systematic DLTS study of the related trap levels has learnt that they are determined both by the type of metallisation and by the starting material used. Both interface and bulk traps have been observed. The interface traps are related to the metallisation strain and interdiffusion of metal and Ga atoms [107]. The sputter related damage has been shown to anneal at 200°C, so that it can be fairly easily removed.

4.3.5 Summary Radiation Defects in GaAs

It can be concluded first of all that there exists a close connection between the grown-in point defects and the radiation-induced ones. However, the identity of some important traps like EL2 – or the family of EL2-like defects – is still not unambiguously established yet. This is related to the lack of resolution of the most powerful defect-spectroscopy techniques like EPR and DLTS. Nevertheless, one can conclude that the most prominent radiation defects are simple point defects or

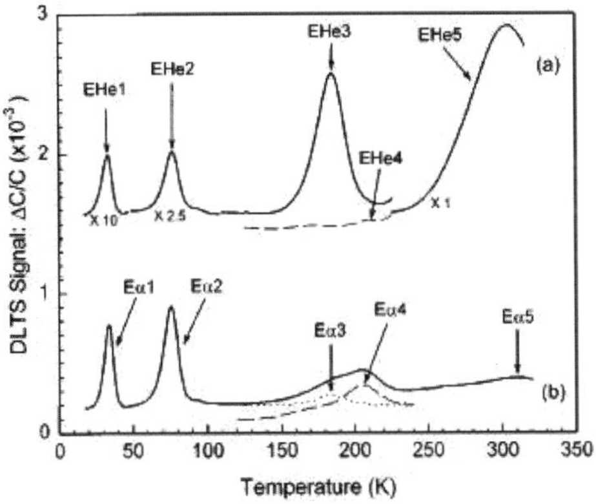

Fig. 4.13. Small signal DLTS spectra of (a) 5 keV He-ion-bombarded and (b) 5.4 MeV α-particle-irradiated n-GaAs doped to 10^{16} cm^{-3} with Si. The dotted and broken lines are the deconvoluted spectra obtained after transforming metastable defects. [After Auret and Goodman [101]]

point-defect complexes, with little involvement of dopant atoms. This is particularly true for n-GaAs, while in p-GaAs some impurity-point defect complexes have been identified. From this correlation between the native and the radiation defects, one can also infer that there will be a strong impact of the growth technique of the starting material on the radiation hardness of devices built in it. This will be illustrated in the next sections dealing with the damage factors and the device response during irradiation.

4.4 Damage Factors and NIEL

As outlined in Sect. 2.3, it is common to describe the degradation of the transport parameters in a semiconductor by empirical damage factors or constants. To a certain extent, these factors allow to model and predict the material and device response in a radiation environment. In principle, it is expected that these factors scale with the NIEL, which is the generally accepted standard for expressing the displacement damage equivalent dose (see Sect. 2.2). Whether this correlation also applies for GaAs will be investigated next.

4.4.1 Carrier Removal and Mobility Degradation in GaAs

Exposure to high-energy particle or photon irradiation causes a degradation of the electrical parameters of GaAs layers. The main effects are a reduction of the carrier density termed carrier removal and a reduction of the mobility, caused by displacement damage. The reduction of the carrier density is a net result of carrier trapping by radiation-induced trap levels and charge compensation by created defects with opposite character [32]. For very high fluences, even type inversion occurs [1]. The mobility degradation is caused by Coulombic scattering at charged centers, which include the created traps. The carrier removal rate $-\partial n/\partial \Phi$ for not too large fluences is constant, leading to an empirical law [108]:

$$n = n_0 \,(1 - a_n\, \Phi) \tag{4.2}$$

with Φ the particle fluence, n_0 the starting carrier density and a_n a damage constant related to the removal rate through: $K_n = n_0 x a_n$. Similarly, for the mobility, derived for example from Hall measurements on exposed epitaxial layers or bulk samples, one can write:

$$\frac{1}{\mu} = \frac{1}{\mu_0}(1 + b_\mu \Phi) \tag{4.3}$$

with μ_0 the initial mobility and $K_\mu = b_\mu/\mu_0$ the mobility damage factor. The damage constants a_n and b_μ have been derived from irradiations of simple "resistors" or Hall structures and yielded the following relationship with starting doping density, both for bulk and epitaxial samples, for fast neutron exposure [108]:

$$a_n = 7.2\text{x}10^{-4}\, n_0^{-0.77} \qquad (\text{cm}^2) \tag{4.4}$$

$$b_\mu = 7.8\text{x}10^{-6}\, n_0^{-0.64} \qquad (\text{cm}^2) \tag{4.5}$$

The corresponding carrier removal rate is represented in Fig. 4.14 and shows an increase for higher starting doping density. A value ranging from 6 to 10 cm^{-1} is derived for a doping density typical for the channel of a MESFET or of a HEMT.

The damage parameters have been studied for epitaxial GaAs for other particle irradiations as well [109 - 112] and also for ion-implanted resistors [110]. A striking result is that the damage factor for carrier removal is more than 1 decade larger for implanted resistors compared with epitaxial (or bulk) ones (Fig. 4.15). The damage factor shows a superlinear dependence on doping density, suggesting that higher implanted GaAs is more radiation tolerant, than at lower implant doses.

For high resistivity changes (> 25 %) strong non-linearities of the carrier removal and the mobility with neutron fluence can be observed [113]. These have been analysed in detail, taking account of carrier trapping and release, dictated by the shift in the Fermi level. In good approximation, the shift of E_F is linear with the neutron fluence [113].

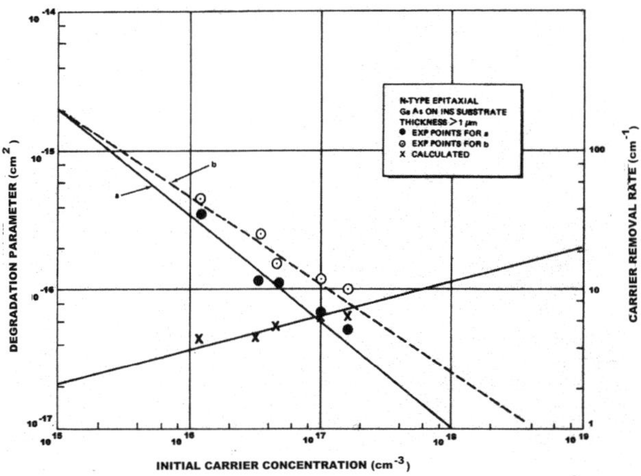

Fig. 4.14. Degradation parameters a_n and b_μ and carrier removal rate versus initial carrier concentration for epitaxial n-type GaAs. [After Behle and Zuleeg [108]]

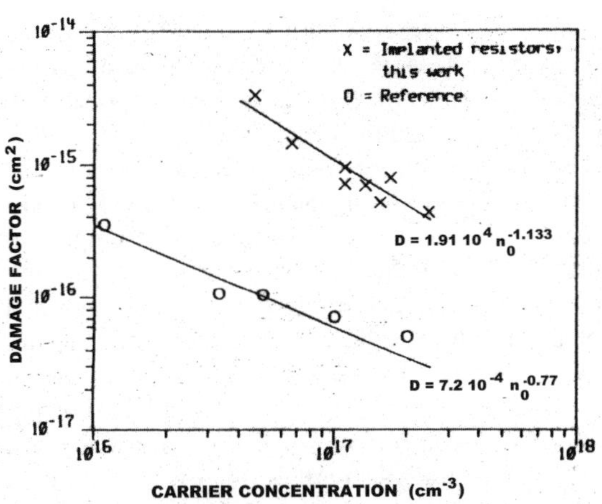

Fig. 4.15. The resistor damage factor versus carrier concentration under neutron bombardment for implanted and epitaxial uniformly-doped resistors. [After Campbell et al. [110]]

Gamma irradiation of GaAs generates similar carrier removal and mobility degradation, as demonstrated in Fig. 4.16, versus total dose D [114]. For the carrier removal, the following empirical relationship has been obtained [114]:

$$\Delta n = 9.92 \mathrm{x} 10^5 \; D^{1.17} \, \mathrm{cm}^{-3} \tag{4.6}$$

The carrier removal rate was found to be independent of the initial carrier concentration n_0.

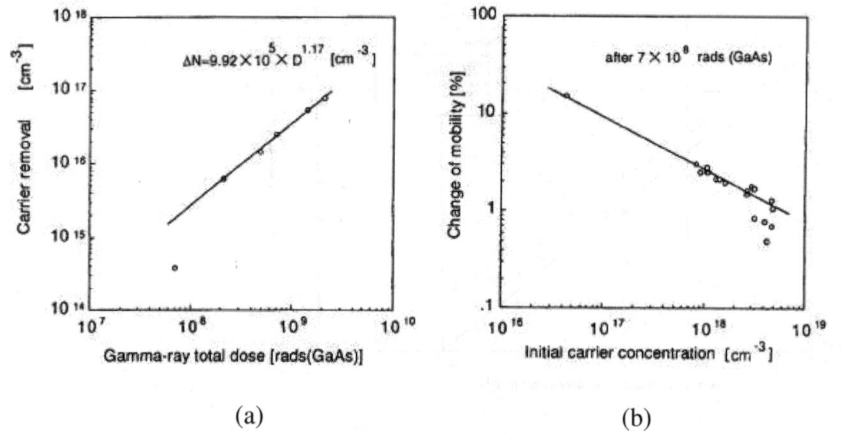

Fig. 4.16. Carrier removal with γ-ray irradiation (a) and density dependence of mobility degradation with γ-ray irradiation. [After Nishiguchi et al. [114]]

The damage factors defined above provide a good empirical description of the macroscopic displacement damage for fairly large total doses or fluences. However, it has been observed on several occasions that certain electrical/optical bulk properties of GaAs improve for moderate radiation fluences [41, 52, 115 - 117]. This has been found for the maximum mobility, the minority carrier lifetime, the carrier concentration, the exciton lifetime and the PL intensity [52]. Such an improvement becomes clear especially after annealing the irradiated samples at 550°C and is ascribed to a so-called radiation-induced 'ordering' effect, suggesting a restructuring of the lattice. The energy deposited during irradiation may induce a configuration modification, which can result in a lower energy state of the crystal. Another possibility is the stimulated gettering of impurities and defects at surface or interface sites. An example of this ordering is given in Fig. 4.17 [41], showing the maximum low-temperature mobility (occurring at ≈90 K), in function of the neutron fluence. An increase of μ_{max} is observed until a fluence of 10^{13} neutrons/cm^2, for irradiated and annealed samples. No increase is noted for unannealed samples.

Figure 4.18a demonstrates that for the high fluence range, the empirical law of (4.3) is in first approximation valid. In the same fluence range, it has also been found that the irradiation-induced gallium vacancy acceptor concentration, detected by PL, increases linearly with fluence (Fig. 4.18b) [41]. These two facts establish a clear correlation between the radiation-induced degradation of a macroscopic parameter (μ_{max}) and a primary radiation defect in n-type GaAs.

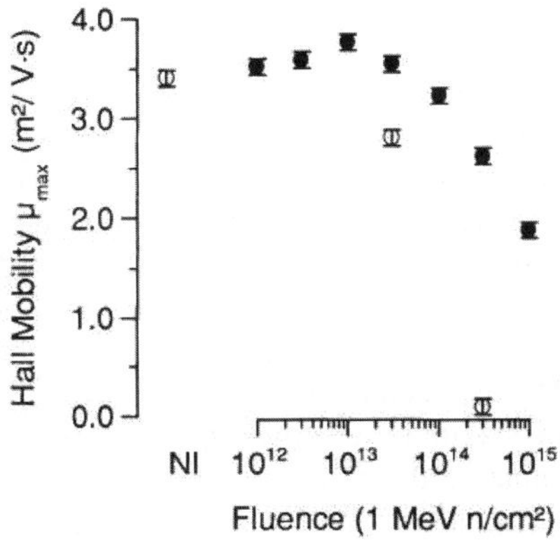

Fig. 4.17. μ_{max} as a function of fluence. For annealed samples (closed circles), it goes through a maximum at 10^{13} neutrons/cm^2. This increase did not occur for the samples which were irradiated but not annealed. NI means not-irradiated. [After Jorio et al. [41]]

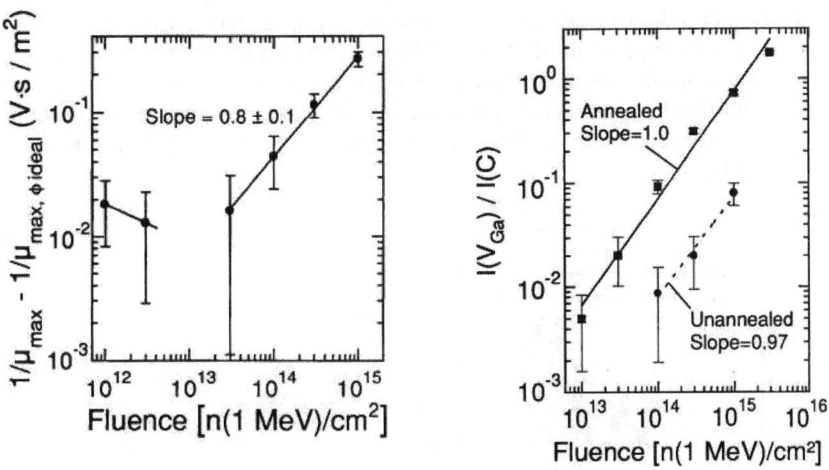

(a) (b)

Fig. 4.18. (a) $(1/\mu_{max}-1/\mu_{max,\Phi ideal})$ as a function of fluence. The measured slope of 0.8±0.1 in the degradation regime for large neutron fluences suggests that (4.3) holds in first instance. (b) Normalised PL intensity $I_{PL}(V_{Ga})/I_{PL}(\text{carbon})$ as a function of fluence. V_{Ga} is observed in the non annealed samples between 10^{14} and 10^{15} neutrons/cm^2, whereas in the annealed samples at Φ between 10^{13} and $3\text{x}10^{15}$ neutrons/cm^2. [After Jorio et al. [41]]

4.4.2 Correlation between Resistance Damage Factor and NIEL

The NIEL can be calculated for GaAs, taking account of elastic and non-elastic collisions. The calculations can be extended to include heavy ions in the Coulombic limit using the Monte Carlo code SRIM (Stopping and Range of Ions in Solids) [118] and have for example been compared with experimental damage factors for the resistance. Summaries of the different results obtained on n- and p-type GaAs and for a wide range of particles and energies are given in Fig. 4.19, showing a linear correlation with the non-ionizing energy deposition (NIEL) [112]. However, the energy of the primary knock-on atoms (PKA's) corresponding to Fig. 4.19 is in fact quite low (50 to 200 eV), while the maximum transferred PKA energy in GaAs can be as high as 2 MeV. This is much larger than the corresponding values for Si (300 keV). For such high maximum energies, nuclear inelastic interactions should be considered. For example, in the case of high energy proton irradiations, high energy inelastic nuclear interactions (PKA's) become important above 10 MeV and dominate for energies higher than 60 MeV. Experimental evidence is provided by the results shown in Fig. 4.20, for high-energy proton exposures, showing the normalised resistance change of a GaAs resistor, as a function of fluence. While the initial slope is approximately constant and can be described with the resistance damage factor K_R, for higher Φ, a superlinear degradation is found [112]. The result is depicted in Fig. 4.21, showing a good agreement between NIEL and the (linear) resistance damage factor, if account is made for the non-local energy deposition of energetic PKA's, which penetrate much deeper in the substrate than the extent of the implanted or epitaxial resistors. It should also be remarked that the proton energy dependence of the carrier removal rate and the mobility are similar, as derived from Hall measurements, at least for protons up to 63 MeV [111].

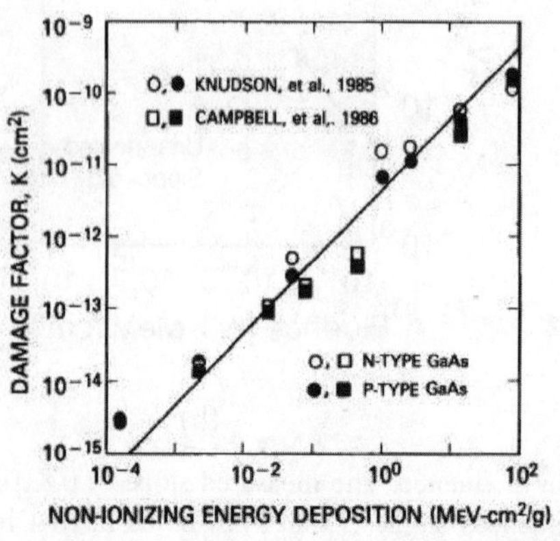

Fig. 4.19. A plot showing the linear dependence of the damage factors for GaAs JFETs and resistors on the non-ionizing energy deposition. The data comes from *Knudson et al.* [109] and *Campbell et al.* [110]. [After Summers et al. [112]]

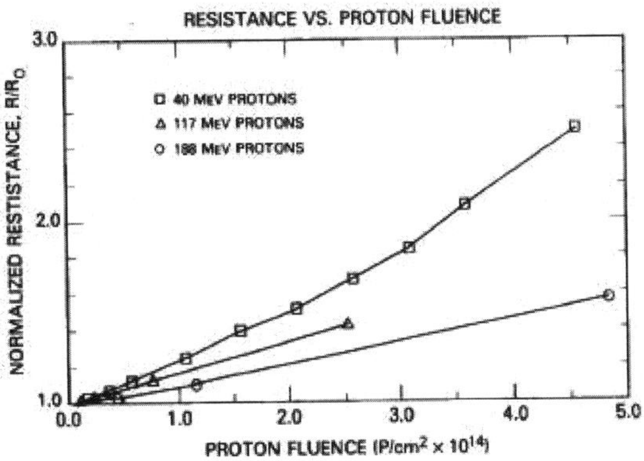

Fig. 4.20. Variation in reduced resistance for ion implanted n-type GaAs resistors as a function of proton fluence for 40, 117 and 188 MeV protons. [After Summers et al. [112]]

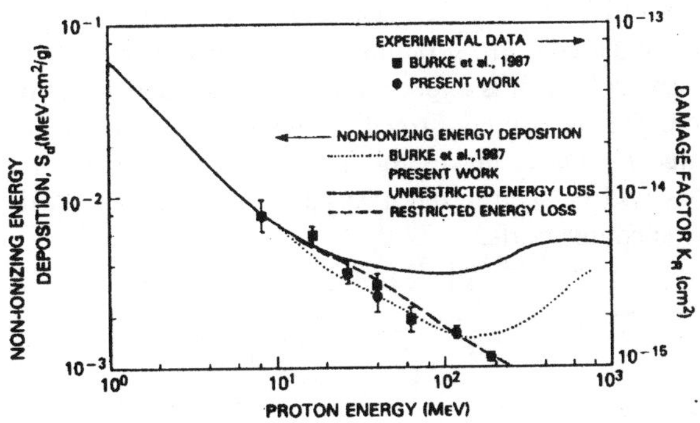

Fig. 4.21. NIEL versus proton energy according to different approximations, compared with the experimental damage factors for implanted resistors. The restricted energy loss curve results when the energy of the residual recoils is arbitrarily limited to 500 keV, i.e., recoils having a range less than the active depth of the resistor. [After Summers et al. [112]]

4.4.3 Lifetime Damage Factor and NIEL

As explained in Sect. 2.3, the carrier lifetime is an important parameter for the operation of p-n junction based devices. The corresponding damage factor, defined by (2.20) has been determined for GaAs, yielding values in the range 8.5x10^{-7} cm^2/s for electron irradiation [119]. In highly Se-doped n-type GaAs, K_τ was found to vary from ~4x10^{-6} cm^2/s/e$^-$ for a doping density of 3.5x10^{16} cm^{-3} to 12x10^{-6} cm^2/s/e$^-$ for a doping concentration of ~10^{18} cm^{-3}, where a saturation of the damage factor occurred [120]. This behaviour is in contrast with what has been reported before for other donors, which showed a constant K_τ [120].

The lifetime damage factor in n-GaAs has been determined for a wide range of energies and particles by *Parenteau et al.* [121], using time-resolved PL spectroscopy at 77 K. The results are summarized in Fig. 4.22. Note that K_τ for electron irradiation is about 10^{-6} cm^2/s, in agreement with the above mentioned numbers.

A parameter which is directly related to K_τ is the diffusion length damage factor K_L, defined by [122, 123]:

$$\frac{1}{L^2} = \frac{1}{L_0^2} + K_L \Phi \tag{4.7}$$

The diffusion length is a key parameter in the operation of solar cells, as it determines the short circuit current I_{SC}. Values for K_L in the range 2x to 7×10^{-8} per electron have been found in the literature for both n- and p-type GaAs [123].

Figure 4.23 shows the doping density dependence of K_L for a few semiconductor materials often used for space solar cells and corresponding with 1-MeV electrons [122]. A striking feature in Fig 4.23 is the relatively small variation of K_L with doping level for GaAs, compared with Si or InP. This implies that the performance degradation of a GaAs solar cell is expected to be fairly independent on the base doping concentration. This is also confirmed by measurements [123]. At first sight, the higher K_L value for GaAs makes this material less favorable for space solar cells. However, as will be seen in the next paragraph, GaAs cells do perform better than their silicon counterparts.

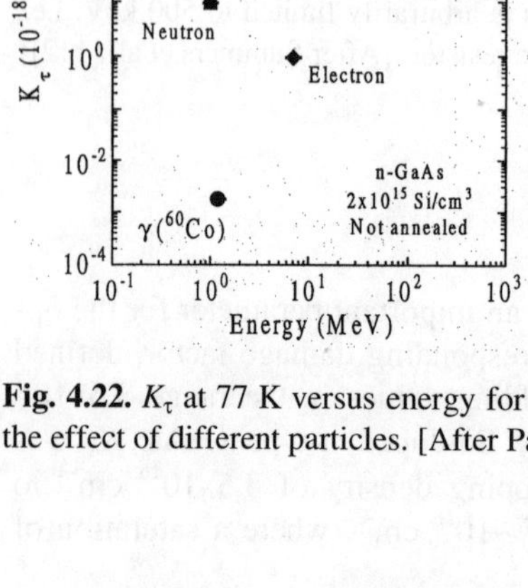

Fig. 4.22. K_τ at 77 K versus energy for n-GaAs doped to 2×10^{15} cm^{-3}. The figure indicates the effect of different particles. [After Parenteau et al. [121]]

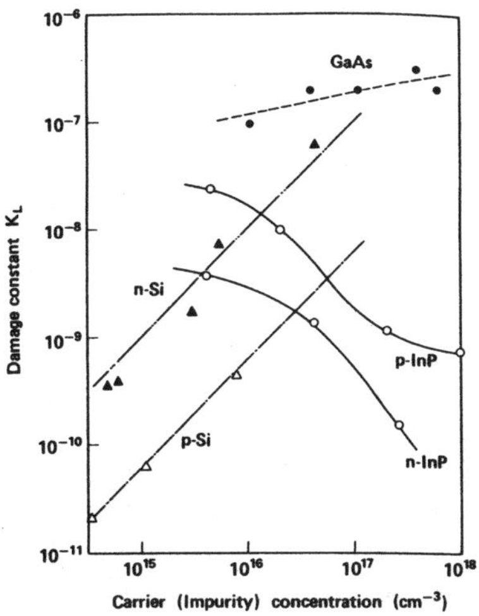

Fig. 4.23. Changes in damage factor for diffusion length K_L in InP, GaAs and Si determined with 1-MeV electron irradiation as a function of material carrier concentration. [After Yamaguchi and Ando [122]]

Note finally that in heavily ion-damaged GaAs subpicosecond recombination lifetimes can be obtained [124 – 126]. This makes the material suitable for photoconductive switches. In this way, picosecond electrical pulses driven by very short laser pulses can be generated for example.

4.4.4 Correlation with Microscopic Damage

The successful correlation established between the NIEL (or TRIM simulations) and the macroscopic damage for not too high particle energies (see e.g. Fig. 4.21) opens up strong perspectives for the prediction and simulation of radiation damage in III-V electronics and enables to reduce considerably the amount of radiation testing. Nevertheless, such a correlation is not that obvious, since the material properties are affected in the first place by stable radiation defects, which are only fractions of the total amount of Frenkel pairs initially formed. Furthermore, the defects responsible for the degradation must not necessarily be simple (primary) defects but can be complexes or larger aggregates. Therefore, for a full understanding of the problem, one should identify the critical defects, which is not an obvious task.

As mentioned above, one of the first successful identifications of critical radiation induced defects was the gallium vacancy, correlating well with the mobility degradation [41]. In addition, the correlation between the radiation induced formation of V_{Ga} and the NIEL has been extensively studied by *Khanna et al.* [2, 40, 73, 121, 127 - 128], using PL, which is a relative technique. The signal peak associated with V_{Ga} is resulting from electron-hole recombination through a donor (D) and a nearby V_{Ga} shallow acceptor (close to the valence band) [14]. Its formation has been investigated for a broad range of substrate doping densities (n-GaAs: 10^{15} up to 10^{18} Si/cm^3) and irradiating particles: electrons, neutrons, protons, lithium, deuteron and oxygen. A comparison with ^{60}Co γ-irradiation was also performed. In general, the spectral PL intensity after irradiation reduces with increasing fluence. This points to the creation of deep(er) trap centres, which cause a non-radiative GR of electron-hole pairs and thus compete with the PL optical recombination transitions. However, annealing at 550°C removes most of this deep -level damage and increases again the PL signal [2, 40, 121].

From these studies, a good correlation between the relative density of the primary defects V_{Ga} and the NIEL was derived as a function of particle fluence (Fig. 4.24 [121]). Also TRIM, which basically calculates the number of vacancies formed along the trajectory of the projectile yields a qualitative agreement. Interestingly, the V_{Ga} was observed for high dose γ-irradiations as well, indicating that indeed displacement damage is formed due to the energetic secondary Compton electrons. It has furthermore been observed that the formation of the gallium vacancy is strongly dependent on the starting material.

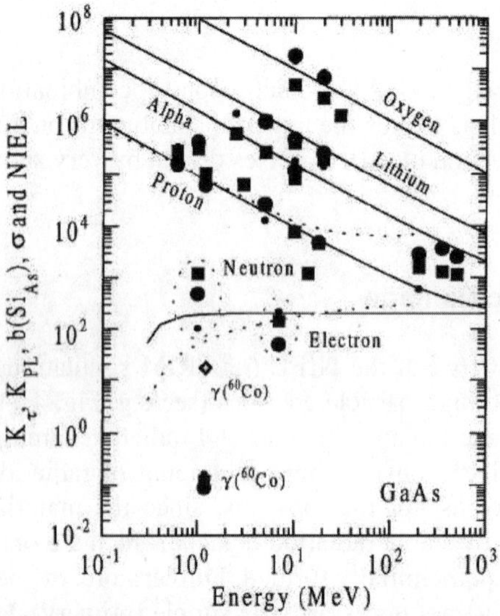

Fig. 4.24. K_τ (large circles), K_{PL} (small circles), the Si_{As} introduction rate $b(Si_{As})$ (squares), NIEL (diamonds and dashed lines) and Rutherford scattering cross-section (solid lines; see (2.8)) versus energy. The circled points to the left of the word "neutron" refer to 1 MeV neutron irradiation, and the square point at the right to 14 MeV neutron irradiation. The circled points left of the word "electron" refer to 7 MeV electron irradiation. All the points above 200 MeV refer to proton irradiation. [After Parenteau et al. [121]]

Looking at Fig. 4.24, it is clear that at high energies, the Rutherford mechanism, predicting a $1/E$ dependence (E the particle energy) is no longer followed. On the other hand, NIEL (or TRIM) follows well the measured damage factors for low to moderate energies, but overestimates the damage at high energies [2, 73, 121, 127 - 128]. In contrast to the interpretation of *Summers et al.* [112], it is believed that this higher damage rate above ≈10 MeV for protons could be caused by a secondary displacement cascade, associated with high energetic PKA's. From the neutron irradiations, it is derived that they are more damaging than electron exposures and give rise to larger defect complexes clustered along the neutron track, which most likely incorporate the V_{Ga} and Si_{As}. These could for instance be defects related to the well-known U-band, which is not formed in electron or gamma-irradiated GaAs and anneals at approximately 450-500°C [40]. These thermal recombination centres are deeper in energy and, therefore, quench the PL signal.

While the PL intensity of the V_{Ga} line is a useful damage monitor for n-GaAs, this is not the case for p-GaAs as no related PL signal is detected after irradiation and anneal [42]. This probably points to the lower stability of the neutral vacancy in p-GaAs. In that case, it has been proposed to use the PL intensity of the C_{As}-line ($I_{PL}(C_{As})$) as a monitor, assuming that the concentration of C_{As} is not affected by the irradiation [42]. The reduction of I_{PL} after irradiation is then a measure for the induced deep levels, which compete with the shallow acceptors for carrier recombination. Using this parameter, a reasonably acceptable correlation between K and NIEL has been found for p-GaAs, as evidenced by Fig. 4.25.

However, it has been observed that the correlation is a (quasi) linear function of I_{PL} in case of 7 MeV electron irradiation, while a (quasi) quadratic dependence is found for proton, neutron and heavy ion irradiations in p-GaAs [42].

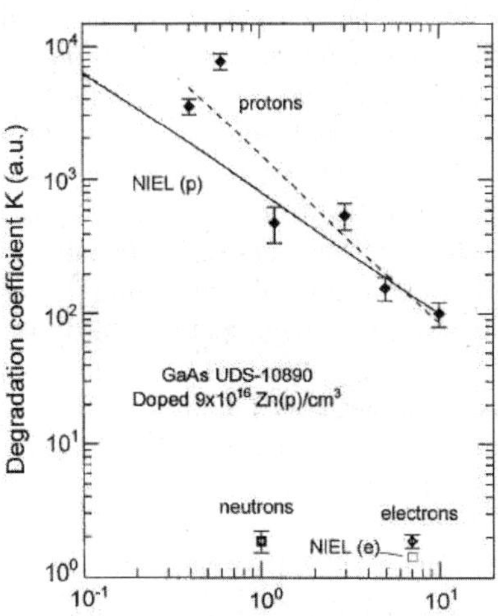

Fig. 4.25. K versus energy for p-GaAs. [After Khanna et al. [42]]

The difference is interpreted as being due to the formation of complex defects for proton irradiation involving shallow dopants, while electron exposures only produce simple point defects in first instance. In n-GaAs, a quasi-linear correlation is observed for all particles studied. Compared with γ-irradiation, a significant discrepancy (i.e., a factor 30) is found between the calculated NIEL and the experimental damage factor (see e.g. Fig. 4.24). This discrepancy is attributed to the weaker bound complexes formed by gamma irradiation compared to electron irradiation [42]. It also indicates that gamma (n- and p-) or electron (p-) irradiations are not useful for damage simulations due to (heavy) ions or neutrons in GaAs.

However, it has been observed that the correlation is a (quasi) linear function of I_{PL} in case of 7 MeV electron irradiation, while a (quasi) quadratic dependence is found for proton, neutron and heavy ion irradiations in p-GaAs [42]. The difference is interpreted as being due to the formation of complex defects for proton irradiation involving shallow dopants, while electron exposures only produce simple point defects in first instance. In n-GaAs, a quasi-linear correlation is observed for all particles studied. Compared with γ-irradiation, a significant discrepancy (i.e., a factor 30) is found between the calculated NIEL and the experimental damage factor (see e.g. Fig. 4.24). This discrepancy is attributed to the weaker bound complexes formed by gamma irradiation compared to electron irradiation [42]. It also indicates that gamma (n- and p-) or electron (p-) irradiations are not useful for damage simulations due to (heavy) ions or neutrons in GaAs.

4.4.5 Summary Damage Factors and NIEL in GaAs

The damage factors for the most relevant electrical material parameters are well documented in the literature and can be used to model/predict the radiation response of more complex device structures. One should keep in mind, however, that there exists a significant difference between the degradation of epitaxial layers, compared with bulk (substrate) material. Most state-of-the-art devices, and also GaAs solar cells, employ a stack of epitaxial layers with different compositions. On the other hand, GaAs radiation detectors are built on a SI substrate, so that in this case, bulk material parameters should be used for the modeling. It has also been observed that the lifetime or diffusion length damage factor is fairly insensitive to the doping density of the material.

The good correlation with NIEL for not too high particle irradiations indicates that NIEL is a useful tool for radiation damage studies in GaAs. In addition, a clear correlation with the created point defects has been established for low to moderate energies. For energies higher than 100 MeV, NIEL predicts a too high damage effect. In addition, the scatter between the different damage factors becomes pronounced. The latter could be related to the fact that at higher energies, beside the point defects also complex centres contribute to the electrical degradation, each having a different electrical impact.

4.5 Impact on GaAs Devices

In this paragraph, the impact of irradiation on the operation of a few classes of GaAs devices will be described. First, the behaviour of Schottky barrier diodes and nuclear-radiation detectors is discussed. The gate of a HEMT or a MESFET consists in fact of a metal-GaAs contact [107, 129], so that it is of basic importance to study its degradation. Secondly, the focus will be on GaAs space solar cells.

4.5.1 Schottky Barriers and Radiation Detectors

The impact of particle bombardment on the parameters of a metal-GaAs Schottky barrier has been studied by many groups (see e.g. [50, 105, 130 – 134]). On the technological more relevant n-GaAs material, a reduction of the barrier height ϕ_b is usually observed, resulting in an increase of the leakage current. For an Au barrier, this may result in a lowering from 0.91 eV to 0.7-0.8 eV typically [105]. In addition, the ideality factor in forward operation increases significantly above 1 with fluence, indicating the contribution of a non-ideal generation-recombination component [132 - 134]. At the same time, the diode series resistance is found to increase [50, 133 - 134]. On p-type GaAs, on the other hand, an increase of ϕ_b is typically observed [105].

A bulk GaAs radiation detector basically consists of a Schottky barrier on SI material. A schematic representation is given in Fig. 4.26 [135]. The advantages of using a GaAs detector are the radiation hardness of the material and the fact that due to the larger band gap (1.42 eV), it can be operated at room temperature and above, this in contrast to e.g. a Ge detector, which needs to be cooled at 77 K. In addition, the high electron mobility (>7000 cm^2/Vs) ensures a fast transport of the created electrons, while the hole mobility is significantly smaller, i.e., ~400 cm^2/Vs. It requires on the average an energy of 4.2 eV for e-h pair creation, when an ionizing particle is absorbed by the material.

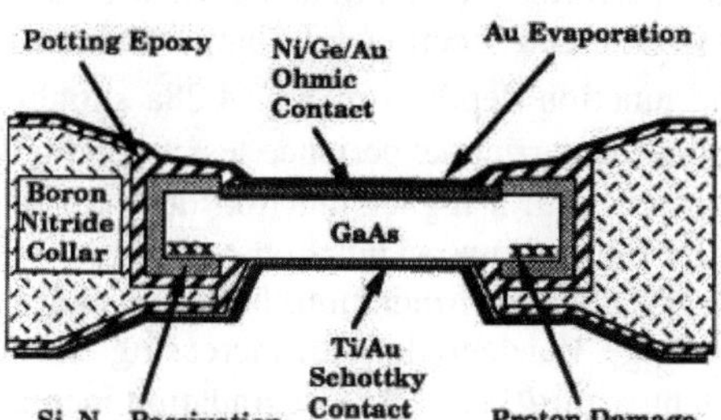

Fig. 4.26. Bulk GaAs detector cross section. [After McGregor et al. [135]]

The detector is operated with the barrier under reverse bias. Ideally, one aims at completely depleting the thickness of the diode, which is typically 100 to 200 μm [135 – 137]. Full depletion ensures a better collection efficiency of the created charge and hence a better resolution. Contrary to what one expects the depletion width W of the barrier changes proportionally with the blocking bias V_R [135, 136]. Full depletion is reached for V_R=V_d (the depletion bias). In order to explain this linear bias-dependence, one has to consider the impact of the deep EL2 donors on the electric field in the material [10].

The parameter that is most sensitive to the radiation damage is the Charge Collection Efficiency (CCE), which also sets the detector resolution and noise [10, 136, 137]. As shown by Fig. 4.27, the CCE scales nicely with the NIEL for the investigated range of energies and particles [137]. It was found that under full depletion, only hole trapping/detrapping contributes significantly to the signal [136]. It has also been demonstrated that mainly the radiation-induced EL2 or antisite defects near mid gap can give rise to trapping/detrapping times with the right order of magnitude [136]. A simple way to monitor the evolution of the EL2-like concentration is by room temperature optical absorption measurements [138]. This enables to derive the ionised fraction of the antisites, which is important for the degradation of the CCE [138].

4.5.2 GaAs Solar Cells

GaAs solar cells, like the one shown in Fig. 4.28a [123], are currently the main source of satellite power. Since GaAs has a high absorption coefficient compared with silicon, the total active thickness is typically only a few μm. The AlGaAs window or cap layer in Fig. 4.28a serves to passivate the GaAs surface and to reduce the surface recombination velocity. The reasons for the current interest in GaAs space solar cells are its high radiation tolerance and high conversion efficiency η. In addition, it has been observed that the radiation damage can be fairly easily annealed at relatively low temperatures, i.e., ~200 °C [123, 139]. The higher cost does not weigh up against the lower weight, especially when deposited on a Ge substrate. Figure 4.28b compares the performance of a Si and a GaAs solar cell under 1 MeV electron irradiation [123]. The GaAs cell clearly outperforms its Si counterpart. For optimal hardness, the junction depth x_j in Fig. 4.28a should be kept at ~ 0.2 μm. For a shallow junction, the carriers experience less trapping due to the radiation induced defects, compared with a higher junction depth. This is because the collection distance is smaller for shallow(er) junctions.

It has been observed that after 1 MeV electron irradiation, both the short circuit current I_{sc} and the open circuit voltage V_{oc} degrade with increasing fluence, while the fill factor (FF) is fairly constant with Φ [123]. The degradation increases with increasing flux, as there is less time for dynamic annealing. This becomes clearer when the irradiations are performed at high temperatures, i.e., 200°C [123]. Little difference in overall performance degradation has been noticed for highly or lowly doped GaAs cells. Similar conclusions can be drawn for the case of proton

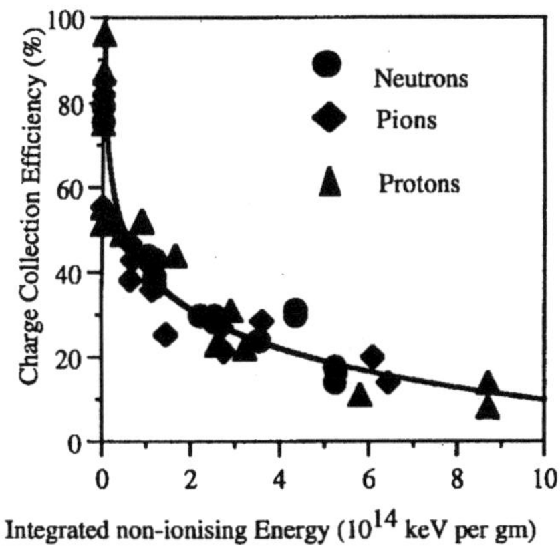

Fig. 4.27. Charge collection efficiency as a function of integrated NIEL for 24 GeV protons, 300 MeV pions and 1 MeV neutrons. [After Bates et al. [137]]

irradiation, although there exists some subtle differences, which are illustrated by Fig. 4.29 [123]. As shown there, the radiation damage in this case depends on the proton energy and hence the range of the ions. In fact, low energy protons (<0.5 MeV) are more damaging than high energy ones, as they deposit maximum damage inside the junction of the GaAs solar cell. This also explains why a silicon cell performs better in that case [123]. This deficiency of GaAs solar cells is, however, mitigated by the fact that a coverglass protection is most effective for low-energy protons. More detailed damage calculations for proton damage in shielded cells have shown that less than 20 % of the damage is created by protons emerging in the cell with energies <0.1 MeV [140]. It is concluded that it will be generally not worth performing tests with protons below ~1 MeV.

A clear correlation has been established between the radiation-induced deep levels observed in DLTS and the performance degradation of GaAs solar cells [123, 141, 142]. The defects cause a reduction of both V_{oc} and I_{sc}, where the latter parameter is determined by the minority carrier diffusion length, i.e., τ_p. A number of deep electron and hole traps were observed. However, the most detrimental ones were recombination centres close to mid gap: an electron trap at E_c-0.71 eV and a hole trap at E_v+0.71 eV [123, 141, 142] It was found that these levels only partially annealed at 200°C, while the other more shallow traps could be annihilated at such low temperatures. The origin of these levels is unclear, but they are probably different from antisite or EL2-like defects. The same deep levels were detected for electron and proton irradiated devices.

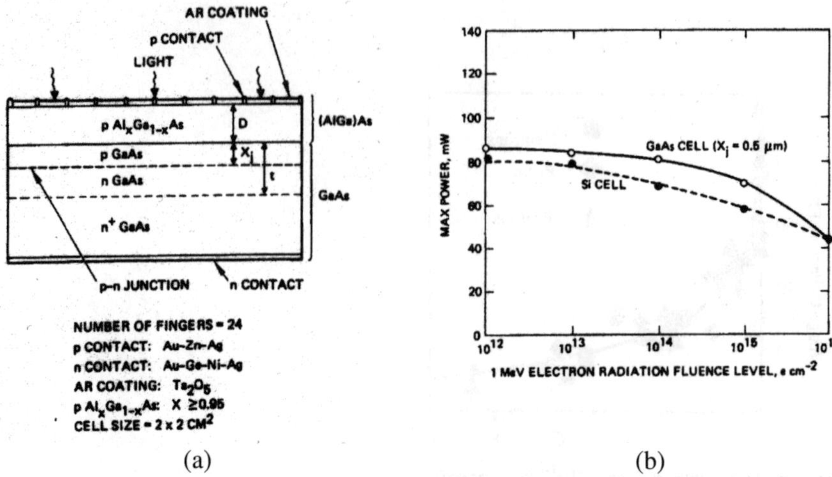

Fig. 4.28. (a) AlGaAs/GaAs solar cell design and (b) GaAs solar cell degradation characteristics after 1-MeV electron damage (x_j=0.5 μm). [After Loo et al. [123]]

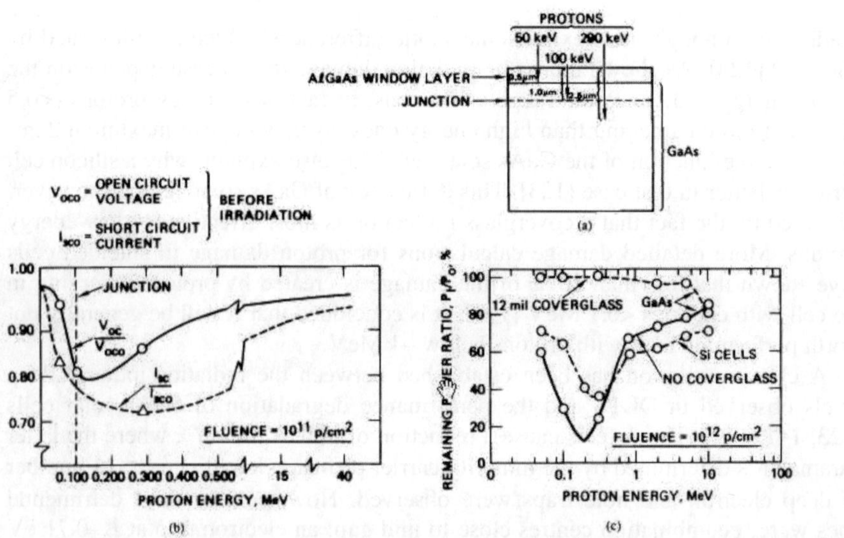

Fig. 4.29. Proton radiation damage as a function of proton energy: (a) proton penetration depth, (b) V_{oc} and I_{sc} versus proton energy, (c) remaining power ratio versus proton energy. [After Loo et al. [123]]

A number of groups have modeled the radiation response of GaAs solar cells [142 – 147], in order to improve its hardness. One result is shown in Fig. 4.30 [147]. It is seen there that the radiation response hardly depends on the choice of the starting doping concentration, which is also confirmed by experiments [123,

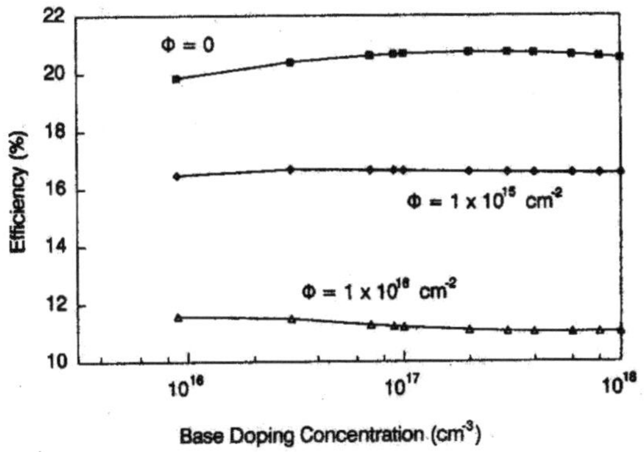

Fig. 4.30. Calculated efficiencies as a function of base doping for GaAs cells before and after 1 MeV electron irradiation at two different fluences. [After Augustine et al. [147]]

142] or by the K_L factor of Fig. 4.23. This stands in contrast with InP cells, where a high starting concentration is more favorable, while the opposite holds for a silicon device [147].

4.6 General Conclusions

GaAs has certainly proven to be a material of choice for space or radiation-hard applications. One of the reasons is that the basic radiation defects are point defects or pairs of primary defects, at least for n-GaAs. Unlike the case of silicon, impurity-related radiation defects are only formed under high flux or high temperature irradiation conditions, which mobilises the As_i in n-GaAs - technologically far more important than p-GaAs. In the latter case, dopant-point defect complexes form more easily. However, for space applications, one should consider the results of low-flux experiments to be more representative for the actual operation conditions. In addition, most of the radiation defects anneal at rather low temperatures, with stage III (~200°C) being the most important one for the room-temperature stable centres. An important remark is that the starting material may contain already high concentrations of point defects, which can interact with the radiation-induced ones. This is also one of the reasons why the identification of the radiation defects is less established than in silicon.

In spite of this, the impact of radiation on the basic material properties is well documented in the literature and many damage factors have been determined accurately. Most of them correlate well with the NIEL for not too high energies. In addition, it has been demonstrated that PL spectroscopy is a useful microscopic damage monitor, whereby V_{Ga} serves as a point-defect marker in n-GaAs, while the PL intensity of the C_{As} line shows the necessary correlation with the damage factors or NIEL in p-GaAs. These macroscopic or microscopic damage factors can be used for predicting the degradation of certain components and circuits, which will be discussed in Chap. 7 and 8. However, it has also been shown that γ- or electron exposures are not a good replacement for (heavy) ion irradiations in GaAs. This is amongst others related to a more complex damage (U-band, L-band) generated in the material.

5 Space Radiation Aspects of Silicon Bipolar Technologies

5.1 Introduction

Bipolar Junction Transistors (BJT) have important applications in analog or mixed signal ICs and BiCMOS (Bipolar Complementary Metal-Oxide-Semiconductor) circuits because of their current-drive capability, linearity, low noise and excellent matching characteristics. Furthermore, their microwave performance compares favorably with respect to CMOS, explaining the use in GHz telecommunications applications and low-cost system-on-chip (SOC) solutions. The implementation of a SiGe heterojunction base will further push the penetration of BiCMOS in the microwave IC market. In addition, BJTs are frequently used in space systems, including operational amplifiers, comparators and voltage regulators, in order to accomplish analog functions. Early generations of BJT based circuits mainly suffered from radiation-induced leakage currents associated with the degraded field-oxide isolation regions. However, solutions for these problems have been implemented in present-day technologies, so that other degradation mechanisms have become more important. In this chapter, an overview will be given related to radiation damage in submicron BJTs and HBTs. First the different type of device architectures (vertical, substrate, lateral BJT) will be briefly defined and the basic degradation mechanisms described. In Sect. 5.3, focus is on the radiation effects in vertical n-p-n BJTs, while Sect. 5.4 covers the radiation degradation in lateral and substrate BJTs. Sect. 5.5 is devoted to the radiation response of SiGe HBTs, while in the last section, some conclusions will be drawn.

5.2 Device Structures and Basic Radiation Effects

The device structure of different types of silicon BJTs will be described. The architecture of a SiGe HBT has already been given in Figs 3.22 and 3.23 of Chap. 3. In addition, some general notions regarding radiation damage in bipolar transistors will be discussed.

5.2.1 Device Structures and Definitions

Different types of BJT architectures are used for specific applications. They are illustrated in Fig. 5.1, for p-n-p devices [1]; complementary structures exist for the n-p-n counterparts. One can distinguish between vertical (V), lateral (L) and substrate (S) transistors. For vertical and substrate devices, the current flow is mainly in the vertical direction. Lateral devices are characterized by a current flow parallel to the surface. A schematic top view of a vertical device is given in Fig. 5.2, defining the emitter and intrinsic base. The extrinsic base is adjacent to the intrinsic one and is more highly doped to reduce the contact and base series resistance. Originally, a crystalline silicon emitter was used, contacted by Al metallization.

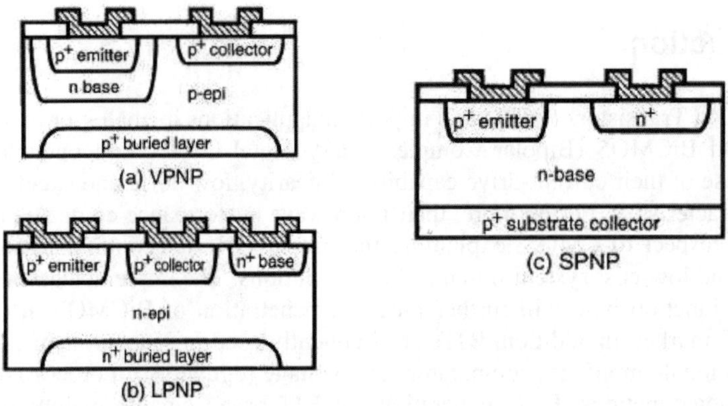

Fig. 5.1. Qualitative cross sections of (a) vertical (VPNP), (b) lateral (LPNP) and (c) substrate (SPNP) transistors. [After Schrimpf [1]]

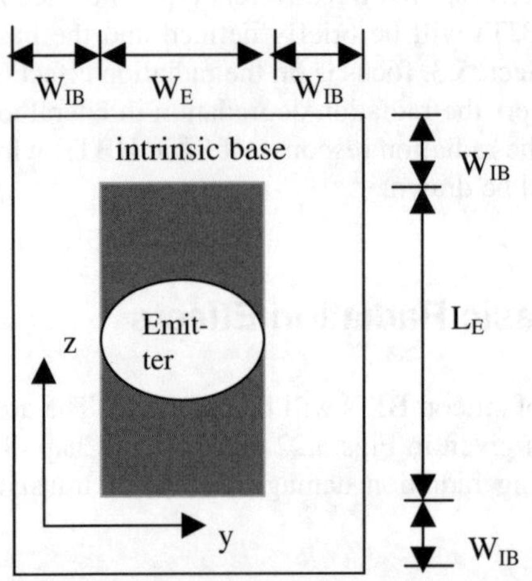

Fig. 5.2. Schematic top view of a BJT. W_{IB} is the width of the intrinsic base. L_E and W_E is the length and the width of the emitter, respectively.

A polycrystalline (poly) emitter is more and more replacing the crystalline emitter, as this is amenable for scaling to the submicron regime. This allows a reduction of the base thickness in the first place. At the same time, a higher base doping density is used in order to prevent base push-out and to reduce the base resistance. Both single and double polysilicon technologies have been developed. In the latter case, a polysilicon base and emitter contact of both doping type is implemented. An important part of the structure is the oxide covering the surface of the base-emitter junction, often called sacrificial, screening or spacer oxide. In commercial technologies, the oxide is grown before the implantation of the base and, therefore, suffers from ion implantation damage. In addition, it is generally exposed to a high temperature annealing, necessary for the dopant activation in the base and in the (poly) silicon emitter. The resulting oxide quality is less than in ordinary MOS gates, leading to a higher than normal native density of interface traps (range 10^{11} cm^{-2}). As will be outlined further, this has serious consequences for the radiation hardness of modern BJTs.

5.2.2 Radiation Damage Mechanisms

In bipolar devices, three main damage mechanisms can be distinguished [2]. First, there is the positive oxide trapped charge, which modifies the surface potential at the Si-SiO_2 interface. As a result, p-type doped silicon will become depleted, or even inverted, depending on the doping density and the amount of trapped hole charge, while n-doped regions become accumulated. Depletion of the p-type base (n-p-n) or emitter (p-n-p) will increase the (surface) recombination current and, hence, the non-ideal base current I_B. This in turn leads to a reduction of the static current gain $\beta = I_C/I_B$, with I_C the collector current. Inversion of the p-type surface layer can lead to an increased leakage current along the field oxide isolation regions. At the same time, the breakdown voltage of the transistor can be reduced, as summarized in Table 5.1 [2]. Similar as for a MOS device, the charge yield strongly depends on the quality (hardness) of the oxide and of the bias (field) across it. For state-of-the-art BJTs, a low oxide field is generally present under normal operation, while at the same time high charge yields, typical for soft oxides have been observed (see Chap. 6 on MOS devices).

A second degradation source is surface recombination in the base-emitter depletion region. This can be related to the density of interface traps D_{it} through the equation:

$$s_r = 0.5 v_{th} \sqrt{\sigma_n \sigma_p} \pi k T D_{it} \tag{5.1}$$

with s_r the surface recombination velocity, v_{th} the thermal velocity, kT the thermal energy and σ_n,σ_p the capture cross section for electrons and holes, respectively. As can be noted from (5.1), a radiation-induced increase of D_{it} will generate a proportional increase in the surface recombination velocity and in the corresponding recombination base current. It should be noted, however, that in contrast to MOS devices, the relevant D_{it} is the one corresponding to mid gap energy (if $\sigma_n=\sigma_p$) and, hence, to a maximum recombination rate. For MOS structures, it is the whole energy range from mid gap up to the relevant band edge that plays a role in the device performance.

Table 5.1. Summary of total dose mechanisms for bipolar devices. [After Johnston and Plaag [2]]

Mechanism	Energy Dependence Particle/Radiation	Bias Dependence during Irradiation (low field)	Effect on Devices
Surface Recombination Velocity	slight	weak	gain reduction
Oxide Trapped Charge	slight	strong	inversion gain reduction increased leakage decreased breakdown
Bulk Damage	strong	none	gain reduction

A third degradation mechanism is bulk (displacement) damage which is particularly pronounced for high energy particle irradiation and to a lesser extent for ^{60}Co γ's. In the latter case, the displacements in the silicon lattice are produced by secondary energetic Compton electrons. This is not relevant for X-ray exposures. The main effect is a reduction of the minority carrier lifetime, resulting in a linear degradation of the reciprocal current gain with particle fluence Φ [3 - 5], given by:

$$\beta^{-1} = \beta_0^{-1} + K_\beta \Phi \tag{5.2}$$

K_β is the damage factor and β_0 is the initial static current gain. As shown previously (Fig. 2.20) an excellent agreement has been found between the experimental K_β and the NIEL of irradiated silicon BJTs [5].

The main degradation mechanisms and their impact on the BJT characteristics are summarized in Table 5.1 [2]. At the same time, the qualitative dependence on the energy of the impinging particle/photon and on the oxide field is indicated.

5.3 Degradation in Vertical (n-p-n) BJTs

The different degradation mechanisms for vertical n-p-n BJTs are discussed. Similar effects are expected for VPNP, although these devices are much harder than their n-p-n counterparts. The basic reason is that in a VNPN the oxide-trapped charge induced depletion of the p-type region is in the "lowly" doped base, while it occurs in the heavily doped p^+ emitter for a VPNP. As will be seen, total-dose (oxide-trapped charge) effects are the main source of damage in modern scaled BJTs. Due to the narrow base widths and the high base doping densities used, displacement damage is in many cases negligible or of second order.

5.3.1 Phenomenology of Total Dose Damage

The primary effect of total-dose degradation of vertical n-p-n transistors is the increase of the base current with dose (Fig. 5.3a) and a corresponding reduction of the current gain (Fig. 5.3b) [6 - 8]. The base current increase ΔI_B is particularly pronounced for small base-emitter voltage V_{BE}, causing a shift of the peak β to higher V_{BE}. In addition, a total-dose-induced increase of the collector current ΔI_C has also been noted [8, 9], as shown in Fig. 5.4. It is, however, strongly technology dependent [8] and only observed for certain types of poly-emitter BJTs, under reverse V_{BE} irradiations.

The excess base current increases superlinearly with dose D, for not too large D (Fig. 5.5) and is higher for larger emitter perimeter/area (P_E/A_E) ratios [8]. For a sufficiently large dose, the excess base current saturates (Fig. 5.6) [10]. This saturation value of ΔI_B is independent of the dose rate during the exposure, or of the irradiation bias. It is only a function of the V_{BE} during measurement [10]. Implanted or polysilicon emitter n-p-n BJTs show qualitatively similar degradation [8].

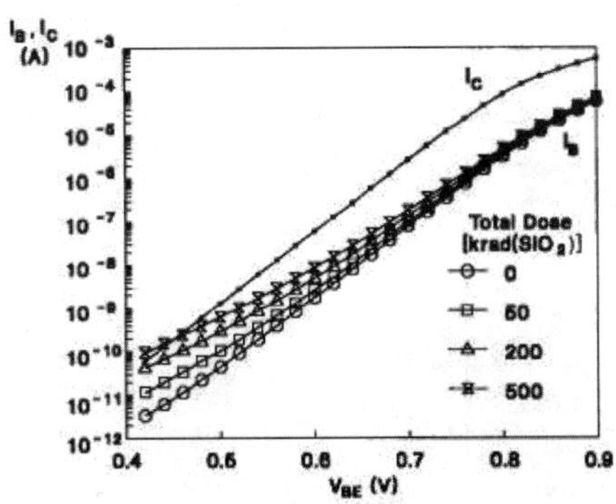

(a)

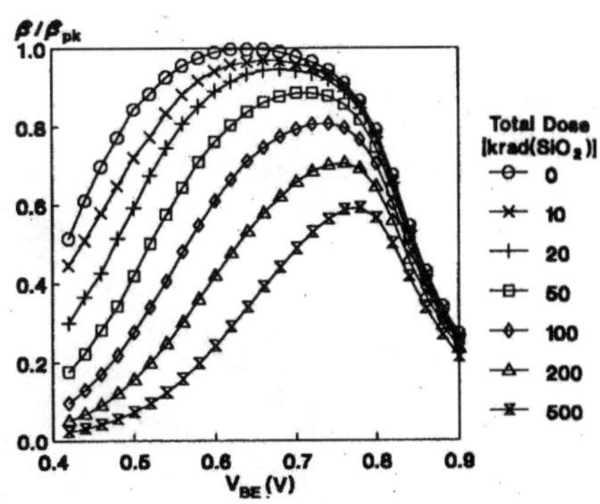

(b)

Fig. 5.3. (a) Typical Gummel curves for a bipolar transistor exposed to ionising radiation. (b) Typical dc current gain (β) degradation in a bipolar transistor exposed to ionising radiation. β_{pk} is the peak pre-irradiation current gain. [After Nowlin et al. [8]]

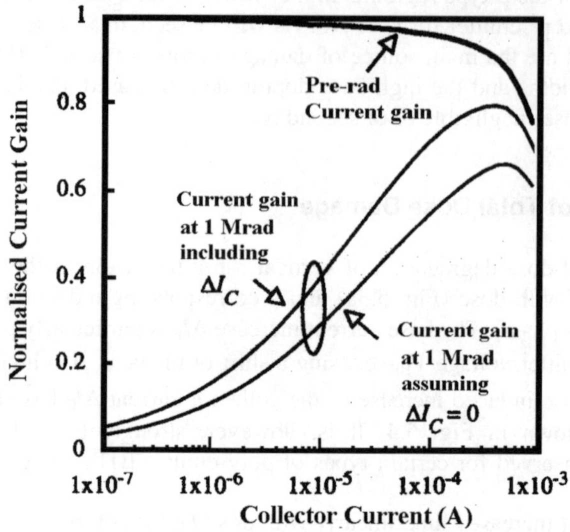

Fig. 5.4. Measured common-emitter current gain, normalized to the peak pre-irradiated value, versus the collector current. Shown are the pre-irradiated current gain plotted against the pre-irradiated collector current, the current gain at a total dose of 1 Mrad(SiO_2) plotted against the actual measured collector current at that dose, and current gain at a total dose of 1 Mrad(SiO_2) plotted against the pre-irradiated collector current. [After Wei et al. [9]]

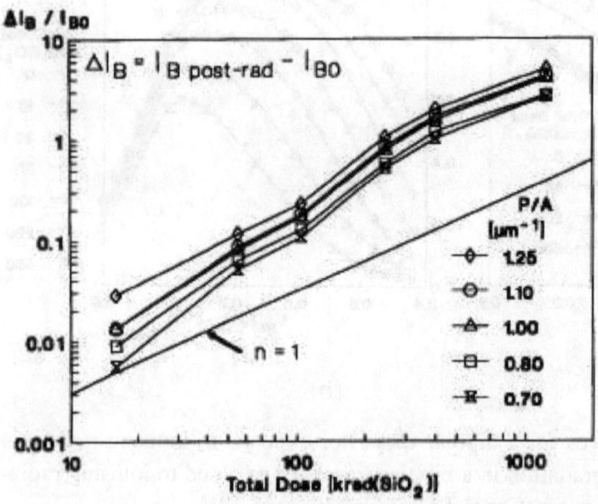

Fig. 5.5. Total-dose and device-geometry dependencies of the excess base current measured at V_{BE}=0.7 V. I_{B0} is the pre-rad current. [After Nowlin et al. [8]]

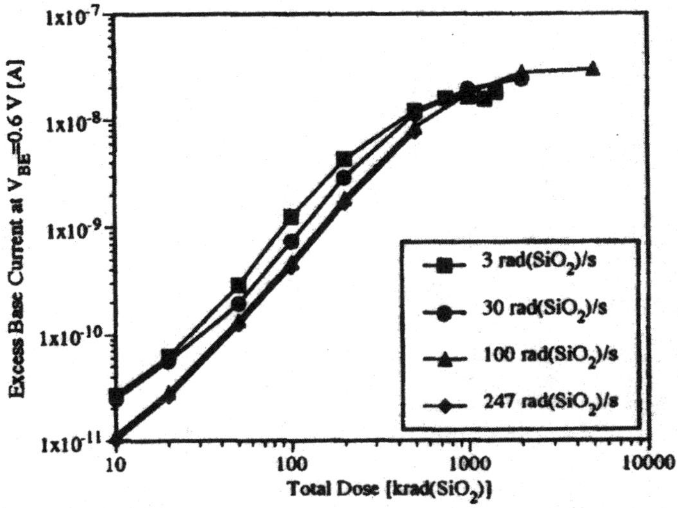

Fig. 5.6. Excess base current at V_{BE}=0.6 V versus total dose at four different dose rates. [After Kosier et al. [10]]

Perhaps the most striking feature in Fig. 5.6 is the dose rate dependence of the excess base current, indicating a *higher* degradation for *low* dose rates [6 - 8, 10 - 11]. This is in contrast with what is known for MOS devices, where the damage generally increases with dose rate. The effect is more clearly evidenced in Fig. 5.7, where the low dose rate response saturates at ≈10 rad(SiO_2)/s for that particular type of devices [12]. In addition, the low dose rate effect is worst for zero bias irradiation, while reverse bias operation is worst case for high dose rate degradation in Fig. 5.8 [8]. The high dose region sets in around 150 rad(SiO_2)/s, whereby the excess base current becomes independent of this parameter [8].

This low dose rate effect has large practical consequences for hardness assurance of modern bipolar space applications. In contrast to the case of the MOS system, the convenient standard laboratory tests at high dose rate do not provide a conservative estimate of the component and circuit degradation, but *underestimate* what can be expected for space radiation conditions. The alternative is to perform testing of BJTs at unpractical and lengthy (expensive) low dose rates. According to the MIL-STD-883D Test Method 1019.4, the low dose rate response can be simulated by applying a high temperature anneal, following a high-dose rate exposure. For bipolar transistors, this test method fails, since, as expected, the degradation anneals out, as evidenced by Fig. 5.9 [8]. In other words, a high dose rate exposure plus high temperature anneal does not reproduce the enhanced degradation observed at low dose rates. Note also in Fig. 5.9 that at room temperature the annealing proceeds very slowly.

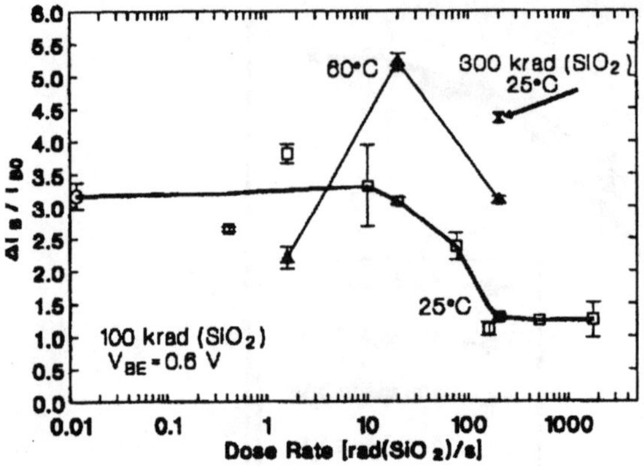

Fig. 5.7. Dose-rate response of the excess base current after irradiation to 100 krad(SiO_2) at 25 and 80°C. Also shown is a 200 rad(SiO_2)/s, 300 krad(SiO_2), 25°C data point representing a factor-of-3 overtest. [After Nowlin et al. [12]]

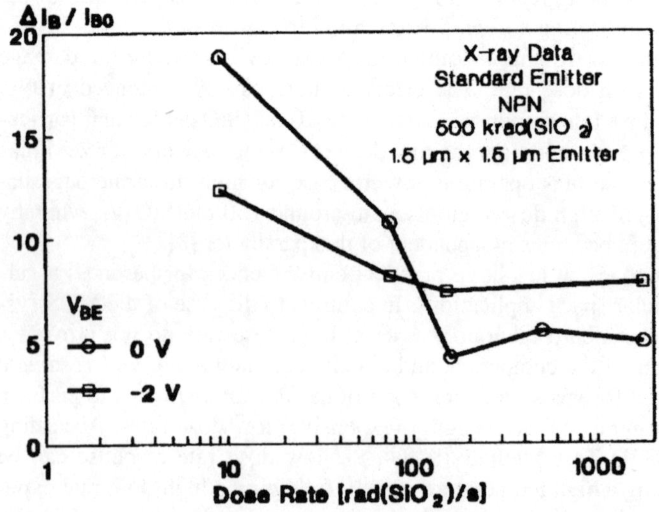

Fig. 5.8. Dose-rate dependence of the total-dose response of standard-emitter n-p-n devices irradiated in an X-ray source. [After Nowlin et al. [8]]

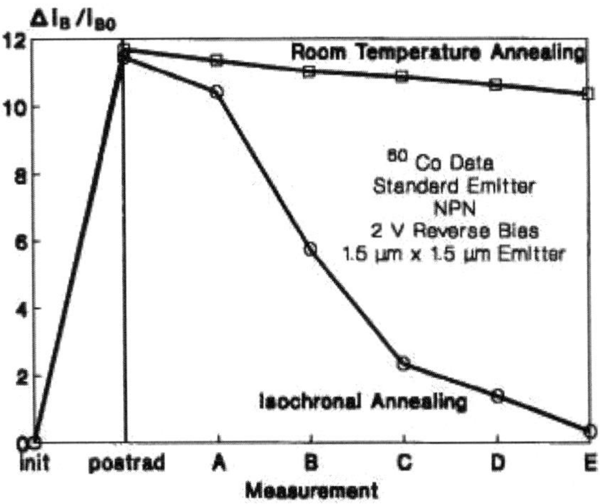

Fig. 5.9. Room-temperature and isochronal annealing response of reverse-biased, standard-emitter n-p-n transistors. All measurements were taken after exposure to 500 krad(SiO_2) (post-rad from ^{60}Co at 240 rad(SiO_2)/s, and after the isochronal anneal). The parts were annealed for 30 minutes at (A) 60°C, (B) 100°C, (C) 150°C, (D) 200°C and (E) 250°C. The room temperature parts were characterized at the same time as the isochronal annealed parts. [After Nowlin et al. [8]]

A final observation with respect to the enhanced low dose rate sensitivity (ELDRS) is that it is significantly impacted by a pre-irradiation elevated-temperature stress (PETS) [13]. Such a stress may occur during the bonding of the devices in packages or during a burn-in test. It can partly explain the strong technology and even batch dependence found in practice for the ELDRS effect. The ELDRS decreases significantly with longer thermal pre-treatment at 175°C [13]. Nevertheless, within some total-dose and annealing-time range, true dose rate sensitivity was observed for the linear circuits studied. Although no exact physical explanation is available as yet, it is speculated that the interaction of hydrogen with defects in the bulk of the base oxide or at the base-emitter junction may play a key role.

This sensitivity to a thermal pre-treatment can have serious implications for radiation hardness testing and assurance in linear bipolar technologies and circuits for space applications. The current ASTM hardness assurance test method requires the use of either a low dose rate or elevated-temperature irradiation [14]. However, the observed interaction between pre-irradiation thermal cycling and ELDRS suggests that one should also test whether the IC is sensitive to PETS or not [13].

5.3.2 Basic Low Dose-Rate Degradation Mechanisms

It is obvious from the above that owing to the different dose-rate dependence, the total-dose degradation in BJTs follows a different mechanism than in MOS devices. An additional factor that can play is that the oxide field during exposure is much lower in BJTs than in MOS transistors (typically ±1 MV/cm). Finally, as mentioned above, the screen oxide covering the base-emitter surface is generally rather thick (50 to 500 nm) and of poor(er) quality [8]. This calls for a radiation testing of such oxides under low bias. From the mid-gap voltage shift shown in Fig. 5.10, it is clear that a larger hole trapping yield under zero bias follows for such field-oxides at low dose rates [8]. The extracted positive oxide charge in Fig. 5.11b nicely tracks the behavior of the excess base current of Fig. 5.11a, emphasizing the important if not dominant role played by the oxide trapped charge N_{ot} in the degradation of the BJTs [11].

The total dose damage in spacer, screen or sacrificial oxides, covering the base-emitter junction of a BJT has been studied in more detail by Fleetwood et al [15, 16], using Thermally Stimulated Current (TSC) and C-V measurements. It has first of all been pointed out that the charge yield in such ion implanted and annealed spacer oxides is very high (close to 100%) compared with hardened oxides (≈0.1 to 1 % range). Also the role of electron trapping has been highlighted. A model has been proposed, which is schematically depicted in Fig. 5.12 [16], and which can explain why slightly more holes are trapped at low dose rates. Key in the model is the assumption that holes are trapped in delocalized trapping centres with a metastable nature. These can be related to ion implantation induced displacement damage, so that good candidates are the so-called E'_{δ} centres detected by ESR [15]. The release of holes from these centers occurs on a time scale of seconds to hours. It has been shown that such centres are more abundant in soft oxides compared with hard ones. They also anneal at moderate temperatures (~50°C) [15]. It has in fact been observed that the ELDRS in linear bipolar circuits has an activation energy of 0.37 eV [17], consistent with the activation energy for

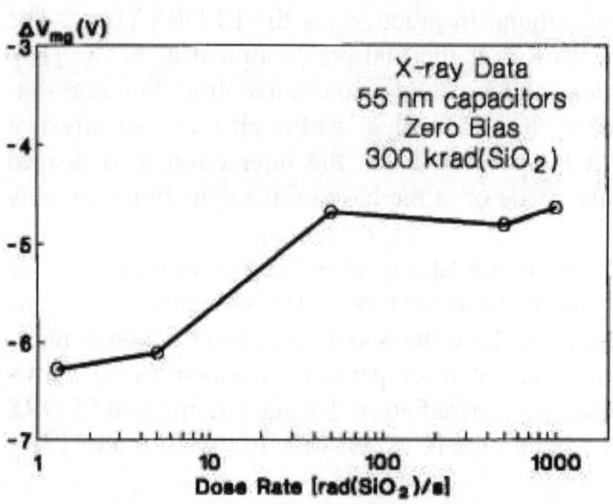

Fig. 5.10. Dose-rate dependence of the mid-gap voltage shift ΔV_{mg} in field-oxide capacitors with 55 nm field-oxides at 0 V bias. [After Nowlin et al. [8]]

the diffusion of molecular hydrogen in SiO_2. This implies that the diffusion of molecular hydrogen or other hydrogen related species may play a key role in the depassivation of metastable trap sites in the base oxide and the passivation of interface-trap precursors at the Si/SiO_2 interface at or near the base/emitter.

The total dose damage in spacer, screen or sacrificial oxides, covering the base-emitter junction of a BJT has been studied in more detail by Fleetwood et al [15, 16], using Thermally Stimulated Current (TSC) and C-V measurements. It has first of all been pointed out that the charge yield in such ion implanted and annealed spacer oxides is very high (close to 100%) compared with hardened oxides (≈0.1 to 1 % range). Also the role of electron trapping has been highlighted. A model has been proposed, which is schematically depicted in Fig. 5.12 [16], and which can explain why slightly more holes are trapped at low dose rates. Key in the model is the assumption that holes are trapped in delocalized trapping centres with a metastable nature. These can be related to ion implantation induced displacement damage, so that good candidates are the so-called E'_δ centres detected by ESR [15]. The release of holes from these centers occurs on a time scale of seconds to hours. It has been shown that such centres are more abundant in soft oxides compared with hard ones. They also anneal at moderate temperatures (~50°C) [15]. It has in fact been observed that the ELDRS in linear bipolar circuits has an activation energy of 0.37 eV [17], consistent with the activation energy for the diffusion of molecular hydrogen in SiO_2. This implies that the diffusion of molecular hydrogen or other hydrogen related species may play a key role in the depassivation of metastable trap sites in the base oxide and the passivation of interface-trap precursors at the Si/SiO_2 interface at or near the base/emitter.
An alternative model suggests that electron trapping in shallow traps can explain the enhanced hole trapping at low dose rates in bipolar screen oxides [18].

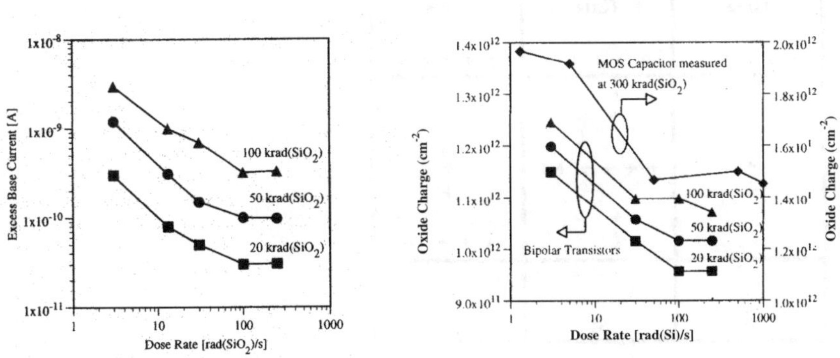

(a) (b)

Fig. 5.11. (a) Excess base current at V_{BE}=0.6 V versus dose rate and (b) oxide charge versus dose rate at various total doses. [After Wei et al. [11]]

At higher irradiation temperatures, the thermally activated release of the trapped electrons causes a reduction of the trapped hole yield at low dose rates and essentially removes the effect. Investigations have pointed out that the non homogeneous fringing field at the edge of the base-emitter junction also plays a role and should be taken into account [15, 19]. The effect is shown to be more pronounced for p-n-p compared with n-p-n transistors.

Another striking difference compared with the degradation of MOS type devices is the observed impact of repeated irradiation and anneal cycles, as shown in Fig. 5.13 [20]. In between irradiations, the samples were annealed at 275°C in N_2 for a sufficient long time in order to restore their pre-irradiation current gain. While for MOS devices a hardening effect is reported, Fig. 5.13 clearly shows a continuous degradation. The degradation is most pronounced after the first irradiation and saturates after a few cycles. In other words, after one (or more) irradiation and anneal steps, the radiation induced degradation becomes worse, for the same total dose. The trend is independent of the BJT geometry, or of the type of emitter (standard implanted or poly). P-n-p devices seem to be less susceptible compared with n-p-n. In order to explain this effect, surface stress measurements have been performed on similarly oxidized and metallized reference wafers [20]. The results of this analysis indicate that the interfacial compressive stress relaxes with increasing number of cycles. The interpretation given to these facts is that for lower compressive stress at the Si-SiO_2 interface, more interface traps are being created for the same total dose, thereby producing a larger excess base current and current gain degradation. The nature of the interface stress, i.e., compressive or tensile, depends on the used fabrication process and will be strongly influenced by the implementation of advanced processing modules for scaling down geometries.

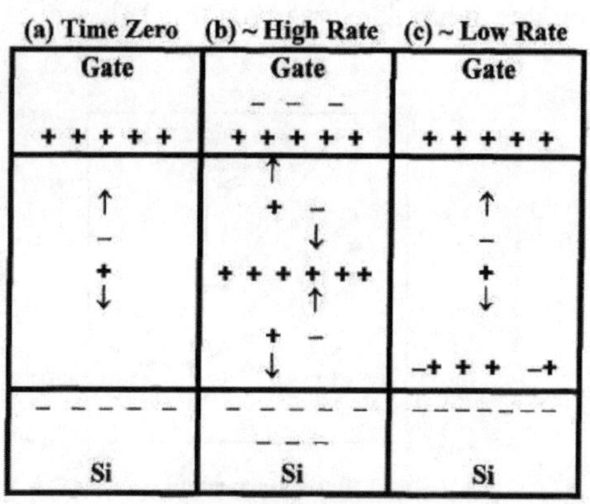

Fig. 5.12. Schematic illustration of electron and hole transport for three cases: (a) at the start of irradiation before trapped charge begins to build up, (b) during high-rate exposure, when holes are captured by metastable traps in the bulk of the SiO_2, causing space charge effects, and (c) during low-rate exposure, when some holes in metastable traps are emitted and transported to the Si/SiO_2 interface. [After Fleetwood et al. [16]]

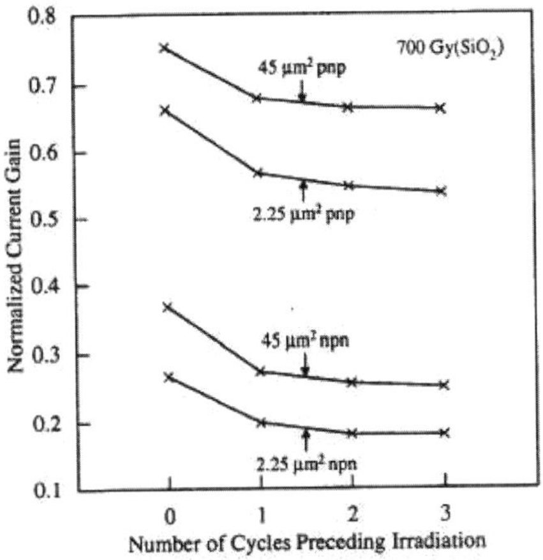

Fig. 5.13. Effect of radiation and anneal cycles on the radiation-induced current gain degradation after a total dose of 70 krad(SiO_2). The gain degradation increases with an increasing number of cycles independent of the emitter geometry or polarity. All data were measured at $|V_{BE}|$=0.6 V. [After Witczak et al. [20]]

5.3.3 Charge Separation in BJTs

Total-dose damage has two main effects in a bipolar transistor: first, the creation of oxide trapped charge changes the surface potential $\Psi_s(y)$ at the interface of the spacer oxide and base-emitter junction. For an n-p-n transistor, the p-type base region becomes more depleted as illustrated in Fig. 5.14 [21]. This enhances the peak surface recombination rate, which corresponds to the lateral position where the electron and hole densities at the surface are approximately the same. In order to determine the exact lateral position of the peak surface recombination rate, two-dimensional device simulations are indispensable [1, 10, 21 - 23]. A typical result is given in Fig. 5.15, for a certain V_{BE} and for different densities of oxide trapped charge N_{ot} [21].

A second contributing factor is the increase in the surface recombination velocity, given by (5.1), through the increase of the density of mid-gap interface states D_{it}. Combining both factors, one can derive the surface recombination rate at a certain lateral position y, defined in Fig. 5.14 [1, 21]:

$$R_s(y) = \frac{n_i s_r exp(\frac{qV_{BE}}{2kT})}{2cosh[\frac{q}{kT}(\Psi_s(y) - \frac{V_{BE}}{2})]} \tag{5.3}$$

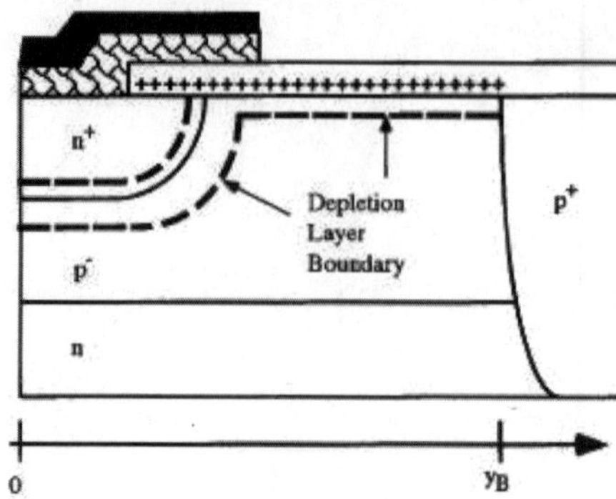

Fig. 5.14. Schematic cross-section of the base-emitter junction of a n-p-n transistor showing the effect of oxide charge on the depletion layer. [After Kosier et al. [21]]

whereby $\Psi_s(y)$ is a function of N_{ot}. The band bending due to N_{ot} can be written in the depletion approximation [21]:

$$\Psi(N_{ot}) = \frac{qN_{ot}^2}{2\varepsilon_s N_s} \tag{5.4}$$

with N_s (in cm^{-3}) the doping density at the surface of the intrinsic base. The peak recombination rate occurs for $\Psi_s(y)=V_{BE}/2$, so that:

$$R_{speak} = \frac{1}{2} n_i s_r exp(\frac{qV_{BE}}{2kT}) \tag{5.5}$$

Note that the peak recombination rate corresponds to an ideality factor n_{if} of exactly 2. The radiation-induced excess base current ΔI_B then follows from an integration over the whole base width in Fig. 5.14 (y=0 to y_B) of the sharp peaked function (5.3) (see Fig. 5.15). Inspection of (5.5) points out that both N_{it} and N_{ot} effects are closely connected and affect the excess base current in an interactive and far from linear way. Therefore, the excess base current shows initially a superlinear increase with dose (Figs 5.5 and 5.6) and eventually saturates for large total dose (Fig. 5.6) [10]. This is opposite to the case of a MOSFET, where ΔV_{ot} and ΔV_{it} effects are additive, so that they can be fairly easy separated (see Sect. 2.3.1). This facilitates tremendously the physical modeling of total-dose effects.

A similar kind of charge separation can be achieved for irradiated BJTs [21]. This is obtained in the following way. Looking more closely to the excess base current characteristics of Fig. 5.16, it is clear that for low total dose, two parts can be distinguished, corresponding to an ideality factor $1<n_{if}<2$ at low V_{BE}, while for

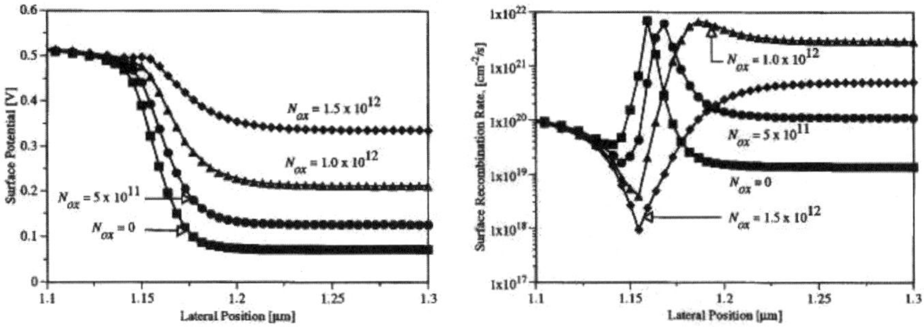

(a) (b)

Fig. 5.15. PISCES simulation results for the base-emitter junction, located at y=1.16 μm. V_{BE}=0.5 V with varying oxide charge in units of cm^{-2}. (a) Surface potential versus lateral position and (b) surface recombination rate versus lateral position. [After Kosier et al. [21]]

high V_{BE}, n_{if}=2. The transition voltage V_{tr}, defined by the crossing point of the two linear fits in Fig. 5.16, is related to a critical oxide charge density, given by [21]:

$$N_{ot} = \sqrt{\frac{2\varepsilon_S N_S}{q}\left[\frac{kT}{q}\ln\left(\frac{N_S}{n_i}\right) - \frac{V_{tr}}{2}\right]} \tag{5.6}$$

and corresponds to the point where the oxide charge band bending of (5.4) is always equal or larger than $V_{BE}/2$. At this point, the maximum recombination rate is no longer occurring at the surface, but subsurface, as can be derived from the simulations of Fig. 5.15 and shown more explicitly in Fig. 5.17 [10]. Given the sharply peaked nature of R_s, the surface recombination will no longer contribute significantly to ΔI_B even if D_{it} shows a further increase with total dose. In other words, the excess base current will no longer be affected by s_r, but by the base recombination lifetime, which is relatively insensitive to total dose (especially for X-ray exposures). This means that ΔI_B will saturate and furthermore exhibits an n_{if}=2. The PISCES simulations of Fig. 5.18 [10] further illustrate this.

On the other hand, the lower n_{if}-value for lower total dose is explained by the fact that R_s is not constant, but varies with lateral position y along the base. Integrating (5.2) yields some effective (average) n_{if} value in that case. Another important consequence is that, while for low total dose and V_{BE} ($n_{if} < 2$) the excess base current scales with the device perimeter (Fig. 5.5), for n_{if}=2, the saturated excess base current will increase proportionally with the area of the transistor (Fig. 5.19) [23]. In conclusion, V_{tr} can be used as a measure for a certain oxide charge density in the spacer oxide.

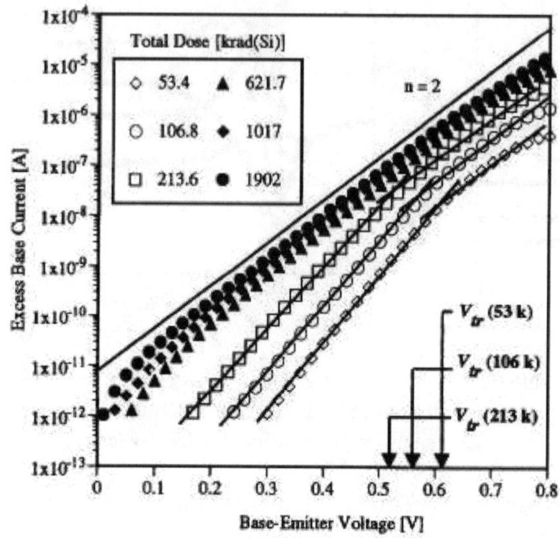

Fig. 5.16. Excess base current versus base-emitter voltage. [After Kosier et al. [21]]

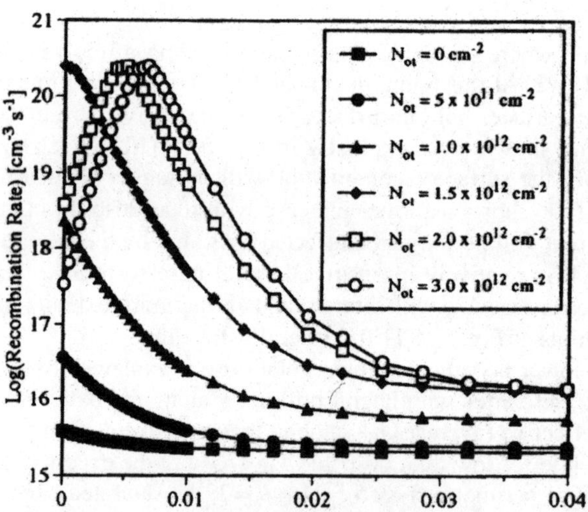

Fig. 5.17. PISCES-simulated recombination rate versus position normal to the interface for V_{BE}=0.5 V and various amounts of positive oxide charge. [After Kosier et al. [10]]

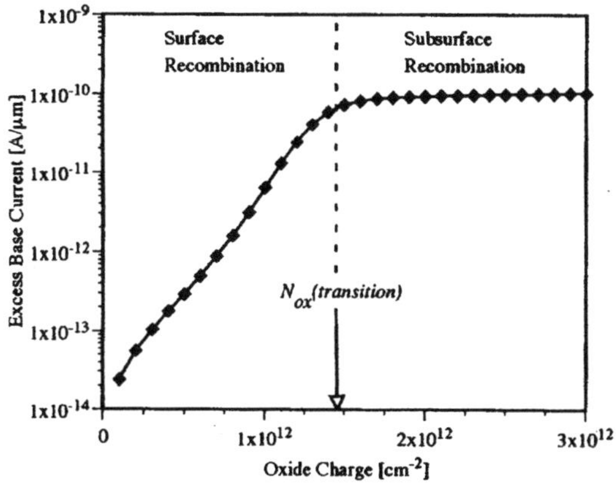

Fig. 5.18. PISCES-simulated excess base current versus positive oxide charge at V_{BE}=0.5 V. [After Kosier et al. [10]]

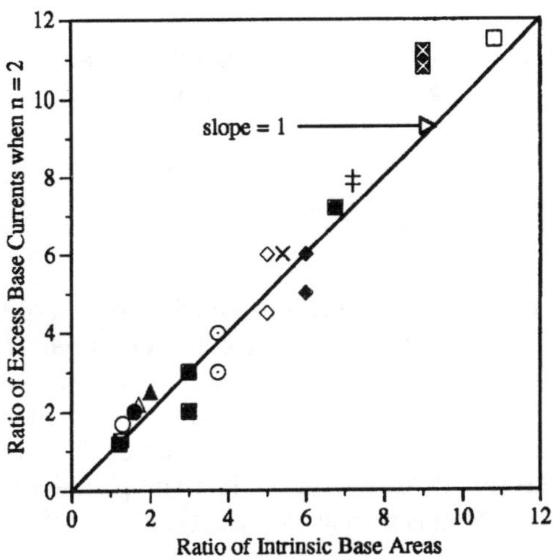

Fig. 5.19. Ratio of excess base current when n_{if}=2 versus ratio of intrinsic base areas. [After Kosier et al. [21]]

A value for the density of interface traps can be derived from the intercept of the exponential fit to the ΔI_B curves, corresponding to lower V_{BE} in Fig. 5.16, i.e., n_{if}<2. According to (5.3) or (5.4) for example, the intercept is proportional to s_r. It is clear from Fig. 5.16 that the intercept and, hence, s_r increases significantly with dose (Fig. 5.20). An alternative approach to determine the interface trap density is to use a gated diode in forward operation [21], although the results may not be identical, like shown in Fig. 5.20. Because of this, preference is given to the intercept method.

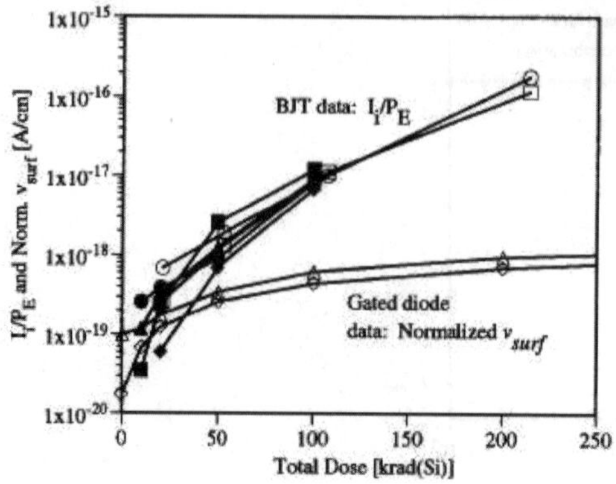

Fig. 5.20. Scaled intercept current and normalised surface recombination velocity versus total dose. [After Kosier et al. [21]]

Based on the proposed analytical model, it is concluded that the key factor is the oxide trapped charge, since, combining (5.4) and (5.5), one obtains [1, 23]:

$$\Delta I_B \sim R_S \sim \exp\left(\left[\frac{N_{ot}}{\sqrt{2}L_D N_s}\right]^2\right) \tag{5.7}$$

with L_D the extrinsic Debye length. As mentioned above, for a certain type of poly emitter vertical transistors, a radiation-induced increase of the collector current is also found (Fig. 5.4) [9]. With the help of PISCES simulations, it was demonstrated that the origin of this increase is an enhancement of the effective emitter area, by an oxide trapped charge induced inversion of the p-base surface. This is illustrated in Fig. 5.21, showing the electron density in the emitter and p^- base region, for large N_{ot}. The excess collector current is then given by:

$$\Delta I_C = A_{EX} J_{EX} \tag{5.8a}$$

with A_{EX} the area of the "extrinsic" base (i.e. adjacent to the emitter) [9] and J_{EX} the corresponding current density. By using A_E as emitter area and GN_{IN},GN_{EX} as the Gummel number for the "intrinsic" and for the "extrinsic" base, respectively, the upper bound for the effective emitter area is given by [9]:

$$A_{E,eff,max} = A_E + \frac{GN_{IN}}{GN_{EX}} A_{EX} \tag{5.8b}$$

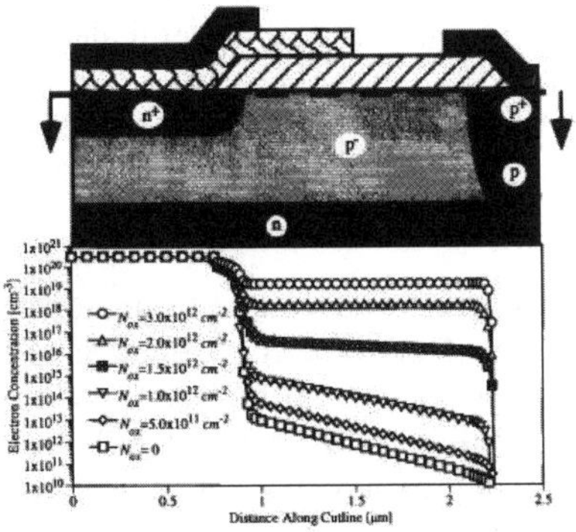

Fig. 5.21. PISCES simulated electron concentration along the oxide-silicon interface at V_{BE}=0.6 V versus increasing oxide charge. [After Wei et al. [9]]

5.3.4 Hardening Guidelines for Vertical BJTs

Based on the above insights, a few guidelines for practical bipolar technology and circuit hardening can be formulated [8, 15]. These recommendations are:

i) to increase the surface doping density in the intrinsic base. As a result, a higher N_{ot} is necessary to deplete (or even invert) the surface ((5.6)) and according to (5.7) reduces significantly the total-dose-induced excess base current.
ii) to reduce the thickness of the screen or spacer oxide. This lowers in first instance the accumulated N_{ot}, the key factor in the damage process.
iii) to use dedicated sacrificial ion implantation oxides over the intrinsic base region, which are subsequently removed and/or to reduce the thermal budget to which the sacrificial oxide is subjected. A good quality spacer oxide can be grown/deposited in a later stage of the processing. This, of course, makes the bipolar (BiCMOS) process more complex and expensive.
iv) to optimise the design by using minimal P_E/A_E ratio devices, whenever possible. This is particularly useful for not too high total doses, below the saturation point of the excess base current. There, the ΔI_B is proportional with the device perimeter.
v) to design circuit operation for high(er) I_C (peak gain) operation. Higher V_{BE} operation corresponds to less ΔI_B and current gain degradation. Furthermore, the peak β shifts to higher base-emitter voltages upon irradiation (Fig. 5.3). Drawback is the higher power consumption which is often incompatible with the low power/low voltage applications of BiCMOS technologies.
vi) to reduce the intrinsic surface area between the base contact and the emitter. In fact, downscaling bipolar and BiCMOS technologies fulfills perfectly this requirement. It is, therefore, anticipated that scaled technologies will be more ra-

diation tolerant, as they also have higher base doping densities and thinner spacer oxides. The use of a heterojunction device may also be a useful alternative [24].

5.3.5 Hardening Assurance and Testing

A particular problem of hardness assurance for bipolar devices and technologies for space is the assessment of the low dose rate effects in a cost-effective and meaningful way. One pathway, which has been pursued, is to replace radiation testing by reverse-bias hot-carrier degradation testing. In the past, it has been noted that in certain cases, there exists a similarity between both types of degradation, which mainly produces a current gain reduction, caused by excess base current [25, 26]. Unfortunately, comparative studies have pointed out that the degradation mechanism is quite different in both cases [22, 23, 27]. As shown above, total-dose damage occurs mainly through the creation of oxide-trapped charge, while hot-carrier degradation proceeds by the creation of interface traps near the emitter-base junction [27], resulting in an excess base current with an ideality factor of 2.

An alternative is the use of high-temperate ($\geq 60^{o}$C), high dose rate irradiations. According to Fig. 5.7 [10], the resulting degradation is of the same order as the low dose-rate room temperature exposures. This is considered a viable method for assessing low dose rate damage in bipolar components and circuits [28]. However, as pointed out by *Shaneyfelt et al.* [13], one key aspect to account for in the first place is the sensitivity of the components to thermal cycling. It could explain the large differences in low dose-rate sensitivity observed in practice from technology to technology.

An important point to mention is also that the worst case testing at low dose rates is for zero bias, i.e., with all terminals grounded (see e.g. Fig. 5.8 [8]).

5.3.6 Bulk Damage in Vertical Transistors

Some studies have also been undertaken to investigate the occurrence of displacement damage in modern BJTs by exposure to high-energy particle irradiation [25, 26, 29 - 31]. This is motivated not only in view of the radiation environment but also for technological reasons. For example, the use of e- (or ion-)beam writing for deep submicron lines may be accompanied by the creation of radiation damage [29, 30]. This exposes the wafers to displacement damage, which can, however, anneal during subsequent thermal processing. For space applications, proton damage is a potential problem.

The degradation of 0.5 μm BJTs fabricated in a single-poly BiCMOS technology has been studied in [31], with particular emphasis on the Gummel plots, static current gain and low-frequency (LF) noise. The latter parameter is important for analog (linear) applications (amplifiers, voltage regulators) and in addition for nonlinear microwave applications, since the LF noise can determine the high-frequency phase noise. One MeV electron irradiation was compared with 59 MeV proton exposures (unbiased). Some clear differences have been noted in the LF noise behaviour, as illustrated by the results of Fig. 5.22. In both cases, the static characteristics have degraded (mainly the base current) [31]. For the LF $1/f$ noise, on the other hand, e^- irradiations have little effect (Fig. 5.22a), while protons cause a marked increase of the flicker noise, at high V_{BE} (Fig. 5.22b). The latter effect is believed to indicate the creation of displacement damage in the base-emitter junction of the BJT, while for electron irradiations, total dose effects on the surface recombination current are most likely predominant. This behaviour is similar as for X-ray or γ-exposures. It should also be noted that with respect to γ-irradiations, a dose-rate effect has been noted in the LF current noise of a bipolar operational amplifier [32]. An increase of the noise was found for increasing dose rate. The main contribution came from a GR noise component.

Most of the studies so far have indicated that the impact of displacement damage is marginal compared with the total-dose effect. However, it has been shown that total-dose irradiation may indirectly affect the silicon substrate by reducing the active p-type base dopant concentration [33, 34]. The effect is most pronounced for irradiation temperatures around 100°C [34]. This raises concern for some applications where the circuit temperature may be raised well above room temperature during operation. It is also a typical irradiation temperature recommended for high-dose testing of the low dose-rate sensitivity. As for the low dose-rate effect, the dopant neutralisation is most pronounced for small irradiation biases, resulting in the surface under depletion. An example is given in Fig. 5.23 [34], showing the density of neutralised acceptors versus the irradiation bias. For simulating the phenomenon, radiation tests were performed on MOS capacitors, with the same type of oxide covering the base-emitter surface. To determine the change in acceptor density at the surface, the improved charge separation technique, mentioned in Sect. 2.3.1 was utilised [35]. The device simulations represented in Fig. 5.24 show that the deactivated dopants near the surface result in an enhanced base current and, therefore, in a reduced static current gain. The enhanced I_B is related to an increase in the surface recombination velocity, as the density of electrons and holes is becoming more comparable, due to the boron neutralisation. It has in addition been observed that the collector current increases, resulting from a higher emitter injection efficiency. This is again related to the reduction of the base doping concentration after exposure.

The origin of the boron neutralisation is related to the interaction with hydrogen, released from the irradiated gate dielectric and electrode. This can occur according to two reaction mechanisms [34]: boron passivation, given by:

$$B^- + H^0 + h^+ -> (BH)^0 \qquad (5.9a)$$

The second reaction leads to acceptor compensation, whereby a positive hydrogen ion is ionically bond to B, according to:

$$B^- + H^+ + e^- + h^+ -> (BH)^0 \qquad (5.9b)$$

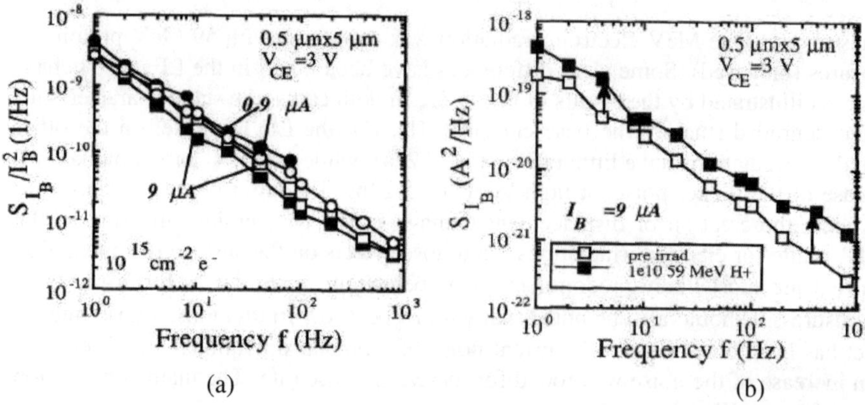

Fig. 5.22. LF noise spectra before (open symbols) and after (closed symbols) a 1 MeV 10^{15} cm^{-2} electron irradiation (a) and a 10^{10} cm^{-2} 59 MeV proton irradiation (b), for a 0.5 µmx5 µm vertical n-p-n BJT. [After Simoen et al. [31]]

The drift of H^+ ions into the Si depletion region can thus also lead to electron-hole recombination through acceptor compensation, where the electrons are supplied from the silicon bulk in order to maintain charge neutrality. In (5.9a), the complexing of of B^- and H^0 requires the presence of holes, whereas in (5.9b) the Coulombic attraction of B^- and H^+ can occur in a region free of carriers. By considering the two reaction paths of (5.9), one can fairly easily understand the trend observed in Fig. 5.23, showing maximum neutralisation for a bias in depletion [34].

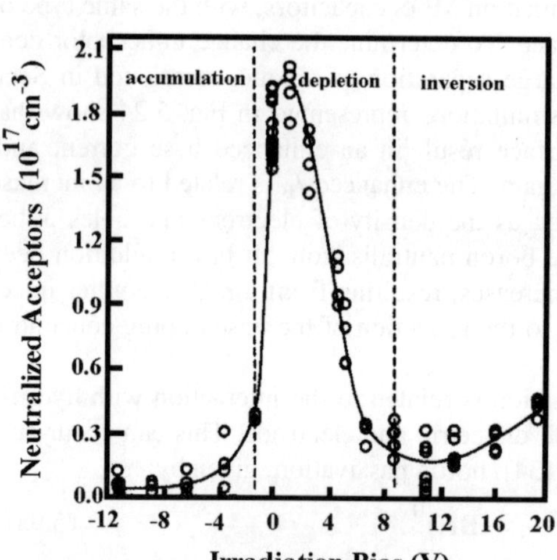

Fig. 5.23. Effect of irradiation bias on the concentration of acceptors neutralised near the Si surface. The acceptor neutralisation is most severe for irradiation biases in depletion. [After Witczak et al. [34]]

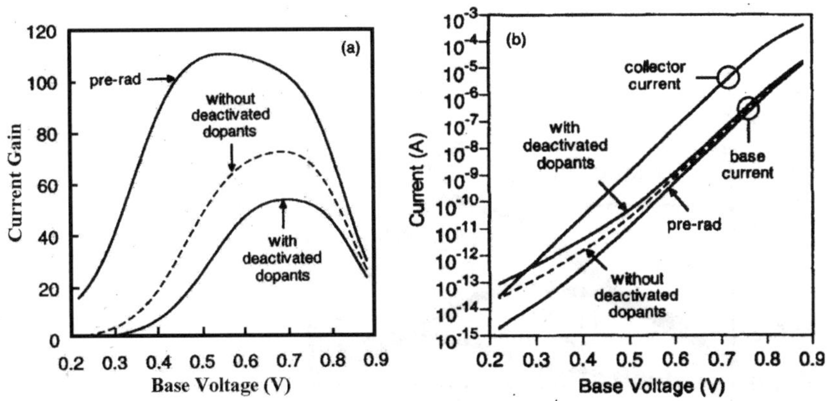

Fig. 5.24. Effect of acceptor neutralisation on the simulated (a) current gain characteristics and (b) current-voltage characteristics. The acceptor neutralisation enhances radiation-induced current degradation by increasing the base recombination current. [After Witczak et al. [34]]

5.4 Degradation in Lateral Transistors

The impact of total dose degradation on lateral p-n-p transistors is described and a physical explanation is provided. In addition, operation conditions will be pointed out which will reduce the impact of total-dose damage on lateral transistors.

5.4.1 Phenomenology

Although vertical p-n-p BJTs are significantly more radiation hard than their n-p-n counterparts, there are indications that certain types of p-n-p transistors in fact show more pronounced radiation-induced damage, particularly at low dose rates. This was derived from experiments on bipolar linear circuits and components [36 - 40]. Therefore, a systematic comparison of different types of p-n-p BJTs, depicted in Fig. 5.1, has been undertaken [41 - 44]. It is clear from Fig. 5.25 that the LPNP shows the most pronounced normalized current gain degradation, compared with the S- or V- type p-n-p transistor [41]. In addition, the low dose rate effect is higher for the LPNP (Fig. 5.26). A typical feature for the degradation of the LPNP is the reduction of the collector current after total dose exposure, given in Fig. 5.27. The post-irradiation base current in Fig. 5.27a shows an ideality factor n_{if} between 1 and 2 for a $V_{BE}< 0.7$ V and $n_{if}=2$ for $V_{BE} >0.7$ V [41]. The excess base current of Fig. 5.28a increases with lower dose rate and follows roughly the same trend as the interface and oxide trapped charges in test capacitors, representative for the oxide above the base-emitter junction (Fig. 5.28b).

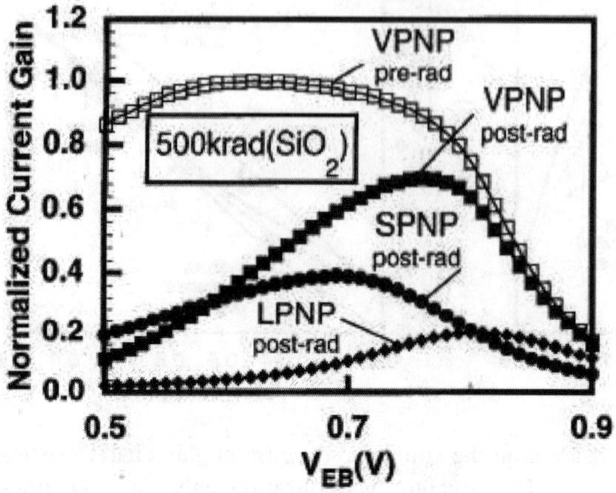

Fig. 5.25. Normalised current gain vs. V_{BE} for lateral (LPNP), substrate (SPNP), and vertical (VPNP) transistors for a total dose of 500 krad(SiO_2). Dose rate for VPNP devices is 158 and for LPNP and SPNP devices is 167 rad(SiO_2)/s. [After Schmidt et al. [41]]

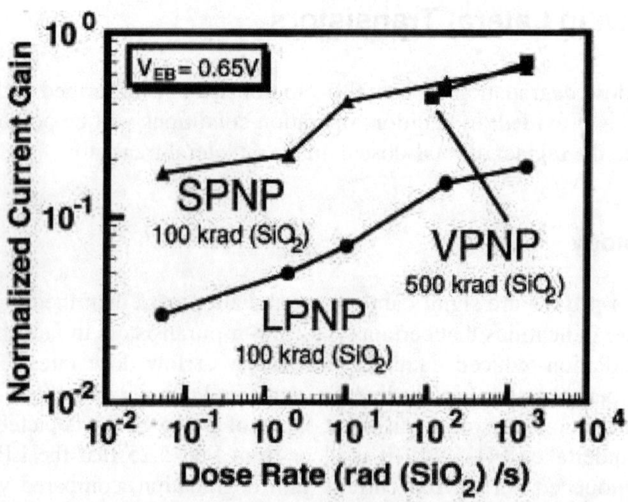

Fig. 5.26. Normalised current gain versus dose rate for lateral (LPNP), substrate (SPNP), and vertical (VPNP) transistors. The total dose for the LPNP and SPNP devices was 100 krad(SiO_2), and the total dose for the VPNP device was 500 krad(SiO_2). [After Schmidt et al. [41]]

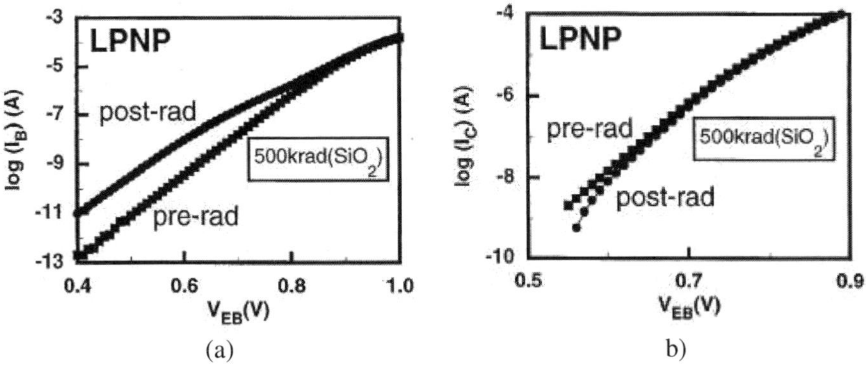

Fig. 5.27. Gummel plots (a) log I_B and (b) log I_C vs. V_{BE} for a LPNP at a dose rate of 167 rad(SiO_2)/s to a total dose of 500 krad(SiO_2). [After Schmidt et al. [41]]

An important factor in the degradation of lateral p-n-p's is the emitter doping [44]. Highly doped emitters show reduced excess base current for the same total dose, compared with lightly doped emitters (Fig. 5.29). A gate control over the LPNP surface can drastically reduce the total-dose degradation [45]. The least degradation is obtained for a dynamical biasing, whereby the transistor is under depletion during the sleep or standby modes and has an accumulated base surface during operation.

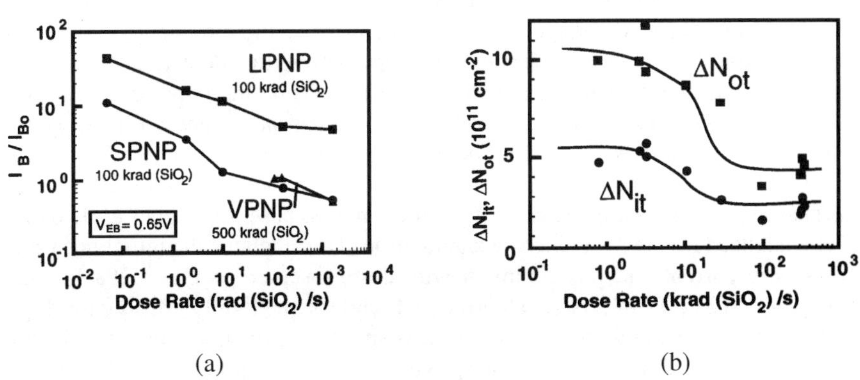

Fig. 5.28. (a) Excess base current $\Delta I_B/I_{B0}$ versus dose rate for the LPNP, SPNP and VPNP transistors. The ^{60}Co data at 0.5 rad(SiO_2)/s have been reduced by the factor 2.73 for comparison with the X-ray data. (b) Change in oxide-trapped charge ΔN_{ot} and interface trap ΔN_{it} density in test capacitors, representative of the oxide above the base-emitter junction in the LPNP transistors. [After Schmidt et al. [41]]

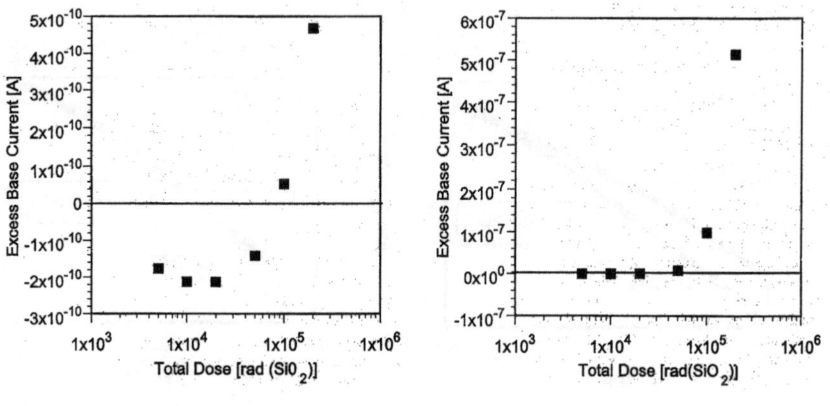

(a) (b)

Fig. 5.29. Excess base current versus total dose for V_{BE}=0.6 V, a dose rate of 83.3 rad(SiO_2)/s for (a) heavily- and (b) lightly-doped emitter. [After Wu et al. [44]]

5.4.2 Physical Mechanisms and Modeling

In order to understand the degradation behavior in an irradiated p-n-p BJT, one has to consider the impact of positive oxide trapped charge on the depletion (inversion) in the base and emitter, like shown in Fig. 5.14 for a n-p-n structure. For a p-n-p BJT the n-base will become accumulated, while the emitter gets depleted with increasing N_{ot} (total dose). It should be kept in mind that depleting the surface region will enhance the surface recombination rate, while the opposite is true for accumulation. This implies that the surface recombination current will enhance in the p-type emitter and is reduced in the base. However, as illustrated in Fig. 5.30, one should in addition account for the radiation-induced increase of the density of interface traps [41]. The net result is that in the n-type base, the higher N_{it} is more or less balanced by the N_{ot} induced accumulation. This explains the initial negative excess base current and the minimum with dose observed in Fig. 5.29a for the heavily doped LPNP [44]. No such effect is found for the lightly doped case (Fig. 5.29b). In the p-type emitter side, both effects should in principle interact with and enforce each other, similar as for the p-type base in an n-p-n device. The effect will be more pronounced for a lowly doped emitter compared with a highly doped one [44].

Two extra degradation mechanisms have to be considered for these p-n-p transistors [41]. The first one is related to the accumulation of the n-type base, which becomes an n^+ base. As shown by the PISCES simulations of Fig. 5.31, more electrons will be back-injected in the emitter through this effect, yielding an increase of the base current. A second effect of the accumulated base surface is that the holes will be pushed deeper in the substrate, leading to a longer 'effective' base. At the same time, more holes will recombine along their path, since they are moving in a heavier doped buried n^+ region, characterized by a lower recombination lifetime. This is represented by the PISCES results of Fig. 5.32 [41] and explains the reduction of the collector current shown in Fig 5.27b. The strongly different radiation tolerance for L and S devices is related to the predominant direction of the hole flow: in the L transistor, it will essentially occur along the surface, while for the S device, only a fraction of the hole current will be lateral. The buildup of oxide-trapped charge forces the holes to flow deeper in the base, explaining the susceptibility of LPNP to total-dose degradation [41].

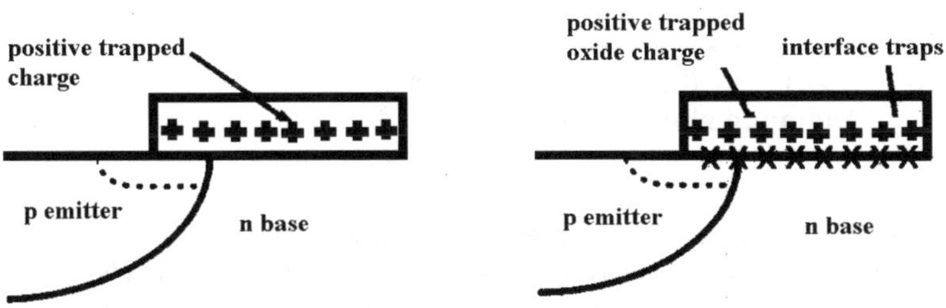

Fig. 5.30. Illustration of positive trapped oxide charge (a and b) and interface traps (b) over the emitter-base junction and the resulting expansion of the depletion region into the emitter. [After Schmidt et al. [41]]

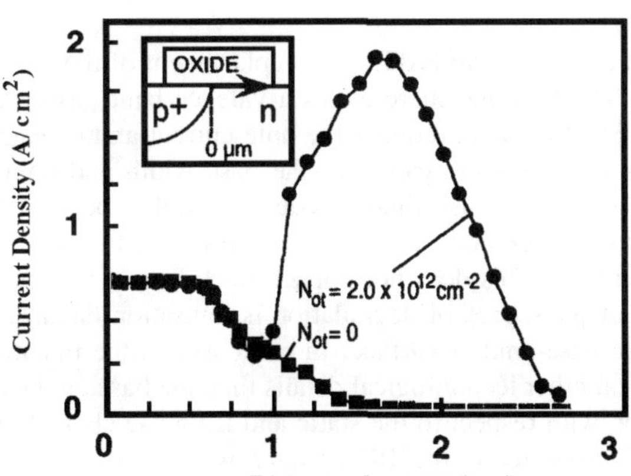

Fig. 5.31. SPISCES simulation of electron current density near the base-emitter junction that shows the effect of excess base current due to back-injected electrons. The cut-line (represented by the arrow) is just under the $Si\text{-}SiO_2$ interface below the base-emitter junction with the origin at the junction. [After Schmidt et al. [41]]

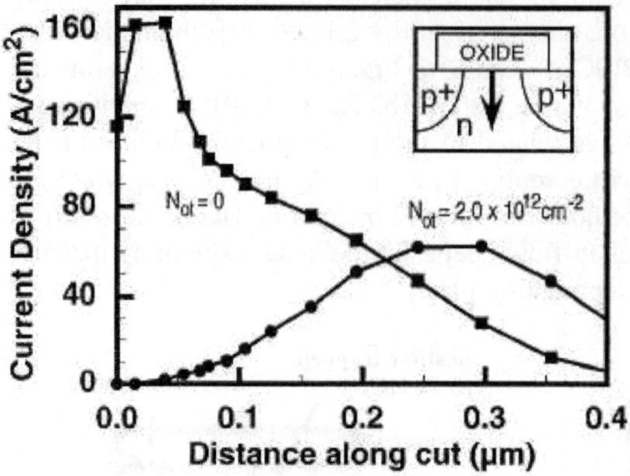

Fig. 5.32. SPISCES simulation of the hole current density midway between the base-emitter and the base-collector junctions at the Si-SiO_2 interface. This illustrates how N_{ot} alters the path injected holes must travel en route to the collector. The origin of the cut-line (represented by the arrow) is at the Si-SiO_2 interface. [After Schmidt et al. [41]]

5.5 Degradation in SiGe HBTs

The main difference between a BJT and an HBT is the replacement of the silicon p-type base by a SiGe base. The fact that there exists a valence band offset between the Si emitter and the $Si_{1-x}Ge_x$ base, reduces the hole current and enhances the current gain. This allows for a further scaling of the base width and doping concentration. Consequently, the base transit time is reduced and the speed of the devices increases. Both abrupt and graded base Ge concentrations, in the 15 to 20% Ge range, are used in practice. This does not impact much from a viewpoint of radiation hardness, as the major source of degradation is ionisation damage in the spacer oxide, covering the base-emitter surface. In other words, the radiation damage will depend strongly on other technological details than the base engineering. This will be demonstrated with respect to the static and LF noise characteristics. In addition, one should characterise the RF or microwave response of Si HBTs before and after irradiation, as this is one of the important application fields of BiCMOS technologies.

5.5.1 Degradation of the Static and Low-Frequency Noise Characteristics

The study of ionisation damage tolerance in advanced, fully developed BiCMOS technologies, with SiGe HBTs has been undertaken mainly by the group of *Cress-*

Cressler, employing room temperature [46 - 48] and 77 K γ-irradiations [47], and neutron [49] and proton exposures [24, 50]. The structure of the HBTs is represented in Fig. 3.22, whereby both Si and SiGe epitaxial transistors have been fabricated on the same wafers. This technology is expected to be intrinsically radiation hard, because of the use of trench isolation, of the thin oxide/nitride emitter-base spacer and of a high doping concentration at the surface of the epitaxial base region.

The total-dose degradation of the SiGe HBTs is summarized in Figs 5.33 to 5.35, for room temperature exposures up to 10 Mrad(Si). As expected, an increase of the non-ideal recombination base current is found for low I_C in Fig. 5.33, resulting in a lowering of the current gain for low collector current (Fig. 5.34). However, the shift of the peak β is only 20 % after 10 Mrad(Si) total dose, which is much more tolerant than many so-called radiation-hardened bipolar technologies. This follows more specifically from the comparison in Fig. 5.36, which gives the change in the base current at a V_{BE}=0.6 V as a function of the dose rate, for different types of BJTs [48].

The response of the current gain as a function of total γ-dose shows three regions in Fig. 5.35: below 20 krad no degradation whatsoever is observed. For intermediate doses (≤1 Mrad(Si)), β increases ! This is believed to correspond to an increase of the collector current [47]. It is demonstrated there that the increase in I_C is not due to a radiation-induced inversion of the base-emitter surface or a radiation-induced heating, but rather related to a change of the lifetime properties of the emitter-base junction. Beyond 2 Mrad(Si), β decreases.

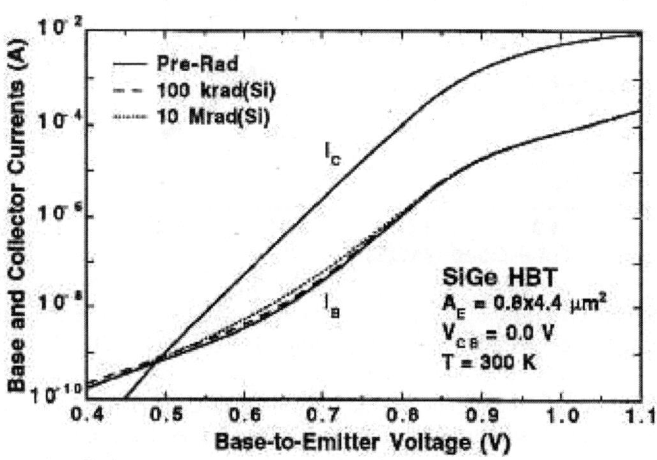

Fig. 5.33. Typical Gummel characteristics for a SiGe HBT (emitter area A_E=0.8x4.4 μm^2) both before and after radiation at room temperature. [After Babcock et al. [47]]

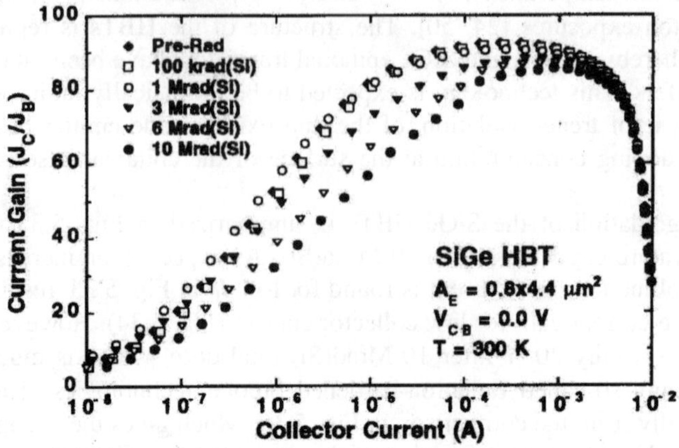

Fig. 5.34. Pre- and post-radiation current gain (β) versus the collector current at 300 K for the SiGe HBT (A_E=0.8x4.4 μm^2) shown in Fig. 5.33. [After Babcock et al. [47]]

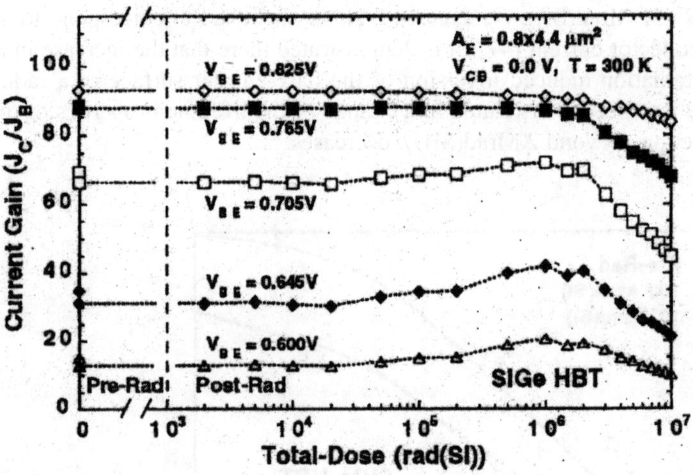

Fig. 5.35. Current gain at 300 K of the SiGe HBT shown in Fig. 3.22 as a function of total radiation dose at constant base-to-emitter voltage V_{BE}. [After Babcock et al. [47]]

From the area/perimeter dependence of the current gain degradation it is concluded that for doses below 2 Mrad(Si) the changes are governed by changes in the emitter-base bulk properties. For higher doses, perimeter effects start to be discernable, pointing towards a positive charging of the spacer oxides [47]. There is also evidence that charging of the trench oxides occurs for high doses, which enhances the Miller and/or base-to-emitter capacitance [47].

A striking feature in Fig. 5.36 is the absence of any clear low dose-rate effect. It is believed that this is related to the robust architecture of the SiGe HBTs studied and more specifically the narrow spacer oxide/nitride composite stack. In addition, the heavily doped extrinsic base layer (>10^{19} cm^{-3} peak doping) directly underneath the spacer effectively confines the ionization damage of the base-emitter spacer and helps prevent degradation of the base current ideality factor under ionising irradiation. It has in fact been observed that a so-called anomalous low dose-rate effect occurs (Fig. 5.37) [48], whereby a reduction of the base current is seen for low V_{BE} and low total dose. It is speculated that this could be due to a transformation of the pre-irradiation deep-level traps to less effective shallower levels. Moving the recombination centers further away from mid gap lowers the surface recombination rate and vice versa [48]. It has in fact been found that this base-current reduction at

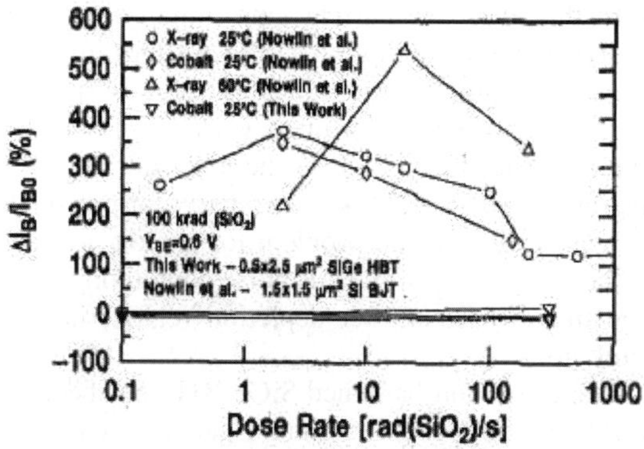

Fig. 5.36. Comparison of base current change with data from existing literature. [After Banerjee et al. [48]]

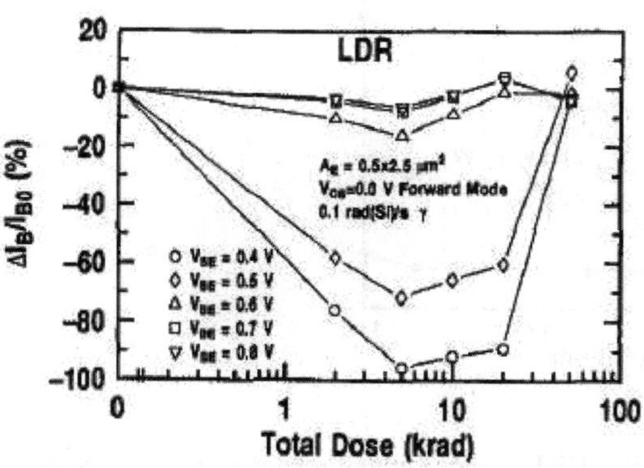

Fig. 5.37. Normalised base current change for the SiGe HBTs in forward mode (i.e., the collector at negative bias with respect to the base) for low dose rate. [After Banerjee et al. [48]]

low total dose also occurs for high(er) dose rates, although it is less pronounced. This points to the fact that two competing GR mechanisms are active in these devices under irradiation conditions

The impact of vertical and lateral scaling of the HBT technology has been explored in [50]. It was concluded that the 0.18 μm technology generation for example shows a much stronger degradation than the 0.5 μm robust technology described above [46 – 48]. A further hardening requires an optimisation of the emitter-base spacer. Nevertheless, in the corresponding BiCMOS technology, it are the MOSFETs which are the weaker parts, suffering significant degradation at far lower total dose.

The radiation response after 77 K exposure (important for certain space applications) is shown in Fig. 5.38. In this case, the current gain initially decreases slightly, but recovers for doses >1 Mrad(Si). This is related to a slight improvement of the post-radiation I_B characteristics and a slight increase in I_C [47]. It has finally been observed that after irradiation, the low frequency 1/*f* noise increases and GR noise components can be introduced (Fig. 5.39). The latter are related to the occurrence of Random Telegraph Signals (RTSs) in the base current of the transistor [46, 47]. The noise increase is a potential problem for analog or mixed mode applications of BiCMOS. The occurrence of the GR noise could point to the creation of displacement damage in the base-emitter depletion region, i.e., the formation of radiation-induced bulk traps.

The impact of displacement damage on non-hardened SiGe BJTs has been extensively studied by *Ohyama et al.* [51]. The resulting current-gain damage factor for neutron irradiation is represented in Fig. 5.40, showing a marked impact of dis-

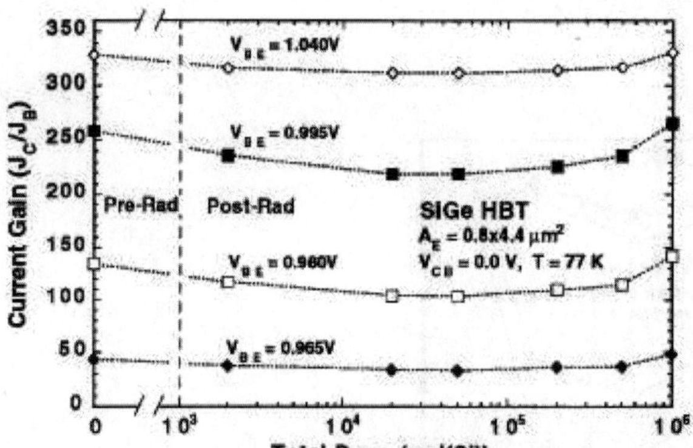

Fig. 5.38. Current gain at 77 K of a SiGe HBT as a function of total dose radiation at constant V_{BE}. Note that β is larger at 77 K than at 300 K (Fig. 5.35) for this device, which has been optimized for cryogenic operation. [After Babcock et al. [47]]

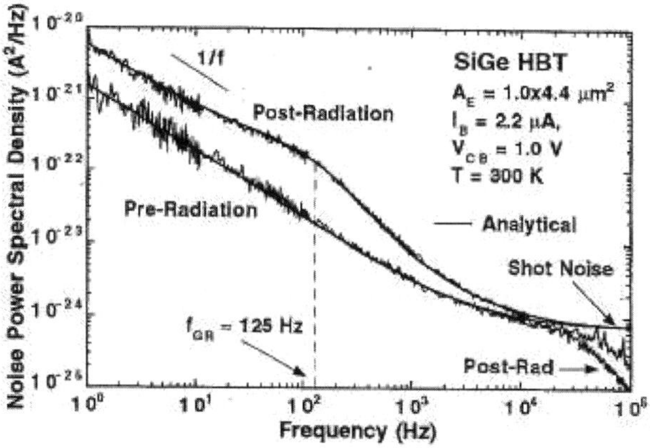

Fig. 5.39. Equivalent input referred base noise power spectral density S_{IB} measured pre- and post-radiation at 300 K for a SiGe HBT (A_E=1.0x4.4 μm^2) showing a radiation induced GR centre. [After Babcock et al. [47]]

placement damage in the base-emitter depletion region. On the other hand, hardly no effect is seen for the commercial 0.5 μm technology studied by *Cressler's* Group [49]. Similar results were obtained for silicon BJTs made on the same wafer, emphasising the robustness against displacement damage effects. Additionally, the Ge doping profile used in the base, i.e., a box or a graded profile (see Fig. 3.23) has little impact on the radiation response [49].

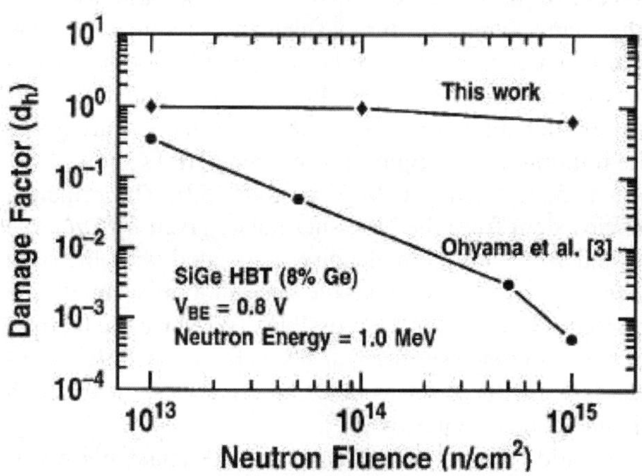

Fig. 5.40. Damage factor as a function of neutron fluence for a SiGe HBT with 8 % Ge content. [After Roldán et al. [49]]

5.5.2 Degradation of the RF Characteristics

Since one of the main application fields of HBTs is RF telecommunications systems, it is important to study the impact of irradiation on the high-frequency behaviour. The RF figures-of-merit of a transistor are [52]: the cutoff frequency f_T, the maximum oscillation frequency f_{max}, power gain, total emitter and base resistance r_{BE} and noise parameters (e.g., the minimum noise figure NF_{min} and the phase noise). The cutoff frequency f_T is the frequency at which the short-circuit common emitter ac current gain extrapolates to unity. Physically f_T can be related to the device profile via the total emitter-to-collector delay time. The maximum oscillation frequency f_{max} is defined as the frequency at which the maximum available gain extrapolates to 0 dB. It can be related to the device parameters as follows:

$$f_{max} = \sqrt{\frac{f_T}{8\pi C_{CB} r_B}} \tag{5.10}$$

with r_B the base resistance and C_{CB} the collector-base junction capacitance.

The degradation of SiGe HBTs in the GHz regime has been investigated for gamma [53, 54], neutron [49] and for proton irradiations [52]. According to Fig. 5.41, negligible degradation is observed for the cutoff frequency f_T, while a stronger degradation was found for the maximum oscillation frequency f_{max}, which was more pronounced for the HBTs (compare Fig. 5.41a and b). According to (5.10) this degradation is either due to a change in the total base resistance r_B or in the total base-collector capacitance C_{CB}, for constant cutoff frequency. Analyzing the *S* parameters in more detail shows that r_B increases after irradiation, although this is not related to an increase in the sheet resistance of the SiGe base layer. The constancy of f_T, on the other hand, points to the fact that the lifetime and transit (transport) parameters in the intrinsic transistor are not affected by the irradiation. This is confirmed by Early voltage measurements, which do not indicate the creation of bulk traps in the base-emitter depletion region after neutron exposure [49].

It was also found that the minimum noise figure for the same HBTs shows little degradation up to a fluence of 5×10^{13} cm^{-2} 46 MeV protons [52]. The principal noise sources at RF frequencies stem from the base shot noise, given by $2qI_B$, the collector shot noise ($2qI_C$) and from the thermal noise associated with the total base-emitter resistance $r_{BE}=r_B+r_E$. In order to make the base shot noise small, the current gain of the transistor needs to be as high as possible. To reduce the thermal noise, the small signal total base-emitter resistance r_{BE} needs to be small. It is concluded that the excellent hardness of the noise is due to the hardness of the high frequency (HF) current gain and cutoff frequency [52].

The impact of gamma irradiation up to 1 Mrad(Si) on the HF phase noise was reported as well [54]. Also this parameter was shown to be quite robust against radiation. The reason for this hardness is related to the minor increase in the LF $1/f$ noise and the device non-linearities. Additionally, it was demonstrated that there is a device non-linearity cancellation mechanism, which maintains a low phase noise up to 1 Mrad(Si). A direct comparison between SiGe and GaAs HBTs dedicated

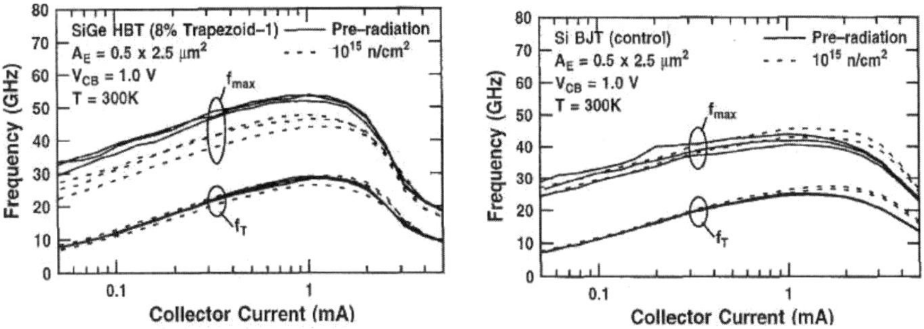

(a) (b)

Fig. 5.41. Maximum oscillation frequency and cutoff frequency as a function of collector current for the SiGe HBT (a) and the Si BJT (b), before and after irradiation. [After Roldán et al. [49]]

for RF operation, proves that both technologies are total-dose hard up to a level of 1 Mrad(Si) [53]. Given the cost, manufacturability and yield/reliability advantages offered by a SiGe HBT technology for the same performance, it is clear that this is the technology of choice for RF system-on-chip solutions for space

5.6 Conclusions

From the above, one can conclude that the main total dose damage mechanisms of modern bipolar transistors and circuits are quite well established and understood. The current gain degradation is mainly caused by the radiation-induced excess base current, which in turn is generated predominantly by the positive charging of the spacer oxide covering the base-emitter junction. The specific nature of this oxide together with the low electric field present explains the higher degradation at low dose rates, compared with high dose rate irradiations. Identification of this mechanism also leads the way to improve the hardening of bipolar technologies. At the same time, it is expected that downscaling will enhance the radiation resistance. One critical point may be the hardness assurance testing, although performing the irradiations at high temperatures (around 100°C) can solve this matter. Another merging issue is the effect of doping neutralisation by radiation-released hydrogen from the gate. This problem could be solved by hardening the spacer oxide, e.g., by the implementation of a nitrided oxide layer close to the $Si\text{-}SiO_2$ interface (see also Chap. 6). Nitrogen is known to provide a barrier for the diffusion of hydrogen, which could prevent H^+ or H^0 to reach the silicon substrate.

SiGe BJTs have shown to offer competitive if not better radiation resistance compared with so-called hardened silicon BJTs. For these devices, suitable for microwave and RF applications, no low dose-rate effect was found. In fact, for low total doses even the opposite has been observed: a reduction of the base current at low base-emitter voltages, implying an increase of the current gain. In addition, also the high-frequency figures-of-merit were found to behave excellently under various radiation environments. A comparable hardness has been demonstrated as for GaAs HBTs. For similar performance, it is clear that SiGe HBTs are the technology of choice for RF system-on-chip applications.

6 Radiation Damage in Silicon MOS Devices

6.1 Introduction

Since the early 90ties, a strong tendency exists to replace hardened electronics for space by so-called "Custom's Of The Shelf" components (COTS), fabricated in a mainstream commercial CMOS technology. One strong motivation for this strategy is the expectation that scaled technologies should inherently be more radiation hard. This relies mainly on the fact that along with the electrical channel length L of the MOS transistors, also the thickness of the gate dielectric t_{ox} is reduced, for every new technology node. Consequently the amount of positive trapped-hole charge during irradiation, will reduce accordingly. This follows from the expression for the threshold voltage (or flat band voltage) shift corresponding to radiation-induced hole trapping [1]:

$$\Delta V_{ot} = \Delta V_T = \frac{q}{\varepsilon_{ox}\varepsilon_0}[b_h(t_{ox} - 2h_1)]\frac{t_{ox}}{2} \tag{6.1}$$

with b_h the fraction of the volume density of holes created throughout the oxide which becomes trapped to form fixed positive charge (hole yield). The parameter h_1 is the distance from the Si-SiO_2 interface into the oxide, where trapped holes can recombine with electrons tunneling from the substrate or from the gate. It is approximately equal to 3 nm for reasonable times t between irradiation and measurement (n varies according to ln t). Equation (6.1) is valid for positive gate bias during the exposure and shows a quadratic dependence of ΔV_{ot} on t_{ox}. For thin and ultra-thin oxides smaller than 2xh_1 (6 nm) essentially no net hole trapping will occur during irradiation, which is indeed found in practice [2]. In case a hole gets trapped near the interface under positive gate bias, it will immediately recombine with an electron tunneling from the substrate, thereby removing all fixed charge. Moreover, a similar trend has been found for the formation of interface traps in thin oxides, whereby $\Delta D_{it} \sim t_{ox}^{\kappa}$ with κ usually between 1 and 2 for fairly thick oxides and $\kappa >> 1$ for $t_{ox} < 10$ nm [3].

Based on these observations, one could come to the conclusion that deep submicron CMOS technologies are radiation hard - particularly for low dose rate space applications, allowing some recovery of trapped charge during exposure - and no extra measures need to be taken anymore. The reality is of course somewhat more complicated than that, as will be reviewed below. Not only charge trapping but also interface state generation during and following irradiation is of concern. Particularly for scaled devices, this can lead to a non-homogeneous and more severe device degradation, as will be shown in Sect. 6.2. In addition, deep submicron processing utilizes a number of aggressive steps, which expose the gate dielectric to radiation or electrical damage - one can think of plasma etching, e-beam, ion and X-ray lithography, etc. This can generate either direct radiation damage or latent damage in the oxide which acts as precursor for radiation induced traps. These aspects will be discussed in Sect. 6.3. Although ionising damage may be of concern for certain applications, reliability and lifetime issues are of more importance for CMOS devices and circuits. The need for a reliable gate dielectric for scaled technologies has triggered the development of alternatives, like nitrided and/or reoxidised nitrided oxides. Their response to radiation will be briefly discussed in Sect. 6.4. The trend to use more exotic gate dielectrics is addressed in Chap. 9. For ultra thin oxides, some new radiation-induced phenomena occur, which may be an issue for large total doses or heavy ion irradiations (Sect. 6.5). Device isolation is still the Achilles' heel with respect to the radiation hardness of MOS transsistors. The fact that for deep submicron technologies the for decades commonly used LOCal Oxidation of Silicon (LOCOS) is being replaced by so-called Shallow Trench Isolation (STI) puts the question of radiation hardness back on the blackboard. This and related issues will be dealt with in Sect. 6.6. Because of the importance of silicon-on-insulator (SOI) for deep submicron CMOS, a brief discussion of SOI is given in Sect. 6.7. For an introduction into the basic radiation damage mechanisms in MOS devices, the Reader is referred to some books in the field [4 – 5].

6.2 Impact of Scaling on the Radiation Response

Scaling of a CMOS technology first of all implies a reduction of the minimum line width defined by lithography. In the past, only optical lithography was used in a production environment. However, since the minimum feature size has become smaller than the wavelength of the visible (ultra-violet) light, other techniques, like e-beam or X-ray lithography are considered as viable solutions for the sub 100 nm era. Besides a scaling of the channel length *L* (i.e., the distance between the source and the drain of the transistor), other technological features have to scale along. If the length is scaling with a factor k_s>1, the same applies for the gate oxide thickness, for the device spacing (isolation length), the junction depth, etc. However, in order to maintain an acceptable gate control over the current flow in the channel, one has to deal with the so-called short-channel effects, like drain-induced-barrier lowering (DIBL), device punch-through and so one. In addition, also hot-carrier degradation and reliability become important issues, if the supply voltage and the maximum electric field are not properly scaled. Whether there is also an impact of scaling on the radiation hardness will be dealt with in this section. Whenever necessary, some basic aspects of the MOS device physics involved will be explained.

6.2.1 Gate Length Dependence

Early studies on the impact of radiation on short channel effects were rather controversial: while in some reports a change was observed [6 - 8], other investigations did not reveal a noticeable effect [9 - 10]. Later it was found that it strongly depends on technological details, whether or not the short channel behavior is affected by irradiation [11 - 12]. In a first attempt to model this issue, *Huang et al.* [7, 8] considered the charge-sharing concept, depicted in Fig. 6.1, in order to calculate the impact of positive fixed charge on the flat band and threshold voltage of a short channel nMOSFET. They found that the corresponding change becomes:

$$\Delta V_{\mathrm{T}} = \Delta V_{\mathrm{fb}} = \frac{\Delta Q_{\mathrm{ot}}}{C_{\mathrm{ox}}} \frac{L}{L - W_{\mathrm{d}}} \tag{6.2}$$

with C_{ox} the oxide capacitance per unit of area and ΔQ_{ot} a certain amount of trapped oxide charge (C/cm^2). As long as the device length L is much larger than the width of the source/drain depletion region W_{d}, the induced flat-band voltage shift is length independent. An increased short channel effect is expected from (6.2) for shorter devices. A stronger V_{T} shift has indeed been noticed for certain CMOS technologies, as shown in Fig. 6.2 [12]. In addition, the subthreshold leakage current (off-state leakage) also increases more severely for shorter n-channel transistors (Fig. 6.3). This means that the standby power consumption is higher for irradiated short channel transistors than for long devices. Studying in more detail the components contributing to the threshold voltage shift, i.e., the creation of positive oxide charge (ΔV_{ot}) and of negative interface charge (ΔV_{it}) in case of an nMOSFET reveals that it is mainly ΔV_{ot} which increases for smaller L, while ΔV_{it} is relatively insensitive to this parameter [12, 13]. In line with this ΔV_{ot} effect, the short p-channel counterparts show a more pronounced negative shift of the threshold voltage. It has also been observed that Q_{ot} exhibits an asymmetry when the irradiation is performed with a non-zero drain bias [14]. The differences are most pronounced in the subthreshold regime and for smaller gate lengths. The asymmetry in the forward-reverse characteristics is explained by assuming after irradiation a non-uniform charge distribution in the oxide along the channel.

Two-dimensional numerical simulations [14, 15] have studied the charge distribution. By solving the Poisson and continuity equations in the oxide and the semiconductor, it is possible to demonstrate that for an nMOSFET, the oxide-trapped positive charge is slightly higher near the drain when irradiated in the triode regime, while the opposite is true when irradiated under saturation [15]. This is explained by the sign and magnitude of the electric field in the oxide. It results in a large lateral non-uniformity of the trapped charge, for irradiations in saturation ($V_{\mathrm{DS}} \gg V_{\mathrm{GS}}$). In that case, after exposure the threshold voltage near the drain is higher than near the source and, furthermore, depends on the drain bias, i.e., V_{T} reduces with increasing V_{DS} [15]. Overall, the charge non-uniformity increases with gate voltage and with total dose. For p-channel devices, both numerical simulations and measurements have shown that more positive charge is trapped

lations and measurements have shown that more positive charge is trapped near the drain, irrespective of the operating condition during the exposure. This is because the electric field in the oxide is more positive close to the drain, for all drain voltages [15].

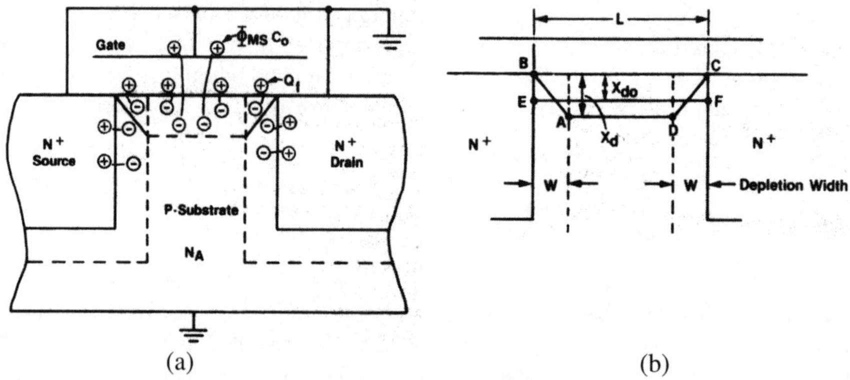

Fig. 6.1. Cross-sectional view of a MOS model: (a) Pictorial view of charge sharing concept. (b) Geometry used for the calculations. [After Huang and Schrankler [7]]

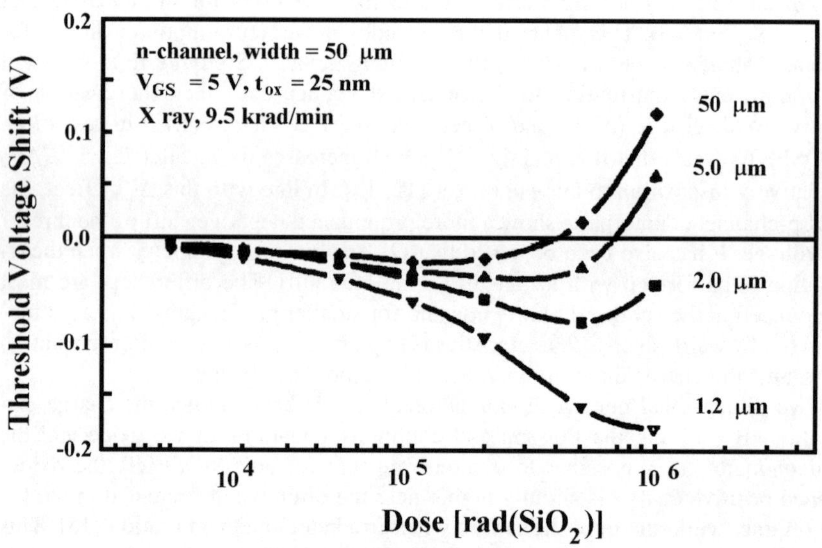

Fig. 6.2. Threshold voltage change ΔV_T for n-channel transistors irradiated to 1 Mrad(SiO_2) at 5 V with lengths of 50 μm to 1.2 μm and widths of 50 μm. [After Shaneyfelt et al. [12]]

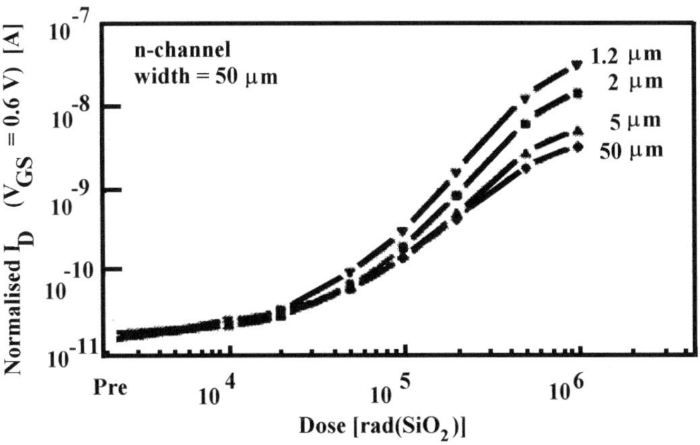

Fig. 6.3. Drain-to-source current at V_{GS}=0.6 V for n-channel transistors. The current has been normalized by *L*/*W* for transistor size. [After Shaneyfelt et al. [12]]

The post-irradiation annealing of trapped positive charge under 5 V gate bias appears to be similar for long and short devices, although the net V_T shift is markedly different (Fig. 6.4) [12]. The rebound observed in Fig. 6.4, whereby the threshold voltage of the nMOSFET becomes larger (i.e., more positive) than before irradiation is less for a short compared with a longer channel device. This may be an advantage for shorter devices for the low dose rate environments, typical for space. However, the increase in off state leakage and power consumption may be a serious drawback, particularly for present-day low power/low voltage applications. Based on the low-frequency (LF) noise results shown in Fig. 6.5, it can be concluded that for short channels, there is an enhanced oxide trapped charge buildup [12]. This comes from the fact that the LF noise magnitude is proportional to the so-called near interface oxide traps, also called "slow" or "border" traps [16]. The stronger radiation induced increase in the normalized noise (normalized for the bias voltage and the device area *A* !) observed for the short channel transistors in Fig. 6.5 is an additional serious drawback for analogue applications of scaled technologies. The fact that more oxide trapping occurs for shorter lengths demonstrates that (6.2) is too simple an approximation to describe the length dependence, since ΔQ_{ot} also varies with *L*.

In the literature, some reports can also be found on the radiation response of narrow-width devices (see e.g. [17]), a topic that is not further discussed here.

6.2.2 Lateral Non-Homogeneous Damaging

As mentioned in the previous section, an additional factor contributing to the enhanced degradation of short devices is the non-homogeneous nature of the damage along the channel, whereby clearly different densities of interface and oxide trapped charge have been found near the source and drain junctions [18 - 20]. This has been revealed by applying a number of dedicated techniques with lateral resolution along the Si-SiO_2 interface, based on a modified charge pumping technique [18], the series resistance extraction of an *L*-array [19, 20] and the study of the so-called gate-induced drain leakage (GIDL) current [19, 20]. The latter technique is discussed in Sect. 6.2.3

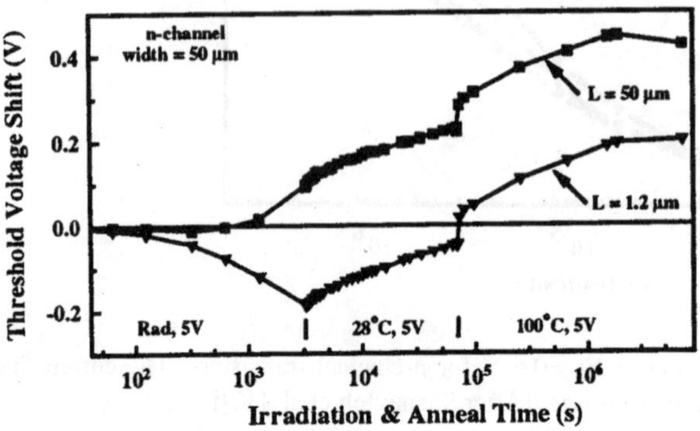

Fig. 6.4. ΔV_T for n-channel transistors with lengths of 1.2 and 50 µm, irradiated to 500 krad(SiO_2) and annealed at 28°C and 100°C. During irradiation and annealing, the gate bias was +5 V. [After Shaneyfelt et al. [12]]

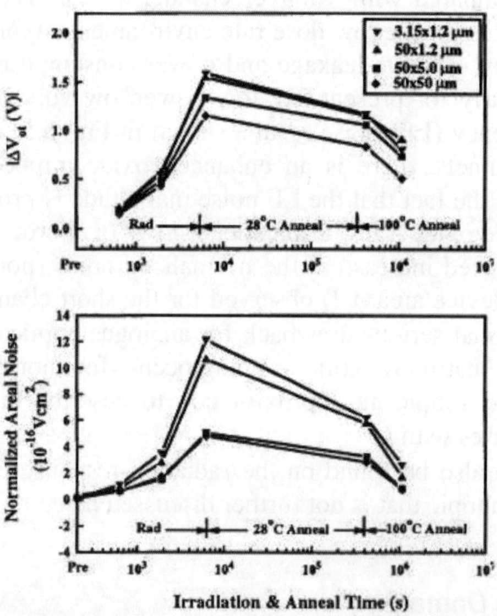

Fig. 6.5. ΔV_{ot} and normalized areal noise power as a function of irradiation and anneal times for n-channel transistors. [After Shaneyfelt et al. [12]]

Figure 6.6 again illustrates the length dependence of the negative threshold voltage shift for nMOSFETs with different 18 nm gate oxides: fluorinated oxide, standard control dry oxide and a hardened oxide [20]. At the same time, the transconductance g_m is reduced after irradiation, whereby the largest effect is seen for the shorter devices, and, for control oxides. Note that in Fig. 6.6 the control and the fluorinated devices show a similar tendency, although the magnitude of the V_T change is different. The rad-hard MOSFETs show the opposite trend, i.e., increased degradation for the shorter lengths. The latter is in line with previous observations for non-hardened oxides [8, 12].

The channel resistance R_{chan} versus the channel length for the different gate oxides, before and after exposure to 1 Mrad(SiO_2) X-rays, is given in Fig. 6.7. After irradiation R_{chan} is in first order given by [19]:

$$R_{chan} = G_{chan}^{-1} = \frac{L_{eff} - \Delta L}{W_{eff}} (\mu_{n,p} C_{ox} [V_{GS} - V_T])^{-1} \tag{6.3}$$

with $\mu_{n,p}$ the electron (n) or hole (p) effective low-field mobility, L_{eff} and W_{eff} the effective (or electrical) length and width of the transistor. L_{eff} is equal to the designed (or mask) channel length minus a channel length reduction δL which is technology dependent, but constant for devices with different L on the same wafer

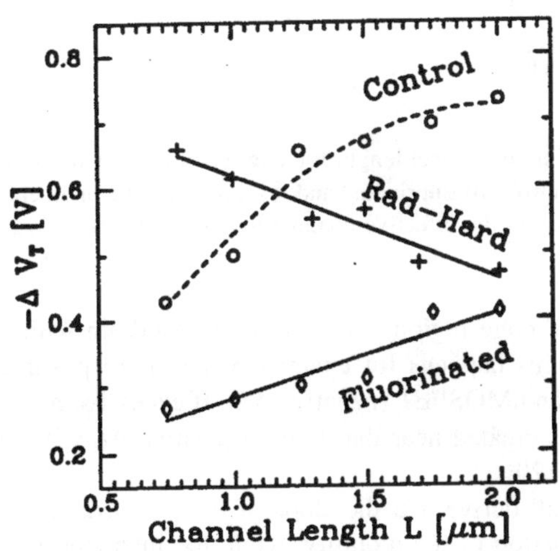

Fig. 6.6. Channel length dependent threshold voltage shifts in irradiated nMOSFETs. During irradiation, a gate bias of +5 V with respect to the substrate was applied. t_{ox}=18 nm. [After Balasinski and Ma [20]]

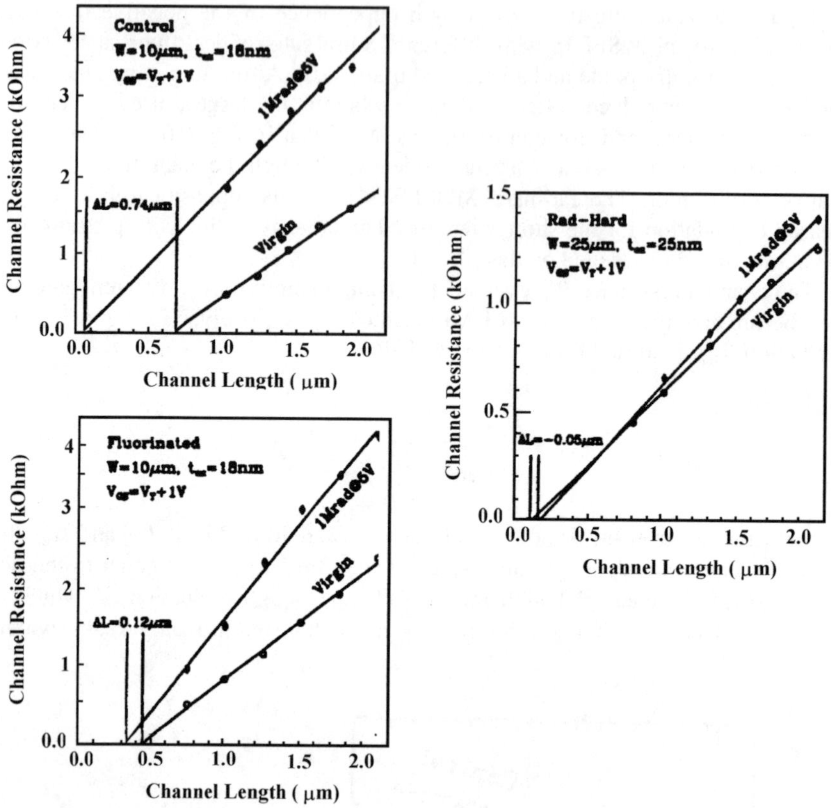

Fig. 6.7. Channel resistance as a function of channel length for virgin and irradiated MOSFETs with different gate oxides: (a) control, (b) fluorinated and (c) rad-hard. The intercepts at the horizontal axis reveal the changes in the effective channel length. [After Balasinski and Ma [20]]

The ΔL in (6.3) corresponds to the edge region, adjacent to source/drain with a lower V_T than the bulk channel. This happens for example when more positive charge is trapped near the drain of an nMOSFET (negative ΔL). If excessive negative charge (e.g. interface states) is created near the drain, a positive ΔL will be found, resulting in a shorter L_{eff} [19, 20].

From Fig. 6.7 one can first of all derive that the slope of the post-irradiation curve is larger, indicating a degradation of the mobility, while the intercept with the *x*-axis has shifted by an amount ΔL, which strongly depends on the gate oxidation conditions. For the control and fluorinated case a positive ΔL points to an excessive negative charge near the source/drain, while a higher positive charge is expected for the radiation-hard devices (negative ΔL). The origin of this excessive charge near the source and drain regions can be studied by using a modified charge-pumping analysis, which provides the lateral distribution of the interface and oxide traps (Fig. 6.8) [20]. It is clear that the control and fluorinated nMOS-

FETs give rise to a lower trapped hole charge near the drain, while the opposite is observed for the radiation hard devices. In contrast, the density of negatively charged (acceptor) interface traps increases significantly when the drain is approached for control/fluorinated devices, while a more or less flat profile is found for the hard transistors (Fig. 6.8b). As a net result, more positive trapped charge is present for the hardened nMOSFETs, giving furthermore rise to a larger negative shift for the shorter device lengths in Fig. 6.6. This is due to the fact that for smaller gate lengths, the contribution of the edge regions to the overall degradation becomes more important. A similar argument holds for the control and the fluorinated splits. Computer simulations point out that the peak transconductance is extremely sensitive to the magnitude and position of the effective oxide charge near the junctions [20].

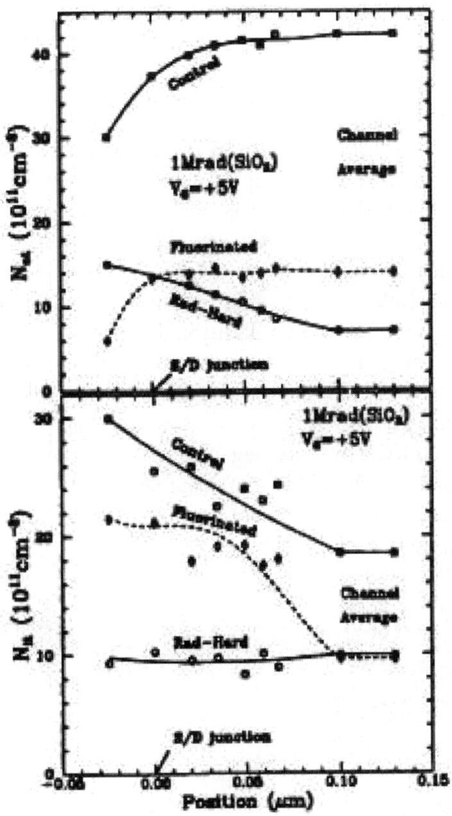

Fig. 6.8. Lateral distributions of (a) oxide trapped charge, and (b) interface traps in the vicinity of the source or drain junction after 1 Mrad(SiO_2) irradiation at V_{GS}=+5 V. [After Balasinski and Ma [20]]

The origin of this non-uniformity is unclear, since different mechanisms can be involved. As already seen above, one contributing factor is the difference in oxide field near the drain during the irradiation, related to the presence of the built-in field of the junction. Furthermore, the chemical oxide properties may be different close to source and drain, due to the presence of ion implanted dopants (As, B,...) in the oxide. In addition, there can be local variations in mechanical strain at the Si-SiO_2 interface, which is known to affect the generation of interface traps. Finally, there can be residual unannealed displacement damage in the oxide generated by previous device processing, i.e., coming from e.g. plasma etching and ion implantation.

For technologies using Lowly Doped Drain (LDD) structures covered by spacer oxides, additional radiation-induced lateral non uniformities in the degradation behavior can be created [21]. Omitting the LDD generally removes or lowers this type of degradation

6.2.3 Gate-Induced Drain Leakage (GIDL)

GIDL has emerged as a new constraint in miniaturizing MOSFET devices. Because it influences the drain current in the off state, it is of particular concern in low power circuits and DRAMs [22]. The GIDL current is observed when a gate bias is applied which accumulates the substrate, thus a negative V_{GS} for an nMOSFET, attracting the holes to the interface. A field induced p^+-n^+ junction is hence created near the drain. Due to the large field existing there, tunneling of carriers from the p^+- to the n^+-region occurs, giving rise to an additional leakage current (GIDL) as illustrated in Fig. 6.9 [22]. The GIDL current is basically independent of the device length L [23] and increases exponentially with the absolute value of the gate overdrive voltage $|V_{GS}-V_T|$ as in Fig. 6.9.

After irradiation, the GIDL curve is shifted to more negative (n-) or more positive (p-channel) V_{GS} (Fig. 6.9). This shift is parallel for an n-channel device and is similar but in opposite direction, as the shift in the subthreshold I_D-V_{GS} curve, shown in Fig. 6.10a. This follows from the fact that [22]:

$$\Delta V_T = q(N_{itacceptor} - N_{ot}^+)/C_{ox} \tag{6.4.a}$$

$$\Delta V_{GIDL/given I_D} = q(-N_{itdonor} - N_{ot}^+)/C_{ox} \tag{6.4.b}$$

while, finally:

$$\Delta V_{mg} = -qN_{ot}^+/C_{ox} \tag{6.4.c}$$

for the mid-gap voltage shift, which corresponds to neutral interface traps (zero interface charge). $N_{itdonor}$, $N_{itacceptor}$ is the density of interface traps with donor and acceptor character, respectively. The former lie in the lower half of the band gap

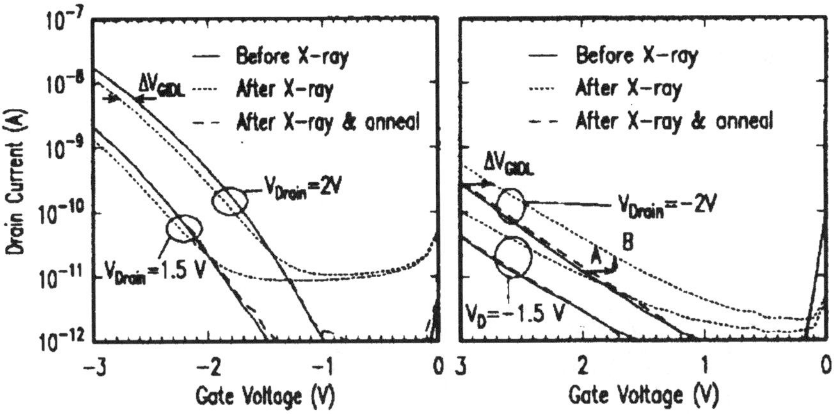

(a) (b)

Fig. 6.9. Typical effect of a 0.5 J/cm^2 X-ray irradiation on GIDL in (a) n- and (b) p-channel MOSFETs with W/L=20 μm/0.65 μm, measured at V_{DS}=±1.5 V and ±2 V. In the n-channel device, the main effect of irradiation is a small parallel shift □V_{GIDL}. In the p-channel device, irradiation shifts the $I_D(V_{GS})$ characteristics - due to trapped positive charges and positively charged interface states (A) - and increases the interface-state-assisted tunnel current (B). [After Acovic et al. [22]]

and are positively charged if the surface potential is below the level, while the acceptor states lie in the upper half and can be negatively charged or neutral, depending on the surface potential. In the case of GIDL (nMOSFET) the surface potential is near the valence band and the donors are charged; the opposite holds when the MOSFET is operated in the subthreshold regime, where the acceptor interface charge is swept from zero to negative for increasing V_{GS}. The difference in shift between the GIDL and subthreshold I_D curve can be explained by this difference in underlying interface charge [22]. From this, it is concluded that in principle, the GIDL curve can be used in a similar way as the subthreshold characteristic to extract the oxide-trapped charge near the drain after irradiation [19, 20, 22, 23]. Such a technique has first been developed to study interface-state generation by hot carriers in small MOSFETs.

For the p-channel devices, the shift of the post irradiation curve in Fig. 6.9b is not parallel and also larger than for the nMOSFETs [22]. This could point to the non-uniformity of the oxide trapped charge, already discussed above [20]. In addition, for low absolute gate bias, there is evidence for an excess GIDL after irradiation, which is more pronounced for the p-channel devices. It is believed that this is due to an additional generation mechanism, whereby interface states assist the tunneling of carriers (ITAT=interface-trap-assisted tunneling) [22, 23]. This causes the curves to shift vertically, while oxide trapped charge in first instance produces a horizontal shift, as indicated in Fig. 6.9b. The larger impact of ITAT for p-channels suggest that probably a higher density of interface traps close to the conduction band is created by irradiation, compared with close to the valence band, i.e. $N_{itacceptor}$> $N_{itdonor}$ [22].

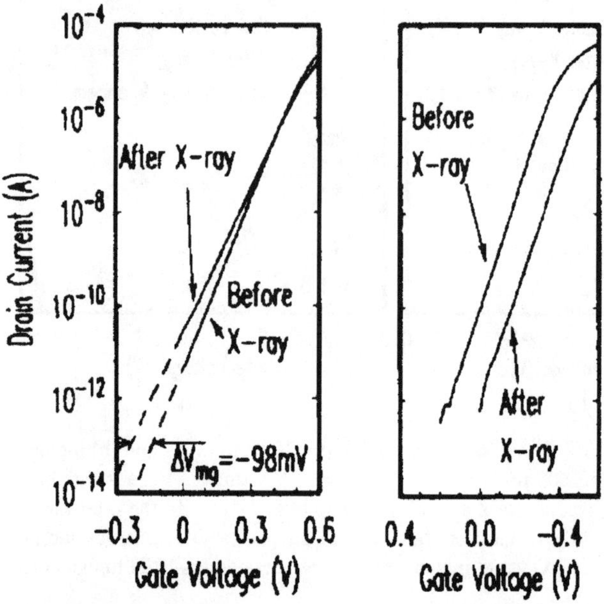

Fig. 6.10. Channel current in the MOSFETs of Fig. 6.9, at V_{DS}=±0.1 V. The dotted lines are extrapolated. In the n-channel device (a) the shift of V_T is 35 mV, and the subthreshold swing S increases from 78 to 92 mV/decade. In the p-channel device (b), ΔV_T=-211 mV and ΔS=3 mV/dec. [After Acovic et al. [22]]

Annealing studies point out that the recovery proceeds faster for the oxide charge near the drain, as derived from GIDL measurements, compared with the charge in the central channel part [23].

6.3 Processing Induced Radiation Damage Effects

Already in the early days of submicron technologies it was realized that a number of processing steps could produce some form of radiation damage, which may not be removed totally by subsequent annealing [24]. This was shown to be the case for plasma or reactive ion etching (RIE), e-beam and X-ray lithography [25 - 27] and ion implantation through the gate oxide. While it is not clear yet whether e-beam, ion-beam or X-ray lithography will ever be used in a mass production environment (maybe somewhere in the sub 100 nm era), plasma and reactive ion etching (RIE) are surely common practice. Other advanced processing steps have emerged for deep submicron technologies, which again raise concerns with respect to radiation damage. Some of them will be critically reviewed.

6.3.1 Plasma Damage

Typical submicron CMOS processes employ much plasma-enhanced chemical vapor deposition (PECVD) and RIE steps, which could produce electrical stress and damage to the thin gate oxides. As a result, interface states and hole and electron traps are created by the Fowler-Nordheim (F-N) tunneling current flowing during the exposure to the plasma, which in a later stage can cause time-dependent dielectric breakdown (TDDB) and long term reliability problems. In addition, it is expected that also the radiation hardness of (sub)quarter micron CMOS could be at stake [28]. So far, little studies have addressed this issue. Nevertheless, from the cumulative probability plots of Fig. 6.11, it is immediately clear that plasma damaged devices show a large spread in the threshold voltage after irradiation, which is on the average also shifted more than for a non plasma-exposed wafer. Associated with the large spread and shift in V_T is a similar spread in the subthreshold swing S. The correlation between the two parameters is represented in Fig. 6.12 [28]. This suggests an enhanced hole trapping and interface-state generation in plasma damaged devices, therefore, seriously jeopardizing the radiation tolerance.

6.3.2 Rapid Thermal Annealing (RTA)

It is well-known that the total dose hardness of an oxide depends critically on the thermal history after oxide growth. For example, furnace annealing of the gate oxide at temperatures above 925°C can significantly degrade the radiation tolerance [29 - 35]. It has also been observed that the addition of a small amount of O_2 to an

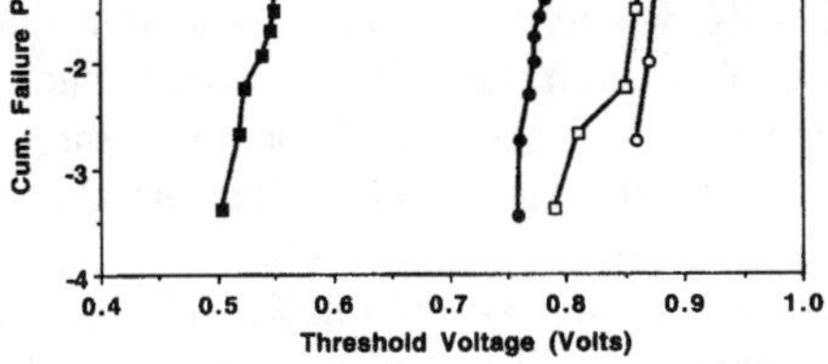

Fig. 6.11. Cumulative probability plots of threshold voltage (V_T) distribution of control and plasma-damaged samples before and after a total dose irradiation to 1 Mrad (SiO_2). nMOS devices were biased at V_{GS}=5 V and V_{DS}=V_{BS}=0 V during the irradiation. Large negative V_T shift indicates positive charge trapping in the plasma-damaged oxide. [After Yue et al. [28]]

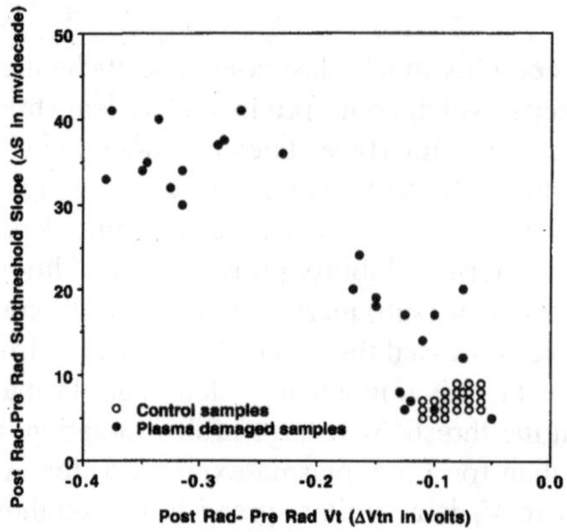

Fig. 6.12. Correlation between radiation induced threshold voltage shift (ΔV_T) and subthreshold swing shift (ΔS) with both control and plasma damaged samples. Results indicate enhanced hole and interface traps in plasma-damaged devices. [After Yue et al. [28]]

inert post-oxidation annealing (POA) reduces the buildup of trapped positive charge due to hot electrons or ionising radiation [33 - 35]. Unfortunately, there exists usually a trade-off between, on the one hand, optimizing the quality of the oxide with respect to native fixed charge and interface states and, on the other hand, improving its radiation tolerance. In general, radiation resistant gate oxides have poorer initial quality and vice versa [35].

Rapid Thermal Annealing (RTA) is routinely used in submicron processing, amongst others for the activation of dopants after ion implantation. RTA is favorable to maintain a shallow depth of the source and drain junction since the diffusion of the doping atoms is limited by the short duration of the high-temperature step. As mentioned before, ULSI technology scaling requires a reduction of the junction depths, which puts severe constraints on the allowable overall thermal budget. RTA is a powerful tool to cope with these limitations. During RTA, the temperature of a wafer is ramped in a short time (seconds) from room temperature to a value between 850 and 1050°C typically, where it is held for some annealing time (a few seconds) and then quickly cooled again. One may wonder if such a drastic procedure will affect the radiation hardness, especially since only the top surface of the wafer is heated significantly and there may exist a substantial temperature gradient across the wafer surface [35, 36].

The results of Fig. 6.13 point out that RTA has a strong impact on the oxide trapped charge in furnace-grown 18 nm oxides, but little influence on the radiation-induced interface states. At the same time, the device-to-device spread of ΔV_{ot} across a wafer is large(r), especially for higher RTA temperatures, whereby the shifts are worst at the wafer periphery. This may be partially related to the non-uniformity of the temperature during RTA, which can be controlled to within 20°C for the case studied [36]. The impact of the temperature on the radiation response is better illustrated in Fig. 6.14, showing the largest negative V_T shift for the highest RTA temperature.

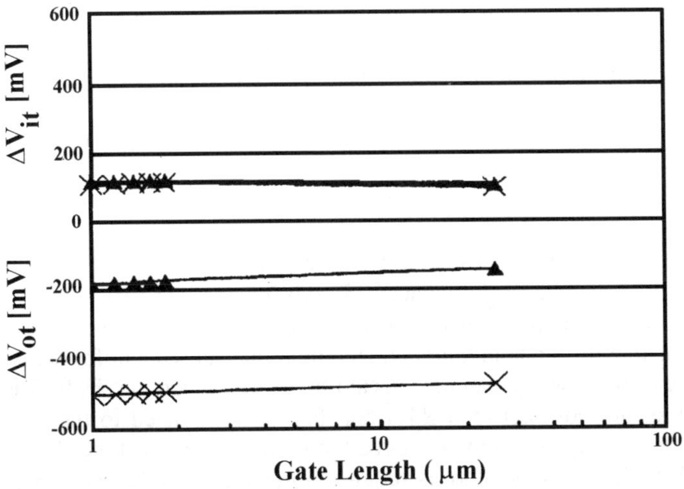

Fig. 6.13. Interface trap (ΔV_{it}) and oxide charge (ΔV_{ot}) contributions to the V_T shift after 1 Mrad(SiO_2) versus gate length of nMOS transistors from wafers with 1050°C RTA (x) and without RTA (triangles). t_{ox}=18 nm, covered by a poly-Si gate. [After Flament et al. [36]]

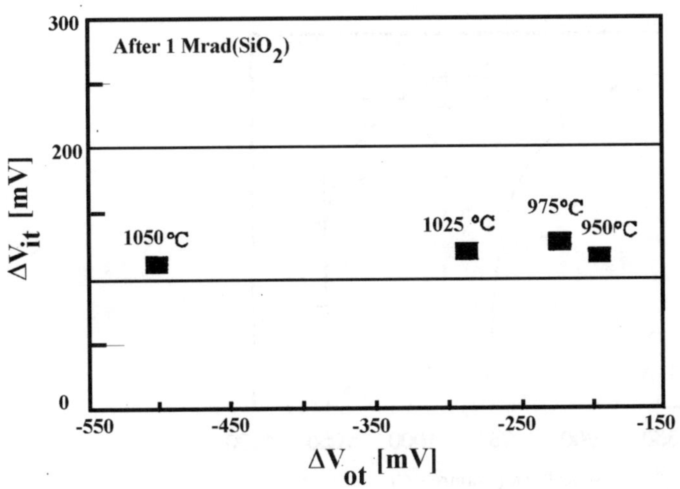

Fig. 6.14. ΔV_{it}-ΔV_{ot} plot for various RTA conditions in nMOS transistors. [After Flament et al. [36]]

In Fig. 6.15, the effect of an RTA post oxidation anneal on the V_T shift is compared with a PO furnace anneal. In both cases, negligible impact is observed up to a certain maximum temperature, while the hardness drops sharply beyond that. For RTA, this maximum occurs around 975°C [36]. Figure 6.15 implies that for a small temperature variation across a wafer during RTA, a strong spread in radiation response can occur for annealing temperatures beyond that threshold. A similar behavior was noted before for PO annealed 22-25 nm RT oxides grown at 1075°C [35]. To pinpoint the mechanism underlying the observations, the Arrhenius plot of Fig. 6.16 is very instructive. The same defect creation activation energy E_a of ≈3 eV was found for both RTA and FA devices, strongly suggesting that the intrinsic oxide properties are responsible for the hardness degradation behavior.

The 3 eV activation energy comes close to the activation energy for the out-diffusion of interstitial oxygen to the adjacent Si substrate (≈2.54 eV), whereby oxygen vacancies are left behind near the interface. The latter centres have been shown to be responsible for the hole trapping, leading to the subsequent formation of *E'* paramagnetic centres, which can be detected by room temperature EPR. The shift in temperature threshold between RTA and FA is then related to the time necessary for sufficient oxygen atoms to leave the oxide. Since the time during RTA is restricted, a higher temperature is required to create a similar amount of oxygen vacancies. The dispersion observed across a wafer can be understood partly by assuming a temperature gradient. However, an additional factor (thermal stress ?) needs to be invoked, in order to explain the large magnitude of the spread.

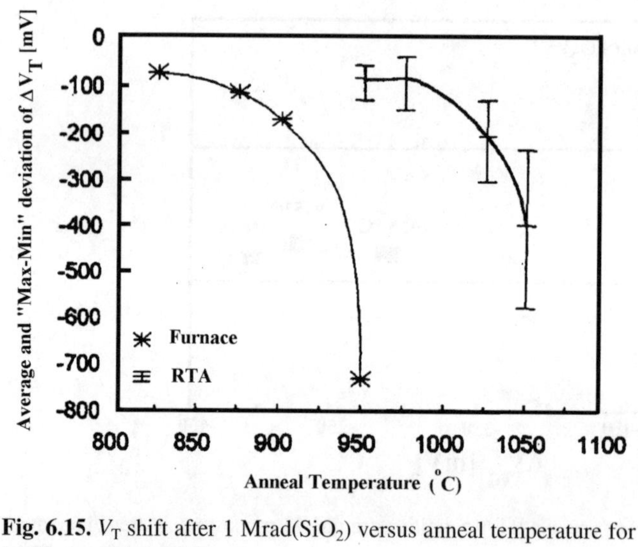

Fig. 6.15. V_T shift after 1 Mrad(SiO_2) versus anneal temperature for nMOS transistors. [After Flament et al. [36]]

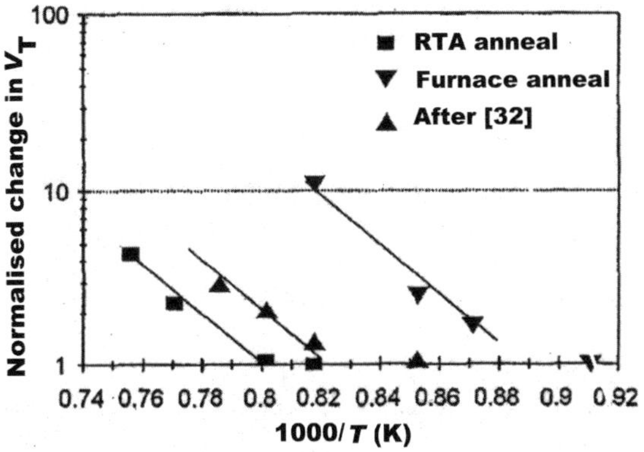

Fig. 6.16. Arrhenius plot of V_T shift after 1 Mrad(SiO_2) for nMOS transistors. t_{ox}=18 nm. [After Flament et al. [36]]

6.3.3 Gate Material and Contacting

Optimizing submicron CMOS requires in many cases the use of a double flavor polysilicon (i.e., partially n^+ doped and partially p^+ doped) gate material, in order to balance the parameters (V_T, g_m,...) of the n- and p-channel devices. For nMOSFETs the standard n^+ poly is applied to the gate, while a surface channel pMOSFET is obtained if a B doped p^+ polysilicon gate is fabricated. This is in contrast to the classical n^+ single-poly pMOSFET, which has rather a buried-channel character and, therefore, a worse control over the short-channel effects. The use of a double-poly technology has a further implication, namely, one has to switch over to silicidation instead of Al metallisation of the gate. The latter would otherwise lead to a p-n junction in the polysilicon interconnect lines.

The implementation of a p^+ poly gate brings about two major changes: first, there is the workfunction difference, which is about equal to the silicon band gap (≈1.12 eV). Second, the doping of the polysilicon is usually achieved by a B ion implantation, which might introduce B in the oxide, especially during the subsequent activation anneal of the dopants. These two factors can change the hardness of the oxide [37]. A thorough investigation pointed out that from an oxide-trapped charge viewpoint p^+ poly gate devices actually show a better radiation response over a wide range of processing conditions, i.e., B implant dose. This is represented in Fig. 6.17a. The n^+ poly devices received after the B implant a phosphorous doping from a $POCl_3$ source. A slightly different trend is observed with respect to the radiation-induced formation of interface states in the upper half of the band gap. There seems to exist an optimum implantation dose for B implanted n^+ poly devices (Fig. 6.17b), which were processed to allow a comparison with their

p^+ counterparts and to separate out the effect of the B implantation from the change in work function difference on the radiation behavior. It was also observed that a maximum ΔV_{it} was obtained for zero field during the exposure [37].

The net result of ΔV_{ot} and ΔV_{it} is strongly in favor of the p^+ poly compared with the commonly used n^+ material, so that p^+ pMOSFETs are significantly harder than n^+ pMOSFETs in older technology generations. The presence of the B in the oxide plays a certain role in the hardening, as found from the B implanted n^+ capacitors. It is speculated that B related (electron) traps could enhance electron-hole recombination during irradiation, so that less holes are available for subsequent deep trapping in oxygen vacancies near the interface. The trend for increasing hole trapping with increasing B dose could point to the presence of increasing residual ion implantation damage. The better behavior at higher doses for the n^+ poly may come from a further damage annealing during the subsequent P-diffusion doping used. The increase of interface-state formation with dose is most likely related to the presence of Si-B bonds, which can break up to leave behind a silicon dangling bond (DB) [37].

There is also interest in replacing polysilicon by amorphous silicon, deposited at lower temperatures [38]. It has been shown that this affects the formation of interface states following ionising irradiation: a significant reduction of D_{it} has been reported for amorphous gates. The improved hardness is assigned to the role of the mechanical stress associated with the amorphous gate. It is believed that a higher compressive stress is created in the oxide, after the gate dopant drive in, compared with polysilicon gates. On the other hand, annealing of polysilicon gates above 875°C lowers the hardness of the capacitor [32]. For sufficiently high temperature anneals, the morphology of the polysilicon grains changes markedly, resulting in a different local stress.

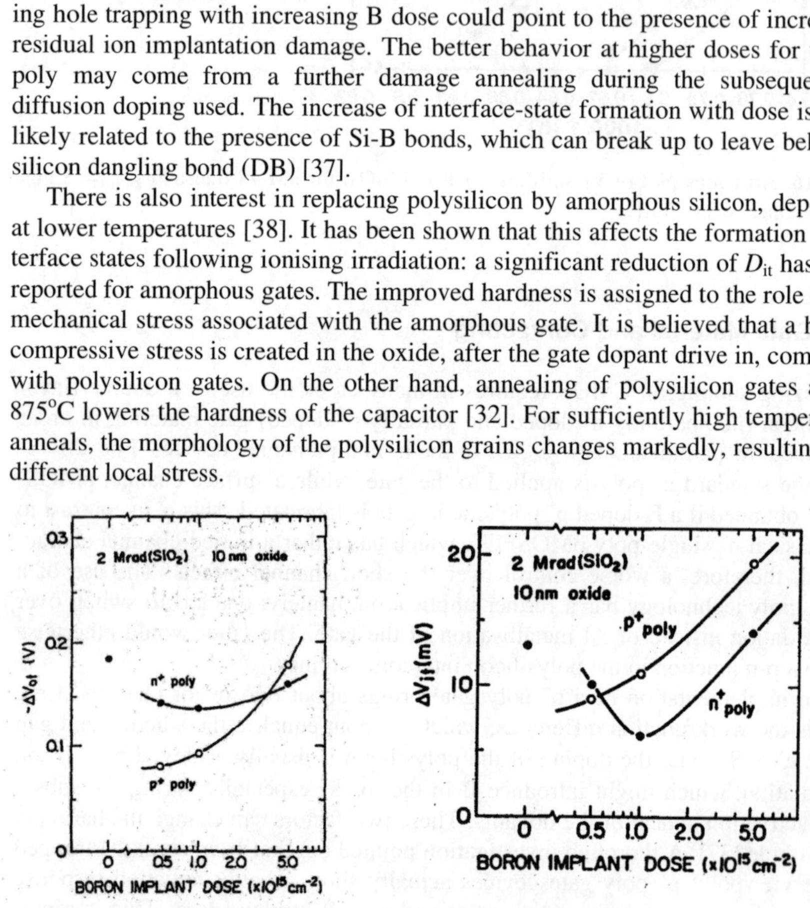

Fig. 6.17. Voltage shifts due to oxide trapped charge (a) and interface states (b) as function of the B implantation dose for p^+ and n^+ poly capacitors with 10 nm oxides. The data represent the maximum voltage shifts with respect to the applied bias. [After Yoshii et al. [37]]

For sub-0.1 μm technologies, more exotic gate materials like amorphous SiGe, characterised by a mid-gap work function or even metal gates are candidates. The choice of the gate material will closely depend on the future high-*k* materials, which will be used as gate dielectric. The impact of such a gate stack on the radiation response is not clear and requires further studies. Some additional information will be given in Chap. 9.

The role of mechanical stress on interface state formation and hole trapping has been studied in detail by *Kasama et al.* [39, 40], for different silicide-gate electrodes. It was found that compressive stress reduces both charging mechanisms and is in principle independent on the silicide material, for the same annealing conditions. The effect is most pronounced for interface-state generation and thin oxides. On the other hand, if the gate metallisation or back-end processing produces tensile stress, an increased interface state generation can result upon irradiation. This may show up as a strong gate length dependence [11].

6.4 Alternative Gate Dielectrics

This section only discussed doped oxides (Sect. 6.4.1), nitrided and reoxidised nitrided oxides (Sect. 6.4.2), and NO_2 or nitrous oxides (Sect. 6.4.3). The more exotic high-*k* gate dielectrics are briefly addressed in Chap. 9.

6.4.1 Doped Oxides

The integrity of the gate oxide is sensitive to the post-oxidation annealing temperature and ambient. Generally, annealing in an inert ambient is employed, whereby the use of a dilute fraction of O_2 can increase the total dose hardness [35]. Furthermore, it has been found that the addition of a small amount of impurities during oxidation (Cl [41], F [42, 43],...) can have a beneficial effect on the radiation response. It turns out that for most additives, an optimum concentration exists, which implies that the gate oxidation conditions need to be well-controlled. An example is given in Fig. 6.18 for 18 nm F implanted gate oxides [43], demonstrating that the best response with respect to interface-state generation is obtained for a dose around $2x10^{15}$ cm^{-2}. It is believed that the presence of F or Cl at the interface changes the local bond strain distribution. A possible mechanism leading to the reduced strain is the interaction of a F atom with a strained Si-O bond, forming Si-F bonds and a non-bridging oxygen bond, resulting in a local strain relaxation. When excessive amounts of F (or Cl) are incorporated, too many non-bridging O-bonds are formed, thereby canceling the positive effect of strain relaxation [43].

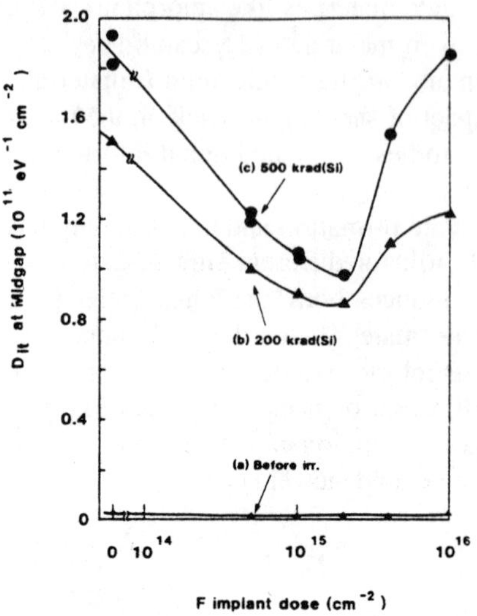

Fig. 6.18. X-ray radiation-induced mid-gap interface trap density as a function of fluorine implantation dose. Gate bias was kept floating during the irradiation. t_{ox}=18 nm. (a) Before irradiation; (b) after 200 krad(Si) and; (c) after 500 krad(Si). [After Nishioka et al. [43]]

6.4.2 Nitrided (NO) and Roxidised Nitrided Oxides (RNO)

The rate of scaling the device feature size is much stronger than the reduction in operating voltage. As a consequence, electrical fields in the oxide and near the drain become progressively higher, subjecting the MOS devices to hot-carrier effects. This endangers the integrity and the lifetime of the oxide, so that there is a need for more reliable, hot-carrier hard dielectrics. Early studies indicated that nitrided oxide (NO) could replace standard oxides for scaled technologies, as a harder, more reliable alternative [44 - 48]. Nitridation of the oxide is usually achieved by annealing it in ammonia (NH_3), using either classical furnace, or rapid thermal annealing [48]. The latter allows a better control of the nitridation conditions and thermal budget.

6.4.2.1 Properties of NO and RNO Dielectrics

Nitrogen incorporation in the oxide increases the dielectric constant of the film, allowing a thicker layer to exhibit the same capacitance as a thinner thermal oxide and thus permitting higher gate voltages before breakdown. Nitrided oxides show furthermore a lower incidence of low field breakdown. There are also some indications that NO has a good potential as a hard(er) gate dielectric, which is not only of interest for space or military applications, but also in view of the application of plasma etching, e-beam or X-ray lithography [49]. Besides improved reliability, nitrided oxides offer some other substantial advantages [47]: it has proven to be a good barrier for dopant diffusion (B, P, O,...). This feature is in fact of crucial importance for the down-scaling of p-channel devices, using BF_2 implantation for the source/drain junctions. In order to prevent boron penetration from the B-implanted p^+-polysilicon gate in deep submicron (0.18-0.13 μm) CMOS, one has to implement a nitrided gate oxide. The N content in the oxide has an important impact on both the dielectric properties and the boron penetration resistance.

However, NO has also some substantial drawbacks, which may limit large-scale implementation. First, it has been noted that NO oxides show a larger fixed oxide and interface trap charge under certain processing conditions (see [47] and references therein). In addition, the presence of a large density of native electron traps (ET) can cause considerable electron trapping under hot-carrier or FN stressing conditions. These native ETs are assumed to be related to the presence of substantial amounts of H or N close to the interface (see below for more details about the microscopic nature). Due to enhanced Coulombic scattering at these additional charge centres, the maximum effective carrier mobility will be smaller in NO compared with standard oxides, resulting in a lower maximum MOSFET transconductance. The reduction is more pronounced for n- than for p-channel devices, due to the presence of the ETs [50]. Since both type of devices are affected, it is a strong evidence that the related near-interface oxide traps have an amphoteric character, i.e., they communicate with both inversion layer holes and electrons. On the other hand, the mobility degradation with increasing gate bias V_{GS} is less pronounced for NO, resulting in a higher transconductance than for standard oxides at high transverse electric fields.

In order to improve the initial characteristics of NO, one can apply an additional reoxidation, either in a furnace [50] or a RTA system [51]. This is often abbreviated as RNO, although ONO or ROXNOX are also used frequently. The reoxidation reduces part of the excess fixed charges and ETs, although not completely, while maintaining sufficient nitrogen concentration near the interface to preserve the advantages of NO. This is illustrated in Fig. 6.19, showing Auger (AES) profiles of NO and RNO oxides [50]. NO oxides contain a nitrogen peak at the Si-SiO_2 interface and at the surface (poly-gate) of the dielectric. For sufficiently heavy nitridations, the N concentration in the bulk of the oxide also increases. Reoxidation removes both the surface peak and the bulk nitrogen, while the interface peak concentration becomes smaller, compared with NO. In most cases, a heavy nitridation and RNO create a high-oxygen-concentration layer at the interface, which plays a crucial role in the hardening of the gate dielectric.

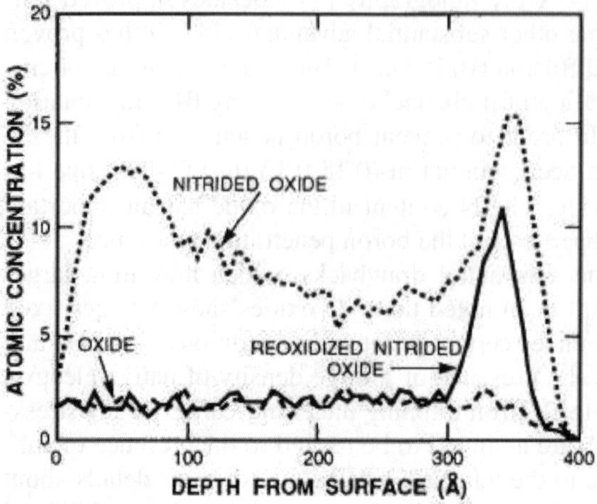

Fig. 6.19. Depth profiles of the nitrogen concentration in oxide, nitrided oxide and reoxidised nitrided oxide, measured by Auger electron spectroscopy. [After Dun and Wyatt [50]]

6.4.2.2 Hardness of NO and RNO Dielectrics

Although NO and RNO have a strong potential for achieving radiation-hard gate dielectrics, the resulting radiation response strongly depends on the processing details and usually requires an optimalisation. An example is presented in Fig. 6.20, showing the impact of the nitridation temperature and time on the radiation-induced flat-band-voltage change due to oxide trapped holes [47]. From these results follows that heavier nitridations (i.e., higher temperatures or longer times) are required to achieve a hardness improvement, while light nitridations yield no improvement or even an increase in the hole trapping and interface-state generation [47, 52]. However, reoxidation of fairly light nitridations may result in a superior dielectric with respect to radiation damage [53]. Several groups have undertaken a systematic study of nitridation and reoxidation conditions on the radiation hardness, for which the interested Reader is referred to [52 - 57].

The gate dielectric degradation by ionising radiation typically consists of hole trapping near the interface by already present neutral traps (with a hole capture cross section σ_p of 10^{-13} to 10^{-14} cm^2) and the creation of interface traps. Electron trapping is usually negligible, since the native ETs have an electron cross section which is typically three orders of magnitude smaller (σ_n~10^{-18} cm^2). This may no longer be the case for NO, where large densities of grown-in ETs exist. Moreover, radiation may create neutral ETs with an electron cross section in the range 10^{-15} to 10^{-17} cm^2, which can become a reliability problem for downscaled MOSFETs, since they trap injected hot electrons [1]. Direct creation of other types of new trapping centres by atomic displacements in the dielectric is rather uncommon.

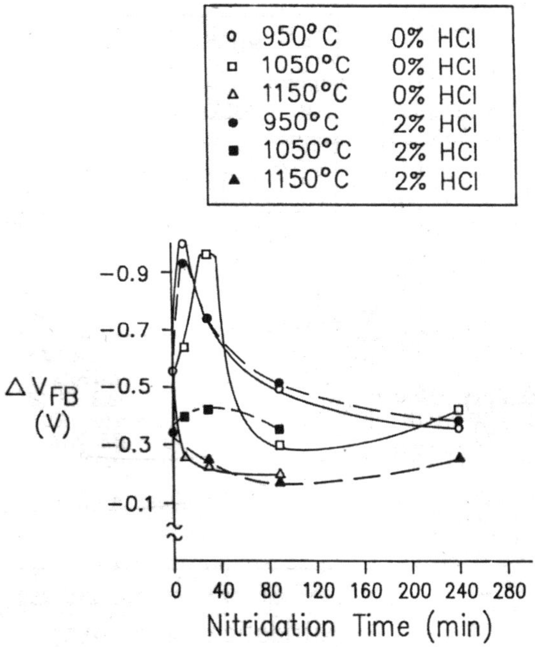

Fig. 6.20. Radiation-induced flat-band-voltage shift vs. nitridation time for 25 nm oxides grown in a dry O_2 ambient. Nitridation temperature and HCl content of oxidation ambient are shown in the inset. Dose is 100 krad. [After Ruggles and Monkowski [47]]

Detailed studies of the hole trapping mechanisms in NO and RNO have been undertaken by a number of groups [58 - 61]. One unusual phenomenon was the observation that for RNO hole trapping was largest for negative gate bias during exposure (Fig. 6.21) [58]. For standard and NO gate dielectrics, maximum radiation-induced hole trapping occurs for positive gate bias with respect to the substrate, whereby the holes are pushed to the interface. The near-interface region is characterized by a deficiency of oxygen and strained Si-O bonds, giving rise to a maximum concentration of neutral hole trapping centres (oxygen vacancies most likely). The fact that RNO behaves differently strongly suggests a different profile of pre-existing hole traps. Analyzing the mid-gap voltage shift leads to the schematic picture of Fig. 6.22, whereby the concentration of hole traps in the oxide peaks towards the gate and silicon interface, with the largest density near the gate interface (region between 3 and 7 nm from the gate) [50, 58]. Another feature, which can be derived from Fig. 6.21, is that the capture cross section of the hole traps is much smaller in NO compared with RNO. This follows from the rapid saturation of the mid-gap voltage shift for RNO, while a continuous increase with dose is found for NO.

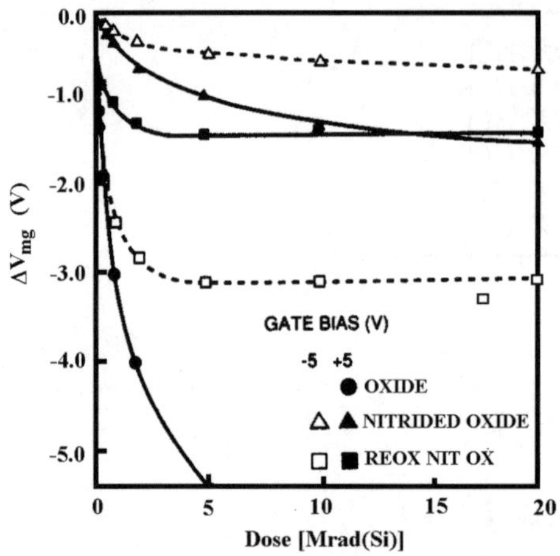

Fig. 6.21. Mid gap voltage shift versus dose for oxide, NO and RNO capacitors irradiated with positive and negative applied gate bias. [After Dunn [58]]

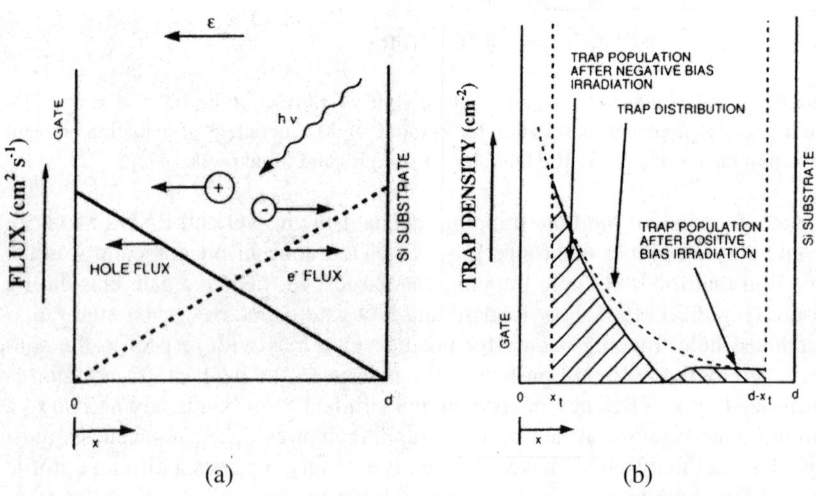

Fig. 6.22. (a) Hole and electron flux in an ideal trapless dielectric during irradiation with a negatively biased gate. (b) Qualitative description of the hole trap population in a reoxidised nitrided oxide following positive and negative bias irradiations. [After Dunn [58]]

Part of the slow positive charging of NO dielectrics can be related to slow donor states, which may be created and ionised during electrical stress or irradiation [59]. Slow donor states are shown to be responsible for a reversible, bias- and temperature-dependent charge relaxation. The number of slow donor states is a function of the nitridation and reoxidation conditions. Using optimized conditions,

RNO dielectrics can be grown with essentially no slow donor states and, hence, with little relaxation [59].

Detrapping studies of positive charge after irradiation have revealed some further interesting observations [60]. Figure 6.23 compares the detrapping behavior of irradiated 36 nm RNO with standard oxides, as a function of the gate bias and this for a 1 hour room temperature 'anneal'. For positive gate bias during the anneal, hardly any detrapping is found for RNO, which is indicated by the unannealed fraction remaining 1.0. This implies that also no interface traps are being created. For the standard oxide, increasing detrapping is found for more positive gate bias, which is in line with the electron tunneling mechanism at the Si-SiO_2 interface. This follows also from the ln(t) time dependence of the unannealed fraction at a constant gate bias [60]. Indeed, for positive gate bias, electrons can tunnel from the substrate in trapped hole centres, which have their maximum density near this interface. As a result, the trapped hole is neutralized by the captured electron, either permanently or reversibly, in the case of a switching oxide trap [62]. For negative gate bias, on the other hand, negligible detrapping is expected and found for standard oxides. For RNO, increasing detrapping is observed for more negative gate bias, confirming that the oxide-trapped charge is mainly in the upper half of the dielectric, close to the gate electrode.

From isochronal detrapping experiments, whereby the temperature is varied for a fixed annealing time (10 min) without application of bias [60], one can derive an effective energy distribution of the discharging hole traps. As can be seen from Fig. 6.24, a completely different spectrum is found for RNO, showing a much more peaked behavior, centered around a higher average activation energy of 0.96 eV. From this it is concluded that the traps which are found in conventional oxides are removed in RNO, whereas a new type of hole trap is introduced by the RNO

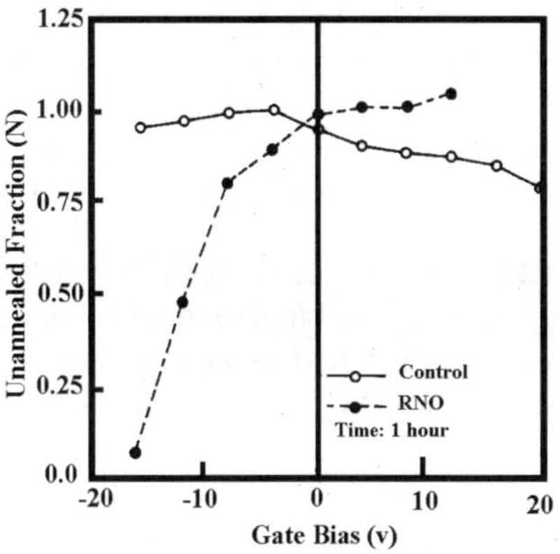

Fig. 6.23. Field detrapping of the radiation-induced trapped charges in control oxide as well as in RNO devices at different gate bias conditions. [After Mallik et al. [60]]

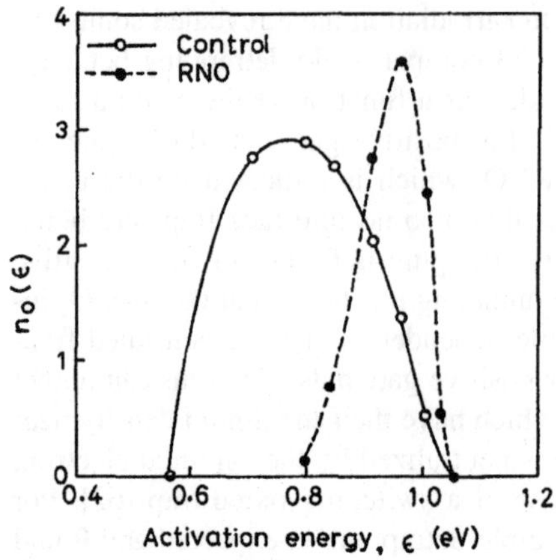

Fig. 6.24. Activation energy distribution of annealing of the radiation-induced trapped charges in control oxide as well as in RNO devices. [After Mallik et al. [60]]

process [60]. Finally, it has been observed that both the density and σ_p of the RNO hole traps are quite different from those in control oxides. The hole capture cross section is significantly larger (7.9 x 10^{-12} cm^2) and the density markedly smaller in RNO compared with a standard oxide [61].

One of the advantages of NO (RNO) is the absence of interface-state generation during and following irradiation or high-field/hot-carrier stress. However, this only applies for optimized processing conditions, as illustrated by Figs. 6.25 [63, 64]. This is further evidenced by the time evolutions of Figs 6.26a and b, corresponding to a control and a RNO oxide [65]. While there is a power law (t^m) dependent delayed D_{it} formation in the control oxide, it does not occur at all for the RNO. Furthermore, exposure to hydrogen after irradiation does not create additional interface traps or a reduction in trapped positive charge, this in contrast to control oxides [65].

Different mechanisms have been proposed for the absence of the D_{it} formation, which are based on the two main models for interface trap buildup [64, 66]. In the electron-hole recombination model, it is assumed that first a hole needs to be trapped in a near-interface oxide trap, which subsequently captures an electron.

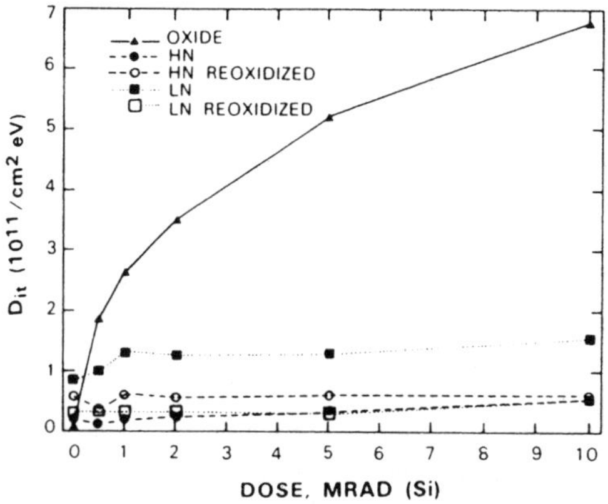

Fig. 6.25. Mid gap interface state densities in MOS capacitors after exposure to 10 keV X-rays. [After Dunn [63]]

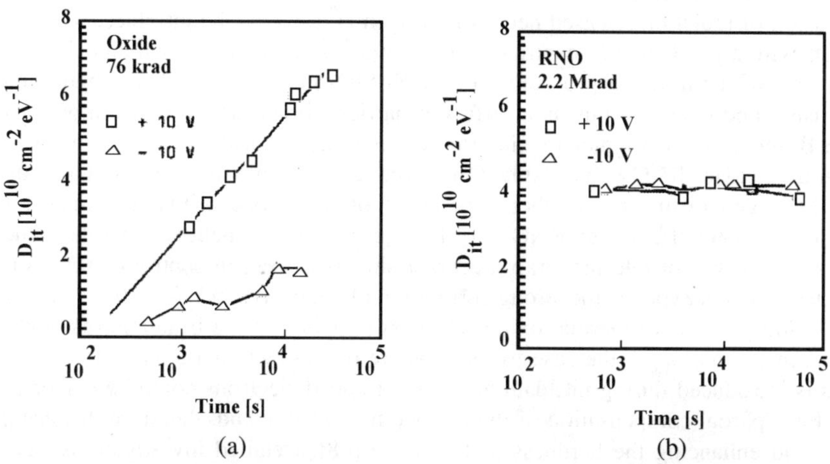

Fig. 6.26. (a) Time evolution of ΔD_{it} for a dry oxide after a total dose of 76 krad(Si). The capacitors were biased at ±10 V after irradiation. D_{it} before irradiation was 10^{10} cm^{-2}eV^{-1} and just after irradiation 8×10^{10} cm^{-2}eV^{-1}. (b) Time evolution of ΔD_{it} for RNO after a radiation dose of 2.2 Mrad(Si). The capacitors were biased at ±10 V after irradiation. D_{it} before irradiation was 2×10^{10} cm^{-2}eV^{-1}. [After Bhat and Vasi [65]]

This gives rise to a bonding reconfiguration which creates silicon dangling bonds at the interface (P_b centers). For light nitridations, the presence of N creates strained Si-O bonds at the interface, which can act as a hole trapping centre and are therefore thought to be precursors of the radiation-induced interface traps [64]. Heavy nitridation or reoxidation shifts the nitrogen peak away from the interface in the oxide and, hence, the corresponding trapping centres, while an oxidation of the interface occurs. In addition, it is believed that the presence of nitrogen in the oxide compensates the compressive stress at the interface. However, as pointed out by *Yount et al.*, one should not overestimate the stress factor for NO and RNO dielectrics; these Authors assign only a minor role to stress in the radiation degradation of the Si-SiO_2 interface [67]. In their view, it is the absence of E' centres (oxygen vacancies) which reduces the hole trapping and probably the reduced D_{it} formation.

Although electron-hole recombination may be partly responsible, it has been shown that hydrogen plays a definite role in the interface-state generation or its absence in NO and RNO [66]. On the one hand, one would expect an effect of the high density of H present in NO [64], if the H-diffusion model were responsible for the radiation-induced D_{it} formation. The result of Fig. 6.27 clearly demonstrates that effectively interface states are being generated in RNO, if exposed to a hydrogen plasma and subsequently subjected to photoelectron injection at fields below the impact-ionisation threshold of $\approx$10 MV/cm in SiO_2. The main difference is that the rate of D_{it} increase is much lower in RNO compared with standard oxides. In fact, the rate of B passivation in the underlying silicon substrate nicely tracks the D_{it} curves, indicating that the same basic phenomenon determines the simple first-order kinetics of both effects. It is believed that the key factor is the diffusion of radiation released neutral hydrogen H^0 towards the interface, where it reacts with a passivated P_b centre (P_b-H) to create a Si dangling bond and H_2. The presence of the nitrogen peak close to the Si-SiO_2 interface retards the hydrogen diffusion and operates thus as a diffusion barrier, similar as for other impurities like B and oxygen. Lighter reoxidation leaves a higher peak concentration (4%) than the heavier RNO (2%), explaining its larger efficiency in retarding/blocking the hydrogen diffusion. Another conclusion of that work is that the interface chemistry is not affected by RNO [64]. In other words, it is believed that the same density and type of interface trap precursor sites is present in control or RNO dielectrics, which explains the strong correlation of the curves in Fig. 6.27.

While in standard thermal oxides electron trapping during irradiation is negligible, this is no longer the case for NO, where it is known that a large density of ETs is introduced during nitridation. In fact, trapped electrons could be responsible for a partial compensation of the trapped hole charge and, therefore, lowering ΔV_{ot} and enhancing the hardness of the oxide [68]. Detailed investigations have demonstrated that, although there is some electron trapping the effect is of second order (10 to 20% of the hole trapping, typically). Figures 6.28 and 6.29 illustrate this. In Fig. 6.28, the isochronal detrapping of radiation-induced charge is plotted versus temperature. Up to point A in the RNO curve, the behaviour is dominated by hole detrapping, which shows a different temperature dependence for both types of oxide. This can be explained by the different energy distribution of the hole traps in the two cases and shown in Fig. 6.24 [60]. This hole detrapping is independent of the magnitude of the field during irradiation, but depends heavily on its polarity (Fig. 6.29). At point A, full compensation occurs between the trapped positive and negative charge for RNO. On the other hand, no sign change is found for control oxides.

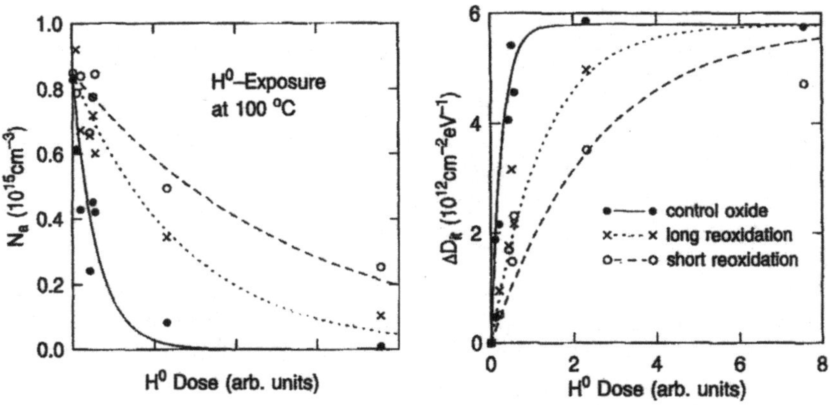

Fig. 6.27. Comparison of the doping density passivation in the depletion region of p-type silicon (a) and of the interface state buildup at the Si-SiO_2 interface (b) as a function of the atomic hydrogen (H^0) dose from a remote hydrogen plasma. Exposures were done at 100°C. In each case, samples with three different oxides are compared. [After Cartier et al. [66]]

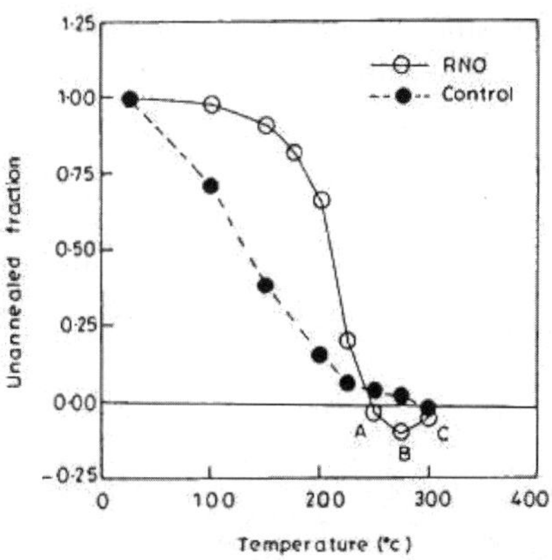

Fig. 6.28. Results of isochronal detrapping experiments following irradiation under floating gate for RNO (36 nm) and conventional dry oxides (33 nm). The values of ΔV_{mg} were -0.7 and -1.66 V for RNO and dry oxides, respectively. [After Mallik et al. [68]]

Further hole detrapping in RNO takes place from A to B in Fig. 6.28. The fact that the unannealed fraction changes sign (super-recovery) in the RNO indicates that electron trapping during exposure does occur. From B to C also the electrons become finally detrapped at higher temperatures, suggesting also a different energy distribution for these ETs. Holes in RNO are completely detrapped around 275°C, while this occurs around 300°C for the electrons. It has also been demonstrated that the fraction of electron trapping is independent of the field polarity during irradiation. This is opposite to the case of hole trapping, which shows maximum intensity in RNO for the most negative bias during exposure (Fig. 6.21) [58]. The amount of trapped negative charge, however, strongly depends on the bias magnitude during irradiation. Highest electron trapping occurs for zero field (floating gate), whereby a region of minimum potential is created by trapped holes, which stimulates electron trapping in the neighborhood. This is not the case for high fields. In addition, the low electron capture cross section of the traps reduces strongly for higher F_{ox}, explaining a negligible electron trapping at hihg gate bias [68].

The role and impact of electron trapping for RNO dielectrics is elucidated in Fig. 6.29, showing maximum impact at zero field and negligible effect for high fields, for both polarities. RNO exhibits highest hole trapping for negative gate bias during exposure and is, therefore, less suitable for such operation conditions, compared with control oxides [68].

Further studies have indicated that nitridation creates neutral hole and electron traps, which can be reduced by reoxidation in a selective way [69], whereby traps with an electron capture cross section of 10^{-17} cm^2 are maintained, while ETs with lower cross sections are removed [70]. One particular reliability issue is the creation of neutral electron traps (NET), following irradiation [1]. This is especially of concern under positive gate-bias operation [71]. However, it has been demonstrated that for RNO, NET formation is successfully suppressed. This is ascribed to the presence of the nitrogen peak close to the silicon interface, which blocks the diffusion of H^+, believed to be necessary for the NET creation.

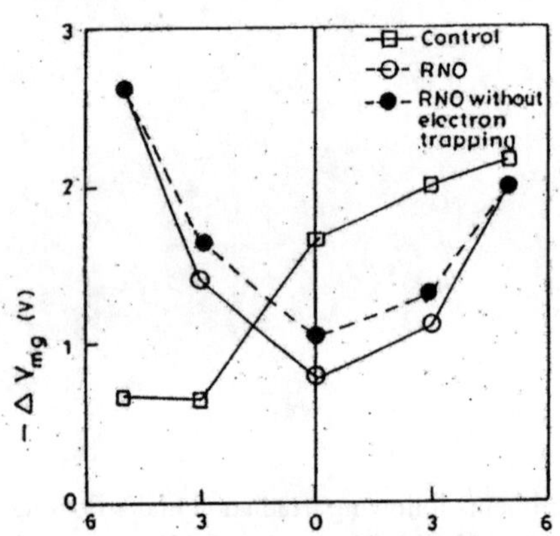

Fig. 6.29. Mid-gap voltage shift as a function of gate bias during irradiation for 1 Mrad(Si). Solid curves represent experimental ΔV_{mg} whereas dotted curve represents estimated ΔV_{mg} if there would not have been any electron trapping. [After Mallik et al. [68]]

The effect of irradiation on transistor mobility and maximum transconductance of RNO devices has been studied in a number of cases [50, 72, 73]. It was found that after irradiation and subsequent low-temperature anneal, a higher inversion

layer mobility is observed for RNO nMOSFETs than before irradiation. Additional LF noise studies indicate that radiation and annealing removes near interface electron traps in RNO, which cause Coulombic scattering and an accompanying reduction of the mobility. The optimum annealing temperature is around 400°C [72]. Similar effects are observed for p-channel devices, although to a lesser extent. This confirms that the near-interface oxide traps responsible for the observations have an amphoteric character, but are much less efficient in trapping holes. After irradiation, the effective mobility of RNO transistors is superior to control nMOSFETs [73] and may even improve after exposure (without anneal). One of the reasons for this improvement is the fact that hole trapping in RNO happens mainly at the gate interface [58], while in contrast, for standard oxides the trapped-hole charge is mainly at the silicon interface. This is related to the profile of the responsible traps, which shows opposite behavior for both types of dielectrics, as shown in Fig. 6.22. The subthreshold slope of irradiated RNO MOSFETs hardly degrades, indicating negligible buildup of interface states [50, 73]. This has been confirmed by charge pumping measurements.
The charge trapping in thermal and nitrided oxides at low temperatures has been studied by *Boesch and Dunn* [74]. From these results follows that RNO has a potentially better total-dose hardness for cryogenic applications, particularly under moderately high-field operation conditions.

6.4.2.3 Microscopy of NO and RNO Dielectrics

In the foregoing, only the impact of nitridation on the macroscopic properties of the gate insulator has been addressed. Microscopic information on the structure of the responsible defects can also be obtained, using the Electron Paramagnetic Resonance (EPR) technique described in Sect. 2.4.1 [75]. So far, the paramagnetic centers represented in Fig. 6.30 have been firmly identified at the interface or in the oxide of irradiated (or high-field stressed) gate dielectrics. At the same time the obtained spin density allows to derive the corresponding absolute concentration to within a factor of 2. EPR can measure trap density profiles on back etched oxides, gradually thinning the sample.

From EPR studies on irradiated films it has been derived that nitridation creates bridging N precursors (c in Fig. 6.30). RNO also shows these radiation induced bridging N traps together with overcoordinated nitrogen centres, represented by d in Fig. 6.30. The impact of the reoxidation time on the different trap species is shown in Fig. 6.31. From this, it is concluded that reoxidation lowers the density of bridging N precursors, which are therefore believed to correspond to the main native electron traps in NO films [75]. Nitridation reduces significantly the number of E' centres (holes trapped in an oxygen vacancy, giving rise to a paramagnetic unpaired electron), while reoxidation produces larger E' EPR signals. However, E' is not the dominant hole trap in RNO films [10]. This follows for example from the defect profiles represented in Fig. 6.32. While the E' centres show a pronounced increase near the silicon interface (position 0 Å) for the 'oxide', a flat profile is found for the NO and RNO case. One might at first sight speculate that the

E' centres in RNO could be (partly) responsible for the radiation-induced hole trapping. However, further experiments show completely different behavior with 'normal' oxygen-vacancy related E' centres [74]. For example, no change is observed in the paramagnetism (density,...) of E' in RNO after the photoinjection of electrons. If these centres were positively charged like in conventional oxides, electrons would be readily trapped with a large cross section, rendering the centre diamagnetic and thus undetectable for EPR observation. This strongly suggests that RNO E' centres are charge neutral and probably single dangling bonds (b in Fig. 6.30) - not vacancies as in thermal oxides.

Figure 6.33 shows the bridging N centre profiles for NO and RNO films; they are thought to be responsible for the significant electron trapping in irradiated NO and to a lesser extent in RNO. This leaves us with the question which centres are responsible for the hole trapping near the gate electrode in RNO? A possible candidate is the overcoordinated N centre, which shows an increasing profile towards the gate [74]. It is expected, however, that this is rather unlikely, given the expected donor nature of the group V nitrogen element. Another important finding is that both the E' hole traps and the N-related defects disappear from EPR observation after a 10 minute hydrogenation in forming gas [75]. This implies that the N-

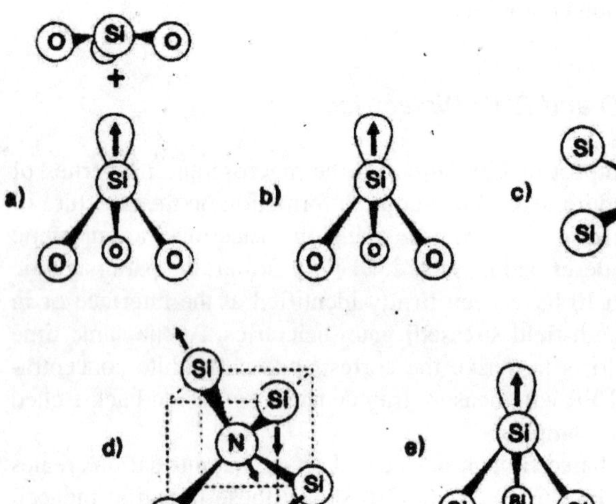

Fig. 6.30. Schematic diagram of EPR defects in irradiated gate dielectric films: (a) positive E' centre associated with an oxygen vacancy; (b) neutral E' centre; (c) bridging nitrogen centre; (d) overcoordinated nitrogen centre (arrows indicate direction of Jahn-Teller distortion) and (e) P_b centre. It is unclear whether the c- and d- defects are bonded only to Si atoms. [After Yount et al. [75]]

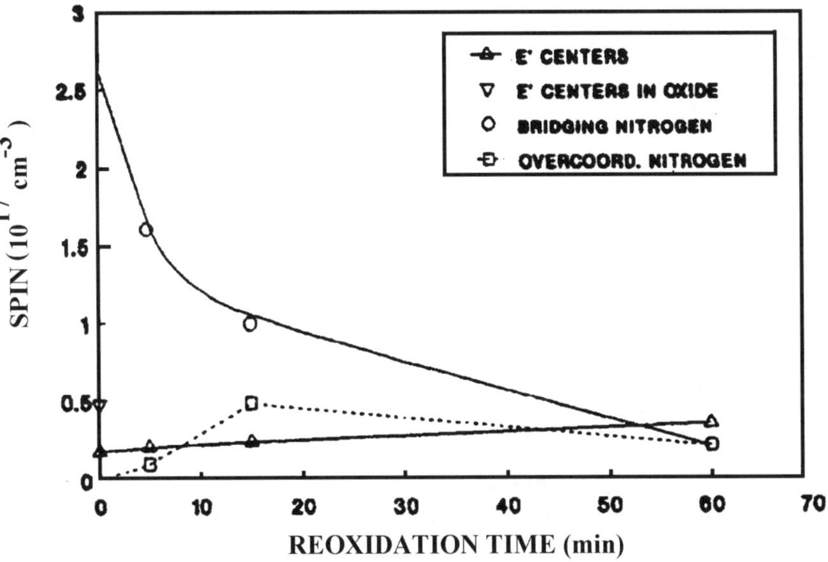

Fig. 6.31. EPR defect density versus reoxidation time. [After Yount et al. [75]]

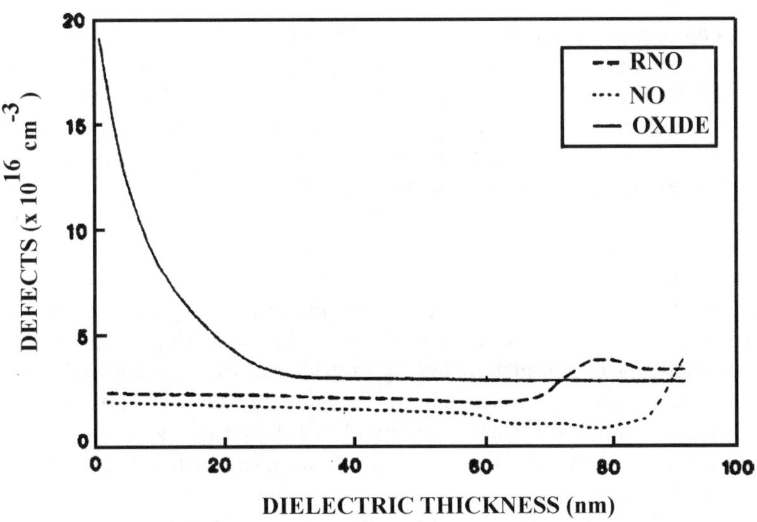

Fig. 6.32. E' distribution in oxide, NO and RNO samples. Distribution curves are approximations based on etch-back data. [After Yount et al. [75]]

related defects quickly react with the H_2, thereby lowering the effective diffusivity (mobility) of hydrogen. This can be a key process in the reduction of interface state buildup and supports the results of *Cartier et al.* [66].

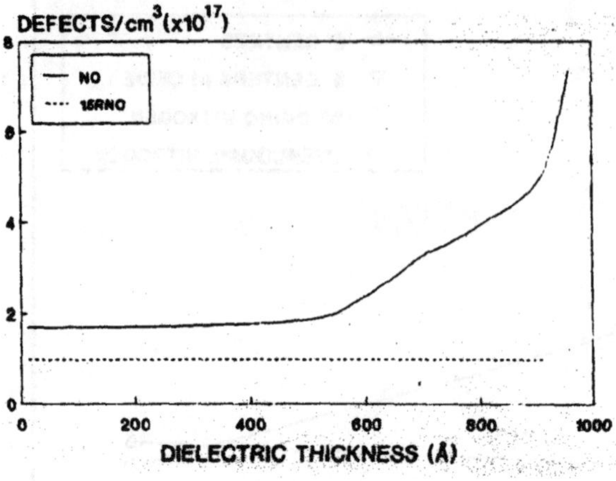

Fig. 6.33. Bridging nitrogen distribution in NO and 15 min reoxidized NO samples. Distribution curves are approximations based on etchback data. [After Yount et al. [75]]

6.4.3 N_2O or Nitrous Oxides

There are strong arguments to grow thin and ultra-thin oxides in an N_2O atmosphere (nitrous oxide or oxynitrides), rather than in NH_3 [76 - 80]. This offers several advantages: first, less hydrogen is introduced compared with NO. Second, the starting quality with respect to fixed oxide charge, electron traps and interface traps is equal compared to conventional oxides. Next, a nitrogen peak is only found at the silicon interface, not at the surface or in the oxide bulk, as seen in Fig. 6.34.

In addition to the advantageous initial quality of the oxynitrides, also the radiation response is improved compared with thermal oxides [76 - 80]. It has been shown that both the oxide-trapped charge and the interface-state generation due to ionisation damage are reduced. An example is given in Fig. 6.35 [77], showing a smaller transconductance degradation for N_2O-nitrided devices. Remark also the length dependence in the figure, whereby a higher degradation is found for the shorter transistors, in line with earlier reports [12, 20] (see e.g. Fig. 6.4). It has moreover been observed that N_2O oxides exhibit an enhanced resistance against NET formation [77].

The impact of the gate dielectric for ≈8 nm gate dielectrics is represented in Fig. 6.36 [79], showing the superior behaviour of N_2O (2-time) oxides. It is mainly the interface-state buildup, which is improved, while the hole trapping is equivalent as for thermal oxides [80]. This points to the fact that the chemistry and structure of the near-interface oxide region, where the border traps reside is clearly different for oxynitrided layers [80 - 82]. It is assumed that less oxygen vacancies are present, while stronger Si-N bonds may replace the strained Si-O bonds.

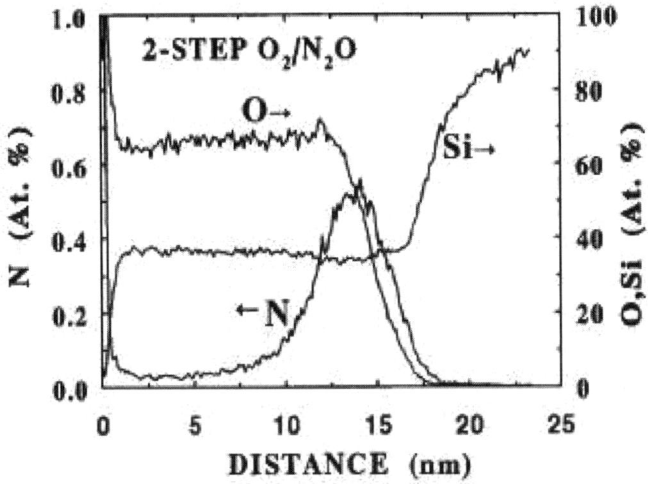

Fig. 6.34. SIMS N, Si and O profiles of a 14 nm thick 2 step O_2/N_2O oxynitride. The N peak is at the Si-SiO_2 interface with a maximum N concentration of ~0.5 at. %. [After Saks et al. [80]]

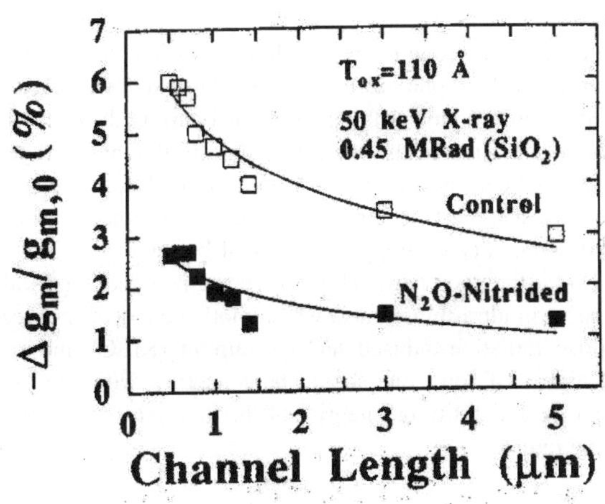

Fig. 6.35. Normalised peak g_m degradation ($\Delta g_m/g_{m,0}$) due to X-ray irradiation as a function of channel length for nMOSFETs with control and N_2O-nitrided gate oxides. [After Lo et al. [77]]

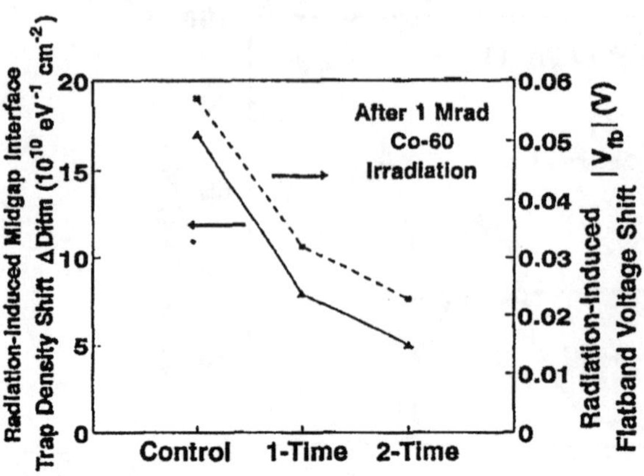

Fig. 6.36. Radiation-induced mid-gap interface trap density shift ΔD_{itm} and flat-band voltage shift $|\Delta V_{FB}|$ for control, one-time N_2O grown and two-time repeated N_2O-grown samples. [After Wu and Hwu [79]]

6.5 Ultra-Thin Oxides

Further downscaling leads to a continuous reduction of t_{ox}. However, when the fundamental tunneling limit around 2-2.5 nm is reached, one has to consider alternative gate dielectrics with a higher dielectric constant, in order to lower the direct tunnel current through the thin gate. For such so-called high-*k* dielectric materials, one can again afford a thicker layer and a correspondingly lower gate leakage current. From a viewpoint of total-dose damage, the tunneling limit has already been trespassed [2]. For a thickness below twice the electron tunneling limit (2x3 nm=6 nm) essentially no trapped holes will be retained in the oxide during or after the irradiation. So far the good news. However, some new radiation-induced problems and drawbacks may arise in ultra-thin oxides, which motivate a deeper investigation. These phenomena comprise radiation-induced leakage current (RILC) and soft or quasi- breakdown (SBD). For high LET ions, one should be concerned with single-event gate rapture (SEGR), while also the reliability or integrity of the thin oxides may be compromised by exposure to ionising radiation.

6.5.1 Radiation-Induced Leakage Current (RILC)

Investigations have shown that irradiated thin oxides can exhibit an enhanced gate leakage current density J_g at low gate biases, before the onset of the classical FN tunneling current, which remains unaffected by the exposure [2, 83 - 86]. An example is given in Fig. 6.37 [2], showing an increase in the gate leakage current for low gate bias. Such an increase is observed after exposure to X or γ-rays [83, 85, 86], high-energy electrons [2, 84], a 1 keV e-beam [87], ion implantation [88] and heavy ions [89 – 91].

There exists a large correspondence with stress induced leakage current (SILC) generation [2, 92 – 95]. Based on the anology, a similar model has been invoked. It is believed that the excess gate current at low field is generated by trap-assisted electron tunneling from the gate to the silicon, through radiation- (or stress-) induced neutral electron traps in the thin oxide. Several experimental arguments point to an inelastic tunneling, whereby the electron looses energy through phonon emission, when tunneling to the trap.

The RILC effect typically occurs for high total ionising doses (several tens of Mrad for X- or γ-rays)). In the first instance, the RILC magnitude increases according to a power law of the radiation dose D [2, 96]. From the slope of Fig. 6.38, an exponent close to 0.9 is derived [2]. This is similar as the interface state buildup with dose in thermal oxides, whereby the exponent is in the range 0.7-0.8 typically. However, for even higher total doses it has been observed that the power law reduces from ~0.92 to ~0.36 [94]. Furthermore, from the different bias impact during irradiation, it is concluded that interface traps do not interfere in the RILC (SILC) phenomenon.

The RILC shows a clear variation with the bias during the irradiation (Fig. 6.39), indicating that the neutral electron trap formation is gate bias dependent. The RILC is most pronounced for zero (low) field during the exposure, yielding a homogeneous neutral trap distribution. For more asymmetric distributions, the trap

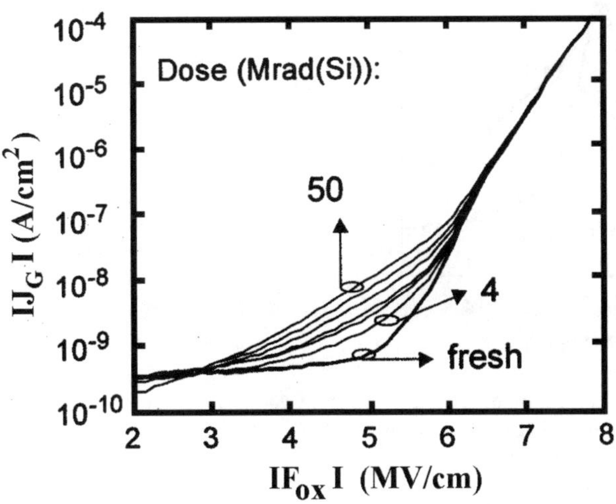

Fig. 6.37. Negative J_G-F_{ox} curves measured before (fresh) and after irradiation for various doses ranging from 4 to 50 Mrad(Si). [After Ceschia et al. [2]]

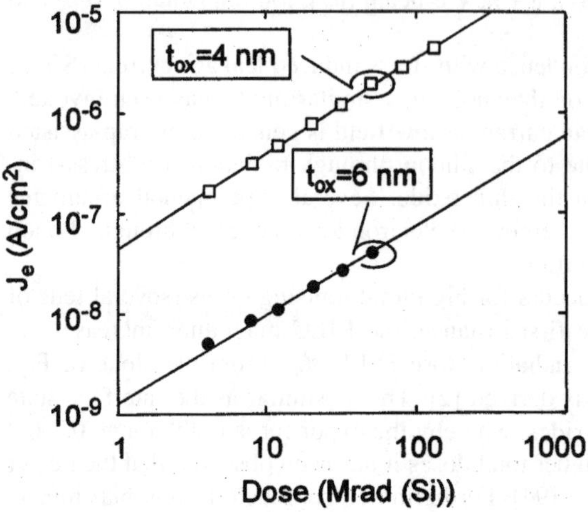

Fig. 6.38. RILC kinetics (negative excess current) for t_{ox}=4 nm (□) and 6 nm (•). [After Ceschia et al. [2]]

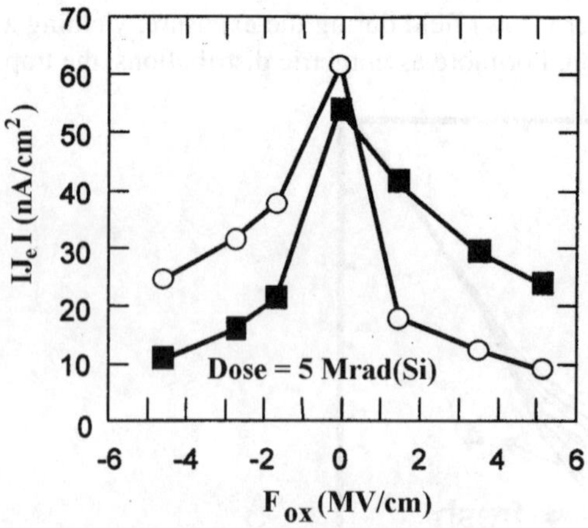

Fig. 6.39. Excess current read at $|F_{ox}|$=5MV/cm for different gate voltages applied during irradiation for t_{ox}=4 nm (o=positive RILC; ■ is negative RILC, i.e. measured with negative bias on the gate). [After Ceschia et al. [2]]

assisted tunneling probability from the gate to the trap, or from the trap to the silicon will be lowered. From the observed field dependence, it is concluded that the underlying traps are not related to displacement damage, but rather require trapping of a charged species, which are released by the irradiation. A possible candidate is hydrogen.

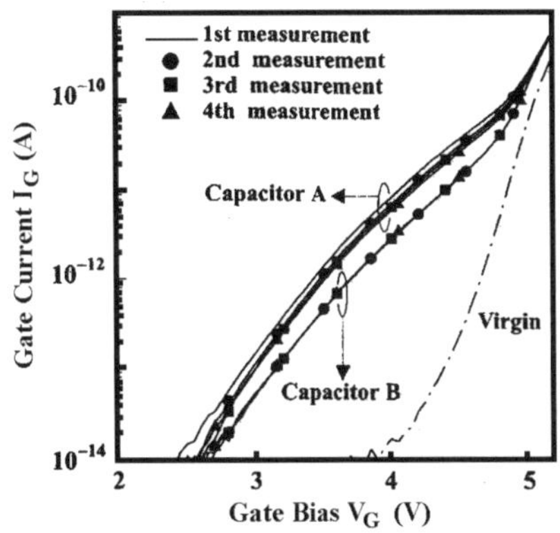

Fig. 6.40. The effect of consecutive I_G-V_G measurements on the RILC of two capacitors exposed to 30 Mrad(Si). Before I_G-V_G measurement, capacitor A was stored for more than 60 h without bias following the irradiation, while capacitor B had been bias annealed for 5×10^5 s under V_G=4 V after the irradiation. [After Ang et al. [98]]

RILC also shows a pronounced transient and annealing behaviour [97]. One typically finds that only after several measurements, a stable RILC is obtained (Fig. 6.40) [98]. In fact, the application of a low gate bias for some time at room temperature leads to a significant reduction (annealing) of the RILC [97, 98]. The annealing follows an ln(t) law and eventually saturates around 10^5 s [98]. This implies that the RILC measurement itself can already generate some recovery, explaining the transient behaviour of Fig. 6.40. This annealing is independent of the polarity of the gate bias.

The effect is greatly enhanced under a hydrogen atmosphere, as evidenced by Fig. 6.41, although no complete recovery is observed [98]. These facts strongly suggest that trapped-holes also contribute to the RILC, besides the creation of NETs.

In addition to bias annealing, RILC can also be removed by a thermal anneal at sufficiently high temperature (Fig. 6.42), i.e., T_A>250°C. Above this temperature, it is believed that NETs are irreversibly annealed. The thermal annealing is more complete for higher temperatures and is more efficient than the bias annealing. On the other hand, the bias annealing can be reversible. It has for example been observed that the application of an unbiased thermal anneal at 150°C after a bias anneal causes a rebound of the RILC characteristics [98]. In other words, some of the oxide traps have been reactivated again. This is in line with the hypothesis that trapped-holes contributing to the RILC may be neutralised by electrons tunneling from the silicon substrate or from the polysilicon gate. In other words, bias annealing actually only deactivates part of the traps, and not really removes them permanently. It moreover explains the higher bias annealing under a hydrogen atmosphere, where H_2 reacts with a trapped hole to form H^+. The latter subsequently drifts towards the anode, removing the positive charge from the bulk oxide [98].

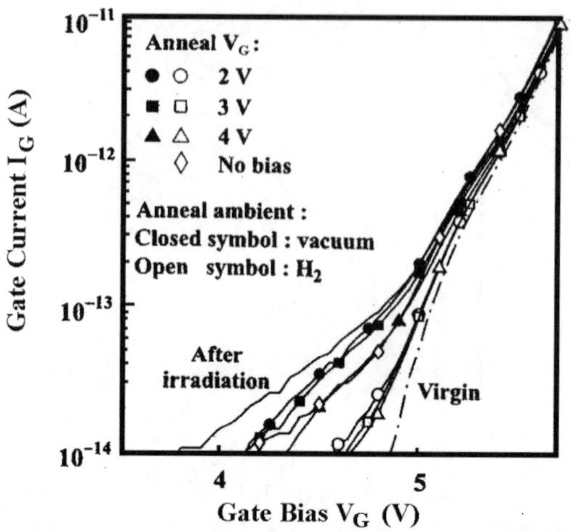

Fig. 6.41. The effect of H_2 exposure on the bias annealing of RILC in pMOSFETs with 4.5 nm gate oxides previously exposed to 50 Mrad(Si). The bias anneal was performed for $5x10^4$ s with the anneal bias as a parameter. The effect of unbiased H_2 exposure on the annealing of the RILC is also shown. For comparison, the corresponding results for bias annealing of RILC in vacuum are also plotted together. [After Ang et al. [98]]

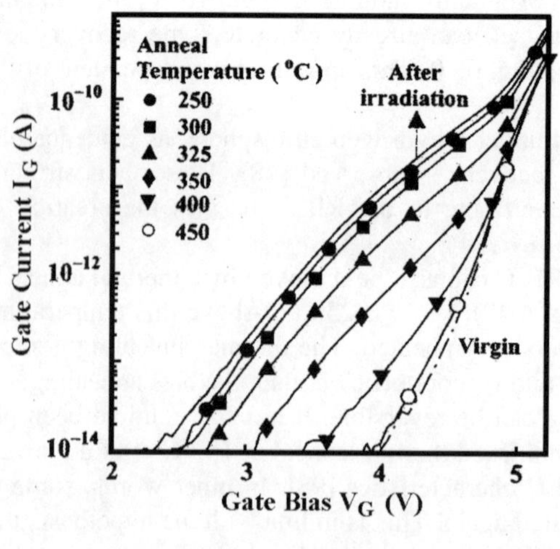

Fig. 6.42. The effect of thermal annealing on the RILC of p-type MOS capacitors with 4.5 nm gate oxides exposed to 30 Mrad(Si). The thermal annealing was performed for 10 min with the anneal temperature as a parameter. [After Ang et al. [98]]

It has been demonstrated that part of the RILC is caused by hole current flow through the gate oxide [95]. In analogy with the electron gate current flow, it is believed that hole trap-assisted tunneling (HTAT) occurs after irradiation. No essential difference is found between the HTAT currents due to inonising irradiation and electrical stress [95]. A maximum contribution is again seen for zero bias during the irradiation for the radiation-induced holes current (RIHC). RIHC becomes more pronounced for lower oxide thickness and higher radiation dose/fluence. As its magnitude is generally quite small, it can only be detected in devices with an area larger than 100 μm^2. It shows similar measurement stabilities, and bias and thermal annealing behavior as the electron RILC. It was proven not to be related to the formation of interface states, which exhibits a pronounced and different dependence on the irradiation bias and polarity [95].

In order to understand better the physics of RILC, one can study the excess leakage current density J_e after irradiation or electrical stress as a function of the oxide field. In Fig. 6.43 two slopes are seen in the J_e versus F_{ox} curves [2]. Furthermore, from the plot it is derived that a law similar as for FN tunneling can represent RILC for the two parts of the curve. The kink in the J_e characteristics defines a critical field F_k whereby the leakage current mechanism changes. This F_k depends on the oxide thickness, as shown in Fig. 6.44 [2]. In analogy with the SILC mechanism, it is assumed that RILC originates from a trap-assisted tunneling of carriers through the oxide barrier, as represented in Fig. 6.45. Neutral traps, induced by the stress or the irradiation are thought to be responsible. The kink in the leakage current characteristics corresponds to a change in the barrier seen by the electron in the oxide trap, which is shown in Fig. 6.45. In other words, for fields below F_k, the trapped electron experiences a trapezoidal barrier, while this becomes a triangular barrier for higher fields exceeding F_k. From the magnitude of F_k shown in Fig. 6.44, it is derived that the tunneling process is inelastic, whereby

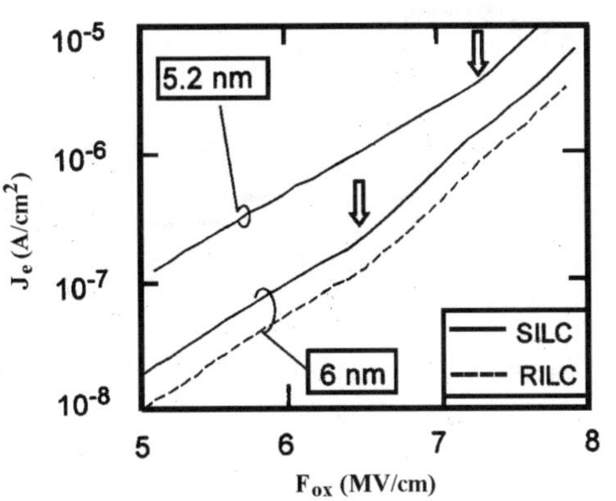

Fig. 6.43. Excess current after irradiation or electrical stress for two different oxide thickness (5.2 and 6 nm). [After Ceschia et al. [2]]

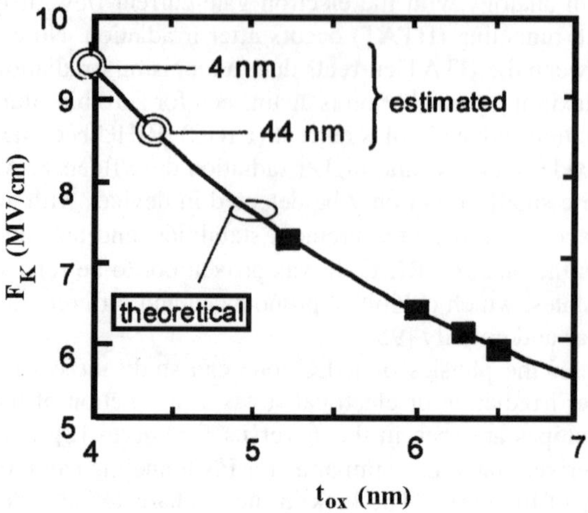

Fig. 6.44. Critical field F_k as a function of t_{ox}. Closed squares represent experimental data, while open dots indicate the expected values for t_{ox} = 4 nm and 4.4 nm. For too thin oxides, the FN tunneling dominates the overall gate current, so that it is difficult if not impossible to extract the RILC contribution. [After Ceschia et al. [2]]

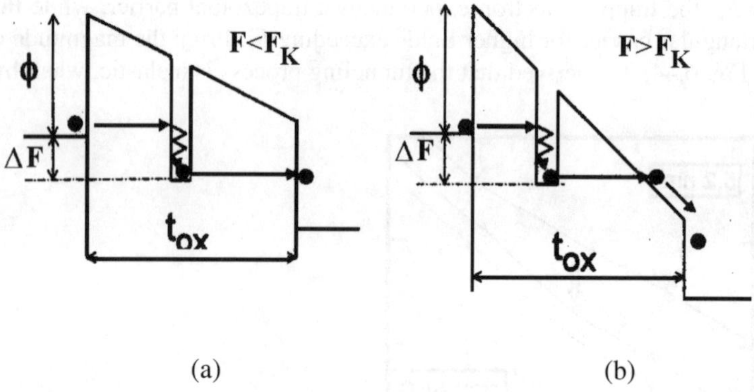

Fig. 6.45. Inelastic trap-assisted tunneling for (a) $F_{ox} < F_k$ and (b) $F_{ox} > F_k$. [After Ceschia et al. [2]]

the carrier looses an energy $\Delta E \approx 1$ eV through the emission of phonons, when it becomes trapped. From a model based on a solution of the Schrödinger equation for a simplified oxide band structure, it has been derived that the underlying NET distribution is spatially symmetrical in the oxide, centered in the middle of its thickness [99].

6.5.2 Radiation-induced Soft Breakdown (RSB)

It has been noted that radiation can also create so-called quasi- or soft breakdown, which is an irreversible (i.e., permanent) damage of the oxide, whereby a local leakage current path has been created in the dielectric. This precedes catastrophic (hard) breakdown where the insulating function of the gate is lost permanently. RSB is characterised by the occurrence of intense fluctuations in the gate current. Often rapid switching in the gate current I_G, also termed a Random Telegraph Signal (RTS), is observed, like illustrated in Fig. 6.46 [89].

The soft- or quasi-breakdown is only observed after high LET ion irradiations, indicating that only dense electron-hole tracks can produce these conductive multi-defect paths across the oxide. It follows amongst others from the fact that in fairly thick oxides both RSB and RILC can be created by high LET ionising irradiation and not by electrical stress for example [89]. It has been found that for LET > 40 MeVcm^2/mg a large RSB current with a characteristic RTS noise occurs after high-fluence irradiation in 4 nm oxides [89]. If any threshold fluence exists for the RSB, it has been shown to be smaller than 10^5 ions/cm^2. No critical field exists to ignite RSB although a bias applied during the irradiation enhances the RSB current intensity. In Fig. 6.47, the excess gate current versus dose is represented, showing empirically [89]:

$$I_{RSB} = KD^{\nu} \tag{6.5}$$

with the exponent ν close to 1. The gate voltage dependence of the excess SBD current is empirically given by:

$$\Delta I_G = aV_G^b \tag{6.6}$$

whereby the coefficients a and b are related to each other. For electrically stressed thin oxides, this relationship is given by [100]

$$b = -0.78.\log(a) - 3.27 \tag{6.7}$$

This has in good approximation also been found for ion doses below 1 krad(Si), while at higher doses the data dispersion becomes quite large and the agreement becomes less good.

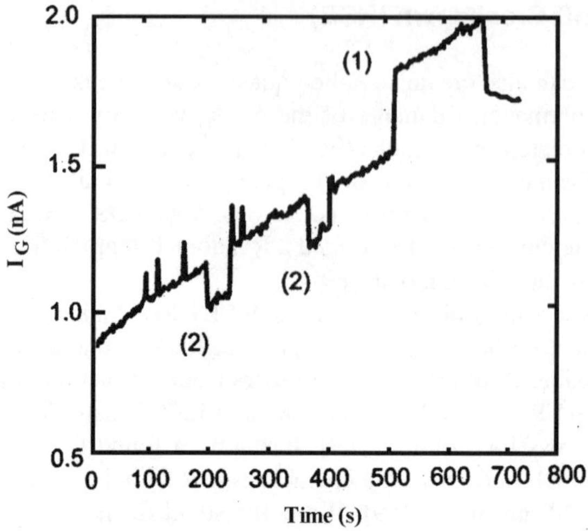

Fig. 6.46. Gate current during irradiation with Ni at V_G=+2.3 V. [After Ceschia et al. [89]]

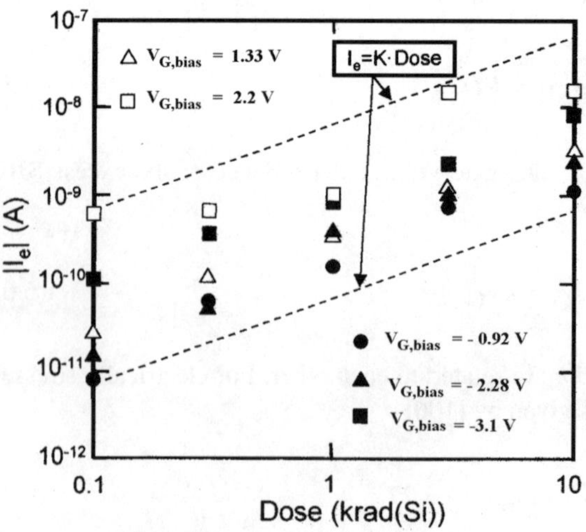

Fig. 6.47. Excess gate current I_e (read at a V_G=-1.8 V) measured in 3-nm capacitors after irradiation with 257 MeV I ions as a function of the radiation dose. [After Ceschia et al. [91]]

6.5.3 Single Event Gate Rupture

Scaled technologies imply a thinner gate and higher oxide fields during operation. F_{ox} values in excess of 5 MV/cm may be expected for technologies approaching 0.1 μm feature sizes [101 - 103]. Therefore, concern has risen that single event gate rupture (SEGR) may become the dominant catastrophic failure mechanism for space-based electronics at the 256 Mbit level and beyond. However, studies have indicated that for thin oxides the SEGR susceptibility in fact improves considerably [101]. The basic reason is that the breakdown field increases in thin dielectrics since less defect creation occurs through hot carriers. Only for devices operating at fields well in excess of 5 MV/cm, SEGR can become problematic. There exists in fact a steep threshold or critical field below which no SEGR can be induced. This field F_{crit} is inversely proportional with LET. SEGR is a single ion event and not a cumulative phenomenon [102]. This follows from the fact that even for extremely low fluences, SEGR can be induced above F_{crit}. Comparing thermal and nitrided oxides, no substantial difference was found with respect to gate rupture susceptibility for 5 nm layers [102]. In addition, no angle of incidence dependence was found for SEGR [102], confirming an earlier hypothesis that when the conducting path diameter obtains the same size as the gate oxide thickness, the angle dependence should disappear. Empirically, F_{crit} can be modeled as [104]:

$$F_{crit} = \frac{F_0}{1 + LET/B_0} \tag{6.8}$$

Sexton et al. [102] proposed a simple analytical model for the empirical parameters F_0 and B_0, providing a good fit with the experiments. This model confirms the two-step nature of SEGR, namely, first, a conducting pipe is created by ionisation damage through the impinging ion and second, the energy stored in the capacitor is discharged through this pipe. In principle, one can define a cross-section for SEGR, similar as for other single-event phenomena, following the procedure described in [105].

6.5.4 Reliability of Irradiated Thin Oxides

A final concern with respect to thin gate dielectrics is the impact of ionising radiation on its integrity or quality [85, 106, 107]. When using X-ray lithography for example, the processed gates can be exposed to a total dose, which may affect their reliability. Figure 6.48 shows an example of the time-to-breakdown t_{BD} distributions of MOS capacitors with a 4.4 nm oxynitride layer before and after X-ray irradiation [85]. It is clear that the t_{BD} is reduced after the exposure. In other words, the critical density of traps necessary to trigger breakdown in the gate oxide is lower in irradiated capacitors. This means that the traps created during the irradiation participate in the formation of the percolation path formed in the gate insulator at hard breakdown. It has been demonstrated, however, that this is not related to the positive trapped charge or the radiation-induced interface states [106]. It is rather believed that NETs play a decisive role in deteriorating the oxide reliability after irradiation.

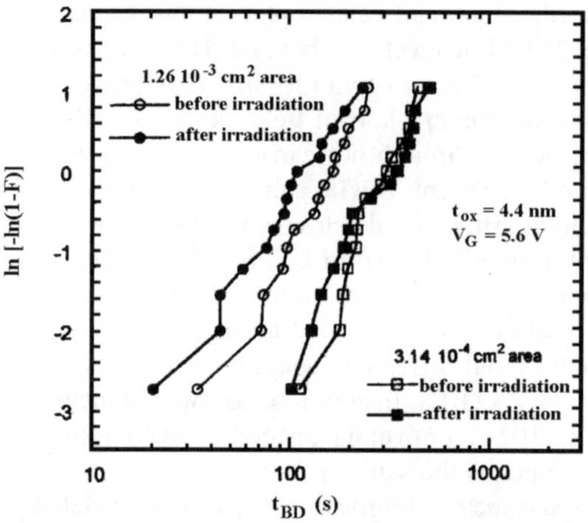

Fig. 6.48. Time-to-breakdown distributions of MOS capacitors with a 4.4 nm oxynitride layer before and after X-ray irradiation. [After Houssa et al. [85]]

6.5.5 Summary

It is clear that thin gate oxides are resistant to high doses of ionisation damage. Consequently, deep submicron CMOS components and circuits are potentially suitable for many space or nuclear applications, without the need for special hardening measures. Nevertheless, while the effect of positive charge trapping can be neglected, this may not be the case for the increased gate leakage current and soft breakdown associated with heavy-ion or high-dose irradiations. Although RILC is not a catastrophic failure, the result will be an increase of the standby power consumption, which may be detrimental for the power management of space missions. Nowadays, the focus is more and more on alternative gate dielectrics with a much higher permittivity [108]. However, in many cases, an ultrathin intermediate oxide layer is needed in order to preserve good interface properties. This layer will govern the electrical properties of the gate stack, so that for future sub-0.1 μm technologies, it is crucial to understand the failure physics of such thin oxides.

6.6 Device Isolation

For more than three decades, isolation between active devices was achieved by growing a thick field oxide, using the so-called LOCal Oxidation of Silicon (LOCOS) concept, depicted in Fig. 6.49 (top). However, the lateral encroachment of the thick field oxide, giving rise to the bird's beak limits the scalability of LOCOS-based device isolation, so that for 0.25 μm and beyond, alternatives like shallow trench isolation (STI) become attractive (Fig. 6.49c). As the thickness of

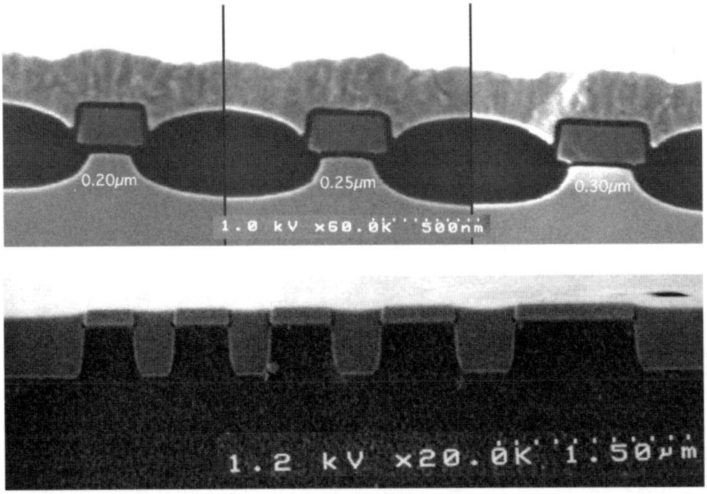

Fig. 6.49. Illustration of an advanced LOCOS approach, called Poly Encapsulated LOCOS or PELOX (top) and suitable for commercial technology nodes down to 0.25 μm and a shallow trench isolation (bottom) commonly used in submicron CMOS technologies.

the field oxide is much larger than of the gate oxide, much more hole trapping will occur there, leading to appreciable device and circuit degradation. In this section, recent advances in the radiation degradation of LOCOS and STI isolation will be summarized.

6.6.1 LOCOS Isolation

A typical cross section of a field oxide used in a 0.7 - 0.8 μm CMOS technology near the gate edge is shown in Fig. 6.50 [109]. Due to the thickness variation of the field oxide, the degradation will also be non-uniform and a distinction needs to be made between the contribution of the different regions. In a first approximation, one can consider the field oxide with a polysilicon layer on top of it as a parasitic capacitor or transistor adjacent to the active devices [109, 110]. To understand the degradation of field oxides, one has to rely on a combination of device simulation and appropriate analysis techniques, like a modified charge pumping method [110]. The parasitic leakage current associated with a field transistor is composed of different contributions, which are summarized in Fig. 6.51. The main source of degradation is associated with the bird's beak region. There, a higher density of trapped charge and a higher creation of interface traps is found (Fig. 6.52). This is ascribed to the higher mechanical stress in the bird's beak region. In fact, also before irradiation, a larger D_{it} exists typically there. It has been observed that similar border traps occur in the different parts of the field oxide [12].

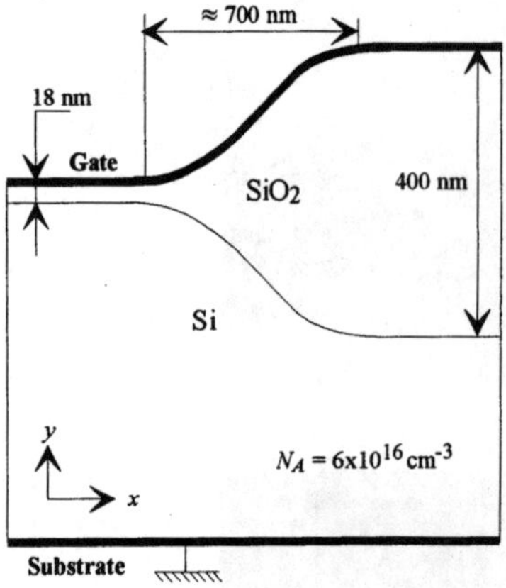

Fig. 6.50. Schematic LOCOS structure. [After Brisset et al. [109]]

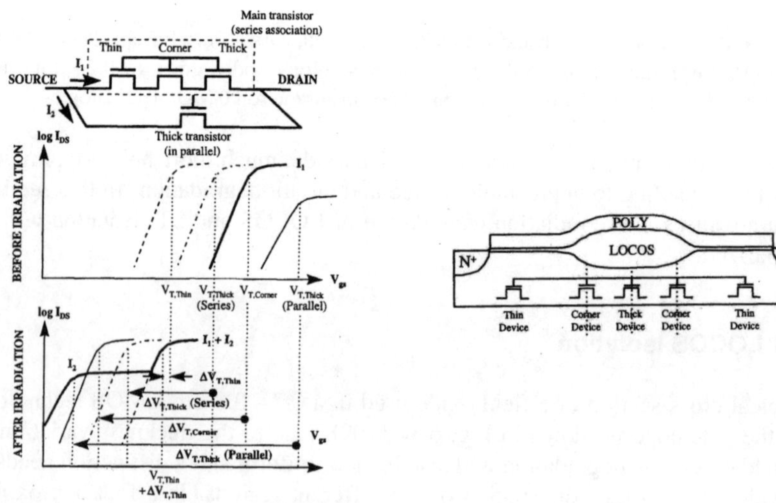

Fig. 6.51. Schematic illustration of total dose effects on $I_D(V_{GS})$ characteristics of an overlap LOCOS transistors. The main transistor corresponds to the series association of the so-called thin, corner and thick devices, shown in the right figure. The thick transistor, in parallel with the main structure, represents a parasitic structure turned on by irradiation. [After Flament et al. [110]]

In order to harden the field oxide, different approaches can be taken. Generally, impurities have been in-diffused or implanted, to create electron traps in the thick oxide. Electron trapping during irradiation will partially compensate for the hole trapping and, therefore, lowers the net negative flat-band or threshold-voltage

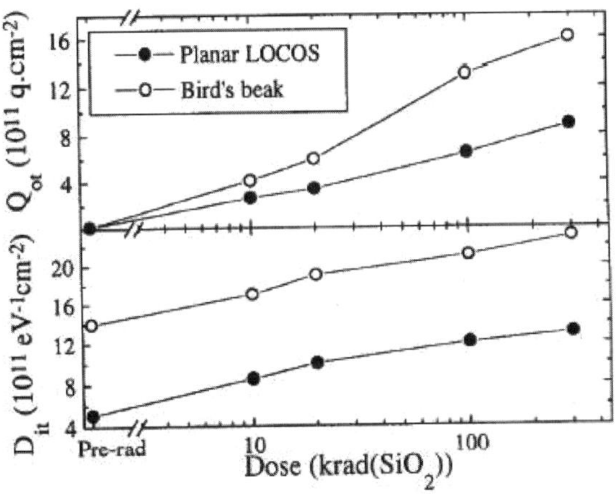

Fig. 6.52. Total dose evolution of Q_{ot} and D_{it} in bird's beak and planar LOCOS regions of the overlap transistor. [After Flament et al. [110]]

shift of the parasitic field transistor. Such dopants can be N, or P. Interest also exists for F-doped field oxides for hardening purposes [111]. Figure 6.53 shows clearly the improvement observed using F-doped field oxides. As a result, the parasitic subthreshold edge leakage for regular nMOSFETs is reduced significantly. The basic reason is that the field V_T is less degraded by the irradiation [111]. An optimum dose of around $5x10^{14}$ cm^{-2} exists, for that particular processing. The improvement is observed for both positive and negative gate bias. Similar results have been obtained for irradiated gated diodes, which show a lower increase of the radiation-induced surface-generation related leakage current with increasing F dose [111]. This points to a reduced interface-state buildup.

6.6.2 Shallow Trench Isolation

Commercial (sub) 0.25 μm CMOS technologies routinely use Shallow Trench Isolation (STI) instead of the LOCOS based isolation. STI consists of a shallow trench, typically, <500 nm deep or so, which is afterwards filled by a deposited oxide. According to Fig. 6.54, at least two leakage paths exist - one along the sidewalls of a single device, leading to subthreshold leakage; a second one in-between devices, whereby the parasitic field transistor is turned on. In order to stick as closely as possible to the COTS principle, one could imagine that an STI based commercial technology is hardened by replacing the filling oxide by a hardened type of dielectric. This requires the smallest amount of changes of the commercial technology [112]. However, as evidenced by Fig. 6.55, such an approach

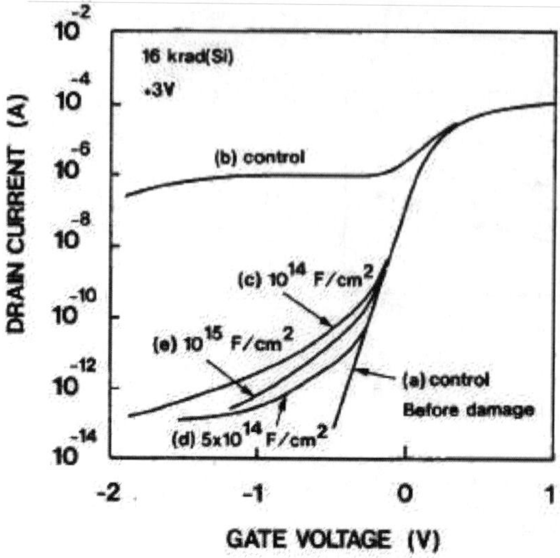

Fig. 6.53. Drain current vs gate voltage for nMOSFETs (L=1.0 μm, W=10 μm) surrounded by LOCOS field oxides before (curve a) and after X-ray irradiation with a V_{GS}= 3 V during irradiation (curve b-e). The curves (a) and (b) represent the control before (curve a) and after (curve b) irradiation, respectively. The field oxide F-implant dose is specified for each device (curves c-e). [After Nishioka et al. [111]]

Fig. 6.54. Arrows indicate two possible parasitic leakage current paths in shallow trench technology. [After Shaneyfelt et al. [112]]

does not work. While in a standard type of technology, the implemented hardened field oxide can withstand total doses up to 1 Mrad(SiO_2), this is not the case in a STI approach. Already at 30 krad (and +5 V on the gate during the exposure) significant subthreshold leakage is noted in Fig. 6.55. In addition, 16 kbit SRAMs fabricated in this technology fail around 80-100 krad(SiO_2). The reason for this unexpected degradation is related to the high electric fields existing in the corners of a shallow trench [112 - 114]. Using 3-dimensional simulations, one can demonstrate that the profiles of the trenches play a crucial role in this. It turns out that a slight overfilling of the trench with oxide, so that its level is higher than that of the active region, shows the best total dose hardness (Fig. 6.56), because it generates a reduction of the corner field and the associated radiation-induced leakage.

Further (or alternative) measures that could be taken consist of an increase of the field implantation and optimising the pull-back of the active regions from the trenches [112]. It turns out that increasing the field implantation dose does not really improve the radiation hardness, especially if the implantation has been done through the filled trenches. Residual ion implantation damage and/or dopants in the oxide are believed to act as trapping centers for radiation-induced charges. It is recommended to avoid ion implantation through the trench oxide for applications requiring total dose hardness. Pull-back of ≈0.3 μm yields better hardness at the expense of packing density and chip area. By carefully optimizing the STI process hardness levels up to several Mrad(SiO_2) can be achieved [112].

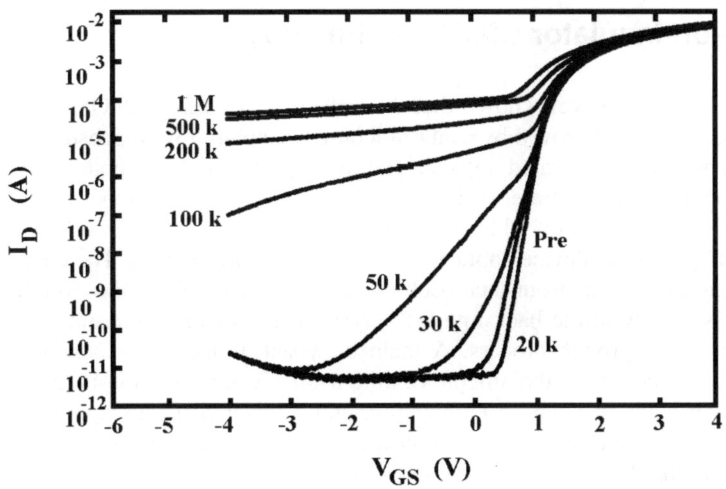

Fig. 6.55. Subthreshold *I-V* curves for n-channel transistors irradiated at room temperature in steps of 1 Mrad(SiO_2) using 10 keV X-rays at 167 rad(SiO_2)/s. [After Shaneyfelt et al. [112]]

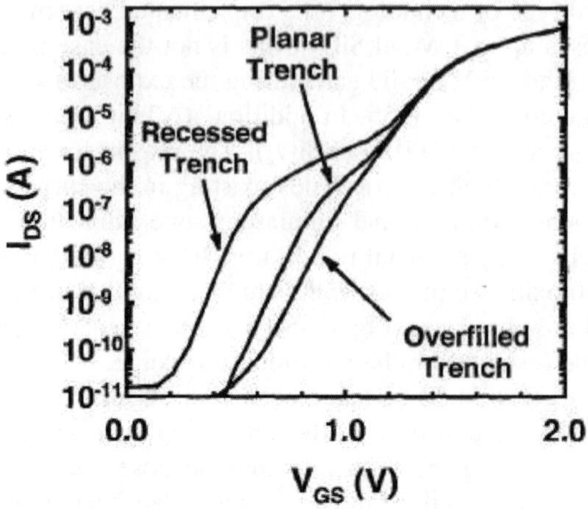

Fig. 6.56. Simulated *I-V* characteristics for three different trench profiles. For the overfilled trench profile, the trench insulator extends 100 nm above the trench corner. [After Shaneyfelt et al. [112]]

6.7 Silicon-on-Insulator CMOS Technology

Since the mid 70-ties it is well known that the radiation hardness of MOS technologies can be strongly improved by using Silicon-on-Sapphire (SOS) substrates. The active devices are fabricated in a crystalline top silicon film, epitaxially grown on a sapphire bulk substrate. Although the first SOS films were reported by *Manasevit and Simpson* in 1964 [115], it took almost a decade before a relative mature material quality for device applications could be achieved [116]. The quality of the silicon film and its front- and back-interfaces are controlling the overall performance. Especially at the back interface crystal defects can be present as a result of the epitaxial growth process. A main drawback is the wafer cost and availability. The reduction of the silicon volume available for radiation-induced carrier generation leads to latch-up elimination, and a lower SEU susceptibility and sensitivity to transient effects. The robustness of SOS technologies resulted for a long time in the dominance of these devices for a variety of space and military applications. Those days the higher circuit manufacturing cost was not seen as a major drawback. Some properties of SOS materials and devices are briefly discussed in Sect. 6.7.1. There is a renewed interest in SOS for microwave applications such as wireless communication.

To improve the carrier mobility, which was much lower than for bulk/CMOS components, alternative SOS-like structures have extensively been studied. The real breakthrough of Silicon-on-Insulator (SOI) technologies, whereby a top silicon film is separated from the silicon substrate by a so-called buried SiO_2 layer, happened in the 1980-ties. A general overview of the most common SOI growth techniques and the used device approaches (thick film, thin film, fully depleted, partially depleted, gate-all-around….) in relation to radiation hardening aspects is given in Sect. 6.7.2. For a more in depth review of SOI materials and devices, the Reader is referred to the textbook by *Colinge* [117]. Important benefits of SOI/CMOS over bulk/CMOS include latch-up immunity, higher packing density, higher speed performance, lower power consumption, lower leakage current and reduced short channel effects. SOI devices can operate in a wide temperature range going from cryogenic electronics at 4.2 K (for satellite applications) to high temperature electronics at 550 K (used for automotives and avionics). Standard SOI technologies are, however, not inherently radiation hard so that a technology or design optimalisation is often required.

SOI technologies are nowadays no longer restricted to niche markets aiming at radiation and/or temperature harsh applications, but are also knowing a breakthrough for commercial applications such as logic, memories and microprocessors [118]. This is especially the case for technologies with deep submicron feature size. The performance benefits of SOI over bulk/CMOS are illustrated in Fig. 6.57, indicating a 20 to 50% higher switching speed or a 2 to 3 times lower power consumption for the same speed. Therefore the speed and power performance of SOI/CMOS is equivalent to 1 to 2 generations of device scaling in bulk/CMOS. In view of the COTS approach for radiation hard applications, as outlined in Chap. 1, it is essential to follow closely the commercial breakthrough of SOI.

Although there is a huge literature on the radiation performance of SOI technologies, only some key aspects will be reviewed in Sect. 6.7.3. In view of its increasing importance, the use of SOI/CMOS for cryogenic electronics will also be

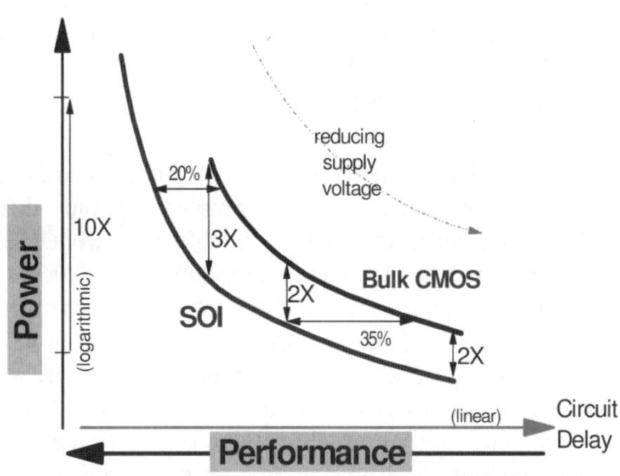

Fig. 6.57. Power drain and access time for bulk and SOI CMOS for supply voltages of 1.2 to 2.5 V. [After Sadana [118]]

be addressed. This section will clearly illustrate that also for SOI a COTS approach turns out to be beneficial. Even more than for bulk CMOS, can the radiation hardness of SOI CMOS be further enhanced by using appropriate design methodologies. This issue will not be addressed.

6.7.1 Silicon-on-Sapphire (SOS)

The starting material is densified Al_2O_3, which is at 2040ºC slowly pulled out of an Ir crucible. After the wafering process of the ingot, a silicon layer is heteroepitaxially grown in a silane ambient. Although there exists a large similarity with the Si growth techniques, SOS material is more expensive and much more difficult to characterise [119]. The fabrication process has a strong impact on both the quality of the silicon-sapphire interface and on the carrier mobility in the silicon film. The mobility is higher in thicker silicon films and can be improved by a higher anneal temperature [120]. The mobility lowers with distance from the Si interface. The depth distribution of the electron mobility versus film thickness and anneal temperature is illustrated in Fig. 6.58. At a given depth, the mobility increases linearly with the film thickness, while the slope itself decreases with increasing depth. A scaling down of the silicon film thickness allows the use of reach-through junctions and reduces short channel effects, however, at the expense of the carrier mobility. Using a solid phase epitaxial and regrowth (SPEAR) [121] or a double solid phase epitaxial (DSPE) growth technique [122] can overcome this dilemma. The main difficulty is to control the defect density at the interface and to limit the Al outdiffusion from the substrate. For an optimised DSPE process, channel mobilities of 500 cm^2/Vs (200 cm^2/Vs) have been reported for n-channel (p-channel) devices in 100 nm thick Si films [123]. For 1.0 μm thick films, an increase of the SPE growth rate above 8 μm/min resulted in mobilities of around 700 cm^2/Vs (275 cm^2/Vs) for nMOSFETs (pMOSFETs) [124]. Thinner films require a further increase of the growth rate to maintain the mobility, i.e., > 10 μm/min for an 0.1 μm thickness.

An important parameter from a radiation hardness viewpoint is the radiation-induced charge trapping at the silicon-sapphire interface, which is not only difficult to characterize but also hard to control and induces a leakage current [125, 126]. The positive charge trapping in the insulating sapphire substrate induces an electron inversion layer in the silicon near the silicon-sapphire interface between source and drain [127]. These back-channel leakage currents, shown in Fig. 6.59, can not be controlled by the gate voltage and thus not turned off. The effect is mainly on the behaviour of the n-channel devices. To harden against back-channel leakage it has been proposed to either heavily dope the silicon film near the interface in order to increase the back-channel threshold voltage or to use a heavy ion implantation to completely 'kill' the back-channel device by reducing the carrier mobility [127]. For the latter approach the difficulty is to control the introduced damage and to avoid complete annealing during subsequent high temperature processing. The amount of radiation-induced charge has been found to be strongly

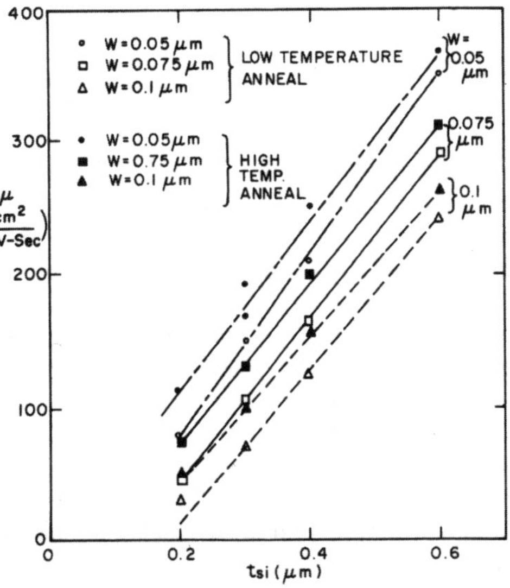

Fig. 6.58. Electron mobility of SOS films with different thickness at 0.05, 0.075, and 0.1 μm from the Si interface. [After Hsu [120]].

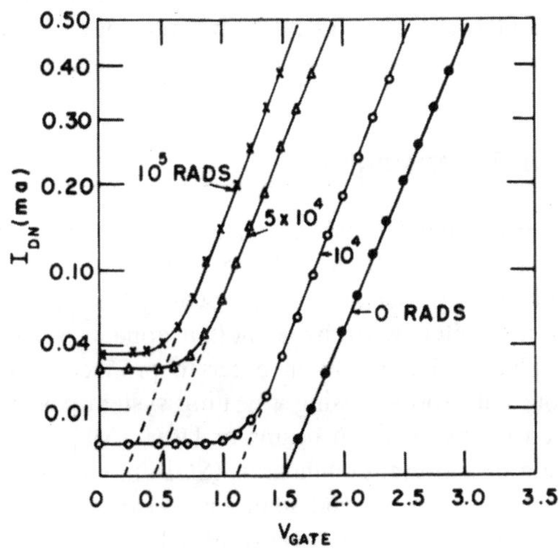

Fig. 6.59. Illustration of the back-channel leakage current of nMOSFET/SOS devices for various total-dose ionisation levels in rad(Si). [After Buchanan et al. [127]]

processing dependent and is influenced by i) the Si epitaxial growth rate, ii) a pre-epi growth hydrogen anneal or so-called hydrogen pre-firing, iii) the epitaxial growth temperature, and iv) the polishing time of the sapphire substrate [128,

129]. More positive charge is found for shorter pre-firing times, increased deposition temperatures and shorter polishing times. In addition to the trapped charge, one has to take into account the impact of radiation on the generation of interface states as validated by DLTS measurements [130].

Although during the last two decades lower cost SOI technologies (see Sect. 6.7.2) have gained in importance and therefore strongly reduced the interest in SOS, it should be remarked that even nowadays there still exist activities on SOS. Beside the standard SOI benefits, SOS has due to its higher thermal conductivity (0.46 W/cm.K for sapphire compared to 0.014 for SiO_2) reduced self-heating effects, a better radiation hardness and a lower parasitic capacitance. SOS is an ideal material for microwave applications [131]. As inductors can be realised, RF front-end electronics on-chip combined with ULSI digital circuits are feasible. Consumer wireless applications in the 1-2 GHz range, fabricated in a 0.8 μm SOS technology, can without problem withstand ionising radiation up to 100 krad(Si) and show single event error rates lower than $1x10^{-9}$ errors/bit-day [132], making them of direct use for satellite communication. By using a sandwich structure consisting of a 10 nm Si buffer layer, a 12nm SiGe layer and a 7 nm Si capping layer on top of the sapphire substrate, a complete suppression of the back gate leakage can be obtained at liquid nitrogen temperature [133]. The SiGe layer induces a valence band offset leading to a confinement of the holes away from the top Si-SiO_2 interface and yielding a decreased surface scattering and a higher mobility. Thin (30-50 nm) strained SiGe layers have successfully been used to fabricate high performance distributed wide-band amplifiers (10 GHz bandwidth), tuned amplifiers (15 dB, 4 GHz bandwidth), 25.9 GHz VCO's and 30 GHz frequency dividers [134].

6.7.2 Silicon-on-Insulator (SOI) Technologies

The problems associated with the back-channel leakage current of SOS structures have strongly triggered the research towards SOI technologies, bringing them their present-day market place. The pioneering technique to fabricate SOI substrates was zone melting recrystallisation (ZMR), whereby a molten zone is scanned across a polycrystalline or amorphous silicon film in order to produce single-crystal material [135]. Molten zones are formed using a heating system based on graphite strip heaters, line-focused lamps, electron beams or lasers. Major problems encountered are the formation of sub-grain boundaries (SGBs), whereby the network of grains is associated with dislocations and sometimes also protrusions. The interface morphology can be dendritic, facetted as well as cellular [136]. The requirement of seed areas making contact with the silicon substrate to initiate the recrystallisation process has a strong impact on the silicon real estate and reduces the achievable packing density.

Over the years a large variety of alternative SOI fabrication techniques has been investigated, such as full isolation by porous oxidized silicon (FIPOS), epitaxial lateral overgrowth (ELO), bonding and etch-back (BESOI), separation by implantation of oxygen (SIMOX) and more recently the smart-cut process. A good overview of these different techniques, including their main advantages and drawbacks, is given by *Colinge* [117]. The nowadays frequently used SIMOX and smart-cut technique are further briefly discussed.

The first SiO_2 buried layer formed by oxygen implantation was reported by *Watanabe and Tooi* [137] in 1966, but *Izumi et al.* [138] first introduced the name SIMOX in 1978. The process consists of a high dose oxygen implantation, typical a few times 10^{18} cm^{-3}, followed by an anneal step. The breakthrough of SIMOX substrates occurred around 1985 due to the industrial availability of a high current ion implanter and the use of a very high temperature annealing (>1250ºC) step. The latter results in the formation of silicon films free of SiO_2 precipitates and a suppression of threading dislocations, although there might still be some silicon islands in the buried oxide [139, 140]. Although SIMOX has a very important SOI market share, there are some doubts about its future for large diameter wafers due to the high temperature annealing. It is also essential to remark that SIMOX is limited to thin film applications.

Wafer bonding and etch-back (BESOI) has for a long time been a very promising technique to compete with SIMOX. Starting with an oxidized wafer, a second wafer is bonded to it by the Van der Waals forces. After increasing the bonding strength by an anneal step (>800ºC) the substrate is thinned down to the desired thickness by grinding and polishing [141, 142]. Both thick and thin silicon films can be fabricated, although the latter are more difficult to control from a viewpoint of uniformity. However, a major drawback has been the rather high production cost associated with the waste of the second wafer and the difficulty to obtain ultra-thin silicon films. An important difference with SIMOX is that the buried oxide (BOX) is thermally grown, which has an impact on its radiation hardness [143].

The smart-cut process is making use of both ion implantation and wafer bonding and has first been proposed in 1995 [144]. The fabrication technique is schematically shown in Fig. 6.60 and consists of the following elementary steps [145]:

- thermal oxidation of a seed wafer (wafer A) in order to form a buried oxide layer similar as for the wafer bond and etch-back technique.
- a hydrogen implantation (range 10^{16}-10^{17} cm^{-2}), whereby microcavities are formed near the projected range of the implantation. For a higher dose blistering will occur. This implantation is the key of the fabrication process.
- bonding of wafer A to another silicon wafer B. The cleaning before bonding plays an important role as it has an impact on the particle density, the surface roughness and the hydrophilicity of the wafer.
- a thermal treatment in the 400-600ºC range leads to an interaction between the cavities so that a splitting is formed, which propagates parallel to the bonding surface. The conditions of this low temperature anneal have to be optimised in order to obtain first a strengthening of the bonding by the formation of Si-O-Si bonds [146] before initiating the splitting process. In the case of hydrophilic bonding a second high temperature (>1100ºC) step is often used to stabilise the bonding interface.

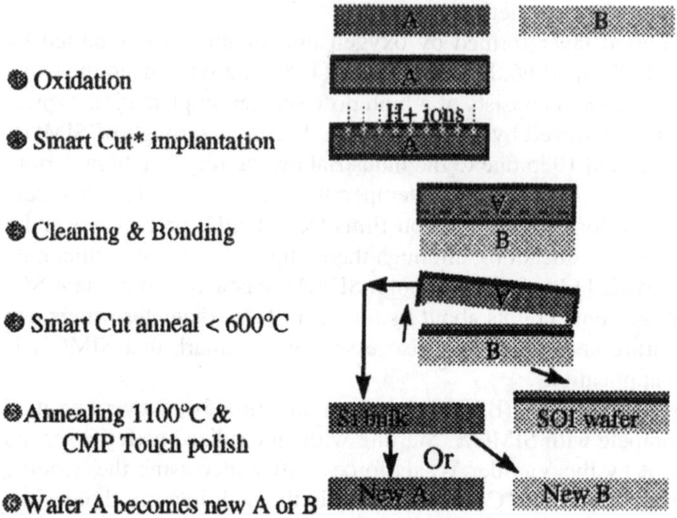

Fig. 6.60. Schematic illustration of the smart-cut process. [After Auberthon-Hervé et al. [145]]

- a polishing step is applied to smoothen the surface and to obtain a silicon surface roughness comparable with the one of standard wafers. A more simple polishing process is applied for wafer A, which is then used again as a seed layer

The smart-cut process has a dominant share of the SOI market because of its relative low fabrication cost, the excellent quality and tight thickness control of the Si films, and the fact that there are no direct limitations with respect to the maximum wafer diameter. The same principle can also be used for the fabrication of other multilayer structures such as Si-Si_3N_4-SiO_2-Si/bulk), Si-Pd_2Si-Si, SiC on oxidised Si and InP or $LiNbO_3$ on Si [147].

From an users viewpoint the SOI devices are further classified according to their operation mode [117]. For the majority of the devices one can differentiate between fully (FD) and partially depleted (PD) transistors and between thick and thin film mode of operation. Depending on the voltages on source, drain, front- and back-gate the depletion regions of the front- and back-interface may touch each other or even overlap. In that case one refers to fully depleted devices. In general, one speaks about thin film devices when the silicon film thickness is smaller than twice the maximum depletion width. In the thick film regime, a neutral region exists between the front- and back-gate depletion layers, so that no coupling occurs between the front- and back-interface. There are electrical performance and reliability differences between FD and PD devices, which are becoming more pronounced for low temperature operation [148]. Special types of transistors are the so-called accumulation mode (AM) devices, such as p^+-p-p^+ or n^+-n-n^+ MOSFETs. They are characterised by a low leakage current and a high subthreshold slope, but are suffering from a more complex conduction mechanism at both room and liquid helium temperature [149]. Compared to enhancement mode FD devices they have a good potential for high temperature operation up to 300ºC, due to their reduced doping [150].

The use of dual-gate devices strongly improves the subthreshold slope, the transconductance and the current drive capability [151]. A special realisation of such a volume inversion device is the Gate-All-Around (GAA) structure proposed by *Colinge et al.* and illustrated in Fig. 6.61 [152]. The same gate oxide is all around the polysilicon gate so that front- and back-gate oxide have the same thickness and quality.

6.7.3 Radiation Hardness of SOI Technologies

The total dose radiation damage of SOI technologies can be caused by charges created in the oxide and at the Si-SiO_2 interface of the gate oxide, the field or sidewall oxides and buried oxides [153]. For the gate oxides there should not be any difference between SOI and bulk devices. In case that the same type isolation technology is used the same is true for the field or sidewall oxides. The creation of edge leakage is illustrated in Fig. 6.62a for a 100 krad(Si) γ–irradiated 0.14 μm n-MOSFET fabricated in a 0.13 μm PD SOI technology with 2.5 nm NO gate oxides and PELOX field isolation. Similar as for bulk CMOS, RILC may be induced in the gate current, for example by a high-energy proton irradiation (Fig. 6.62b). On the other hand, no degradation of the thin oxide or its interface occurs for the conditions reported in Fig. 6.62.

Because of the thickness of the buried oxide and its different initial quality extensive studies have been devoted to the charge build-up at the interface and in the bulk of the BOX [154 - 161]. The threshold voltage shift is related to the hole trapping in the oxide, which requires the simultaneous trapping of a hole and a hydrogen atom [159]. The differences that can be observed for different BOX fabrication techniques are illustrated in Fig. 6.63 for SIMOX and Unibond (i.e., smart-cut) wafers [161]. In early days, the difference between ZMR and SIMOX was as high as 300% [154]. The SIMOX-BOX radiation response strongly depends on the process conditions such as implantation dose, single or double implantation steps and anneal conditions (time and temperature). An in depth discussion is out of the scope of this section.

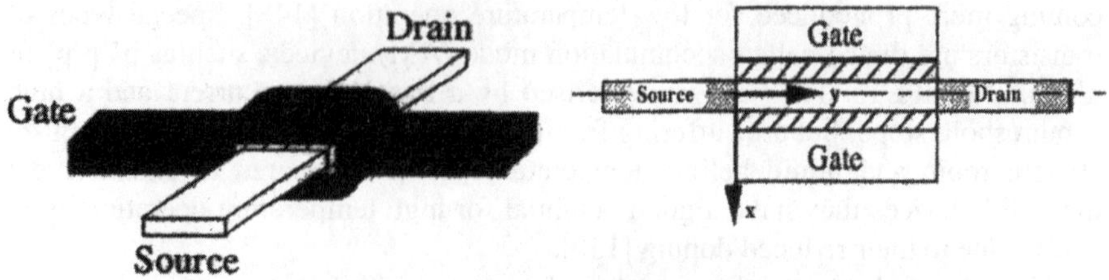

Fig. 6.61. Schematic illustration of the gate-all-around structure. [After Colinge et al. [152]]

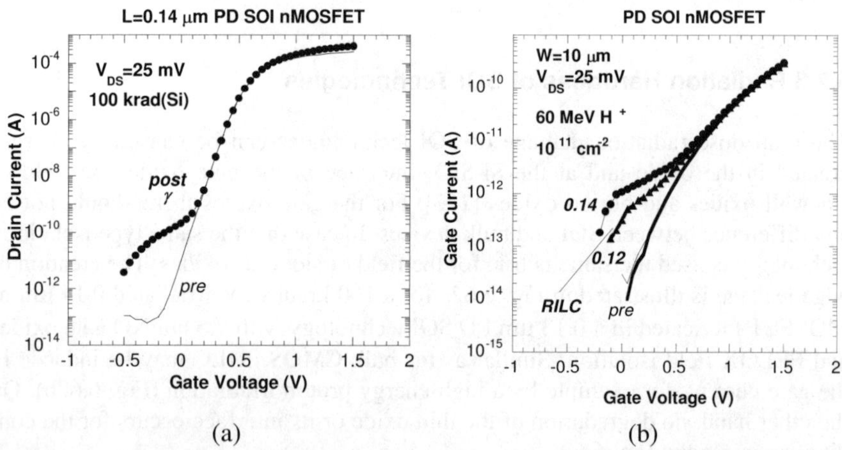

Fig. 6.62. (a) Impact of a 100 krad(Si) γ-irradiation on the I_D-V_{GS} characteristic in linear operation of a 10 μmx0.14 μm PD SOI nMOSFET and (b) RILC induced by a 10^{11} p/cm^2 60 MeV proton irradiation for a L=0.12 and 0.14 μm PD SOI nMOSFET. The devices have been fabricated in a 0.13 μm CMOS technology using a PELOX isolation and a 2.5 nm NO gate oxide

As there is no field oxide present in GAA devices, they should inherently be radiation hard. Initial studies pointed out that GAA nMOSFET only show a minor reduction of the threshold voltage and the transconductance after 10 Mrad(Si) [162 - 163]. The total-dose testing of memory devices based on the GAA structure pointed out that a rebound effect occurs as function of the radiation dose [164]. For nMOSFETs there is first a decrease of the threshold voltage (from 0.51 V before radiation to 0.40 V after 350 krad(Si)) before an increase is observed (0.62 V after 18 Mrad(Si)). The transconductance reduced to 80% of its initial value after 18 Mrad(Si). These results are in good agreement with the rebound effect observed by *Simoen et al.* [165] during their DC and low frequency noise characterisation of GAA devices irradiated up to 30 Mrad, as illustrated in Fig. 6.64. The rebound starts around 3 Mrad(Si) and is explained by the fact that the contribution of the oxide-trapped holes reduces with dose and time, while the interface-state contribution grows steadily causing a monotonous reduction of the maximum transconductance and an overshoot of the threshold voltage. The rebound dose depends on both the technological parameters of the devices and the dose rate used. The impact of the radiation-induced interface-states build-up on the transconductance degradation can be simulated by a model computing the latter as a function of the pinch-off voltage dependence on the drain voltage and the interface trap density [166].

The interest in the use of SOI devices for cryogenic applications has strongly increased [148]. Many of the early observed phenomena associated with the cryogenic operation of SOI devices, such as kink, hysteresis and transient effects, are well understood and can greatly be suppressed by either technology or design optimisation [167]. For detailed information on the different aspects associated with cryogenic operation, the Reader is referred to [168]. Especially for deep submicron feature sizes and nanoelectronics applications cryogenic CMOS has a strong potential [169]. From a radiation hardness viewpoint, it was already in 1988 reported that at cryogenic operation the radiation-induced interface-state generation is suppressed [170 - 171]. This has more recently been confirmed during the 77 K investigation of a radiation hard 0.35 μm CMOS SOI technology [172]. The basis of the so-called hydrogen-model is that radiation induces both neutral hydrogen and electron-hole pairs. The interaction between trapped holes and H^0 leads to H^+ ions, which drift towards the Si-SiO_2 interface and are directly correlated with interface traps [171]. At temperatures below 100 K both the trapped holes and H^0 are immobile so that the interface-state generation is suppressed. Only a subsequent anneal step can initiate this process. However, if electron injection experiments are performed before annealing, then also the N_{it} generation during annealing is suppressed. The models whereby the interface-state generation is only based on trapped holes, i.e., without any role for hydrogen, are not able to explain the observations for their generation during annealing of samples irradiated at low temperatures.

In general, for total dose effects low temperature irradiation is expected to be worse than room temperature irradiation. This is clearly illustrated in Fig. 6.65, giving the front-gate threshold voltage shift for 300 and 77 K irradiation of 0.35 μm CMOS devices [172]. The fact that there is a larger shift for the off-gate bias conditions, which is also seen in Fig. 6.63, is not obvious as it is generally accepted that the worst case corresponds with devices under bias during irradiation. A more detailed study of the electric field distribution within the device structure may shed some light on this point. Experiments performed at 90 K have further pointed out that there is also a dose-rate effect at low temperatures, whereby a lower dose rate gives a lower threshold voltage shift for the same integrated dose [173]. In high-energy physics experiments typical dose rates are of the order of 0.02 krad/hr. As seen in previous chapters, dose rate effects have a serious impact for a variety of semiconductor devices. However, for cryogenic irradiations the available information is limited.

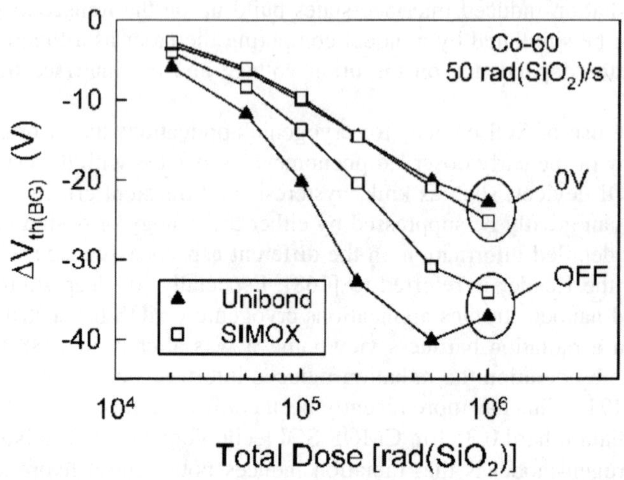

Fig. 6.63. Back-gate threshold voltage shift for nMOSFETs fabricated on Unibond and SIMOX substrates. The transistors were irradiated with Co-60 gamma rays at 50 rad(SiO_2)/s and biased in either 0 V or off configuration during irradiation. [After Schwank et al. [161]]

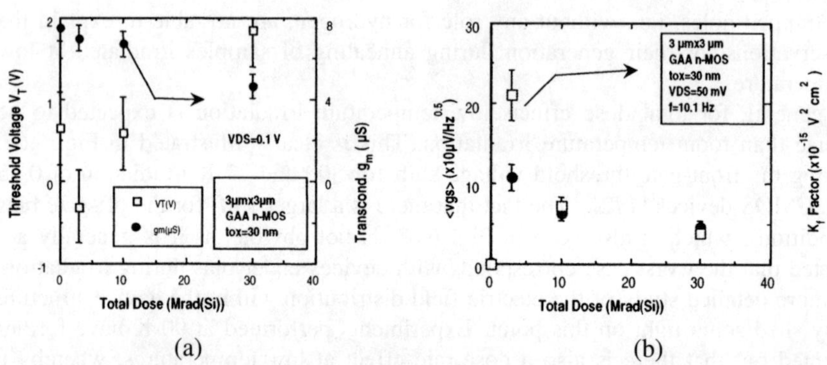

Fig. 6.64. Behaviour of the threshold voltage and maximum transconductance (a) and the average input-referred noise (b) versus total for GAA n-MOS devices. In the latter case a comparison is made with the noise damage factor K_f. [After Simoen et al. [165]]

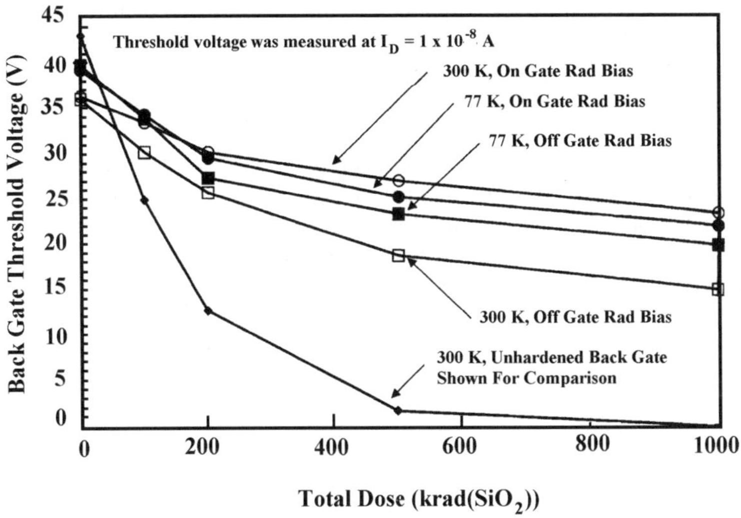

Fig. 6.65. Front-gate threshold voltage shift as a function of the total dose and irradiation condition for both 0 V and off gate biased 0.35 μm SOI-CMOS. [After Jenkins and Liu [172]]

6.8 Conclusions

With the trend of reducing the thickness of the gate dielectric or even replacing it by harder alternatives, whereby N_2O oxynitrides show the best performance, it is expected that scaled submicron technologies can tolerate increasing levels of total dose irradiation. To a certain extent, this also goes for the shallow trench isolation used nowadays, although it requires a careful optimisation of the processing steps. Clearly more work needs to be done in this respect to screen all potential hardness problems and solutions. In line with the COTS philosophy, hardening commercial technologies should be done with as few as possible process modifications, rather than taking measures for extreme hardness levels.

From the above follows that there may exist some 'second order' problems related to the frequent use of certain processing steps like plasma etching, RTA, etc... In many cases, research is developing rapidly, mainly from a viewpoint of manufacturing yield and reliability/lifetime. The hardness issue comes generally last place, but shows many common points of interest with oxide integrity and device lifetime. This means that some of the results and improvements obtained there can be directly translated to the hardness issue.

It is hard to foresee, whether the application of (sub) 100 nm CMOS will bring about some new radiation-induced degradation concerns. Although RILC and SEGR may occur, it is expected that this will only be the case for extreme conditions, which are not typical for the space environment. They are, therefore, at the moment more of academic interest. However, shrinking the device area can have a negative impact on the transient total dose behavior of circuits, like single event upsets and latch-up, since the charge required to switch a node scales along with the technology.

Silicon-on-Insulator is gaining more and more importance as the minimum feature size is entering the 100 nm and below era. These technologies not only have a good radiation performance, but also show a strong potential for cryogenic applications. Therefore, in view of the COTS approach for space applications, it is important to watch closely the advances in this field.

7 GaAs Based Field Effect Transistors for Radiation-Hard Applications

7.1 Introduction

The high electron mobility of GaAs and related III-V compounds renders these materials very suitable for high-speed digital and microwave/millimeter wave applications. The superior operation frequency combined with low high-frequency noise and power dissipation has been exploited for the development of satellite and other telecommunications systems. In this respect, the extreme radiation hardness quoted for these materials is an invaluable plus point: total dose radiation tolerance levels up to 1 Grad(GaAs) have been observed, which is at least two orders of magnitude better than for hardened Si-based technologies. Nowadays, InP is more and more replacing GaAs substrates. However, the use of the semi-insulating (SI) substrates brings about some specific radiation effects, which jeopardize the correct functioning of devices and circuits. Therefore, in Sect. 7.2, some relevant material properties, related to ternary alloys used in a FET and the structure and operation of the Metal-Semiconductor Field-Effect Transistor (MESFET) and the High Electron Mobility Transistor (HEMT), also called Modulation Doped FET (MODFET) will be described. The radiation response and hardening of MESFETs will be presented in Sect. 7.3, followed next by the behavior of HEMTs (Sect. 7.4). Some general conclusions are drawn in Sect. 7.5.

7.2 Material Related Issues and Device Structures and Operation

MESFETs and HEMTs are usually fabricated on SI GaAs substrates grown by the LEC technique, which are known to be more radiation tolerant than Cr-doped material [1]. The semi-insulating nature of the material is a result of the native donor trap called EL2, approximately 0.78 eV below the conduction band, i.e., close to mid gap. It is attributed to an As_{Ga} antisite, as determined by EPR measurements [2, 3] or a complex involving the antisite. In fact, as mentioned in Chap. 4 its exact nature is still a matter of controversy and intense research [3 - 6]. The partial compensation of the deep EL2 donor by acceptor levels from trace impurities introduced during the crystal growth [7] or by the Ga-sublattice related acceptors [2] pins the Fermi level near the mid gap, yielding a semi-insulating substrate. A more detailed description has been given in Chap. 4.

In addition to the III-V substrate, one should also consider the transport and material properties of the epitaxially deposited layers (ternary III-V compounds). Some relevant information will be given in the first sub-section. The structure and basic physics of MESFETs and HEMTs will be described subsequently.

7.2.1 Defects in AlGaAs Layers

In the ternary wide gap AlGaAs layers used in HEMT devices, a similar situation as with EL2 in GaAs exists, where the native DX centers are predominantly present for an Al content larger than 22% [8], showing also radiation hardness. An example is given in Fig. 7.1 for He ion irradiations. No change in the DLTS spectrum is observed after an exposure to 7.6×10^{11} cm^{-2} 5 MeV He ions [9]. These deep donor DX centers are a complex of a donor impurity (D) and an unknown (X) native defect [8]. Like EL2, metastable behavior can be induced by various excitation mechanisms, which provoke long term transients in the device characteristics.

7.2.2 MESFET Structure and Operation

The structure of a MESFET is schematically shown in Fig. 7.2a: in the SI substrate, a highly doped n-layer is fabricated either by Si ion implantation or by epitaxial deposition of a several μm thick layer, which is contacted by the n^+ source and drain region. In many cases, a lowly p-doped buffer layer is deposited on top of the SI substrate before the creation of the channel layer (Fig. 7.2b). It is generally , like in Fig. 7.2b. In Sect. 4.3.2 it was pointed out that the channel technology i.e., epitaxial versus implanted has important consequences for the hardness of the FETs. Typical doping densities are in the range 1 to 5×10^{17} cm^{-3}, depending on the application. For an epitaxial layer, the constant doping level is mostly much lower

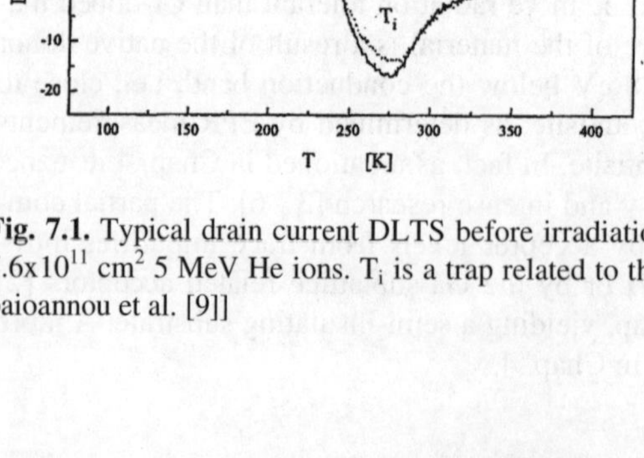

Fig. 7.1. Typical drain current DLTS before irradiation (dashed line) and after a fluence of 7.6×10^{11} cm^2 5 MeV He ions. T_i is a trap related to the hetero-junction interface. [After Papaioannou et al. [9]]

than the maximum doping peak in an implanted device. The reason is that the gate leakage current of the reverse biased Schottky diode should be kept small. For higher surface n-doping densities, the barrier height will be lowered and hence the leakage increased by field-assisted tunneling through the barrier [10]. The application envisaged will also dictate to some extent the use of either an implanted or an epitaxial channel. Finally, on top of the doped n-layer, a metal gate (Au-based) is formed.

In the threshold region of a uniformly doped MESFET, the source-drain current satisfies the relation:

$$I_{\mathrm{D}} = \frac{Z_{\mathrm{W}} \mu_{\mathrm{eff}} \varepsilon_{\mathrm{g}}}{2 d_{\mathrm{n}} L} (V_{\mathrm{GS}} - V_{\mathrm{T}})^2 \tag{7.1}$$

with Z_{W} the gate width, L the length, d_{n} is the channel depth (order ≥0.5 μm), ε_{g} is the permittivity of GaAs, μ_{eff} the effective electron mobility and V_{GS} the gate to source voltage. Using the gradual channel approximation, the threshold voltage V_{T} obeys in first instance:

$$V_{\mathrm{T}} = V_{\mathrm{bi}} - \frac{q d_{\mathrm{n}}^2 n}{2 \varepsilon_{\mathrm{g}}} = V_{\mathrm{bi}} - V_{\mathrm{Po}} \tag{7.2}$$

Hereby is n the channel carrier density (in cm^{-3}), q the elementary charge and V_{bi} the built-in potential of the gate junction. The pinch-off voltage V_{Po} corresponds to the potential necessary to deplete the channel layer. The threshold voltage can according to (7.1) be obtained by extrapolating $\sqrt{I_{\mathrm{D}}}$ to zero V_{GS}, for constant V_{DS} in the linear region.

In case of an implanted profile, a modified theory has been derived [11], starting from the general expression:

$$V_{\mathrm{Po}} = V_{\mathrm{bi}} - \frac{q}{\varepsilon_{\mathrm{g}}} \int_{x_{\mathrm{o}}}^{x_{\mathrm{c}}} x n(x) \mathrm{d}x \tag{7.3}$$

with x_{o} the start of the channel and x_{c} the end, while x_{c}-x_{o}=d_{n} is the channel width.

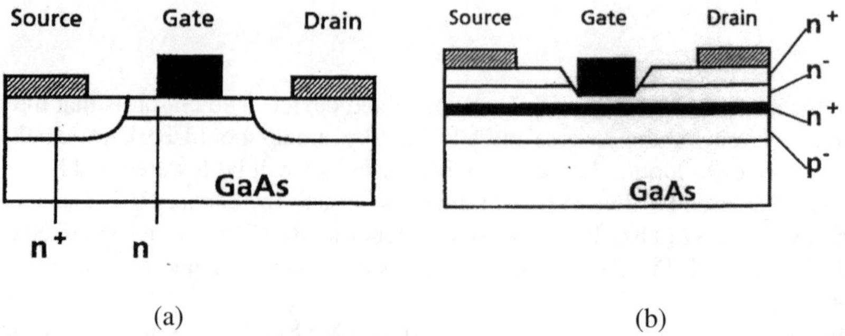

Fig. 7.2. A conventional GaAs MESFET structure (a) and a proposed radiation hard architecture (b). [After Nishiguchi et al. [10]

For a constant profile $n(x)$=constant, (7.2) is found immediately, for x_o=0. The integral has been worked out also for a Gaussian implantation carrier profile [11].

For device and circuit simulations, Eq. (7.1) is too simplistic to describe the full I_D-V_{DS} curve in the whole drain bias range. A suitable empirical model, which has been utilized to simulate the impact of radiation damage is given by [12]:

$$I_D = \beta'(V_{GS} - V_T)^{\kappa} \tanh(\alpha' V_{DS})(1 + \lambda_H V_{DS}) \tag{7.4}$$

β' sets the saturation current (flat level at large V_{DS}), α' models the initial slope (linear region, small V_{DS}) and accounts for the higher mobility in linear operation. The exponent κ is usually assumed to be 2, but can also be considered as a fitting parameter and λ_H is the channel length modulation parameter and models the output conductance of the transistor. It often takes the form:

$$\lambda_H = \frac{\lambda_{Ho}}{1 + G_{Ho}(V_{GS} - V_T)} \tag{7.5}$$

with λ_{Ho} and G_{Ho} fitting parameters.

For analog applications the low-frequency (LF) noise is an important parameter. This even applies for micro- and millimeter-wave non-linear circuits, where the phase noise of an oscillator is determined by the LF noise. Figure 7.3 illustrates the possible sources of LF noise in a GaAs MESFET [13]. It could be (1) the metal-semiconductor interface; (2) the free surface region between source and gate and drain and gate (ungated region), which is frequently covered by a passivation layer (SiO_2 for example); (3) the neutral channel region; (4) the gate depletion region and (5) the channel-substrate depletion region. It has been shown that the dominant source of LF noise is generation-recombination (GR) by trap levels in the depletion region [13]. By studying the impact of the substrate type (SI versus Cr doped) and the effect of the presence of a buffer layer, it was concluded that the main source of

LF noise is the substrate. It is not clear whether the GR noise is predominantly generated in region (4) or (5). On the other hand, it was also observed that the surface contribution to the noise generation is small. Finally it has been found that the input noise voltage varies according to $1/\sqrt{L}$, with L the device length. This is in line with GR noise generated in the depletion region of the MESFET [13]. For L<0.5 μm, the noise varies according to $1/L$.

7.2.3 HEMT Structure and Operation

The layer scheme of a lattice matched Si doped AlGaAs HEMT is represented in Fig. 7.4a [14], while a pseudomorphic HEMT (P-HEMT), where the lattice mismatch between the GaAs substrate and the InGaAs channel layer is larger is shown in Fig. 7.4b [15]. For not too large gate biases, charge transport from source to drain is by the two-dimensional electron gas (2-DEG) existing at the GaAs/(undoped)wide-gap InGaAs interface. The carriers are provided by the highly doped AlGaAs (or InGaAs) donor layer (typically 10^{18} cm^{-3} donors), which is offset from the channel by the lowly doped AlGaAs spacer layer. The fact that the 2-DEG electrons are separated from the charged donor atoms in the highly doped layer reduces strongly the Coulombic scattering contribution to the mobility, so that effectively high mobility transport occurs in the channel. This becomes particularly pronounced for cryogenic temperature operation, where Coulomb scattering is relatively more important. The GaAs buffer layer is undoped but has in practice a slight p-type character (i.e. contains a net acceptor density). At higher gate

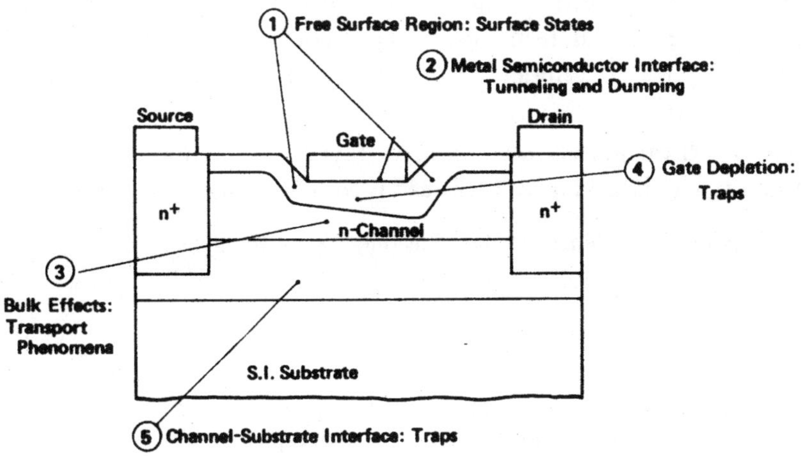

Fig. 7.3. Possible sources of 1/f noise in a GaAs MESFET. [After Su et al. [13]]

biases, also the low-mobility donor layer will contribute to the drain current, which gives rise to a degradation of the effective mobility and hence of the device transconductance g_m. The band diagram of a GaAs HEMT is represented in Fig. 7.5, defining also the relevant device parameters. More complex layer schemes are sometimes used, for example with an extra AlGaAs buffer layer on top of the SI substrate, whereon the p-GaAs channel layer is deposited [9, 16]. This extra heterojunction has been shown to provide improved resistance against ionising radiation-induced photoconductivity (back-gating and transients).

The on voltage of a AlGaAs/GaAs HEMT is determined by a number of parameters. In the depletion approximation, it can be shown that the electric field at the heterojunction interface is given by [17]:

$$F_i = \frac{-V_{off} - (E_F/q) + V_{GS}}{d_n + d_s} \tag{7.6}$$

Hereby is d_n the doped AlGaAs(n) layer and d_s the spacer layer thickness, E_F the Fermi level relative to the bottom of the quantum well and V_{off} is the voltage which annihilates the channel charge, and given by:

$$V_{off} = \varphi_b - (\Delta E_c/q) - V_{A1} \tag{7.7}$$

ϕ_b is the Schottky barrier height and ΔE_c is the conduction band discontinuity at the AlGaAs/GaAs interface. For radiation damage studies, it is generally assumed that the latter two parameters are not affected by the irradiation [9, 17]. This has been verified experimentally for ΔE_c under γ-irradiation, yielding a constant value of ≈0.165 eV up to 40 Mrad(GaAs) [14]. Finally, V_{A1} in Eq. (7.7) is the potential drop across the AlGaAs layer. For a homogeneously doped layer, V_{A1} is given by (7.2) and is equal to minus the pinch-off voltage ($-V_{Po}$). For radiation damage studies, it is only V_G', given by:

$$V_G' = V_{GS} - \varphi_b + (\Delta E_c/q) \tag{7.8}$$

which needs to be considered. For the calculation of E_F, it is generally assumed that only the first quantum level in the 2-D well is occupied [9, 17], which simplifies the analysis considerably. Based on that, it can be shown that the 'threshold' voltage defined in (7.8) is a function of the AlGaAs (n) and the GaAs (N_A) doping density and of the geometry (AlGaAs donor and spacer layer thickness d_n+d_s) [17].

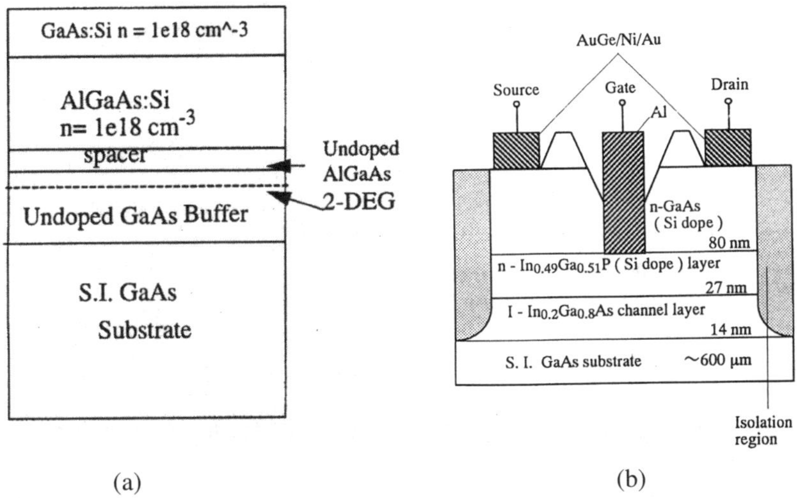

Fig. 7.4. (a) Schematic cross-section of an AlGaAs/GaAs HEMT. [After Subramanian et al. [14] and (b) of an InGaP pseudomorphic HEMT. [After Ohyama et al. [15]]

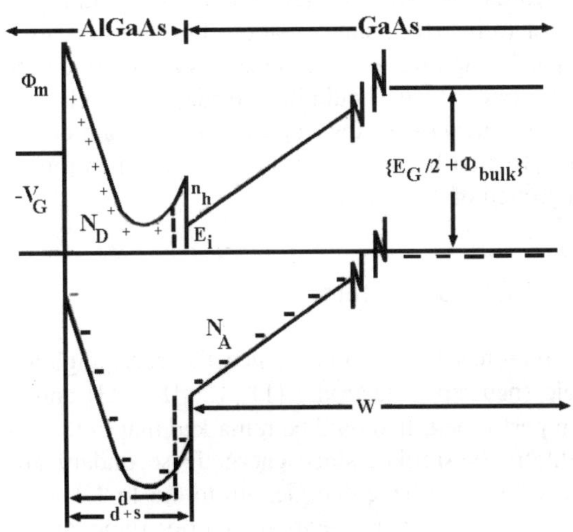

Fig. 7.5. Band diagram of a HEMT structure with Schottky contact, with barrier height ϕ_b, under gate bias V_{GS}. AlGaAs donors and unintentional acceptors (density N_A) are assumed to be completely ionised. Near the threshold, the Fermi level (dash-dot line) lies below the bottom of the 2-DEG well. [After Krantz et al. [17]]

This results finally in a non-linear relationship between the gate bias close to threshold and the surface electron density in the 2-DEG n_S [9]:

$$V_{GS} = \frac{q(d_n + d_s)n_S}{\varepsilon_g} + (\phi_b - (\Delta E_c / q) - V_{Po}) + \frac{kT}{q}\ln[\exp(\frac{qn_S}{D_H kT}) - 1]$$
$$+ \alpha_0 (n_S + N_A^*)^{2/3} \qquad (7.9)$$

N_A^* is the surface acceptor density in the GaAs buffer layer (in cm^{-2}), D_H is the density of states of the 2-DEG and equal to 3.24×10^{17} $m^{-2}V^{-1}$ [9] and $\alpha_0 = 2.5 \times 10^{-12}$ eV $m^{4/3}$. A satisfactory approximation for n_S values below 5×10^{11} cm^{-2} is given by (7.9) [9].

7.3 Radiation Damage and Hardening in GaAs MESFETs

In first instance, a MESFET can be considered as a "bulk" type of device, whereby charge transport occurs through a channel which is at some distance from the Schottky contact interface. It will therefore be mainly degraded by displacement damage in the channel region. Only in the ungated source-gate and drain-gate region, the carriers may be exposed to the surface and, hence, to charging in a passivation dielectric layer if present or to interface state creation. Furthermore, ionisation can also induce some typical trapping and transient response in the SI substrate, so that total dose and dose rate effects should be considered as well, particularly on a circuit level [18]. However, most of the important device and circuit parameters will be predominantly affected by displacement damage. Therefore, a large part of the studies has concentrated on this issue.

7.3.1 Degradation of the Basic FET Parameters

The radiation response of MESFETs has been studied since the early eighties, whereby both high-energy particle (neutrons, electrons) [11, 12, 18 - 23] and γ-irradiations [10, 22, 23] have been performed. It should be remarked that in the latter case, the degradation is qualitatively similar, since energetic secondary and Compton electrons are generated, which can have energies up to 0.6 to 1.1 MeV [20]. The latter again create the necessary displacements for carrier removal and mobility degradation. Also the combined effect of neutron/gamma irradiations showed some interesting features [24].

The degradation of the main MESFET parameters, namely, mobility μ_{eff}, pinch-off voltage V_{Po} and carrier removal is illustrated in Figs 7.6a to 7.6c [20], for different types of devices, after exposure to gamma's and 1 MeV electrons. It is seen that the epitaxial FETs show a lower degradation than the ion-implanted devices, although the doping density is approximately the same ($\approx 2 \times 10^{17}$ cm^{-3}). This is in line with the resistor results of Fig. 4.15 in Chap. 4 [25]. It is generally assumed that the degradation of the pinch-off voltage is proportional to the carrier removal rate and inversely proportional to the original doping density. In other words [11, 20]:

$$\frac{\mathrm{d}V_{\mathrm{Po}}/\mathrm{d}\Phi}{V_{\mathrm{Po0}}} = \frac{\mathrm{d}n/\mathrm{d}\Phi}{n_0 - n} = 1 - a_{\mathrm{n}}\Phi \tag{7.10a}$$

V_{Po0} and n_0 are the original pinch-off voltage and doping density and a_{n} the damage factor for carrier removal, defined in (4.2).
Likewise, for the mobility (4.3) is valid [11]:

$$\frac{1}{\mu} = \frac{1}{\mu_0}(1 + b_{\mu}\Phi) \tag{7.10b}$$

while for the transconductance [11]:

$$\frac{g_{\mathrm{m}}}{g_{\mathrm{m0}}} = \frac{n\mu}{n_0\mu_0} = \frac{(1 - a_{\mathrm{n}}\Phi)}{(1 + b_{\mu}\Phi)} \tag{7.10c}$$

and for the drain current [11

$$\frac{I_{\mathrm{D}}}{I_{\mathrm{D0}}} = \frac{n^2\mu}{n_0^2\mu_0} = \frac{(1 - a_{\mathrm{n}}\Phi)^2}{(1 + b_{\mu}\Phi)} \tag{7.10d}$$

Again, b_{μ} is the mobility degradation factor; the symbols with an '0' subscript are pre-irradiation values.

From Fig. 7.6c it is derived that for an average Compton electron energy of 300 keV within the devices, only a few gamma created electrons will produce atomic displacements, explaining their lower effectiveness apparent in Figs 7.6a and 7.6b. The higher degradation found for implanted FETs is first of all explained by the presence of (shallow) unannealed (Si) implantation damage, which becomes electrically activated by the irradiation [7].

The lower radiation tolerance of an implanted MESFET technology has been further elaborated by *Janousek et al.* [11]. An example of the threshold voltage degradation for a neutron irradiated implanted FET is given in Fig. 7.7 versus fluence. As can be seen the change in V_{Po} (or V_{T}) is much larger than can be predicted by (7.10a) for a uniform doping density and certainly non-linear with fluence.

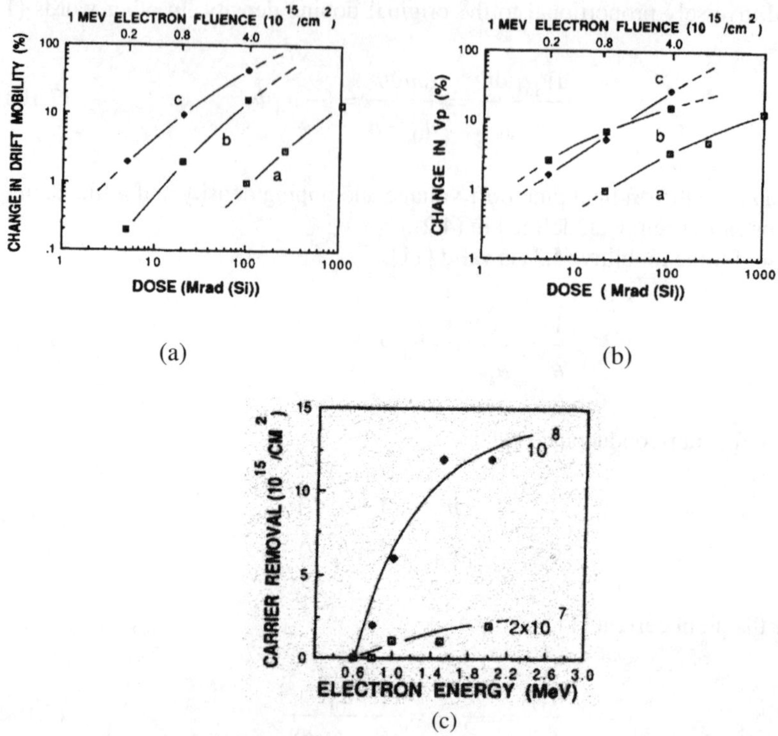

Fig. 7.6. (a) Percent degradation in mobility versus fluence and dose for γ– and 1-MeV e⁻ irradiation of vapor phase epitaxy (VPE) and ion-implanted material. Curve a: VPE, γ (dose abscissa); curve b: VPE, e⁻irradiation (dose and fluence abscissa); curve c: implant, e⁻ irradiation (dose and fluence abscissa). (b) Percent degradation in pinch-off voltage versus dose and fluence for γ and 1-MeV e⁻ irradiations of FATFETs (=50 μm long) on VPE and implanted GaAs. Curve a: VPE, γ (dose abscissa); curve b: VPE, e⁻-irradiation (dose and fluence abscissa); curve c: implant, e⁻-irradiation (dose and fluence abscissa). (c) Carrier removal at 2×10^7 (curve a) and 10^8 rad(Si) (curve b) for different e⁻ energies in FATFETs fabricated on GaAs epitaxial layers. [After Meulenberg et al. [20]]

A model has been developed based on a Gaussian doping profile and including a local doping density dependent carrier removal rate of the form [11]:

$$a_n(x) = a_0 n(x)^{-\nu} \tag{7.11}$$

which is equivalent to a carrier removal rate:

$$a^*(x) = a_0 n(x)^{1-\nu} \tag{7.12}$$

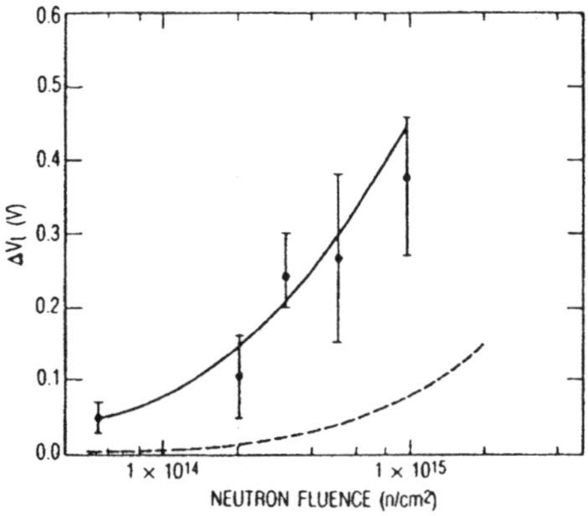

Fig. 7.7. Shift in threshold voltage as a function of neutron fluence (solid line). Predicted shift in threshold voltage using theory for a uniform profile (broken line). The solid dots are the average of the data. [After Janousek et al. [11]]

It is not only shown that the irradiation can change the shape and width of the profile (channel), but in order to explain the measured V_T shift one has to use a carrier removal rate which is larger for implanted FETs than the value of 6 cm^{-1} standardly assumed for 10^{17} cm^{-3} doped n-GaAs [26], with a value of 20 cm^{-1} closer to the observations. This is in line with other results for implanted technologies [11], yielding rates between 10 and 100 cm^{-1}. This higher removal rate can be associated with the nature of neutron-induced defects formed in ion-implanted/annealed material versus epitaxial material. Only for low neutron fluences the threshold voltage of implanted MESFETs can be described by a simple linear relationship with fluence as given by (7.10a). On the other hand, the uniform doping of an epitaxial layer may lead to a substantial hardening of the technology.

The combined effect of neutron and gamma irradiation on implanted MESFETs has been studied by *Chang et al.* [24]. In the standard experiment, the neutron-irradiated sample was first annealed at 150°C in air for 30 min before the application of a γ-exposure to $2x10^7$ rads. This agrees with the onset of the 150-225°C annealing stage III of radiation-induced point defects (simple primary defects; V_{As}-As_i). A second anneal stage occurs at 400-450°C and is thought to correspond to larger "clustered" damage (U band, EL6,...) [27], expected for neutron irradiations, not for γ's. A typical result of this procedure is shown in Fig. 7.8. For moderate neutron doses, a slight improvement of the device transconductance is noted - the same applies for γ-doses up to $5x10^6$ rad(GaAs). This is in line with the results shown in Fig. 4.17 for the Hall mobility [28, 29] of GaAs.

As can be noted in Fig. 7.8, annealing produces a slight recovery of the characteristics. However, a subsequent γ-irradiation gives rise to an enhanced degradation in the high neutron fluence range. This degradation is not caused by the γ-irradiation alone, as experimentally verified. It is concluded that there exists some latent damage (neutral defects) in neutron exposed material which becomes activated (i.e., filled by charge carriers) by the gamma's [24]. For low to moderate neutron fluences, the device degradation will be governed by the carrier removal, since b_μ is negligibly small (mobility degradation factor). For high neutron fluences, on the other hand, where b_μ is no longer negligible, the mobility degradation becomes the predominant degradation mechanism for the transconductance and the drain current (see (7.10)).

Another important observation is that the experimental carrier removal rate is linear with Φ only for fluences up to $\approx 10^{15}$ cm^{-3} (Fig. 7.9), in line with the observations of *Janousek et al.* [11]. For the largest fluences, the removal rate becomes smaller than expected from (7.10a). *Chang et al.* [24] propose a model, which is valid, for the whole neutron fluence range. It not only considers carrier removal but also carrier (donor) generation, in order to explain the reduced V_T change. This is achieved in the following way [24]:

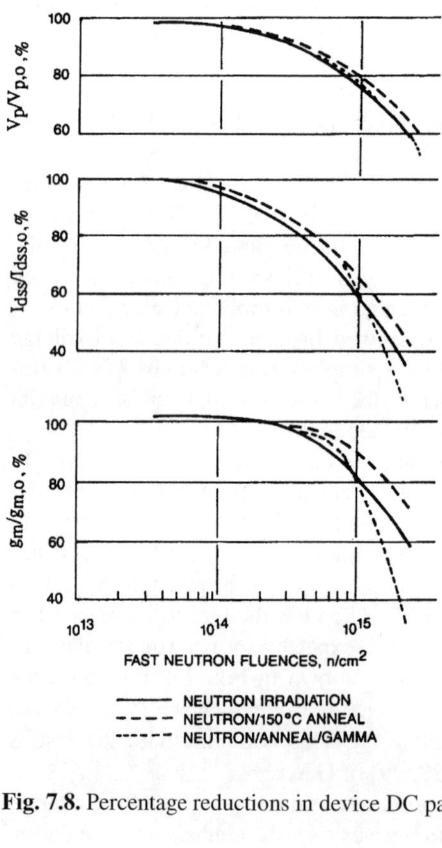

Fig. 7.8. Percentage reductions in device DC parameters. [After Chang et al. [24]]

$$\frac{V_{Po}}{V_{Po0}} = \frac{n}{n_0} = 1 - \frac{a_n \Phi}{(1 + a_n' \Phi^{\kappa})} \tag{7.13}$$

Fitting (7.13) to the data shows that κ=1 and $a_n' \approx a_n$, so that one can write:

$$\frac{V_{Po0}}{V_{Po}} = \frac{n_0}{n} = 1 + a_n \Phi \tag{7.14}$$

and for the carrier removal rate:

$$\frac{dV}{d\Phi} = \frac{a_n}{n_0} n^2 \tag{7.15}$$

According to (7.15) the carrier removal rate depends quadratically on n, which is indeed found experimentally as shown in Fig. 7.10 and in agreement with other results, which predict larger values than 6-10 cm^{-1} for implanted GaAs with a doping density of 10^{17} cm^{-3} [11].

Using the carrier removal rates after neutron and after the combined neutron/γ irradiation, one can distinguish between the contributions of point defects and defect clusters, as shown in Fig. 7.11. While the point defects show a 1-to-1 slope with the number of disappearing carriers, a slope of two (or more) is found for the defect clusters, implying that at least two carriers are captured in these neutron-related centers. This suggests thus larger defects, which are expected to be more thermally stable than the point defects and, therefore, correspond with the 400-450°C annealing stage.

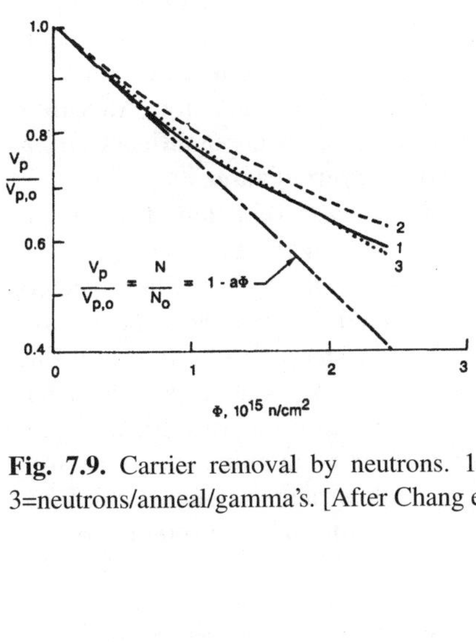

Fig. 7.9. Carrier removal by neutrons. 1=neutrons; 2=neutrons + 150°C annealing and 3=neutrons/anneal/gamma's. [After Chang et al. [24]]

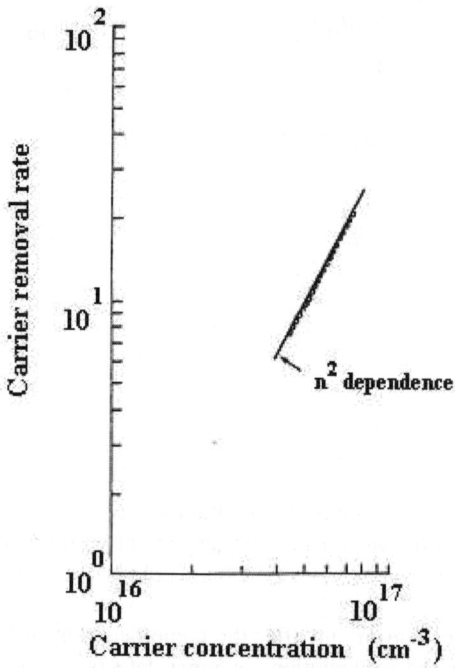

Fig. 7.10. n^2 dependence of the carrier removal rate. [After Chang et al. [24]]

These observations help to understand the difference in effects induced by gamma's (very moderate) and by energetic neutrons (more pronounced) in GaAs MESFETs. The gamma photons and the secondary and Compton electrons are seen as being rather ineffective in producing carrier trapping effects in GaAs other than a moderate number of shallow trapping levels very near the conduction band [24] (and valence band [30]). These shallow trapping levels coincide with the ones corresponding to unannealed implantation damage and are activated thermally at room temperature - they are responsible for the room temperature conduction. Annealing at 150°C enables to stabilize the shallow levels, resulting in no significant difference in the device performance, between pre- and post γ-irradiation.

Exposure to energetic neutrons, however, leads to the production of more permanent defects, which continue to grow in number with increasing fluence and may even coalesce into defect clusters at a sufficient high fluence. These centers are deep enough so that charge carriers trapped in them become permanently lost for conduction. At the same time, these charged (cluster) defects are responsible for a significant mobility degradation through Coulombic scattering, which becomes the dominant degradation mechanism at high neutron fluences. Shallow levels, on the other hand, are at room temperature ineffective in reducing the electron mobility. It has also been shown that beside the shallow radiation-induced defects, one needs to assume two other types of centers, in order to fully interpret the obser-

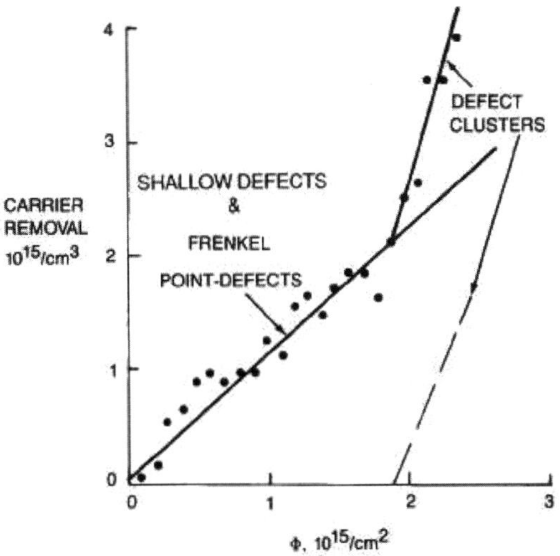

Fig. 7.11. Defect buildup versus neutron fluence. [After Chang et al. [24]]

ved degradation. For moderate neutron fluences, deep-level permanent Frenkel type of point defects are generated linearly with fluence and capable of trapping one carrier. Beyond a certain fluence threshold, in addition, larger aggregates of point defects may be formed, with each cluster capable of trapping two (or more) charge carriers. They also produce the largest effect on mobility [24].

A final remark is that due to this synergistic effect of a combined neutron/gamma exposure, one may wonder how a MESFET device/circuit will behave in a mixed radiation environment.

7.3.2 Low-Frequency Noise and Defect Related Effects

Recently, interest has emerged for the impact of radiation on the LF noise performance of MESFETs [7, 31, 32]. It has been observed that the 1/f noise voltage of a MESFET follows both a $1/\sqrt{L}$ [13] and a $1/\sqrt{Z_W}$ dependence [32]. It was found that the LF noise after γ-exposure increases with accumulated total dose (Fig. 7.12), for frequencies above 1 kHz. In that frequency range, a clear GR noise component develops, which points to the creation of radiation-induced shallow traps or the stimulation of already-present shallow levels, like unannealed Si ion-implantation-related damage [7]. The displacement of a Si donor to an As site introduces shallow acceptor levels [30], which can cause such a high-frequency GR noise. The broad frequency range suggests a distribution of different shallow levels. The LF part of the spectrum << 1 kHz is thought to be dominated by EL2-like deep levels and remains unchanged here [32].

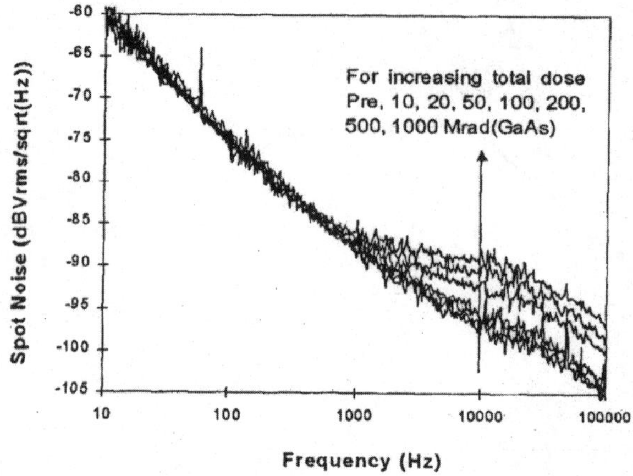

Fig. 7.12. 1x600 μm GaAs MESFET spot noise for various total dose, post-irradiation at test set output. [After Hiemstra et al. [32]]

Another issue is the dose rate effect *during* the exposure, illustrated by Fig. 7.13. It is seen there that the LF noise degrades severely for increasing dose rate [32]. The phenomenon saturates in the range 150 krad(GaAs)/hr (Fig. 7.14a), which is attributed to the fact that the range of stimulated traps (selected time constants) is narrowed down, while at low dose rates, a broad ensemble of time constants (energy levels) is excited. In addition, the concentration of traps is finite, leading to the observed saturation. However, the effect of the dose rate reduces for higher accumulated doses (Fig. 7.14b) [7]. At the same time, noise spikes at low frequencies (range 10 - 100 Hz) have been reported [7], which correspond to the irradiation induced stimulation of the deep level traps. These could be substitutional oxygen donors on an As site, having a similar energy level as EL2, which may 'anneal' under irradiation, and are known to anneal at rather low temperatures [7]. These LF noise peaks increase for increasing dose rates and are believed to be due to a sidegating effect, whereby the deep level traps are excited by the exposure over the whole substrate. Sidegating is the effect where the substrate next to the MESFET can affect its performance, while back- or selfgating is related to the substrate below the channel. No saturation of this detrimental sidegating effect, causing noise peaks, has been reported.

This side- or back-gating effect is directly related to the metastable behaviour of the EL2 defect in the SI substrate, mentioned in Sect. 4.2.1. Absorption of a photon transforms the EL2 centre in a charged metastable state, which leads to long-term transients in the drain current of a FET. These originate from the electrons, which have been excited to the deep traps by the photons and slowly relax back to the conduction band by thermal emission [7, 33 – 35].

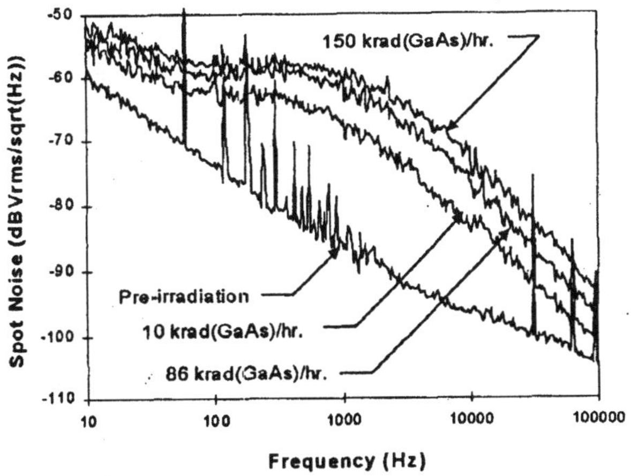

Fig. 7.13. 1x300 μm GaAs MESFET spot noise for various dose rates at the onset of irradiation at test output. [After Hiemstra et al. [32]]

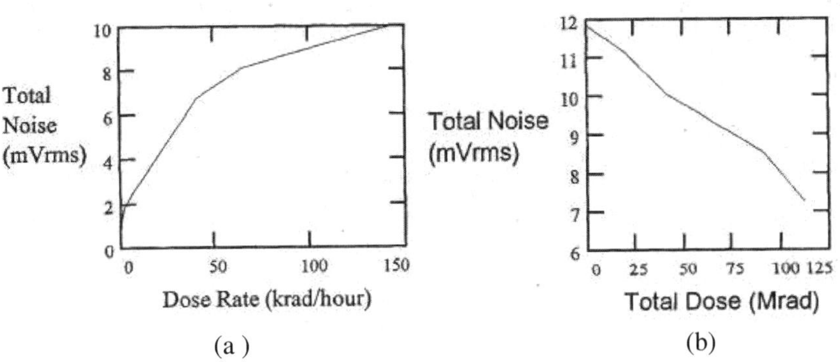

Fig. 7.14. (a) Dose rate dependent total noise and (b) total noise during irradiation versus total dose for a 142 krad(GaAs)/hr dose rate. [After Hiemstra [7]].

In order to improve the dose rate dependent 1/*f* noise, a few measures can be taken. A better performance could be achieved by the implementation of an epitaxially grown buffer layer between the substrate and the implanted active area. An alternative approach involves controlling the state of the semi-insulating substrate by applying a substrate-contacted process, which provides a die backside metallization pattern below the device channels. Application of an appropriate bias eliminates backgating [7].

The impact of cryogenic irradiations in the range 100 K and higher has been studied by Shaw et al. [36]. The a_n and b_μ damage factors have been studied as a function of temperature, for 3 MeV protons. The results are represented in Fig.

7.15a and 7.15b, showing that the carrier removal rate is a factor 2.5 larger for 100 K irradiations, compared with RT. The same applies for the mobility degradation (Fig. 7.15b). Thermal annealing experiments revealed a broad annealing stage at 270 K, while no apparent recovery for T_A<225 K was found (Fig. 7.16) [36]. However, this annealing was only found for the carrier density n and not for the mobility.

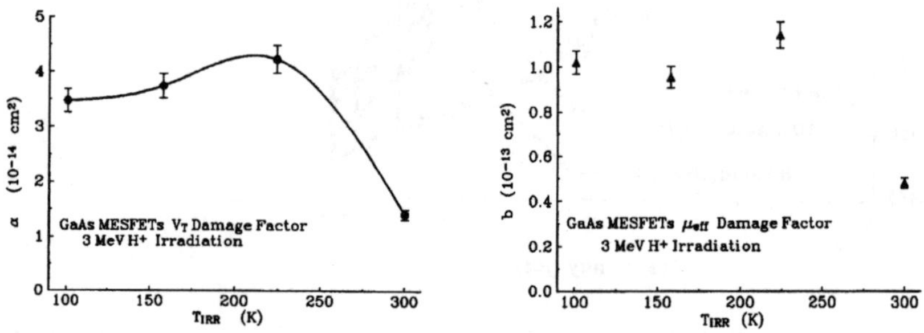

(a) (b)

Fig. 7.15. (a) V_T damage factor versus irradiation temperature T_{IRR} and (b) corresponding μ_{eff} damage factor. There is no strong T_{IRR} dependence for temperatures below T_{IRR}~ 224 K. [After Shaw et al. [36]]

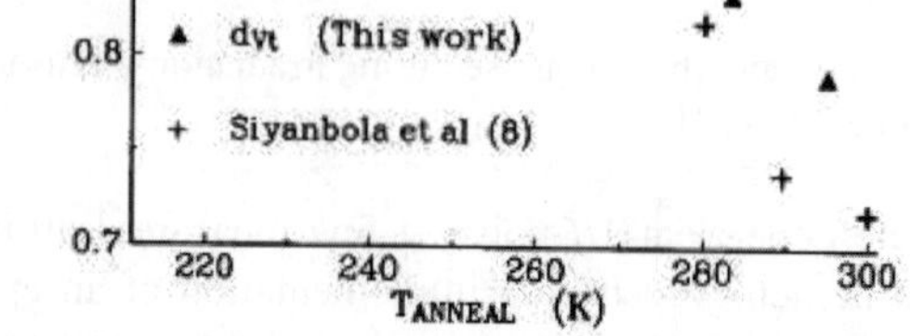

Fig. 7.16. Recovery of the differential change in threshold voltage versus isochronal annealing temperature, plotted with the annealing data of *Siyanbola et al.* [38]. The results are qualitatively similar. [After Shaw et al. [36]]

A comparison of the annealing behaviour of resistors and MESFETs pointed out that the recovery of a resistor above 225 K is approximately half that of the

FET's, which can be explained by the different annealing behaviour of μ and n. The observed annealing stage is in agreement with earlier published work [37, 38] related to the observation of deep level traps and carrier removal, measured by DLTS and C-V, respectively. Two acceptor levels at E_v+0.25 eV and E_v+0.42 eV have been observed after a 120 K 1 MeV proton irradiation of n-type GaAs, showing an annealing stage at ≈280 K. It is suggested that the corresponding defects are related to Ga displacements, possibly a deep state of V_{Ga}, or a mixed divacancy (see Sect. 4.2.2). They are thus shown to be responsible for the carrier removal during low temperature irradiation. Additional charged centers should mainly cause a mobility degradation. Other annealing stages of radiation induced defects have been reported at 180 K, 235 K, 280 K and 520 K [36 - 38].

By studying the transient phenomena in the gate capacitance or the drain current, it is possible to perform a kind of deep-level spectroscopy on an irradiated MESFET [9, 39]. This should enable the identification of the trap levels in the channel region responsible for the device degradation. Beside DLTS-like techniques, it has been demonstrated that the study of the frequency dispersion of the transconductance g_m or of the channel conductance g_D provides an alternative way to investigate radiation-induced deep levels. The advantage is that the study can be performed at one temperature (e.g. RT), whereby the frequency of the small-amplitude excitation signal applied for example to the gate is varied in a broad range. This allows to identify the presence of the U-band at E_c-0.5 eV and EL14 at E_c-0.2 eV after 6 MeV (average energy) neutron exposure [39]. The presence of EL2 (E_c-0.79 eV) was also detected before and after irradiation, showing no change. Furthermore, also the gate to source current may be an important monitor for radiation damage in a MESFET [39]. In general, a reduction of the gate current was observed in both polarities. This indicates that no deep generation centers were created in the depletion region. The behavior suggests a compensation of the donor levels in the channel by radiation induced acceptors, giving rise to a reduction of the tunneling current in reverse bias.

7.3.3 Circuit Related Degradation

Several studies have considered the effect of displacement damage (carrier removal and mobility degradation) on MESFET based circuits [12, 18 – 20, 22, 23, 40]. One of the basic building blocks for digital circuits is a ring-oscillator consisting of a number of inverters. Studies of neutron irradiated inverters show that the noise margins and gain increase after exposure (Fig. 7.17). Especially the high noise margin improves dramatically (Fig. 7.18). The characteristic becomes also more centered (symmetrical) which explains in first instance the improvement of the invertor performance. This also suggests an easy way for radiation hardening of digital MESFET circuits, whereby the initial inverter should have its switching point to the right of the optimum center point.

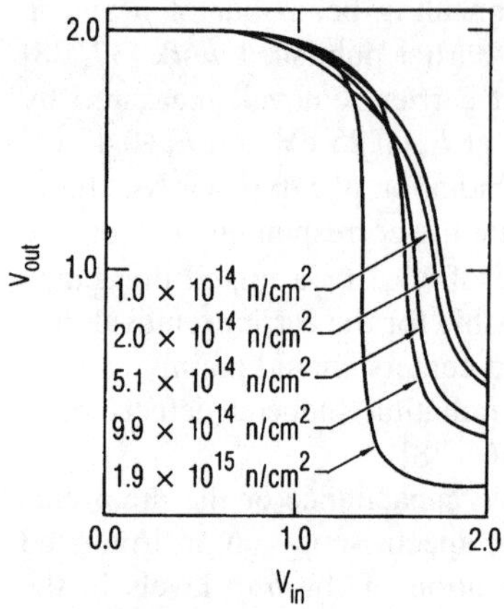

Fig. 7.17. Measured Schottky Diode FET Logic (SDFL) inverter characteristics as a function of neutron fluence. V_{in} and V_{out} are the input and output voltage, respectively. [After Bloss et al. [12]]

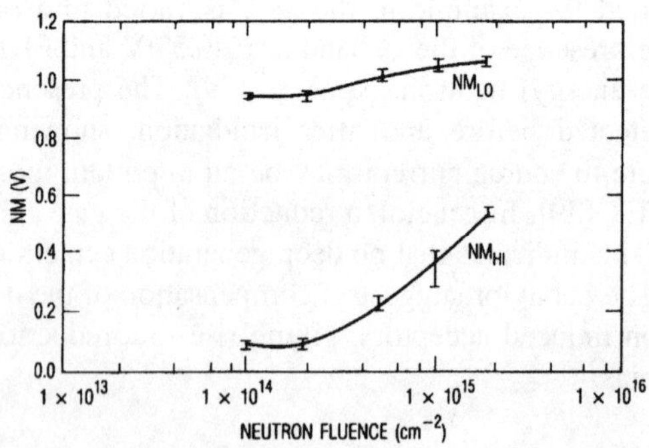

Fig. 7.18. High and low noise margins of SDFL inverters as a function of neutron fluence determined from measured data. Note the increase of the noise margins for higher neutron fluences. [After Bloss et al. [12]]

The fact that the threshold voltage of the FETs becomes more positive causes the observed changes in Fig. 7.17. In addition, the output conductance (in saturation) of the individual FET improves (becomes larger), giving rise to a flatter I_D-V_{DS} curve [32] and steeper inverter characteristics. In other words, the short-channel effect improves upon irradiation. This is related to the fact that the SI substrate becomes more compensated upon irradiation - thus more resistive. This reduces substrate currents and carrier injection, back- and sidegating coupling and kink effects, yielding a higher R_{out}.

It is concluded that although the threshold voltage and the transconductance of individual devices are severely degraded after exposure to $\approx 2 \times 10^{15}$ cm^{-2} neutrons, the logic circuits will still be functional, even under extreme radiation conditions

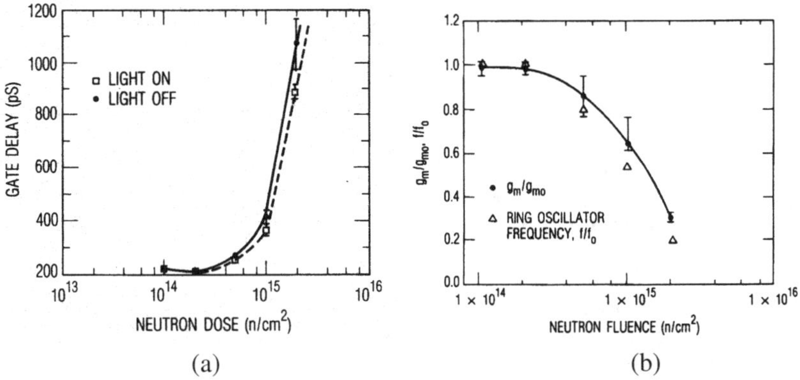

Fig. 7.19. (a) Measured gate propagation delays of a 13-stage ring oscillator as a function of neutron fluence and (b) measured fractional transconductance degradation of GaAs FET and frequency degradation of a 13-stage ring oscillator as a function of neutron fluence. [After Bloss et al. [12]]

[12]. However, the speed of operation will degrade significantly, as seen from the drastic increase of the ring oscillator gate delay in Fig. 7.19a. A slight improvement is seen upon illuminating the circuit; this is because of the ionisation of deep radiation-induced traps by the incident photons. The observed speed degradation follows exactly the observed degradation of the FET transconductance with fluence (Fig. 7.19b) [12].

The high frequency performance of irradiated MMIC circuits (amplifiers) has been reported for example by *Meulenberg et al.* [20]. In that work, a broad-band distributed amplifier, using an ion-implanted technology was compared with a K_a band power amplifier on epitaxial material. Figures 7.20a and 7.20b show the gain degradation in the GHz range for both amplifiers, after a 10^8 rad(Si) 1-MeV electron irradiation. The gain reduction is more pronounced at the high frequency end, while being negligible for the lower frequencies.

It has also been observed that the transient response upon a high dose rate electron (or γ-pulse) is quite different for the two circuits [20]. The implanted amplifiers show a faster, higher-amplitude peak after 10^8 rad(Si) 1-MeV electrons, while the opposite is true for the epitaxial circuit. A number of factors contribute to these phototransients, related to design aspects (distributed finger gate), the operation mode (applied voltage), the passive circuit components and to the material/technology. The origin of the transients is the substrate, whereby the buffer layer in the epitaxial material plays the major role. It is even suggested that pre-exposure to a low dose can be used as a hardening technique against the transient effect for epitaxial technologies [20]. This results in a more resistive (compensated) p-type buffer layer. The same could in principle be obtained by lightly p-type doping the buffer layer. Other suggestions have been the implementation of an extra AlGaAs layer at the back of the structure [9, 16], or using a thinner substrate.

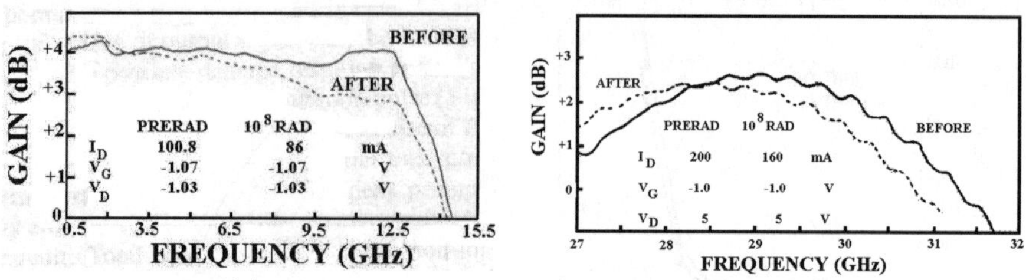

(a) (b)

Fig. 7.20. (a) Performance of MMIC distributed amplifier before and after a 1-MeV electron irradiation (10^8 rad(Si)), biased for pre-irradiation gain flatness. (b) Performance of K_a-band MMIC amplifier before and after a 1-MeV electron irradiation. [After Meulenberg et al. [20]].

The temperature dependence of these transient effects has been studied by Anderson et al. [40], showing that for neutron irradiated MMIC circuits, the induced damage levels provoke long time transients at low temperature operation. A final ionisation related circuit effect is the occurrence of single event upsets for which the Reader is referred to the review by *Zuleeg* [18].

7.4 Radiation Damage and Hardening in HEMTs

This section addresses displacement damage effects on HEMTs. First, the change in the basic parameters is highlighted for high-energy particle irradiations. Next, a few words will be spent on the degradation of the 2-DEG properties under low-energy electron exposures. This is of interest in view of the application of electron-beam (e-beam) lithography, which is necessary for the fabrication of devices with sub-100 nm gate lengths. In a third part, circuit aspects will be discussed.

7.4.1 Degradation of the Basic Parameters

The basic radiation damage mechanisms in GaAs/AlGaAs HEMTs have been studied by a number of groups [9, 14, 17, 41 - 44]. The analysis of the basic displacement damage effect on the threshold voltage is based on (7.6) to (7.9). Starting from there, the radiation induced change can be modeled by [17]:

$$\Delta V_T = \Delta V_{Ga}' \Phi[1 - V_{A1}(a_n n / N_A^2)' / \Delta V_{Ga}] \quad (7.16)$$

with a_n the carrier removal rate in the AlGaAs, $V_{\mathrm{Al}}=n\, d_{\mathrm{n}}^{2}\, 2\varepsilon_{\mathrm{g}}$ the threshold shift, due to the AlGaAs layer and ΔV_{Ga} depends on the GaAs material parameters as described in [17]. Equation (7.16) predicts a linear change of V_{T} with fluence, which is valid in the low fluence range. A slightly modified theory has been developed in [9, 40], whereby only the net acceptor concentration in the GaAs buffer layer is considered. Figure 7.21 demonstrates that the degradation of the threshold voltage of a HEMT is a complex function, composed of different contributions. Figure 7.22 shows the different terms, whereby it is clear that the carrier removal in the AlGaAs layer contributes for only 2.5% in the whole fluence range. The major impact comes from the change in the Fermi level and the change in the field at the interface due to the redistribution of the charge in the GaAs buffer [17].

Typical degradation of the *I-V* characteristics for neutron irradiation is given in Figs 7.23 and 7.24 [43]. The degradation of the drain saturation current is given by a sublinear law [43]:

$$I_{\mathrm{DSAT}} = I_{\mathrm{DSAT0}}(1-\beta_1 \Phi^{\theta}) \tag{7.17}$$

where θ=0.425, for neutron irradiations, this in contrast to the case of He ion irradiation which produces a linear reduction [9]. Such a non-linear law is thought to be caused by a change in conduction mechanism for increasing neutron fluence, whereby the original 2-DEG quantum well transport is overtaken by conduction in the AlGaAs doping layer. This is related to the carrier removal within the 2-DEG upon irradiation.

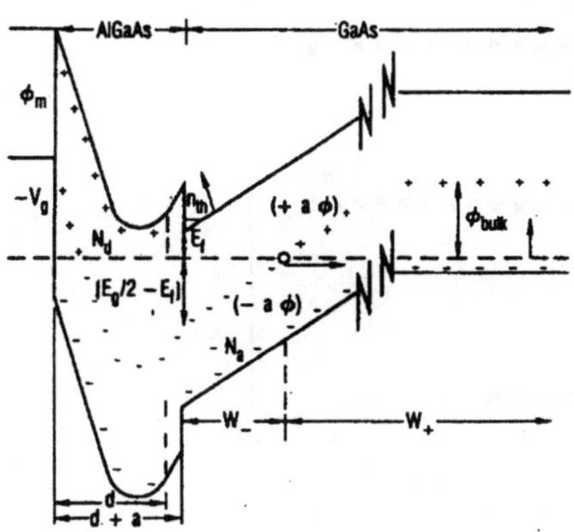

Fig. 7.21. Post-irradiation band diagram of a typical MODFET structure. W_- is the extent of the depletion region in the GaAs in which the traps are below the Fermi level. W_+ is the extent of the region in which the traps lie above the Fermi level. [After Krantz et al. [17]]

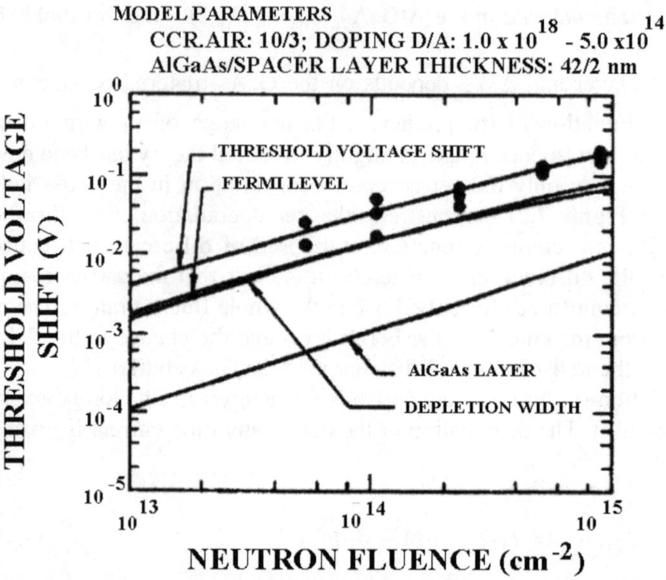

Fig. 7.22. Contributions of the threshold voltage shift versus neutron fluence. For the modeling, a carrier removal rate of 10 cm^{-1} is assumed in the AlGaAs, while a GaAs acceptor introduction rate of 3 cm^{-1} is considered. [After Krantz et al. [17]]

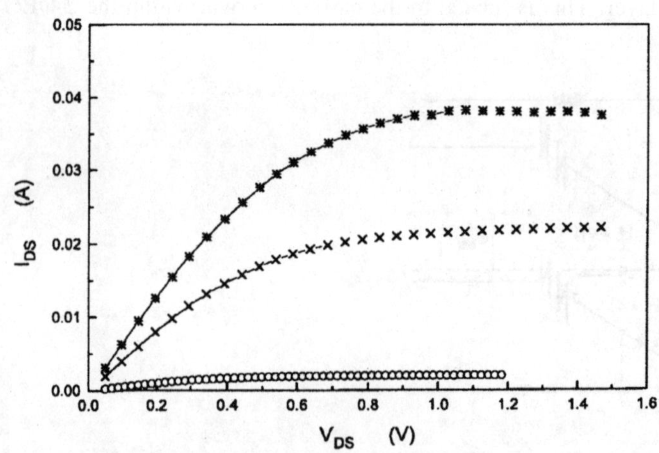

Fig. 7.23. Dependence of I_D-V_{DS} (V_{GS}=0 V) characteristics on the radiation fluence. (*) before irradiation; (x) after a fluence of 10^{15} neutron/cm^2 and (o) after a fluence of $1.7x10^{16}$ neutrons/cm^2. [After Papastamatiou et al. [43]]

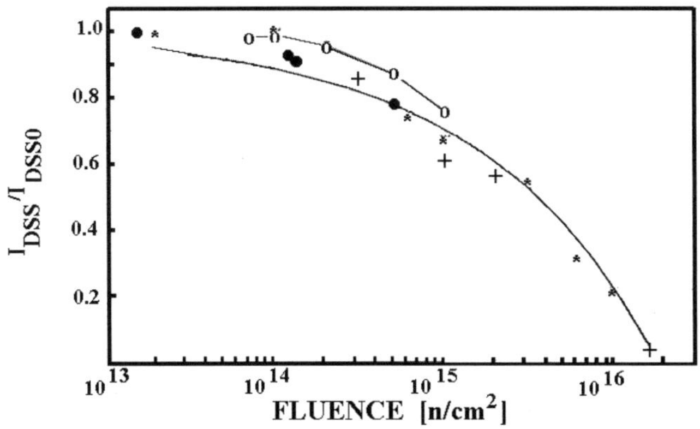

Fig. 7.24. Dependence of the HEMT drain saturation current versus radiation fluence for (*) structure A. (+) structure B with a backside AlGaAs layer and (o) data from [45], and (o) transconductance degradation from [46]. [After Papastamatiou et al. [43]]

The reduction of the 2-DEG concentration is represented in Fig. 7.25. It is found to decrease linearly with fluence (Fig. 7.26), giving rise to a damage constant of $6x10^{-17}$ cm^{-2} [43] (neutrons) or $2.5x10^{-13}$ cm^2 (He ions) [9]. Furthermore, using *C-V* measurements on large area (long) HEMTs enables to determine the carrier profile in the AlGaAs/GaAs heterojunction [9], yielding the result of Fig. 7.27. The heavy ion irradiation used there clearly reduces the carrier concentration in the AlGaAs donor layer and in the 2-DEG. The carrier removal rate following from fitting the threshold voltage model to the experimental data gives rise to values in the range 100 cm^{-1} (neutrons) to $2x10^5$ cm^{-1} (He ions) which is much larger than what is commonly assumed for III-V compounds (i.e. 10 cm^{-1} [17]).

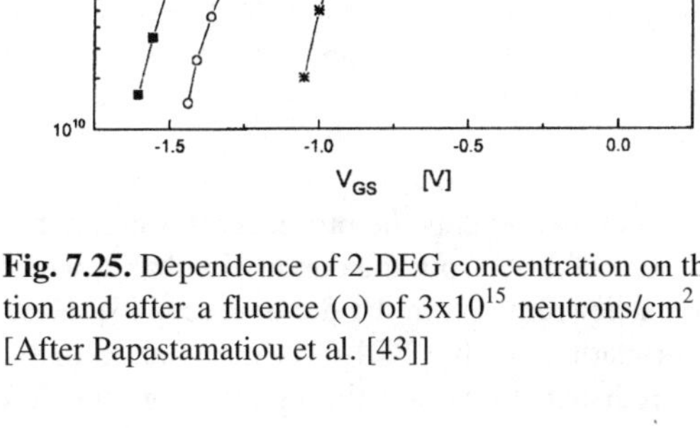

Fig. 7.25. Dependence of 2-DEG concentration on the gate bias (■) before irradiation and after a fluence (o) of $3x10^{15}$ neutrons/cm^2 and (*) $6x10^{15}$ neutrons/cm^2. [After Papastamatiou et al. [43]]

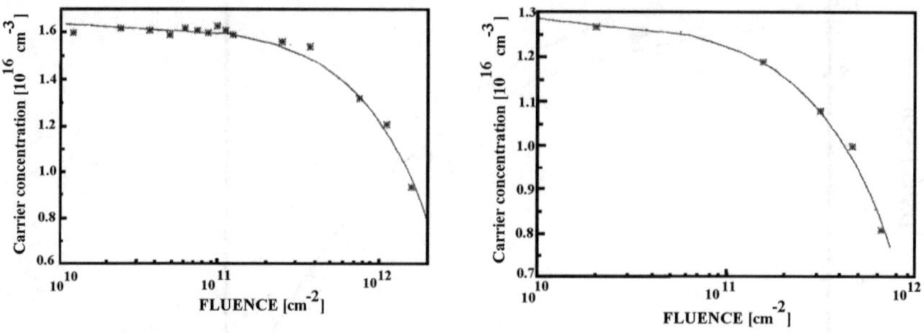

(a) (b)

Fig. 7.26. (a) Dependence of the 2-DEG carrier concentration on the 5-MeV He^+ radiation fluence. The continuous line corresponds to a linear fit. (b) Dependence of the AlGaAs donor layer carrier concentration on the radiation fluence. The continuous line corresponds to a linear fit. [After Papaioannou et al. [9]]

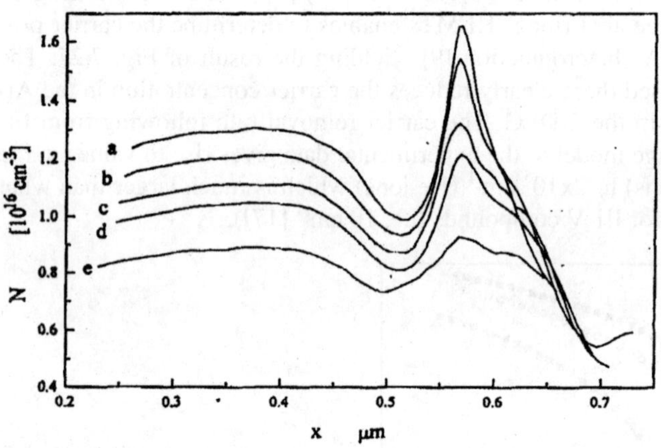

Fig. 7.27. Profile of an AlGaAs/GaAs heterojunction obtained from C-V characteristics (a) before irradiation, (b) after a 5-MeV He^+ fluence of 3.8×10^{11} cm^{-2}, (c) after a fluence of 7.6×10^{11} cm^{-2}; (d) after a fluence of 1.1×10^{12} cm^{-2} and (e) after a fluence of 1.5×10^{12} cm^{-2}. [After Papaioannou et al. [9]]

DLTS of high fluence irradiated structures reveals the presence of a number of deep traps at 0.46 eV (DX), 0.59 eV and 0.62 eV from the conduction band, which are thought to be responsible for the carrier removal in the donor layer [43]. However, the dominant factor in the degradation of the 2-DEG is the change in the buffer layer doping density (charge redistribution) and the creation of interface traps at the AlGaAs/GaAs interface [9, 17, 43]. A similar electron trap at E_c-0.55 eV has been observed in 220-MeV carbon irradiated AlGaAs HEMTs [47, 48], while in InGaP p-HEMTs traps at E_c-0.50 eV (1 MeV electrons) and E_c-0.52 eV (220-MeV carbon) were found by DLTS [49].

The behavior of the mobility is shown in Fig. 7.28. For low 2-DEG densities, the mobility increases due to screening of Coulombic interaction, followed by a peak. At higher carrier densities, part of the conduction takes place in the highly doped AlGaAs, which is characterized by a lower mobility [9, 43]. The 2-DEG mobility degrades almost linearly with fluence, which is correlated with the creation of charged gallium vacancies (acceptors) in the GaAs layer [30].

Another cause for mobility degradation is the creation of charges and the increase in roughness at the AlGaAs/GaAs interface [9, 43]. Evidence for the latter has been obtained in [14]: according to the analysis of the results of Fig. 7.29 a surface state density at the n-GaAs/n-AlGaAs interface of $\approx 10^{11}$ cm^{-2} is created after exposure to 40 Mrad γ's. The induced charges provoque a reduction of the electron mobility by Coulomb scattering. The mobility damage factor β_μ amounts to 0.59×10^{-15} cm^2 (neutrons) [43] and 4.25×10^{-13} cm^2 (He ions) [9]. However, a strong impact of the device structure has been observed [9, 40]. While the addition of a back AlGaAs interface layer doubles the mobility damage coefficient in the case of neutron exposure [40], a reduction of it is seen for HEMTs with a Low Temperature (LT) deposited AlGaAs donor and spacer layer. A better radiation hardness is thus obtained in the latter case (Fig. 7.30) [9]. It is thought that the presence of additional background bulk and interface defects introduced during the low temperature growth causes this improvement, requiring a larger radiation induced charge to compensate for it. The penalty is a lower initial device and material quality.

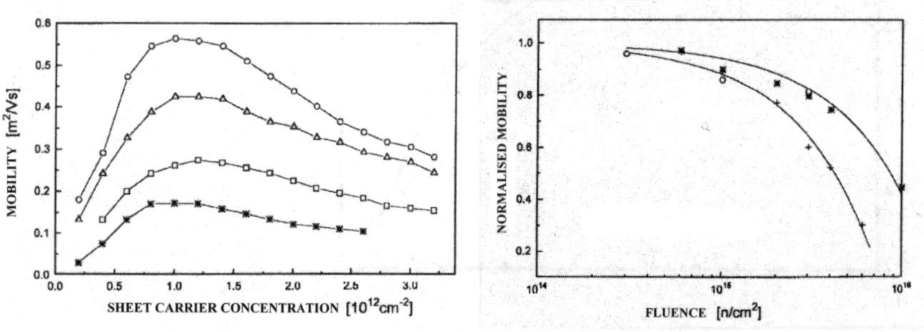

(a) (b)

Fig. 7.28. Typical dependence of the mobility on the sheet carrier concentration: (o) before irradiation and after a fluence of (Δ) 6×10^{14} neutrons/cm^2, (□) 3×10^{15} neutrons/cm^2 and (*) 6×10^{15} neutrons/cm^2. (b) Dependence of the normalized mobilities to their pre-irradiation values, on the radiation fluence (*) standard structure, (+) structure with additional back side AlGaAs layer to reduce the photo-transients, and (o) of GaAs, whose data were obtained from [29]. [After Papastamatiou et al. [43]]

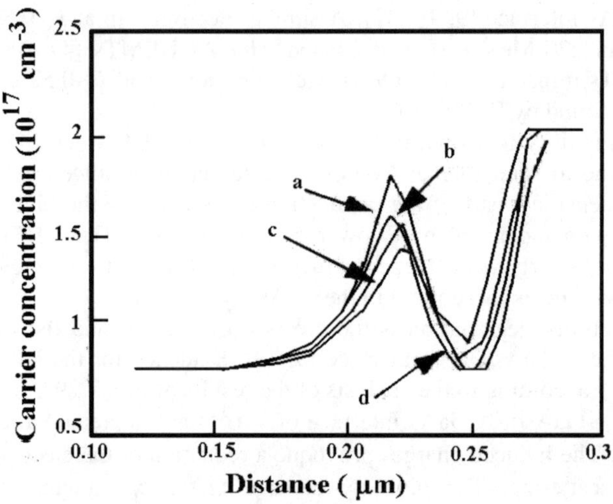

Fig. 7.29. Measured carrier concentration profiles of a GaAs/AlGaAs single heterojunction structure for different gamma ray irradiation doses: (a) unirradiated; (b) 1.0 Mrad (c) 10.0 Mrad; (d) 40 Mrad. [After Subramanian et al. [14]]

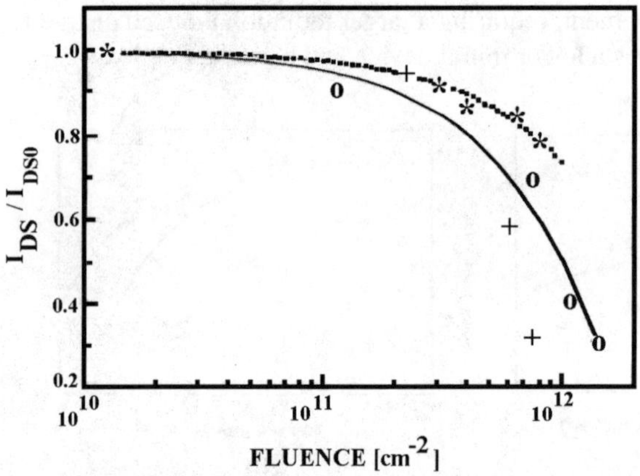

Fig. 7.30. Relative degradation of drain saturation current at V_{GS}=0 V vs 5-MeV He$^+$ fluence for (+) a D-mode HEMT; (b) a commercially available one, and (*) a HEMT with low-temperature donor and spacer layers. The continuous and dotted lines represent fitting curves to the linear degradation model. [After Papaioannou et al. [9]]

An additional diagnosis for radiation-induced defects is the study of the drain current as a function of (low) temperature [9, 42, 43]. An example of the impact of neutron exposure is shown in Fig. 7.31. At large radiation fluences, a drastic reduction of the current is observed, corresponding to a change in conduction mechanism. For curve c in Fig. 7.31, the 2-DEG conduction has vanished and current is flowing through the donor layer. In other words, it behaves as an AlGaAs MESFET. From the I_D-T characteristics information concerning the trapping mechanisms can be extracted, as indicated in Fig. 7.31. A thermally activated current is observed corresponding to two activation energies (0.05 and 0.18 eV). It is furthermore derived that the deeper level shows a donor nature [43].

Beside the channel properties, it has been observed that also the device parasitics are degraded upon irradiation [9, 43]. This is particularly true for the series resistance R_{SD}, which comes from the ungated source to gate and gate to drain regions plus the contact resistance. One can generally write for the total resistance:

$$R_{tot} = R_{SD} + R_{chan} \tag{7.18}$$

with R_{chan} given by:

$$R_{chan}^{-1} = \mu(n_S)\frac{qn_S Z_W}{L} \tag{7.19}$$

R_{SD} depends on the device geometry, doping levels and the fabrication process. Irradiation causes generally an increase of R_{SD} as shown in Fig. 7.32 for different HEMT structures, after He exposure [9]. The change is small for low fluences, but

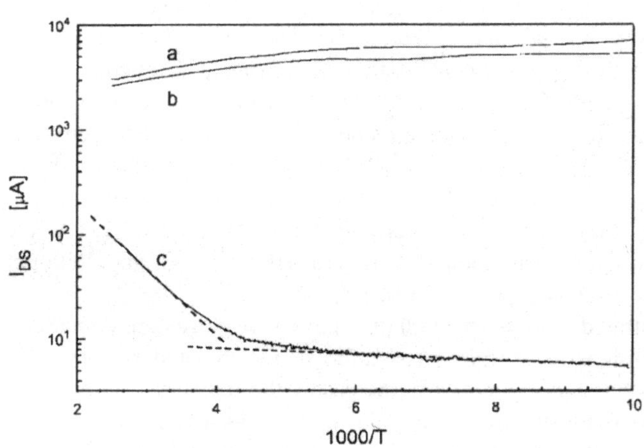

Fig. 7.31. Temperature dependence of the drain current of a conventional HEMT. All curves were obtained at V_{GS}=0 V and V_{DS}= 50 mV: (a) before irradiation; and after a fluence of (b) 3×10^{14} neutrons/cm^2, and (c) 1.7×10^{17} neutrons/cm^2. In (c), the dotted lines result from fitting of the experimental data to a thermally activated drain current with activation energies of 0.18 and 0.05 eV. [After Papastamatiou et al. [43]]

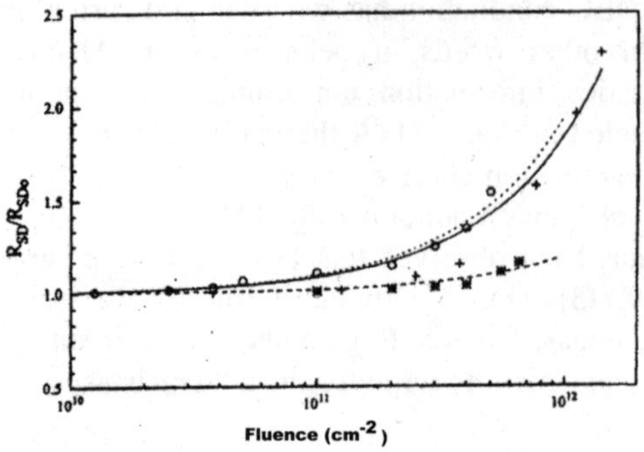

Fig. 7.32. Variation of the normalised series resistance to their values before irradiation vs He ion radiation fluence for (o) a standard structure, (+) a HEMT with a back AlGaAs layer and (*) a LT-HEMT. [After Papaioannou et al. [9]]

once a threshold surpassed, the increase is fast. For neutron exposure, a linear increase with fluence has been reported [43]. Such behavior can not simply be explained by carrier removal only and other mechanisms need to be considered: for example chemical reactions with Au metallization at the ungated surface could play a role [41].

In order to explain the degradation of the threshold voltage a refined charge control model has been developed [9, 42, 43] with three fitting parameters: the thickness of the AlGaAs donor layer d_n, the donor layer pinch-off voltage V_{Po} and the net acceptor layer density N_A^* per unit of area in the GaAs buffer layer. Based on the model, a good fit with the experimental data can be found, in function of the fluence. An example is given in Fig. 7.33 [43]. The resulting N_A^* versus neutron fluence is represented in Fig. 7.34, showing a reduction of the net acceptor concentration. It is a general observation that the slightly p-type GaAs buffer layer becomes more intrinsic after irradiation. Based on the linear relationship between ΔN_A^*, one can define a corresponding damage factor which equals 4.2×10^{-7} for conventional HEMT structures and about 1.4×10^{-6} for HEMTs with an additional buffer layer and this for neutron exposures [43].

It has been demonstrated that also γ-irradiation causes displacement damage in HEMTs [14, 42]. Additionally, surface passivation of the ungated regions with SiO_2 for example can play a definite role in increasing the radiation tolerance for total dose γ's [14]. This is illustrated in Fig. 7.35 for a GaAs/AlGaAs modulation doped structure without (a) and with (b) passivation. The degradation is mainly found for the electron mobility. In the same work [14], it has been reported that the GaAs/AlGaAs interfaces are quite stable and robust against radiation damage. GaAs/InGaAs/GaAs quantum well structures show some impact of the irradiation on the symmetry of the carrier profiles (Fig. 7.36). This is interpreted as being due to a slight rearrangement by In diffusion in the well during irradiation. The peak charge density, however, remains approximately constant, suggesting a negligible increase of the interface charge density, which is already high at the beginning.

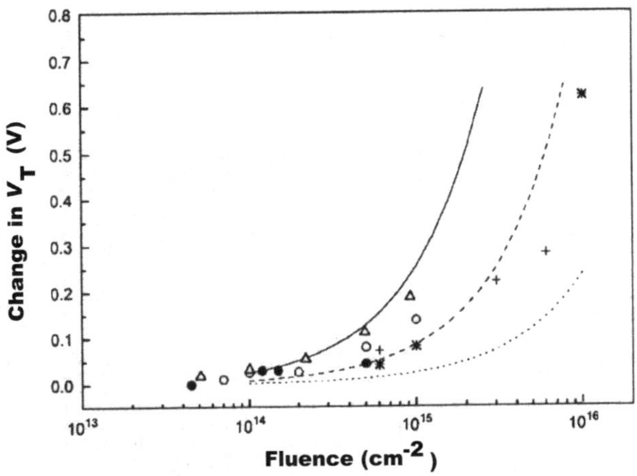

Fig. 7.33. Variation of threshold voltage with the radiation fluence for devices which data were obtained from literature and devices with a conventional structure (*) and (+) and additional AlGaAs buffer layer. The theoretical curves were obtained using a carrier removal rate (dotted) of 30 cm^{-1}, (dashed) 100 cm^{-1} and (continuous) 300 cm^{-1}. [After Papastamatiou et al. [43]]

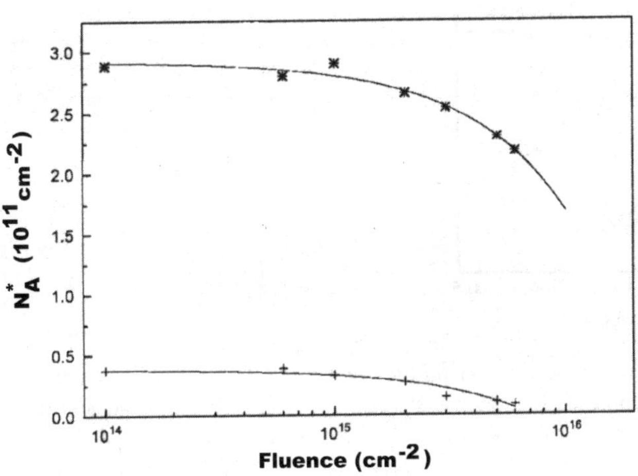

Fig. 7.34. Buffer layer, equivalent net acceptor concentration N_A^* per unit area versus neutron radiation fluence (*) for a conventional HEMT and (+) a HEMT with an additional AlGaAs buffer layer. [After Papastamatiou et al. [43]]

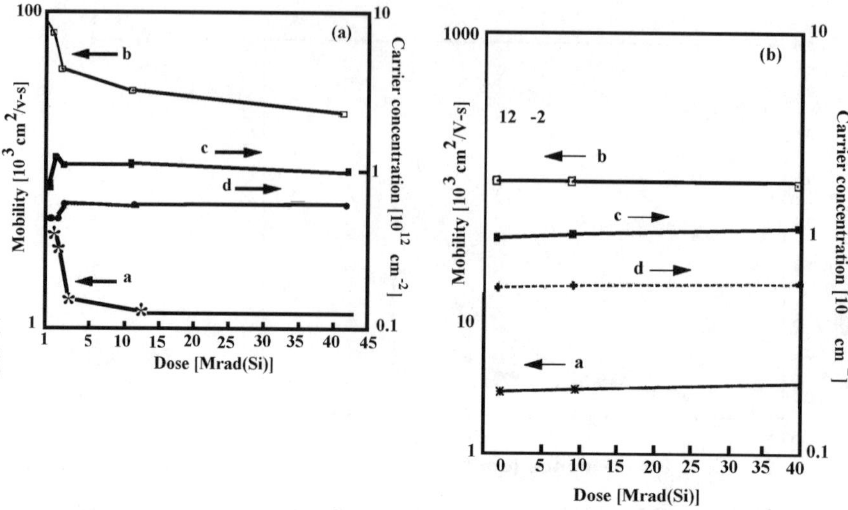

Fig. 7.35. Gamma irradiation dose dependence of Hall mobility at 300 K (a) and 77 K (b) and carrier concentration at 300 K (c) and 77 K (d) of a GaAs/AlGaAs modulation doped heterostructure sample without SiO_2 passivation (A) and with SiO_2 passivation (B). [After Subramanian et al. [14]]

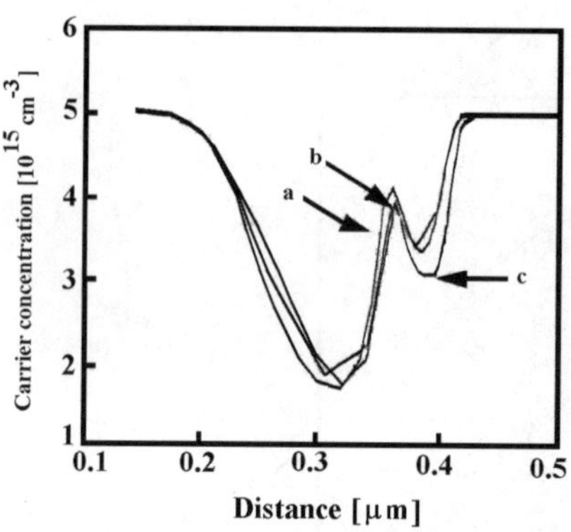

Fig. 7.36. Measured carrier concentration profiles of GaAs/InGaAs/GaAs quantum well structure for different gamma irradiation doses: (a) unirradiated; (b) 10 Mrad and (c) 40 Mrad. [After Subramanian et al. [14]]

Initial results of 1.8 MeV proton irradiation effects on $Al_xGa_{1-x}As/GaN$ HEMTs have also been reported [50]. Such devices raise interest because of their potential for high-temperature, high power microwave applications. It was shown that the drain current and transconductance were seriously degraded after a fluence of 10^{14} cm^{-2}, but could be mostly recovered after a rapid thermal anneal at 800°C [50].

7.4.2 Low-Energy Electron Effects on 2-DEG Properties

E-beam lithography has been frequently used for writing (sub) 100 nm lines in HEMT technology. Typical electron beam energies in the range of a few keV are used, which is much lower than the displacement damage threshold E_d. However, studies have pointed out that degradation of the 2-DEG parameters can occur for low-energy electron- or Ar-ion irradiations [51 – 55]. According to Fig. 7.37a, it is mainly the electron mobility of the channel, which is degraded, while the carrier density suffers less [54]. Figure 7.37b shows the energy dependence of the mobility degradation. It has been demonstrated that the maximum damage occurs for some intermediate electron energies, depending on the thickness of the respective epitaxial layers [52, 54]. For energies below 5 keV, the electrons can not penetrate deep enough to create significant damage, while for energies well above 10 keV, most of the damage is created too deep to affect the 2-DEG behaviour. This depth dependence can also be inferred from Fig. 7.37b.

As to the nature of the responsible defects, it has been concluded that besides Coulombic (charged or ionised) scattering centres also neutral, short range scatterers close to the AlGaAs-GaAs interface contribute to the mobility degradation [54]. In first instance, it is believed that existing defects are being modified (ionised) by the electrons, as their energy is too low to produce an atomic displacement [52]. The thermal stability of the effects has been studied on a few occasions [53, 55]. Partial recovery in the mobility at above 200 K indicates that the defects become thermally mobile (or reconfigure) [53]. Complete recovery is found above 400 K [55], which has been ascribed to the annealing of V_{As}-related complexes close to the hetero-interface [55].

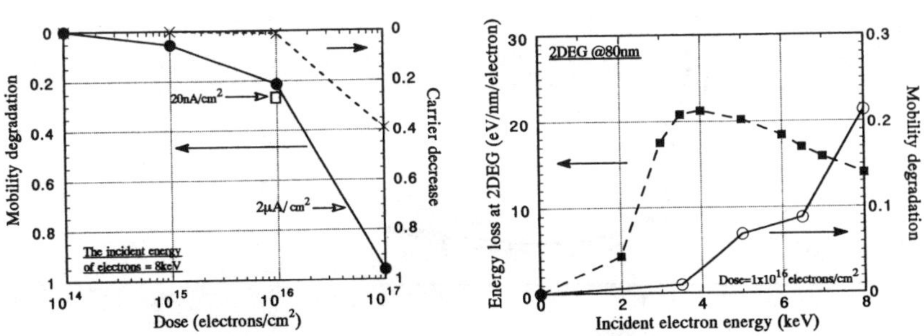

(a) (b)

Fig. 7.37. (a) The electron fluence dependence of the mobility degradation and sheet carrier decrease. The incident energy of electrons was 8 keV. The closed circles represent the mobility degradation when irradiated with the beam current density of 2μA/cm^2 and the open squares, 20 nA/cm^2. The crosses indicate the sheet carrier decrease. (b) The incident electron energy dependences of the mobility degradation (open circles) and the electron energy loss at the 2 DEG region obtained by the Monte Carlo simulation (closed squares). The 2 DEG depth is 80 nm and the irradiated fluence is 1x10^{16} cm^{-2}. [After Wada et al [55]]

7.4.3 Circuit Degradation Aspects

The impact on HEMT based circuits and MMICs has been the subject of a few studies [46, 56]. Inverters still work properly after a neutron fluence of 10^{15} neutrons/cm^2, as can be judged from the result of Fig. 7.38 [46]. The inverter characteristic becomes slightly more steep, while the noise margins drop by about 200 mV for 10^{15} neutrons/cm^2. Figure 7.39 shows the degradation of the propagation delay with neutron fluence, which is more pronounced for 77 K operation. The reason for this larger sensitivity at cryogenic temperatures is the higher mobility and the relatively larger degradation upon exposure [46].

A general observation is also that HEMT devices and circuits are generally more radiation tolerant than MESFETs (see e.g. [56]).

Important for microwave circuits is the degradation of the high-frequency (HF) gain and the minimum noise figure (NF) of the HEMTs (see also Chapter 5). Results for an AlGaAs device after 20 MeV α-irradiation are given in Fig. 7.40 [57]. One can derive a strong increase of the NF with proton fluence, accompanied by a reduction of the current gain. It has also been demonstrated that a partial recovery can be achieved by a 15 min anneal at 150°C [57].

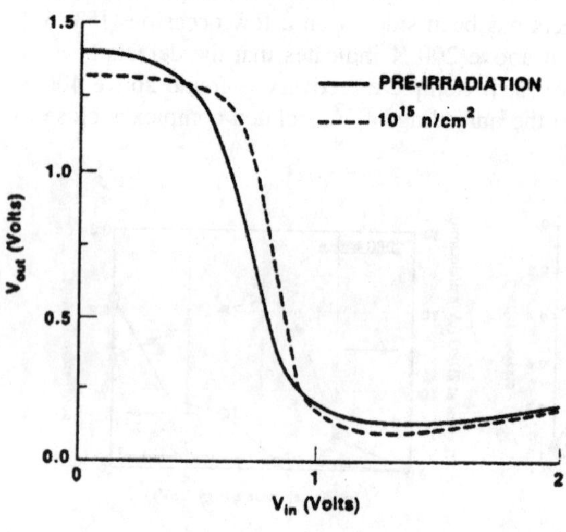

Fig. 7.38. SFFL inverter characteristics comparing pre-irradiation curve to curve following 10^{15} neutrons/cm^2. [After Janousek et al. [46]]

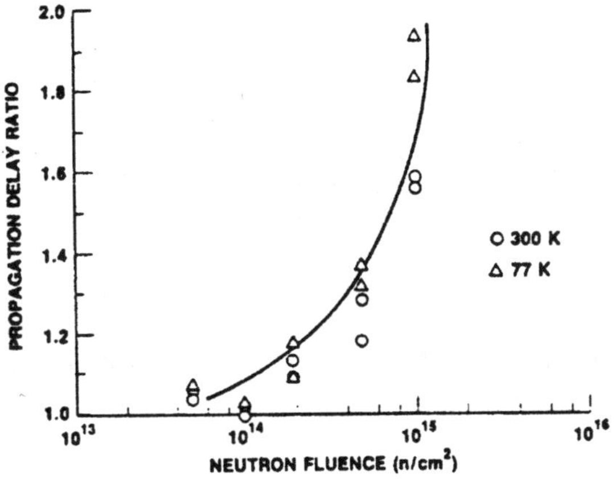

Fig. 7.39. Propagation delay degradation (normalised to the pre-radiated propagation delay) of the HEMT ring oscillator circuits as a function of neutron fluence at 300 K and 77 K. [After Janousek et al. [46]]

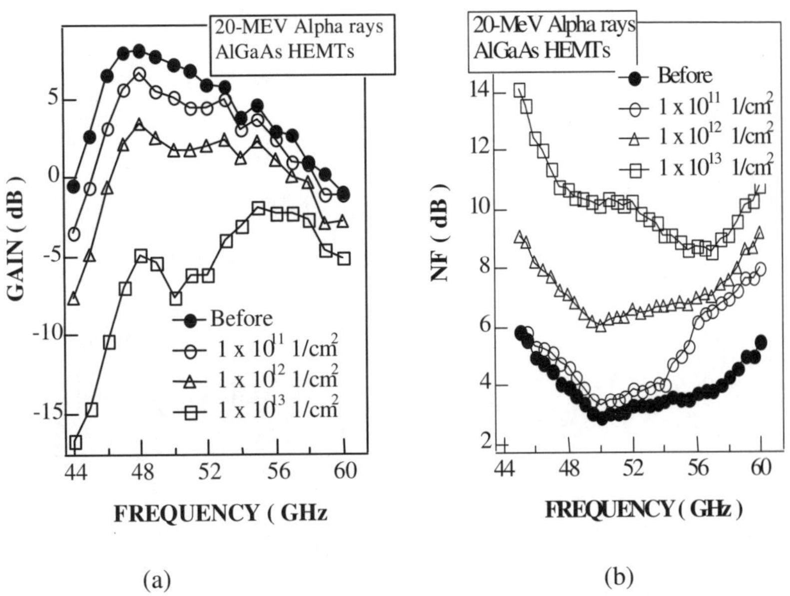

Fig. 7.40. Gain (a) and NF (b) as a function of frequency. [After Ohyama et al. [57]]

7.5. Conclusions

GaAs and related compounds surely have a hardness advantage of several decades compared with Si electronics and are, therefore, the technologies of choice for heavy radiation environments. In addition, a proper choice of substrate, technology and device structure further boost the radiation performance of GaAs MMICs. Less is known for recent InP based technologies, which thus requires additional testing. On the other hand, the fundamental understanding of for example the displacement damage is reasonably well developed and some of the critical radiation defects have been positively identified. Furthermore, the NIEL concept describes quite accurately the damage coefficients with particle energy for not too large energies.

8 Opto-Electronic Components for Space

8.1 Introduction

Photonics systems are ideally suited for space applications for a number of reasons: there is the high bandwidth and speed of operation, the immunity for electromagnetic interference and high reliability, low power consumption and cost and above all, light weight. Given the direct band gap, yielding a high quantum efficiency and the superior radiation tolerance of GaAs and related compounds, III-V opto-electronic components are the technology of choice for applications in a broad wavelength range, going from 700 to 1600 nm, whereby operation at 1300 nm is particularly suitable for fiber optics, since it corresponds to maximum radiation hardness of the mono-mode fibers. In this chapter, the behavior of advanced III-V opto-electronic components is described and problem areas defined. First, a description of the most promising device structures and their operation parameters is presented and this for Light Emitting Diodes (LED) and Laser Diodes (LD), on the one hand, and Photodiodes or Photodetectors (PD), on the other. A brief introduction to optocouplers is also given. Next, the fundamental and material issues related to radiation degradation are pointed out, followed by a discussion of recent irradiation studies of LEDs, LDs, PDs and optocouplers. A summary and the identification of issues requiring further studies conclude the chapter.

8.2 Opto-Electronic Components

This section presents a short introduction to the basic opto-electronic components. In brief, the device structure and its operation principle will be outlined. It should also be emphasised that while the original components were fabricated in bulk III-V material, state-of-the-art and advanced structures rely heavily on epitaxially grown layers, resulting in heterojunctions or (multiple) quantum wells. Current developments are towards the use of even lower-dimensional systems like quantum wires and dots.

8.2.1 Light Emitting Diodes (LEDs) and Laser Diodes (LDs)

Early LEDs were based on amphoteric Si-doped or Zn-doped GaAs, active in the range 850-950 nm [1 - 3]. However, better radiation tolerance was observed for GaAsP based diodes, emitting in the 850 nm region [4]. Good radiation resistance was observed for high radiance GaAlAs (820 nm) and InGaAsP (1300 nm) LEDs also [3]. The improvement of epitaxial growth techniques allows to deposit nowadays complex layer stacks, with sharp interfaces and steep concentration and composition gradients. Modern LEDs are therefore often based on Quantum Well (QW) structures in for example the InGaAs/GaAs/AlGaAs system, active around 875 nm, as represented in Fig. 8.1 [5].

Interest has also developed for the wavelength range around 1300 nm, which is the λ of choice for immunity to radiation induced attenuation of optical fibers [5]. In that case, InGaAsP is the preferred material for fabricating an LED.

The operation of an LED is characterized by the light output L_o versus the applied forward bias. L_0 relies heavily on the forward current flowing through the device, which may have different dependencies on forward bias, as shown in Fig. 8.2. The light output is proportional to the radiative recombination current and degrades if the non-radiative recombination current increases. The latter occurs when deep-level generation-recombination (GR) centers are induced by displacement damage. L_0 can be represented by [3, 4]:

$$L_0 = C_2 \tau \exp(\frac{qV_F}{kT}) \tag{8.1}$$

with τ the carrier lifetime, composed of a radiative (τ_{ra}) and a non-radiative (τ_{nr}) part, i.e.:

$$\tau^{-1} = \tau_{ra}^{-1} + \tau_{nr}^{-1} \tag{8.2}$$

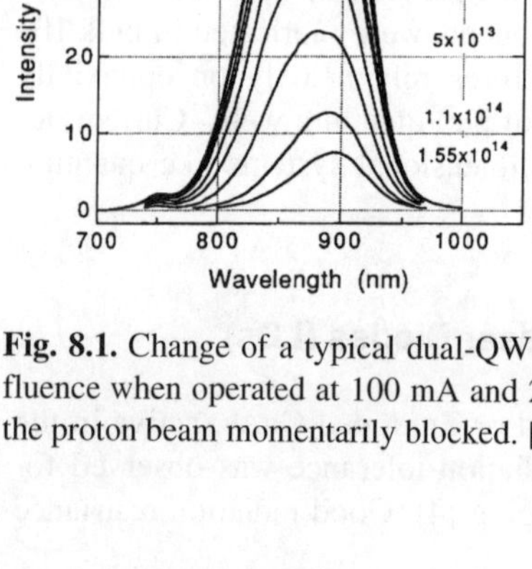

Fig. 8.1. Change of a typical dual-QW LED spectrum as a function of accumulated proton fluence when operated at 100 mA and 20°C during irradiation. The spectra were taken with the proton beam momentarily blocked. [After Evans et al. [5]]

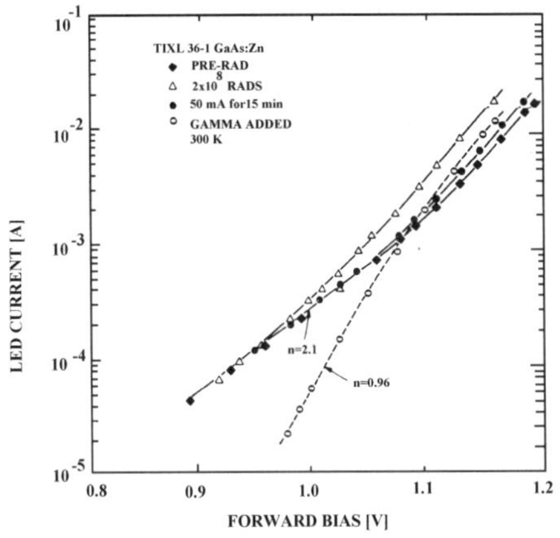

Fig. 8.2. Pre- and post-irradiation current-voltage characteristics of a typical Zn-diffused GaAs LED. The dashed curve is the difference between pre- and post-irradiation I_F-V_F curves. The n_{if} values refer to the proportionality : $I_F \propto \exp(qV_F/n_{if}kT)$. Note that application of forward bias returns the I_F-V_F curve to nearly its pre-irradiation characteristic. [After Barnes [2]]

C_2 is a constant and V_F the applied forward bias. The forward current density J_F, on the other hand, obeys:

$$J_F = \frac{C_3}{\sqrt{\tau}} \exp(\frac{qV_F}{kT}) \tag{8.3}$$

or:

$$J_F = \frac{C_5}{\tau} \exp(\frac{qV_F}{2kT}) \tag{8.4}$$

Equation (8.3) applies if the current is diffusion dominated (n_{if}=1), while (8.4) holds for space charge recombination controlled current (n_{if}=2).

Remark that given the large current levels at which the diodes are operated, a significant power is dissipated in the resistive part of the structure, causing a certain amount of self-heating. Since the band gap of the materials reduces for higher temperature, the central wavelength of the electroluminescence spectrum will shift to longer values if this occurs. This is particularly true for LDs; measuring the spectral line shift, therefore, provides a means to determine the local device temperature. Note also that the creation of non-radiative GR centers during exposure increases the power dissipation, since the band-gap energy is released through the emission of phonons (=heat). This heat loading effect may reduce L_o in a radiation environment, as will be explained in more detail below [6].

For high-power and high-speed (e.g. high data rates) applications, LDs are more appropriate than LEDs [7]. The higher speed of LDs is related to the low carrier lifetimes, which are in the range of 10^{-11} s during the lasing action, i.e., for stimulated photon emission, which is several orders lower than for LEDs (range 1 ns typically) [6]. From this simple fact, one can immediately derive that LDs are expected to be more radiation tolerant than LEDs, if the dominant lifetime degradation is due to displacement in the substrate. This follows from (8.2), where for small initial τ_{ra}, a much larger fluence is required and, hence, a larger density of radiation-induced GR centers to bring τ_{nr} to the same order of magnitude. This trend was indeed found for early-generation components [7].

There exists different types of III-V based LDs, consisting of a single or multiple QWs. A typical structure is represented in Fig. 8.3, featuring a QW active region sandwiched between GaAlAs layers, with higher Al content (higher band gap) [7]. The LD of Fig. 8.3 contains an array of active stripes, wherein the current is confined by an ion (hydrogen for example) implantation. This creates high resistive isolation regions, separating the different arrays. Another possibility is to fabricate a broad-area device, where the active region is a single current-carrying QW stripe, defined by an oxide insulator. A multiple quantum well (MWQ) structure is represented in Fig. 8.4 [8]. The LDs of Figs. 8.3 and 8.4 are horizontal cavity lasers, which emit their photons in the plane of the junction (Fig. 8.4b). Vertical-cavity surface-emitting lasers (VCSELs) emit their light perpendicularly to the junction (Fig. 8.5) [9] and can be directly modulated at rates in excess of 10 GHz. They can be used to transmit signals through space or in multimode optical fibers, thereby showing great potential for optical interconnects. Again, the lateral current confinement can be achieved either by oxidation or by proton implantation.

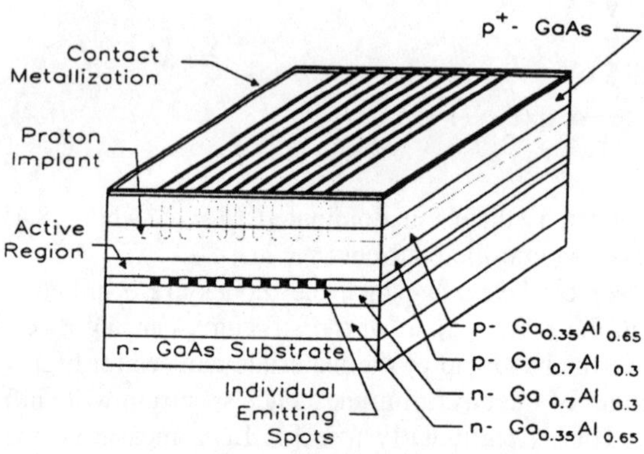

Fig. 8.3. Structure of the semiconductor laser diode array. [After Carson and Chow [6]]

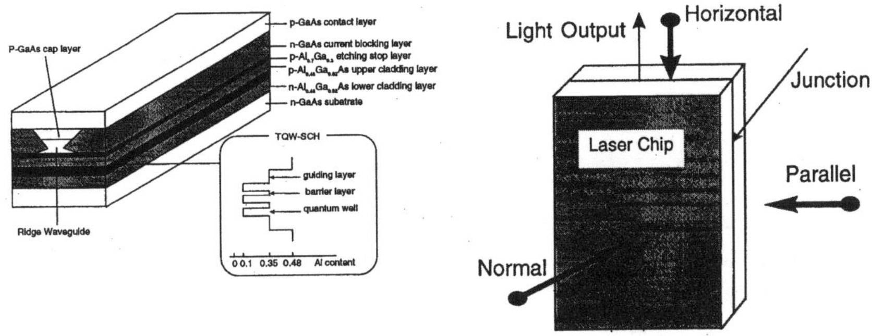

(a) (b)

Fig. 8.4. (a) Schematic structure of the laser diode and (b) of the proton incident directions. [After Zhao et al. [8]]

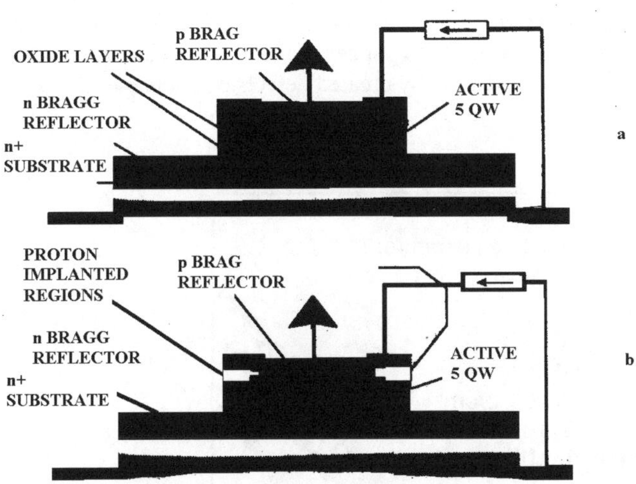

Fig. 8.5. Structure of VCSELs. The resonant cavity is vertical, with the mode confined between two Bragg reflectors made up of quarter-wave stacks of AlGaAs with the Al concentration varying between high and low values. In the center of a one-wavelength-thick undoped region, there are 5 GaAs quantum wells where there is optical gain. Two types of lasers are represented: (a) lasers with the lateral current confinement provided by edge-oxidized layers with very high Al content, and (b) lasers with current apertures defined by proton implantation. [After Paxton et al. [9]]

The light output or optical power L_o of a LD varies with the diode current passing through the junction, as in Fig. 8.6 [8]. Important parameters are the threshold current I_{th} for lasing action, separating the region of spontaneous emission (LED action) from the stimulated emission at high diode current. It can be expressed as [8]:

$$I_{th} \propto \frac{\tau_{ra}}{\tau}[\alpha_c + \frac{\ln(1/\rho_f)}{l_c}] \qquad (8.5)$$

with α_c the cavity loss coefficient (cm^{-1}), assumed to be unaffected by irradiation in first instance [6]. ρ_f is the facet reflectance and l_C the cavity length (cm). The optical power above threshold is given by [8]:

$$L_o = \eta_o (I_{op} - I_{th}) \qquad (8.6)$$

with η_o the slope efficiency in mW/mA. I_{op} is the operation current (mA). For the linear characteristics of Fig. 8.6, η_o is a constant. On the other hand, VCSELs suffer from thermal roll-over, as evidenced by Fig. 8.7 [9]. This is related to the fact that in a VCSEL, a temperature change (by selfheating at higher I_{op}) causes the cavity resonance and gain peak to move to longer wavelengths. Increasing self-absorption of emitted photons for example by created defects can degrade the slope efficiency [7 - 9].

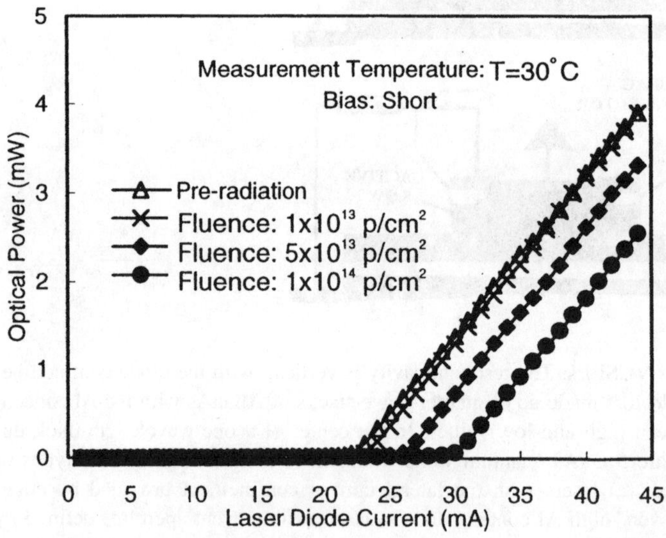

Fig. 8.6. L_o versus I_{op} characteristics for devices shorted during irradiation. [After Zhao et al. [8]]

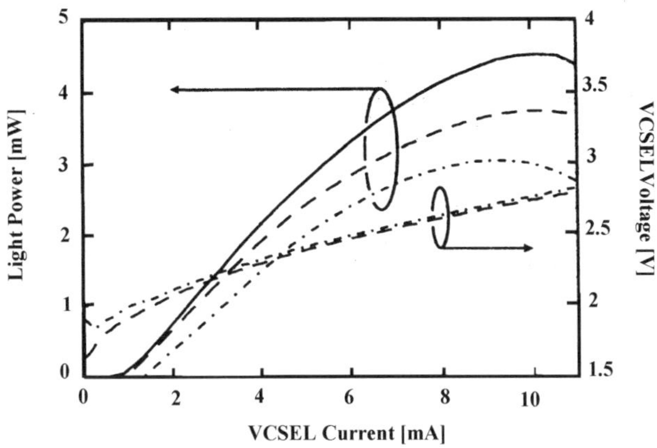

Fig. 8.7. Light-current and voltage-current curves for a VCSEL. [After Paxton et al. [9]]

It is important to remark that for most of the LED and LD structures, the active region is several μm below the surface of the device. This has to be taken into account when calculating the local energy deposition or NIEL going into displacements at the site of the (M)QW junction [5].

8.2.2 Photodetectors

Depending on the spectral region, different types of photodetectors can be used. In some cases, a silicon p-i-n diode or a bipolar phototransistor is preferred, because it can be fabricated monolithically with the necessary signal amplifier. However, direct-gap III-V materials based PDs offer certainly advantages compared with silicon. For example, $In_{0.47}Ga_{0.53}As$ is infrared sensitive, since the band gap is ≈0.75 eV, so that wavelengths in the range 900 to 1600 nm can be absorbed. The layers can, furthermore, be deposited lattice matched on InP SI substrates. A typical structure for such a p-i-n diode is shown in Fig. 8.8 [10].

Another emerging technology is the so-called Quantum Well Infrared Photodetectors (QWIP's), which are large monolithic two-dimensional detector arrays, for wavelengths > 2 μm. In the future, they may replace the existing HgCdTe material, which has problems in detecting long wavelengths >15 μm and also in working in a very low background photon environment [11]. These problems are related to the difficulties in growing highly uniform and defect free stable HgCdTe material. A further advantage is that for multispectral imaging, the peak response wavelength of a QWIP can be tuned by changing the externally applied bias across the AlGaAs/GaAs multistack of different one color QWIPs (Fig. 8.9). Thus QWIPs could provide multispectral imagery by a suitable bias voltage across a given fabricated device. In addition, it has been demonstrated that an LED can be integrated on the same chip, with each detector in the QWIP array. A combined QWIP-LED can be used as a wavelength converter from the mid to far infrared (QWIP) to the near infrared wavelengths emitted by the LED coupled to the QWIP. This technology has the potential of improved performance - because of better quality layers and lower dark currents - ease of production and cost reduction, compared with their present-day competitors for mid to far infrared camera's.

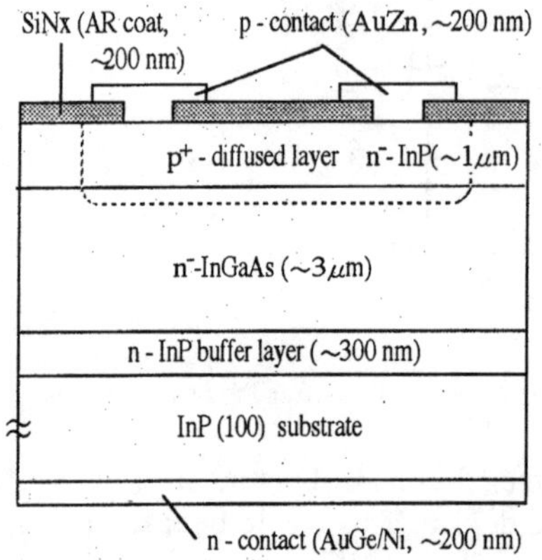

Fig. 8.8. Schematic representation of an $In_{0.53}Ga_{0.47}As$ p-i-n photodiode. [After Ohyama et al. [10]]

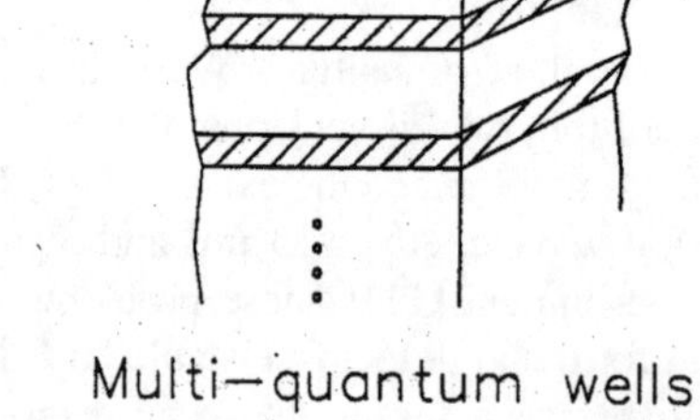

Fig. 8.9. Schematic of nm thick GaAs/AlGaAs multiple layers in a QWIP structure. [After Khanna et al. [11]]

Important operational parameters of a PD are the dark current, the low-frequency noise and noise equivalent power (NEP) and the optical responsivity, which is determined by the charge collection efficiency. Although a PD is a device with a surface active layer (junction), the light is absorbed much deeper in the material. The absorption coefficient $\alpha(\lambda)$ depends on the wavelength of the light. The light intensity in the detector material follows an exponential law with distance from the surface (x=0):

$$I(x) = I(0)\exp[-\alpha(\lambda)x] \tag{8.7}$$

so that $x=1/\alpha(\lambda)$ corresponds to the distance where the light intensity drops to 1/e of its value at the surface. Electron-hole pairs are thus generated at depths much larger than the reverse biased photodiode depletion region, so that the carriers have to diffuse to the active region to be collected. The collection efficiency, therefore, strongly depends on the minority carrier diffusion length $L_n=\sqrt{D_n\tau}$. D_n is the (electron) diffusion constant. The photoresponse will begin to be affected when the lifetime is reduced to the point where $L_n < 3\alpha(\lambda)$ [12]. From this, it is concluded that both photodiodes and LEDs/LDs are in first instance determined by the same material parameter, namely the minority carrier lifetime. This implies that these classes of devices are predominantly prone to displacement radiation damage and less, if at all, to ionisation effects. Therefore, some kind of correlation with NIEL is anticipated, which will be discussed in the next section.

8.2.3 Optocouplers

Optocouplers form an essential part in many photonics systems. They provide a dc isolation between circuit blocks, which increases the reliability of space-born instruments [12]. Optocouplers are generally constructed of two dies, separated by an optically transparent but electrically isolated medium. Information is transferred by light generated by an LED (or LD) and sensed by a photodetector, as represented in Fig. 8.10 [13], showing two different arrangements. In Fig. 8.10b, the isolating medium is for example a silicone coupling compound. The detector is followed by an amplifier stage [12].

The important figure of merit for an optocoupler is the Current Transfer Ratio (CTR) which gives the ratio of the detector current to the LED current and is thus a combined quantity, which not only depends on the sending and the receiving components, but also on the coupling medium. The radiation response of an optocoupler is thus rather complex and generally combines permanent and ionisation/transient radiation effects [12].

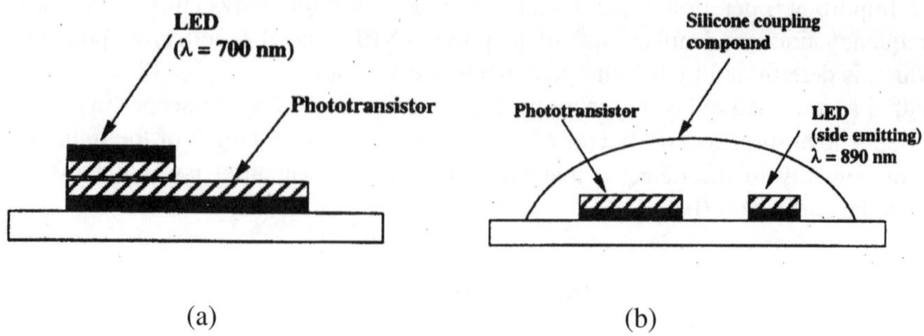

Fig. 8.10. Physical configuration of two different optocouplers. [After Rax et al. [13]]

8.3 Basic Radiation Effects and Material Issues

In Chap. 4, the basic radiation damage mechanisms and defects in GaAs have been described to some extent. For opto-electronic components, beside the degradation of the electrical parameters, one should be concerned with the optical material properties, like the absorption coefficient (see (8.7)) and its wavelength dependence, the reflectance, the refractive index, etc. Therefore, in a first section, the effect of irradiation on some basic optical parameters of GaAs will be described. In a second section, the dominant radiation defects in ternary and quaternery III-V alloys, which are often used as barrier or active layers in LDs, LEDs or MQW structures, will be highlighted. Similar as for the case of $Si_{1-x}Ge_x$, it is expected that the main radiation defects are related to the GaAs ones, but modified by the presence of the third and/or fourth material constituent. In addition, deep-level parameters can be affected by so-called alloy broadening. Finally, the relevant macroscopic damage factors in III-V alloys and the relationship with NIEL (Sect. 2.3.2) will be discussed.

8.3.1 Impact of Irradiation on Optical Material Properties

The impact of radiation damage on basic optical properties of GaAs and III-V alloys will be investigated first. It is shown that irradiation changes the optical absorption, the refractive index, the photoluminescence (PL) and the photoconductivity of these materials, which has important consequences for Photonics Integrated Circuits (PICs) and Optoelectronic Integrated Circuits (OEICs). However, in the case of MQW structures ion damage may be beneficially used to produce so-called quantum well intermixing (QWI), which provides an elegant method for local band-gap engineering of opto-electronic components. At the same time, it opens up the road for novel low-dimensional device structures, like quantum wires and quantum boxes. One critical aspect remains the removal of the non-radiative damage centres after ion exposure.

8.3.1.1 The Absorption Coefficient

The impact of high energy neutron [14 – 16] and proton irradiation [17] on the absorption coefficient of GaAs has been studied for some time. Figure 8.11 shows the evolution of $\alpha(\lambda)$ with fast neutron fluence for a lightly n-type doped GaAs substrate [14]. It is first of all noted that the fundamental absorption edge, occurring at 1.4 eV at room temperature, becomes less steep after neutron irradiation.

Generally speaking, there are two contributions to α, namely the band-to-band (α_{bb}) and the excitonic (α_{ex}) absorption [18]. They can be expressed as:

$$\alpha_{ex} = \frac{1}{\pi}\frac{\Gamma_x}{(E_{ph} - E_G + E_x)^2 + \Gamma_x^2} \tag{8.8a}$$

and:

$$\alpha_{bb} = A_o\left[\frac{1}{2} + \tan^{-1}\left(\frac{E_{ph} - E_G}{\Gamma_{bb}}\right)\right] \tag{8.8b}$$

In (8.8) E_x is the free exciton binding energy, Γ_x and Γ_{bb} are broadening parameters and A_o gives the relative importance of the two contributions. According to (8.8b), the absorption edge is a smeared out step function, with threshold energy the optical band gap E_G. From Fig. 8.11, one can see that an absorption tail below E_G develops with increasing neutron fluence [14, 16]. At E_{ph}=1.5 eV and T=15 K, it has been found that [16]:

$$\alpha = 32.5(\Phi \text{x} 10^{-14})^{0.323} \quad (\text{cm}^{-1}) \tag{8.9}$$

while the corresponding optical gap is 1.519 eV. The change of the absorption edge at room temperature for neutron-irradiated n- and p-type GaAs is represented in Fig. 8.12 [14]. It shows a nearly exponential dependence on photon energy. Since a similar tendency is found after electron irradiation, a possible explanation is a change in α due to the regions of strain near the induced Frenkel defects.
The sub-band-gap absorption is governed by two competing mechanisms: on the one hand, the free carrier absorption diminishes owing to the carrier removal, while on the other hand, α increases with increasing fluence, due to the creation of a higher level of damage, as can be derived from Fig. 8.11. The corresponding absorption coefficient follows a $1/\lambda^2$ or an $E_{ph}{}^2$ dependence, and increases linearly with fluence, as shown in Fig. 8.13 [14]. For sufficiently high Φ, the $E_{ph}{}^2$ absorption starts to dominate; the threshold fluence for this to happen is higher for higher doping density (Fig. 8.13). Note in addition that for the highest neutron fluences,

the α is the same irrespective of the initial doping density or type. It is believed that this absorption is related to neutron-damage spikes, which behave as metallic-like regions dispersed throughout the semiconductor [14].

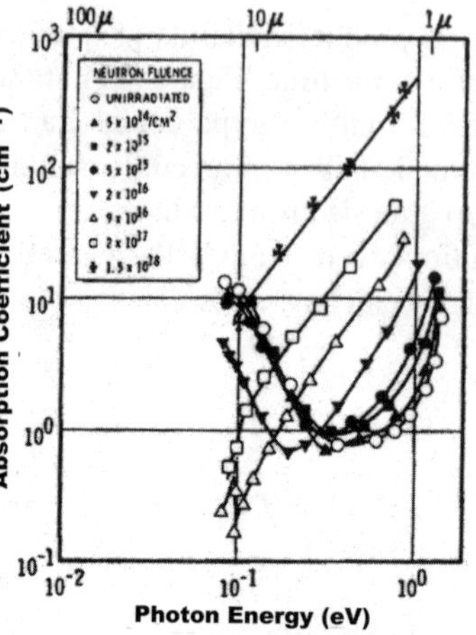

Fig. 8.11. GaAs absorption coefficient at room temperature after fast neutron exposure for lightly n-type material. [After Burkig et al. [14]]

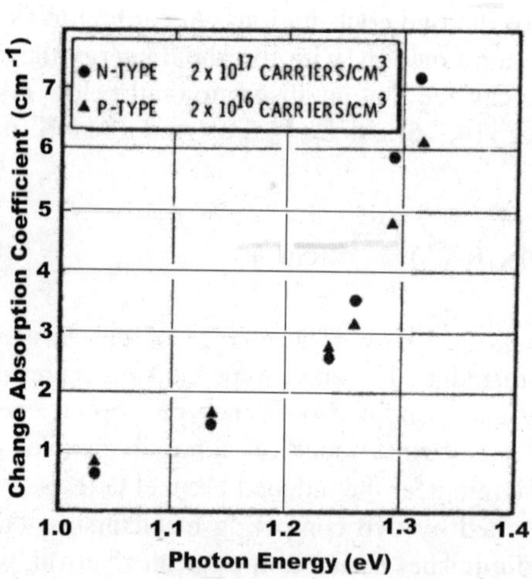

Fig. 8.12. Absorption coefficient change near the absorption edge after neutron irradiation of $5x10^{15}$ cm^{-2} for lightly doped n- and p-type GaAs. [After Burkig et al. [14]]

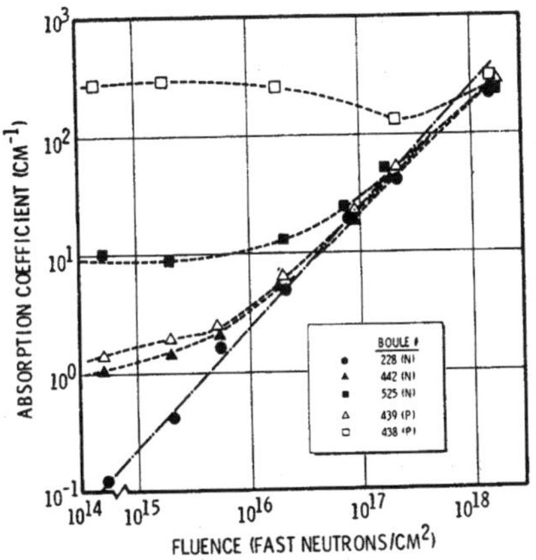

Fig. 8.13. GaAs absorption coefficient at 0.7 eV as a function of neutron fluence. [After Burkig et al. [14]]

Before the dominance of the E_{ph}^2 region, one can rely on the reduction of α, induced by the irradiation, to estimate the carrier removal rate. Assuming that α is proportional to the carrier density n, one arrives at $\Delta n/n_0=\Delta\alpha/\alpha_0$. This enables an optical determination of the removal rate. Values in the range of 8 cm^{-1} for 2x10^{17} cm^{-3} n-type material were found at a neutron fluence of 2x10^{15} cm^{-2} [14]. Similarly, for 2x10^{16} cm^{-3} p-type one arrives at a value of about 4 cm^{-1} after a 2x10^{15} cm^{-2} fast neutron irradiation [14].

In addition to the E_{ph}^2 region, discrete absorption bands have been found, following high-energy neutron or proton irradiations [14, 19]. As mentioned briefly in Sect. 2.4.3 these absorption bands are the basis of Fourier Transform Infrared (FTIR) spectroscopy of optically active radiation-induced defects.

A similar behaviour of the absorption coefficient as after neutron exposures has been observed in low energy proton-irradiated p-type GaAs [17]. This is illustrated by the results of Fig. 8.14, showing a reduction of the transmitted light intensity by ~89 % after a 10^{16} cm^{-2} 300 keV proton implantation. Such implantations are used for the definition of stripe laser diodes. The highly damaged implanted regions provide high-resistive regions confining the active laser diode. However, absorption of the generated laser light in the isolation regions is highly undesirable. Fortunately, it has been demonstrated that the transmission coefficient recovers after thermal annealing, as evidenced by Fig. 8.14. This recovery is characterised by at least two activation energies at ~1.5 and 3.4 eV, corresponding with the annihilation of different radiation defects. At the same time, the resistivity of the regions remains high, indicating that the radiation damage leading to carrier removal is more thermally stable. This fortuitous combination of facts enables to define an implantation plus annealing window for achieving high resistive and transparent proton-implanted isolation regions [17]. The absorption coefficient, which is a measure of the induced defect concentration, increases sublinearly with the 300 keV proton dose [17].

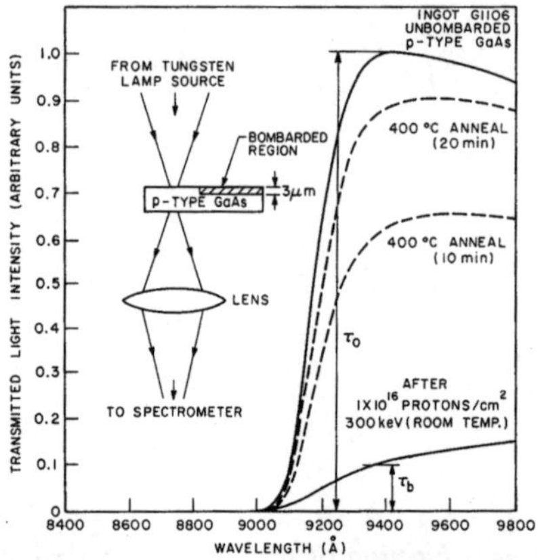

Fig. 8.14. Inset shows the method for measuring the transmitted light intensities τ_0 and τ_b through the unbombarded and bombarded sides of a p-type GaAs sample. Solid lines show typical transmission for an unbombarded sample compared with transmission after bombardment with protons at $\Phi=10^{16}$ cm^{-2} (300 keV). Dashed lines show how the bombardment-induced optical absorption is reduced after annealing at 400°C for 10 and 20 min. The values of τ_0 and τ_b are determined at a wavelength of 936 nm which is approximately the maximum in the transmitted light intensity for the unbombarded sample. [After Dyment et al. [17]]

The impact of electron [20], proton [21, 22] and helium irradiation [23] on the optical absorption of MQW structures has been studied in view of the development of a new generation (very) long wavelength infrared detectors. The detection relies on intersubband transitions in the quantum wells. As can be observed from Fig. 8.15 the absorbance can be completely quenched or depleted for a high fluence particle irradiation. This can be explained in terms of a reduction of the two-dimensional electron gas (2DEG) in the quantum wells due to carrier trapping by radiation-induced defects. The integrated absorbance intensity is indeed proportional to the density of the 2DEG [20]. A reduction of this parameter after irradiation thus points to a reduction of the carrier density in the well. Thermal annealing recovers the intersubband transitions, as evidenced by Fig. 8.16. However, for high fluences no recovery occurs. In the case of electron irradiations, two annealing stages have been found at ~140 and 250 K, respectively. They are believed to correspond to the annealing of certain radiation defects [20].

8.3.1.2 Photoluminescence and Photoconductivity

As discussed extensively in Sect. 4.3.4, the PL spectrum of GaAs is a sensitive monitor for radiation damage. From the example shown in Fig. 8.17 [16], one can derive that the intensity of the donor-exciton (D°-X) transition is much more degraded by fast neutrons than the deep-level transition in the infrared spectral region. Based on these observations, a monitor for the microscopic damage has been established which follows the NIEL over an energy range up to 100 MeV.

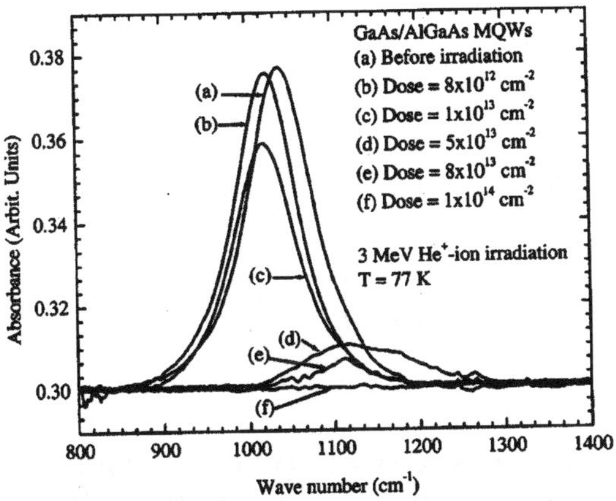

Fig. 8.15. Absorbance spectra of intersubband transitions in unirradiated [spectrum (a)] and 3 MeV He^+-ion irradiated GaAs/AlGaAs MQWs measured at T=77 K. The doses are (b) $8x10^{12}$ cm^{-2}, (c) $1x10^{13}$ cm^{-2}, (d) $5x10^{13}$ cm^{-2}, (e) $8x10^{13}$ cm^{-2} and (f) $1x10^{14}$ cm^{-2}. [After Berhane et al. [23]]

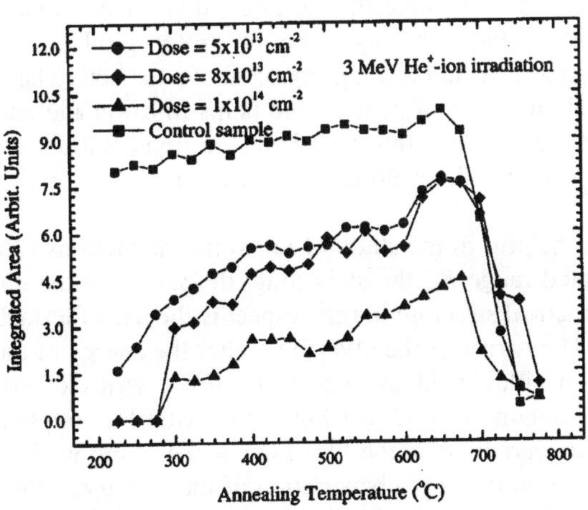

Fig. 8.16. The total integrated area of the intersubband transitions in the control (squares) and irradiated GaAs/AlGaAs MQW samples (circles, diamonds and triangles) obtained as a function of annealing temperature. The annealing time was 15 min at each temperature. [After Berhane et al. [23]]

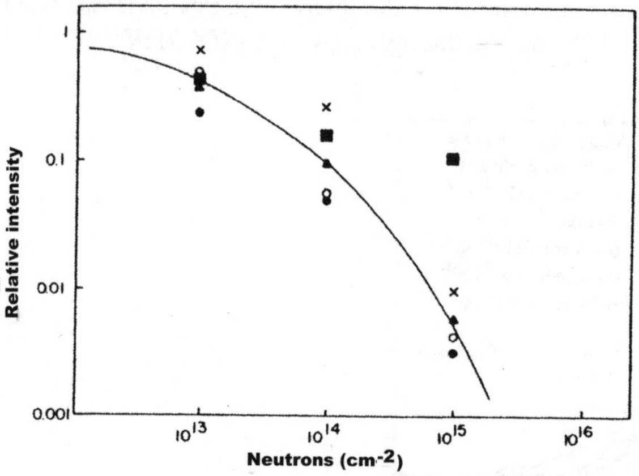

Fig. 8.17. Intensity of the (D°-X) transition (•, ▲, x) for a GaAs sample irradiated by fast neutrons, relative to the unirradiated sample, as a function of neutron fluence. The solid squares refer to the deep level (infrared) transition. [After Carlone et al. [16]]

PL analysis has also been applied to other III-V compounds and alloys in order to evaluate ion implantation damage and its recovery upon thermal annealing [24 – 26]. Generally, the PL intensity drops severely due to the presence of non-radiative recombination centres, while it recovers when these defects anneal. In other words, PL analysis can be used to optimise certain process steps, which are prone to radiation damage. This has been applied for example for the fabrication of distributed Bragg reflectors of vertical-cavity surface emitting lasers by proton implantation [24]. The latter step was used to fabricate an isolating region above the active laser region. This results in an optimal current confinement and helps to lower the current threshold. However, in order to achieve this goal both the implantation and annealing parameters have to be fine-tuned, resulting in no detectable deterioration of the PL.

Alternatively, PL can also be helpful in the study of ion-solid interactions [26, 27], i.e., to determine the projected range R_p, the straggling, the radius of the ion track or the damage profile. A practical structure in this respect is the use of a stack of QWs with varying thickness. The narrower the QW, the higher the energy of the corresponding PL line will be, as illustrated by Fig. 8.18a for a series of AlGaAs/GaAs wells with thickness ranging from 2 to 10 nm [26]. After He^+ implantation, a part of the lines has disappeared, while the PL of the deeper lying wells is unaffected. This yields information on the depth penetration of the damage, which is represented in Fig. 8.18b, as a function of the He^+ energy. Likewise, the radius of a single ion track can be studied by following the change in the PL line position (line shift) as a function of ion dose [27]. This yields important information with respect to the use of Focused Ion Beam (FIB) techniques for the fabrication of quantum wires and dots and more in particular regarding the lateral resolution of the technique.

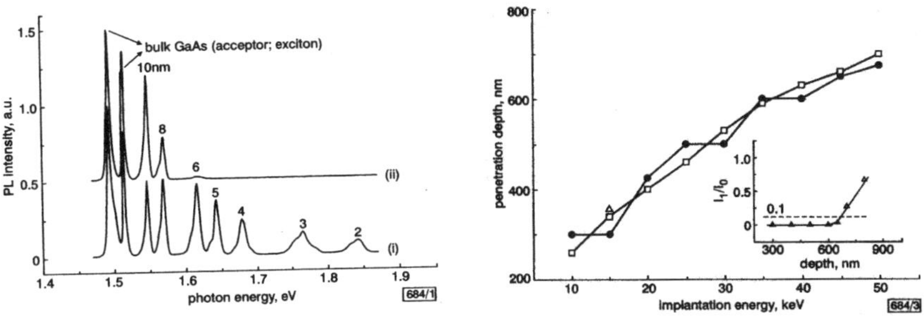

(a) (b)

Fig. 8.18. (a) PL intensity against photon energy before (line (i)) and after a $4x10^{13}$ cm^{-2} 50 keV $^4He^+$ implantation (line (ii)) for a sequence of AlGaAs/GaAs QWs with increasing depth. The 2 nm well lies closest to the surface. (b) Damage penetration depth against $^4He^+$ energy, determined from the PL QW method (•) and TRIM simulations (□). The inset shows the PL intensity ratio I_1/I_0 against depth for E_{ion}=50 keV. [After Gramlich et al. [26]]

The development of lower-dimensional structures like quantum wires (1D) or even quantum dots (0D) may lead to the fabrication of improved, more performing optical components or even new device types. It has been shown for example that quantum dot (QD) lasers have a low threshold current and a high gain [28]. Important question is whether they are also more radiation tolerant than their QW counterparts. This can again be studied by comparing the response of PL to particle exposure. Figure 8.19 compares the PL spectrum of an InGaAs/GaAs QW and QD before and after a 1.5 MeV proton irradiation to different fluences. It is obvious from Fig. 8.19b that the degradation of the QD is less severe than of the QW [28]. The reason for this increased radiation hardness is related to the fact that the total volume of the active region is much smaller than in the QW case. Consequently, the 3D confinement of the excitons in the dots lowers the probability of non-radiative recombination by radiation-induced defects outside the QD. Interestingly, no shift of the PL energy is induced here by the proton irradiation. In addition, there is even a slight increase in the PL intensity of the QDs for 1.5 MeV proton fluences in the $7x10^{11}$ to $7x10^{12}$ cm^{-2} range (Fig. 8.19b), which is related to the 3D confinement effect. The degradation observed at higher Φ is mainly related to the reduced hole mobility, i.e., to the less favorable carrier transport towards the QDs. It can be concluded that in general, devices can be ranked in ascending order of hardness depending on their quantum confinement, i.e., from bulk, over QW to QD structures.

Beside the PL spectrum also the photoconductivity (PC) spectrum may reveal interesting information on radiation damage. As such, PC is closest to the electrical performance of the material, since it measures the conductivity (or current) induced by the photogeneration of carriers in a piece of semiconductor material, as a function of the photon energy. It has been observed that the PC changes in fact more after electron than after neutron irradiation [18]. This is illustrated in Fig. 8.20 for the case of GaAs. From the temperature dependence of the PC it has been derived that.

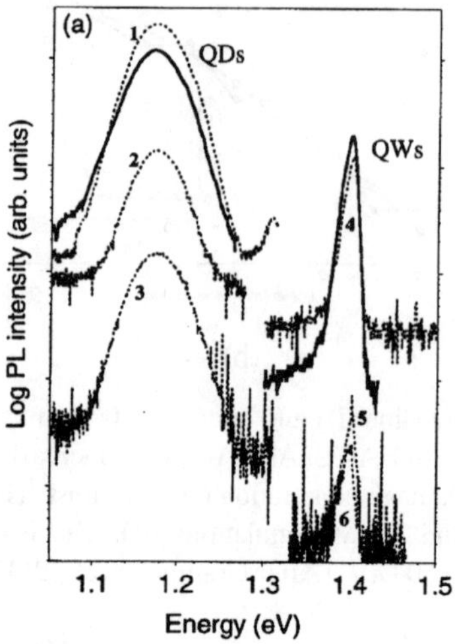

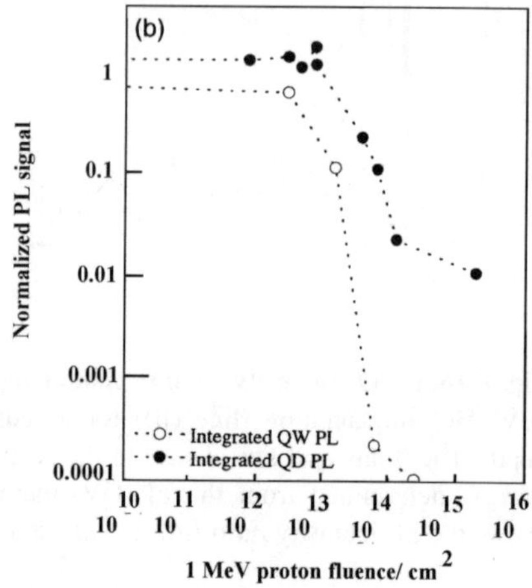

Fig. 8.19. (a) Comparison of PL spectra (measured at 5 K) from InGaAs/GaAs quantum wells and from quantum dots in high surface densities ($2.4x10^{10}$ dots per cm^2) after irradiation with selected proton fluences. The solid lines show the spectra before irradiation. The dotted lines show spectra after 1.5 MeV proton irradiation for fluences of (per cm^2) (1) $7x10^{12}$, (2) $6x10^{13}$, (3) $2x10^{15}$, (4) $3x10^{12}$, (5) $6x10^{13}$, and (6) $2x10^{14}$. (b) Integrated PL emission normalized to the as-grown samples for QW and QDs as a function of proton fluence. [After Leon et al. [28]]

both electron and neutron irradiation induce a shallow donor state, about 40 meV below the conduction band in GaAs

8.3.1.3 Radiation-Induced Intermixing of Heterostructures

It was pointed out in Chap. 4 that ion implantation or irradiation can be used beneficially in some applications. It is particularly useful to create high resistive isolation regions, where current confinement is necessary. It is, therefore, also employed in the fabrication of opto-electronic devices and detectors [29]. Another unique feature is that ion implantation followed by an anneal leads to so-called Quantum Well Intermixing (QWI), which opens up the way for new applications [30]. The basic principle of QWI is illustrated in Fig. 8.21 [31] and relies of the interdiffusion of the Group III dopants Al and Ga across the QW barrier. Consequently, the originally box-like QW potential profile becomes rounded after intermixing the layers, due to the error-function-like diffusion profile of Al or Ga. The change in potential well has a marked impact on the optical properties of the QWs: the optical band gap changes, which results in a change in the refractive index [30]. This is for example a useful property for providing optical confinement, gratings or even laser reflectors. In addition, the band gap increases, since the conduction band becomes higher, while the valence band is deeper (Fig. 8.21). Consequently, the PL energy of an intermixed QW is blue-shifted. Alternatively, PL can be used as a very sensitive monitor of intermixing effects

Intermixing can be achieved by different techniques [30], amongst which ion implantation is certainly the most versatile one. It provides a post-epitaxial-growth method for local and lateral band gap engineering across a single or a multi QW wafer. The basic principle behind the point-defect-induced QWI is the enhancement of the Al and Ga diffusion by the injection/creation of Group III vacancies [32]. These can be injected from the surface under thermal annealing, but are of course also generated along the track of an energetic ion. The flexibility of ion implantation as a QWI method is relying on the fact that the created vacancy profile can be fine-tuned by optimising the implantation energy and fluence and the post-implantation annealing.

Figure 8.22 shows the basic processing scheme for ion-implantation induced intermixing [33]. The role of the post-implantation annealing is first of all to activate the interdiffusion step and secondly, to remove the non-radiative defect centres from the active layers. Possible applications of QWI band-gap engineering are demonstrated by Fig. 8.23, showing the requirements for the optical band gap for different parts of a monolithic integrated photonics circuit [33]. If one for example wants to perform integrated wavelength division multiplexing (WDM) in an optical communications system, it is necessary to integrate lasers, modulators and waveguiding regions on the same epitaxial wafer. This can be accomplished by locally modifying the band gap of the epitaxial layers by QWI.

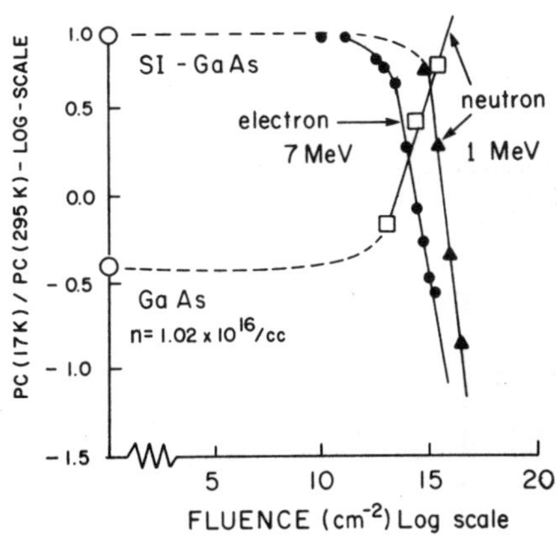

Fig. 8.20. The effect of fluence on the temperature dependence of the photoconductivity of GaAs. [After Khanna et al. [18]].

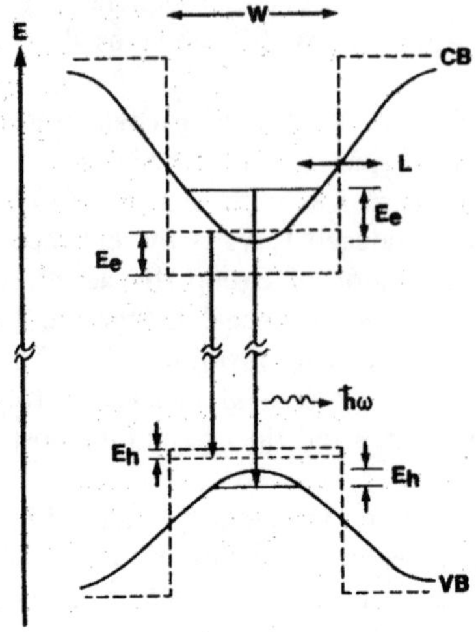

Fig. 8.21. Well profiles before intermixing (square-well profile) and after intermixing (error function profile). The vertical arrows show the exciton luminescence photon energies $\hbar\omega$ (neglecting exciton binding energies) before and after intermixing. The vertical energy scales and horizontal length scales show the potential profile and the electron and heavy-hole confinement energies $E_{e,h}$ for a 6-nm well with a diffusion length of 3 nm. [After Kash et al. [31]]

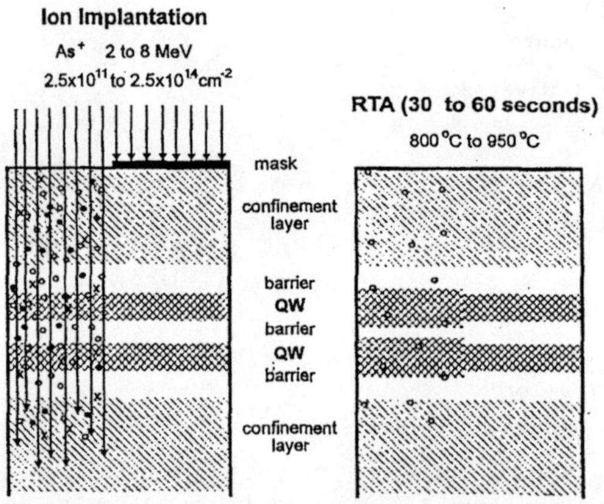

Fig. 8.22. Schematic illustration of the processing steps needed to implement vacancy enhanced QWI: (a) selective vacancy generation through a mask, (b) rapid thermal annealing of the whole wafer. The crosses and open and closed circles represent various defects (vacancies, interstitials, and other complexes) created by the ion implantation. [After Charbonneau et al. [33]]

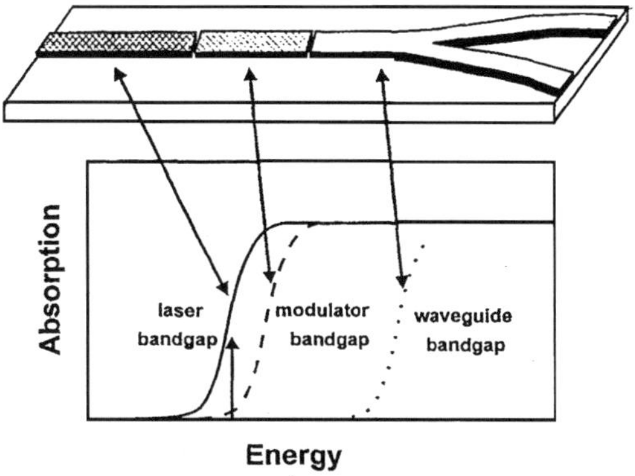

Fig. 8.23. Schematic illustration of the relative band gap positions of laser, modulator and waveguide needed for successful photonics integration. [After Charbonneau et al. [33]]

Implantation with different ions has been proposed for QWI: dopant atoms like Zn or Si [31], neutral atoms (F or B) [30] and lattice constituents, e.g. Al [31, 34], As [33, 35], Ga [34, 36, 37] or dual Ga and As [32]. Especially low-energy Ga ions, produced by a FIB allow band-gap modulations with a high spatial resolution. This ability can be exploited for the fabrication of new low-dimensional structures, like quantum wires and boxes [38 – 43].

Key in the successful application of QWI by the ion implantation technique is the control of the damage in the active layers [37], which reduces the intensity of the excitonic PL for increasing fluence. As a result, the QWI induced blue-shift tends to peak with ion fluence (Fig. 8.24) [44], while the PL intensity drops significantly, due to the accumulation of the radiation damage. Evidence for the presence of radiation-induced extended defects has been provided by TEM analysis, for example [37]. In order to recover the PL intensity, prolonged annealing is required. On the other hand, the intermixing itself saturates rapidly with annealing time. One way to overcome these problems and to enhance the energy shift (the interdiffusion length) is to perform a repeated implant plus anneal cycle [44]. Another possibility is to use light ion implantation, like protons [45 – 48]. They offer the additional advantage that the projected range at the same energy is much higher compared with heavier ions. For certain applications (lasers), penetration depths of at least a few μm are required in order to produce the necessary vacancies across the whole active stack [33, 48]. This requires MeV energies. Moreover, it has been observed that the use of high fluences of keV protons induces high QW energy shifts, without any sign of saturation (Fig. 8.25) [45]. This method has been successfully applied to perform wavelength shifting in QW lasers [46, 48].

Furthermore, it has been shown that the capture of photo-excited carriers is faster in intermixed QWs, compared with a reference QW [47, 49]. This is advantageous to enhance the frequency response and quantum efficiency of intermixed optoelectronic devices. It is believed that the reduction in the capture time is related to the change in the shape of the intermixed quantum well [47].

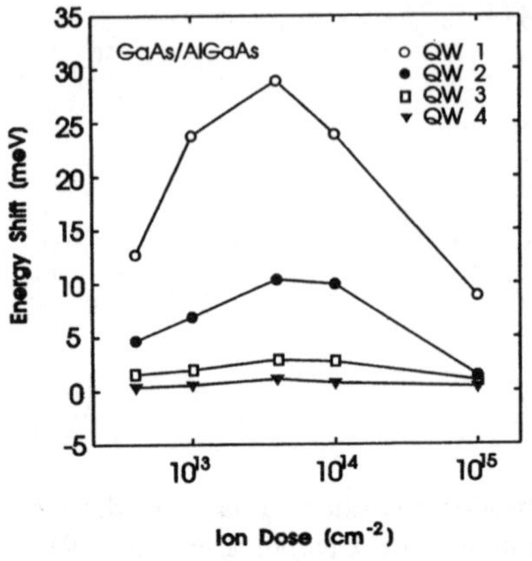

Fig. 8.24. A plot of the energy shifts observed in the photoluminescence peaks for the four quantum wells in the GaAs/AlGaAs sample after ion implantation and annealing. The narrowest well (QW 1) nearest the surface has the largest energy shift, and all the wells demonstrate a peak in energy shift against ion fluence of $4x10^{13}$ cm^{-2}. [After Poole et al. [44]]

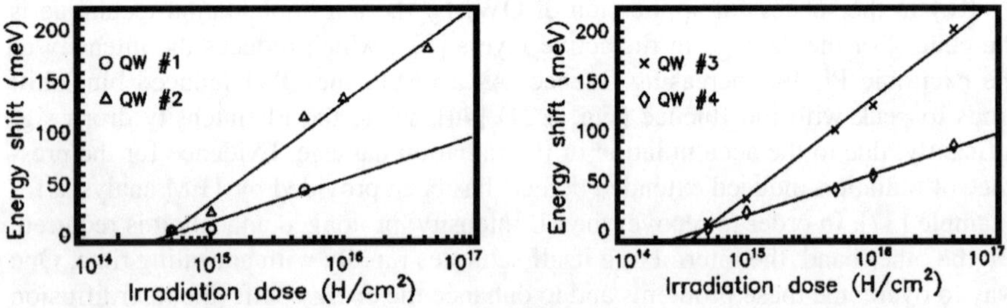

Fig. 8.25. The magnitude of energy shift for each quantum well as a function of proton irradiation fluence for samples annealed at 900°C, 30 s. [After Tan et al. [45]]

The potential of implantation induced QWI has been demonstrated amongst others by the realisation of interdiffused quantum-well infrared photodetectors for color sensitive arrays [50]. A more fundamental use of QWI aims at the study of dopant diffusion (Al, Ga,..) across a heterojunction interface [51, 52].

8.3.2 Radiation Defects and Material Aspects in Ternary Compounds

There are a few material-growth related aspects that need to be considered carefully for GaAs and related photonic ternary and quaternary compounds. For the deposition of QW layers, quite often, use is made of techniques like Metalorganic Vapor Phase Epitaxy (MO-VPE) or Molecular Beam Epitaxy (MBE), which differ from the commonly used Vapor Phase Epitaxial (VPE) growth for the fabrication of HEMTs and MESFETs. One may wonder whether there is a dependence on the growth technique of the radiation response. Initial studies for electron-irradiated material indeed suggest a different behavior [53]. Not only is an additional radiation-induced trap observed for high background doping material, but on top of that the simple linear response of the radiation-defect density with electron fluence is not retrieved. Instead, a complex carrier removal rate, depicted in Fig. 8.26 is observed. This implies that the popular damage-factor concept, which is applicable for bulk GaAs, is not strictly valid for such material. More studies are required to identify the origin of this behavior and to establish an alternative displacement damage monitor.

Another issue is related to the nature of the fundamental displacement centres created in epitaxial compound layers like AlGaAs, InAlAs, InGaAsP, etc., which are often used in photonic devices. In the first instance, one expects simple Frenkel pairs, whereby, like in GaAs, the As vacancy (V_{As}) is one of the predominant stable radiation defects at room temperature [54, 55]. In the past, a number of studies were devoted to electron-irradiation effects in AlGaAs material [54 – 57]. A trap at 420 meV from the conduction band (25% Al), with largest introduction rate has been ascribed to V_{As}. It corresponds to the E2 level in GaAs (see Table 4.1). The corresponding shallower state is expected to lie at approximately E_c-320 meV [55]. Two additional unidentified, deeper levels have been observed in the 10^{18} cm^{-3} doped AlGaAs layer of neutron irradiated HEMTs [58].

The impact of the Al fraction on the electron-irradiation induced deep levels has been studied in more detail by *Zaidi et al.* [54]. As shown in Fig. 8.27, the trap level position of the five radiation related defects in the DLTS spectrum and, hence, the corresponding activation energy, shifts with x. The DX centre, present in un-irradiated material and giving rise to the peak at 140 K remains unaffected by the irradiation. It is believed that the EZ1 centre corresponds to the E2 second V_{As} level in GaAs and is thus related to the arsenic vacancy. The four other levels are thought to correspond with the E3 level in electron irradiated GaAs, i.e., with an V_{As}-As_i pair. This level assignment and interpretation stands, however, in slight contrast with another study [59]. The reason for the splitting of a single level in GaAs in four levels in AlGaAs is related to the presence of the Al fraction, which changes the local atomic arrangement around the same basic Frenkel pair. It has furthermore been shown that some of the levels are pinned to the conduction band edge (donor nature), while other follow the valence band. No hole traps corresponding to the H0 and H1 traps, observed in GaAs (Table 4.1), were detected in irradiated AlGaAs. It was finally observed that the trap introduction rates increase with increasing Al concentration. This is evidenced by the individual and total introduction rates for 1 MeV electron irradiation in Fig. 8.28a and b, respectively. From this, it is concluded that the presence of Al stabilises the Frenkel defects in the As sub-lattice. Calculations indicate that the variation of the introduction rates with x could be explained by 2 or 4 neighboring Al atoms [54].

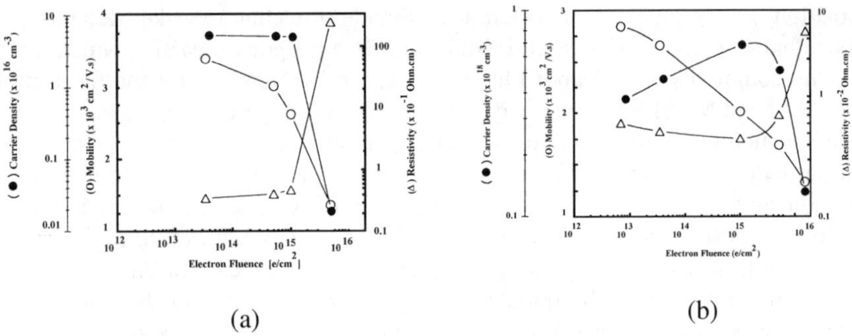

Fig. 8.26. Room temperature carrier density, mobility and resistivity of $n \approx 10^{16}$ cm^{-3} (a) and $\approx 10^{17}$ cm^{-3} MOVPE material as function of the 7 MeV electron fluence. [After Yousefi et al. [53]]

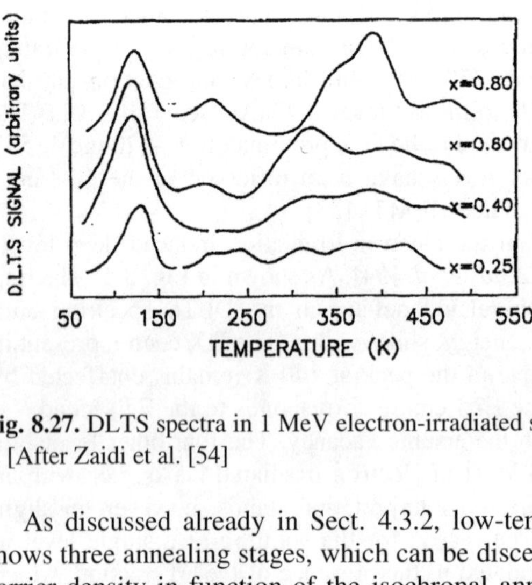

Fig. 8.27. DLTS spectra in 1 MeV electron-irradiated samples of various alloy compositions *x*. [After Zaidi et al. [54]]

As discussed already in Sect. 4.3.2, low-temperature irradiated n-type GaAs shows three annealing stages, which can be discerned in Fig. 8.29, representing the carrier density in function of the isochronal annealing temperature, after 85 K 1 MeV proton irradiation [59]. Stage I occurs around 235 K, stage II at 280 K and stage III at 520 K. The same three annealing stages can be identified for low temperature irradiated AlGaAs in Fig. 8.29. One marked difference is the stage III annealing which is smeared out over a broader temperature range. This can be understood by considering the presence of the Al fraction, which causes local variations in the atomic configuration of the As related defects. Another important finding is that the stage II step is far less pronounced in $Al_{0.22}Ga_{0.78}As$ than in GaAs [59]. From the observations it is concluded that, first, the underlying defects are in the Ga sub-lattice and secondly, more than one Ga atom is involved in the defect centers. Possible candidates are Ga_{As}-V_{Ga} antisite-vacancy close pairs, with a nearby

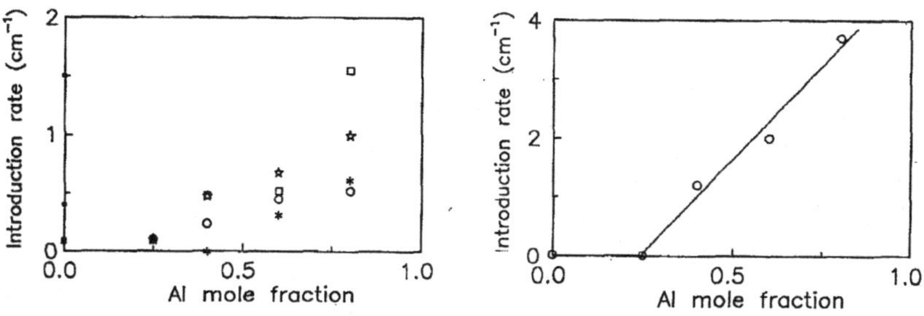

(a) (b)

Fig. 8.28. Variation of the introduction rates versus alloy composition for the various defects EZ2(*), EZ3 (stars), EZ4(squares) and EZ5 (o). The case of the radiation defects in n-type GaAs is indicated by the full circles (a). Variation of the total defect introduction rate vs alloy composition (b). [After Zaidi et al. [54]

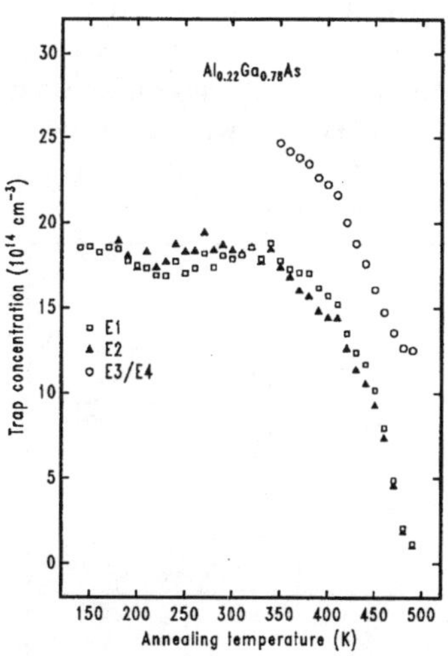

Fig. 8.29. Open triangles and squares: recoveries of the post-illumination 85-K values of the electron concentration, as a function of 15 min isochronal annealing temperature after simultaneous irradiation of the two materials at 85 K to a dose of $4.6x10^{12}$ 1-MeV H^+/cm^2. Solid triangles: recovery of the post-illumination electron density in the GaAs after correction, to remove the influence of the E1 and E2 defects from the GaAs plot. [After Irvine and Palmer [59]]

interstitial As_i, produced by the displacement of one Ga and an adjacent As atom. This also explains why most of the Ga sub-lattice defects are hardly observed in room temperature irradiated GaAs, since their formation/stability temperature lies well below 300 K.

The situation is somewhat simpler for neutron-irradiated bulk AlGaAs: irrespective of the Al content, a broad peak is found in DLTS, which resembles the U-band for GaAs [60]. It has been observed, however, that the activation energy of the band shifts to higher values with increasing Al content, as shown in Fig. 8.30a [60]. The experiments lead to the conclusion that the deep levels are pinned to the valence band. At the same time, the trap introduction rate increases (Fig. 8.30b), which triggers the idea that alloying with Al softens the compound. A similar type of behaviour has been found in neutron irradiated GaAsP LEDs [61, 62]. A defect band at E_c-0.7 eV was observed by DLTS, with an introduction rate of ~1 cm^{-1}. In this case, however, R_T is independent of the Al fraction or the doping concentration [61, 62]. From the photocapacitance quenching observed after illumination with ~1 eV photons, it was concluded that the neutron-irradiation induced defects are related to the EL2 family [61, 62].

Another important aspect for photonic devices is the current transport across AlGaAs/GaAs interfaces, which border the GaAs QW in many structures. It has been observed that the *I-V* characteristics of the AlGaAs/GaAs barriers at sufficiently high temperatures are dominated by trap assisted tunneling, whereby carriers tunnel to traps near the interface, which have been identified as DX centers [55]. They are subsequently emitted to the conduction band with an ionisation energy E_i, which is determined by Poole-Frenkel lowering, i.e., showing a linear

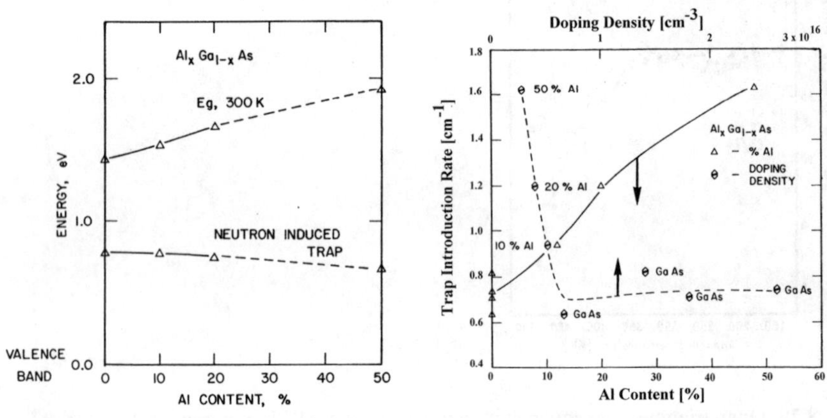

(a) (b)

Fig. 8.30. (a) Thermal emission energy of the dominant neutron-induced trap in n-type AlGaAs plotted relative to the valence band as a function of the Al content. Also shown is the variation in energy gap with Al content. (b) The trap introduction rate for the neutron-induced trap in n-type AlGaAs as a function of Al content and dopant density. Comparison of the curves suggests that the introduction rate is primarily dependent on Al content. [After Barnes et al. [60]]

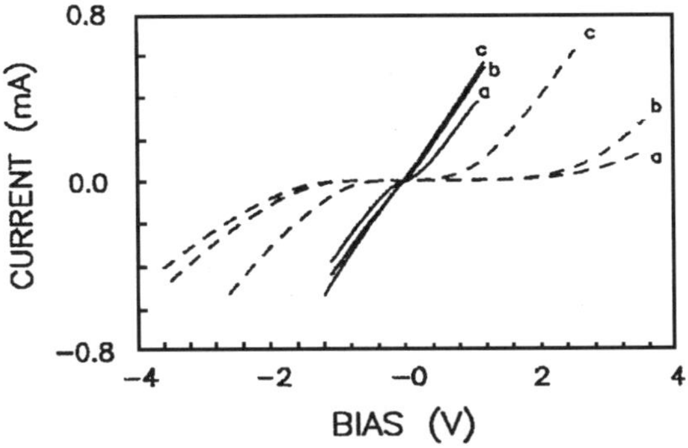

Fig. 8.31. Typical current-voltage characteristics monitored at various temperatures: 80 K (a); 180 K (b), and 300 K (c) before (full lines) and after (dashed lines) 1 MeV electron irradiation. [After Chaabane and Bourgoin [55]]

dependence of E_i on the square root of the electric field or on $\sqrt{V}$. This additional source of dark current is detrimental for the detectivity of QW photodetectors. In order to control the mechanism, one can deliberately introduce traps by high energy particle irradiation. It has been shown that exposure to high fluences (10^{17} cm^{-2}) 1 MeV electrons significantly alters the barrier height from 225 meV (prior to -) to ~300 meV (post irradiation) [55]. This yields a substantial reduction of the dark current (Fig. 8.31) and offers the possibility for defect control in such barriers. The responsible centers are thought to be mainly V_{As} [55].

In 1 MeV electron irradiated $In_{0.53}Ga_{0.47}As$ p-i-n photodetectors, two electron traps have been identified by DLTS after exposure [63], at 0.1 and 0.29 eV below the conduction band (Fig. 8.32). The introduction rate of the deeper level is 0.07 cm^{-1} and shows a thermal annealing stage around 425 K. A similar deep level, at E_c-0.31 eV has been observed in 63 MeV proton irradiated InGaAs [6]. It is believed that this level corresponds to the V_{As} defect. It is furthermore demonstrated that the annealing behaviour of the dark current of irradiated p-i-n diodes follows closely the annealing of the 0.29 eV centre [63]. Deeper electron traps at E_c-0.37 to 0.44 eV have been observed in 20 MeV alpha [64] or 220 MeV C [65] irradiated InGaAs photodiodes. In addition, after high dose γ-exposures deep electrons traps at 0.53 and 0.61 eV from the conduction band have been reported [66]. Neutron irradiation of InGaAsP laser diodes introduces typically two hole traps, at E_v+0.25 and +0.49 eV, respectively [67].

Some studies have also been devoted to GaInP alloys irradiated by 1 MeV electrons [68 - 70]. Analogous electron traps like in GaAs have been observed, although the E1 to E5 levels are shifted in energy [68]. In addition, a dominant hole trap at E_v+0.55 eV was detected which governs the performance of p-InGaP solar cells [69]. This level shows minority-carrier injection-enhanced annealing, which is in fact observed in many III-V based opto-electronic components (see next sections). This effect provides an additional advantage of InGaP-based solar cells for space applications, compared with other types.

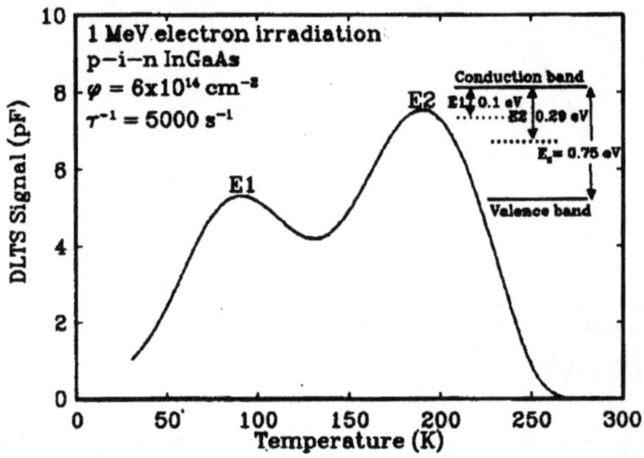

Fig. 8.32. Typical DLTS spectrum measured on the $Ga_{0.47}In_{0.53}As$ material after 1 MeV electron irradiation. [After Walters et al. [63]]

8.3.3 Damage Factors and NIEL

When dealing with displacement damage in photonic devices, one should keep in mind that the active regions are generally not at the surface but may reside several µm (or more) below the exposed surface. In addition, for photodetector devices, the region where photo-generated carriers are collected can extend several tens of µm in the device, e.g. a few minority carrier diffusion lengths. For these reasons, one should correct for the energy loss of an impinging particle with a given energy before it reaches the active region, in order to calculate the correct "local" NIEL, as illustrated schematically in Fig. 8.33 [5].

From (8.1) to (8.4) clearly follows that the recombination lifetime of the active material plays a dominant role in the operation of opto-electronic components. Quite often, the degradation of these devices under irradiation is described by a lifetime damage factor K_τ, which has been defined in Chap. 2 (2.20). A constant K_τ suggests that the reciprocal lifetime increases proportionally with fluence, which was verified experimentally for proton irradiated GaAs LEDs, in the energy range 1 to 500 MeV. The result of this study is summarized in Figs 8.34 and 8.35 [71]. In the proton fluence and particle range studied, the change in inverse lifetime shows a linear relationship with fluence (Fig. 8.34). A value of $1.06\text{x}10^{-4}$ cm^2/s was obtained at an incident energy of 9.0 MeV (corrected energy 8.0 MeV). Comparing this with the calculated NIEL shows a good agreement in Fig. 8.35. This clearly demonstrates that the lifetime degradation is due to displacement damage. Deviations occur, however, for proton energies above 150 MeV, showing lower measured damage factors than predicted by the restricted energy loss model

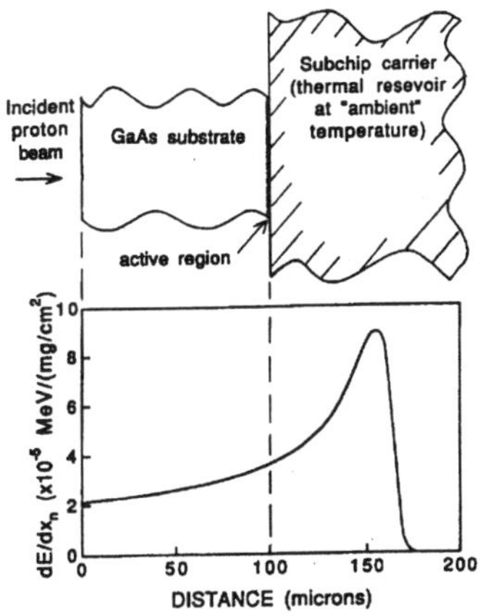

Fig. 8.33. Approximate non-ionizing energy deposition profile for 5.5-MeV protons incident on GaAs. Device substrate thickness and active layer position are depicted above. [After Evans et al. [5]]

developed in [72]. One possible interpretation is that the clustered damage, which is created by highly energetic Primary Knock-on Atoms (PKA's) is less effective in reducing lifetime, than the point defects induced at lower proton energies.

Meanwhile, the use of NIEL as an adequate metric for radiation-damage prediction in III-V optoelectronic compounds has been seriously questioned (see e.g., [73] and [74]). One problem is that for higher fluences than studied by *Barry et al.* [71] the damage evolves no longer proportionally with the fluence but rather follows a power law [73]. Careful experiments on AlGaAs single and double heterojunction LEDs have confirmed the deviations of the experimental damage factors from the calculated NIEL, observed in Fig. 8.34. The same discrepancy is also found for the leakage-current damage factor in proton-irradiated InGaAs photodetectors, as evidenced by Fig. 8.36 [75]. This stands in contrast with the earlier reported good agreement between the NIEL and the generation lifetime damage factor, derived for InGaAs p-i-n detectors [6].

Based on the observations of Figs 8.34 and 8.35, it was originally proposed to use the damage factor in GaAs LEDs as an alternative method for displacement damage monitoring [76], instead of using the resistance damage factor of epitaxial or implanted resistors, JFETs, etc... Given the sensitivity of the light output L_o, this monitor should be particularly suited for the low fluence range (or low γ dose) and is, therefore, complementary to the resistor-based methods. Furthermore, the damage in an LED is purely due to lattice displacements, while negligible ionisation damage occurs. Care has to be taken, that during the measurement of the light output, no self-annealing takes place. It is thus recommended to operate the device at low currents (1 to 10% of the normal operation current).

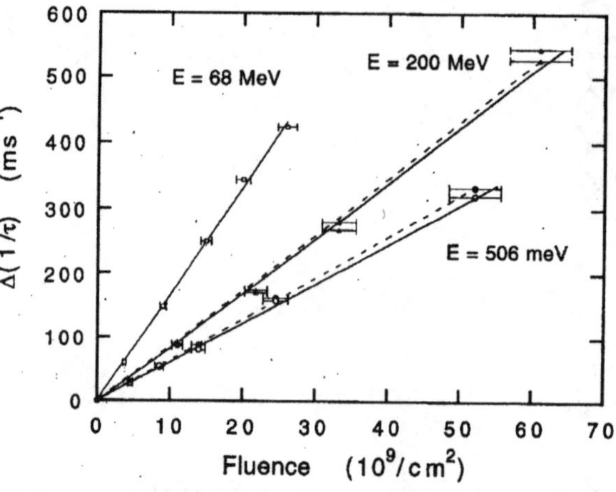

Fig. 8.34. Typical curves of the reciprocal lifetime with proton fluence. For 200 and 506 MeV protons, the two sets of points represent irradiations from the front (solid lines) and from the rear (dashed lines). [After Barry et al. [71]]

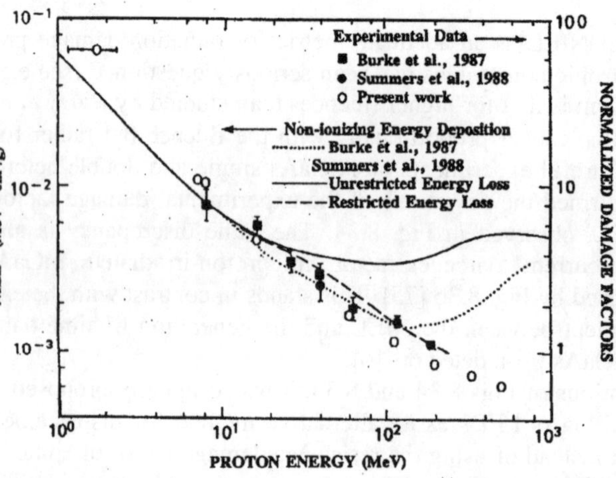

Fig. 8.35. Lifetime damage factors (open circles) fitted at 10 MeV to the calculated NIEL curve versus proton energy of [72]. The right-hand ordinate is normalized at 10^{-15} neutrons/cm^2 in [72], and at 10^{-5} cm/s in this work. [After Barry et al. [71]]

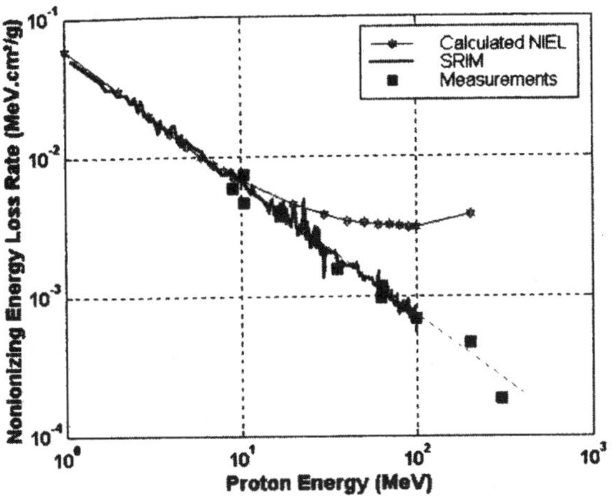

Fig. 8.36. The calculated NIEL in InGaAs compared with the increase of the leakage current damage factor (nA/cm^2/proton/cm^2) after filtering of the highest values and the NIEL derived from SRIM. The two last curves are scaled by the constant of 2.32x10^{-5} nA.g/MeV/cm^2 and 0.5, respectively. The SRIM-NIEL can be fitted by a function as 0.1159 $E^{-0.9433}$. [After Barde et al. [75]]

The light output damage factors (τ_0xK_τ) obtained in the quoted study, were in the range of 6x10^{-7} rad^{-1} (for ^{60}Co), indicating a detection limit of 5 to 10 krad(GaAs) [76]. For 1 to 2 MeV neutrons, a $K_\tau\tau_0$ product ≈10^{-12} cm^2 was established, which is 3 to 4 times higher than the data derived in earlier studies on similar devices. This is related to the higher initial recombination lifetime τ_0 (higher material quality), which was found to be 800 ns for the radiative lifetime, to be compared with previous values of 270 ns. This supports the observation that the better the starting material quality, the poorer the radiation tolerance is [3, 4]. A similar conclusion was reached for neutron-irradiated GaAsP alloys; Fig. 8.37 summarises the lifetime damage constant as a function of the P content [62]. One MeV electron-irradiated GaAs or AlGaAs exhibit a recombination lifetime damage constant in the range of 4 to 14x10^{-6} cm^2/s/e$^-$ [77], whereby K_τ shows a clear dependence on the Se donor concentration.

In summary, it can be stated that the calculated NIEL clearly overestimates the experimental displacement damage at higher proton energies, irrespective of the monitor used. Relying on NIEL for damage prediction during a space mission may underestimate by a factor 3 to 7 the real damage in opto-electronic components [73]. It has, therefore, been proposed to follow an empirical approach, based on the available damage factors at specific energies and by assuming a linear interpolation in between data points [73].

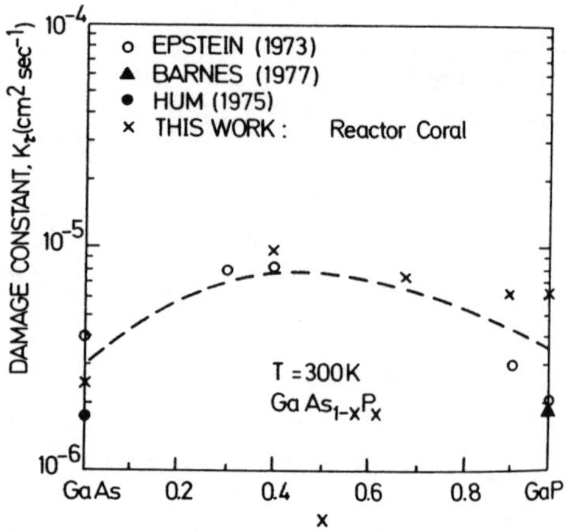

Fig. 8.37. Carrier lifetime damage constant as a function of alloy composition in VPE GaAsP alloys. [After Garcia et al. [62]]

8.4 Radiation Effects in Opto-Electronic Components

The impact of irradiation on the optoelectronic components, which were introduced in Sect. 8.2, will be discussed one by one. As will be seen, for certain applications like optocouplers, the LED may be the weak link with respect to radiation damage. It is, furthermore, clear that the hardness strongly relies on the type of component, i.e., bulk, heterojunction, QW, etc. It will be shown that the more advanced technologies yield harder devices. This relies on the shrinking thickness of the active layers, which renders them less susceptible to the created non-radiative defects. Another point which will be highlighted below is that radiation testing of opto-electronic components is not straightforward. In many cases, the damage factor depends on the operation point. One has to take into account the injection-annealing effect at high operation currents, both during irradiation and testing.

8.4.1 Light Emitting Diodes and Laser Diodes

8.4.1.1 Light-Emitting Diodes

The degradation of the light output L_0 of an LED following irradiation is often expressed in terms of a damage factor [2 - 5], given by:

$$\frac{L_{\Phi}}{L_{\mathrm{o}}} - 1 = \tau_{\mathrm{o}} K_{\tau} \Phi \tag{8.10a}$$

or alternatively:

$$\left(\frac{L_{\Phi}}{L_{\mathrm{o}}}\right)^{2/3} - 1 = \tau_o K_{\tau} \Phi \tag{8.10b}$$

(8.10a) is valid if L_{Φ} is measured in function of the forward bias, and (8.10b) in function of the forward current [3], whereby it is assumed that L_{Φ} is dominated by the diffusion component given by (8.3). Other expressions are available if the *L-I* (or *J-V*) characteristics are governed by space charge recombination (see for example Table 2 in [4]). As shown in Fig. 8.38, (8.10b) reasonably well describes the degradation of a QW LED, in function of the proton fluence [5]. A damage factor of $\approx 4 \times 10^{-14}$ cm^2/proton was found from the slope in Fig. 8.38. It has also been observed that not only the maximum spectral intensity (Fig. 8.1) but also the spectral width degrades upon irradiation (Fig. 8.39).

As shown in Fig. 8.2, also the LED forward *I-V* is modified by the irradiation. Often an increase of the current is found, which indicates the creation of an excess space charge recombination current component. The latter is caused by non-radiative recombination centers, introduced by the irradiation and normally corresponds to an $n_{\mathrm{if}}=2$ ideality factor of the forward diode characteristic, although $n_{\mathrm{if}}=1$ could also occur if the levels form an overlapping band of shallow states [2]. Another important observation is that the characteristics improve under forward current annealing (Figs 8.2 and 8.40) [2]. In other words, operating the irradiated device for a certain time at large forward current either *during or after* the exposure removes part of the damage. This defect annealing may be partially related to self-heating although it is believed that an athermal diffusion mechanism, to be discussed below, is operational.

From (8.10) readily follows that the larger the starting recombination lifetime, the more prone the LED will be to radiation damage. Its light output is thus a sensitive function of the starting material quality and explains why early amphoteric GaAs LEDs where less suitable for space applications than GaAlP or GaAsP counterparts. The damage factor also relies on the material composition, as shown in Fig. 8.41 [4], where a decade difference can be seen in $\tau_0 K_{\tau}$, as a function of the P fraction. Typical values for epitaxial GaAs are in the range of 1-2×10^{-13} cm^2 per 2 MeV electron [3], one decade higher. It also makes clear why in earlier studies, laser diodes turned out to be harder than LEDs. This may no longer be the case for modern QW structures [5], where the opposite can be found. Investigations of state-of-the-art heterojuntion LEDs have confirmed the gain in hardness, compared with diffused or amphoterically doped devices [78 – 80]. In addition, the degradation of heterojunction LEDs is less dependent on the bias during exposure or subsequent post-irradiation testing. In other words, there is less bias-dependent recovery for such devices. On the other hand, *Reed et al.* have shown that single heterojunction LEDs are one decade more susceptible to radiation damage than their double heterojunction AlGaAs counterparts [73]. It has also been demonstrated that QW LEDs hold great promise for rad-hard applications [74]. This relies amongst others on the forward *I-V* characteristics, which are little degraded for example by 1-15 MeV protons or alpha particles. Results on radiation effects in GaN LEDs are available and show a dramatic change in the electroluminescence under e-beam irradiation [81].

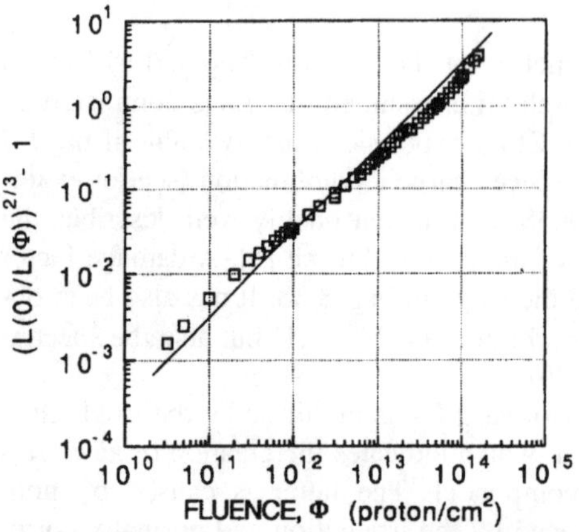

Fig. 8.38. Analysis of strained QW LED degradation while operated at 100 mA and 20 °C corresponding to a $K_\tau\ \tau_0$ product of 4.0×10^{-14} cm^2/5.5 MeV proton. [After Evans et al. [5]]

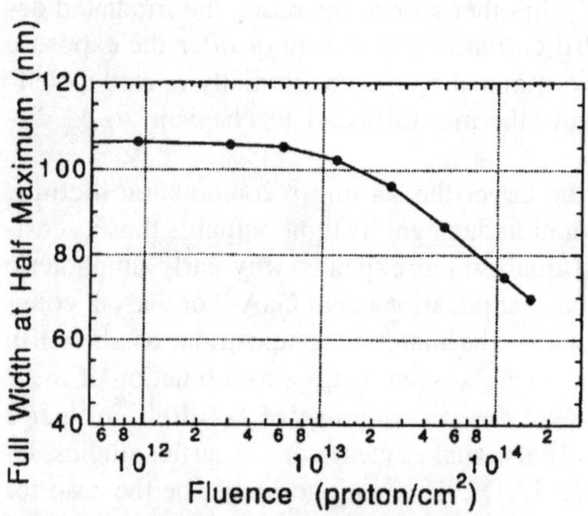

Fig. 8.39. QW LED spectral width versus proton fluence. [After Evans et al. [5]]

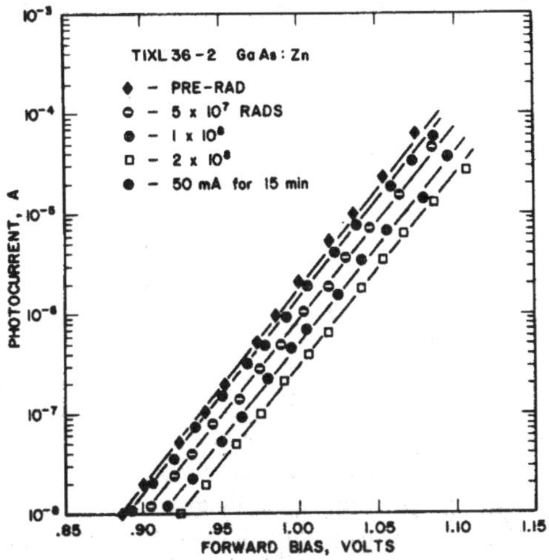

Fig. 8.40. Light intensity (phototransistor collector current) - voltage curve of a typical Zn-diffused LED prior to and following gamma irradiation and forward bias annealing. The true *L-V* characteristic is distorted by the nonlinear phototransistor response. As in the case of the I_F-V_F data, forward bias recovery of the *L-V* characteristic is nearly complete. [After Barnes [2]]

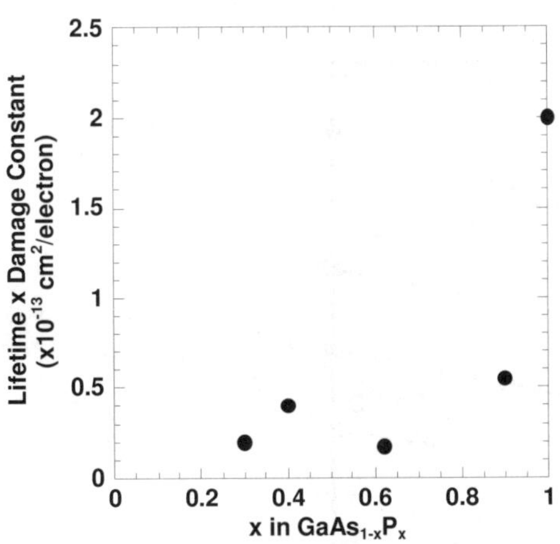

Fig. 8.41. Lifetime damage constant product versus composition. [After Dimiduk et al. [4]]

8.4.1.2 Laser Diodes

As shown in Fig. 8.6, the major degradation of a laser diode is the increase in the current threshold I_{th}. A more or less linear variation of I_{th} with proton fluence is observed in Fig. 8.42 [8]. This radiation-induced change in I_{th} can be modeled by the threshold current damage factor K_{Io} [8]:

$$\frac{I_{th}}{I_{th0}} = \frac{\tau_o}{\tau} = 1 + K_\tau \tau_o \Phi = 1 + K_{Io}\Phi \tag{8.11}$$

I_{tho} is the pre-rad current threshold. From Fig. 8.42 follows that the increase in I_{th} is larger for higher operation temperatures. The degradation is also higher for devices shorted during the exposure, than for those biased with I_{bias}=35 mA. The latter is related to the occurrence of forward-bias annealing during the irradiation. From Figs 8.43 and 8.44, one can derive that the degradation of the normalized optical power is smaller at lower operation temperature and high operation current, irrespective of the bias during exposure. Typical K_{Io} values are in the range 1.5 to 4.6×10^{-15} cm^2/p (200 MeV energy) [8]. It is furthermore assumed to be proportional to the NIEL S, through: $K_{Io}=4.5 \times 10^{-12}$ S. It was also observed that K_{Io}(horizontal) > K_{Io}(parallel) > K_{Io}(normal), whereby the direction of incidence of the proton beam has been defined in Fig. 8.4b.

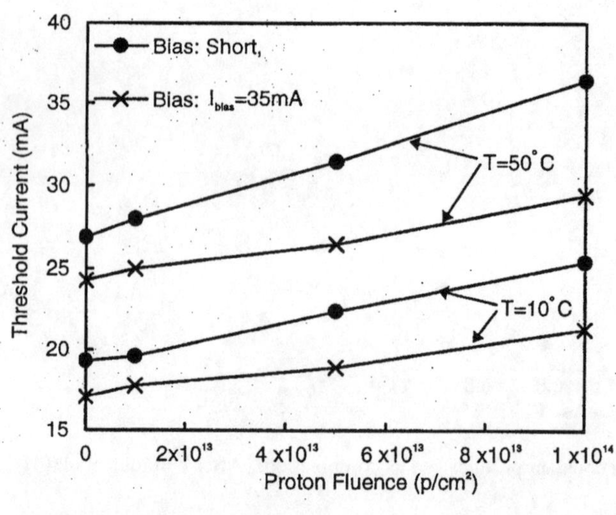

Fig. 8.42. Threshold current vs. proton fluence for devices with different biases during irradiation and measured at different temperatures. [After Zhao et al. [8]]

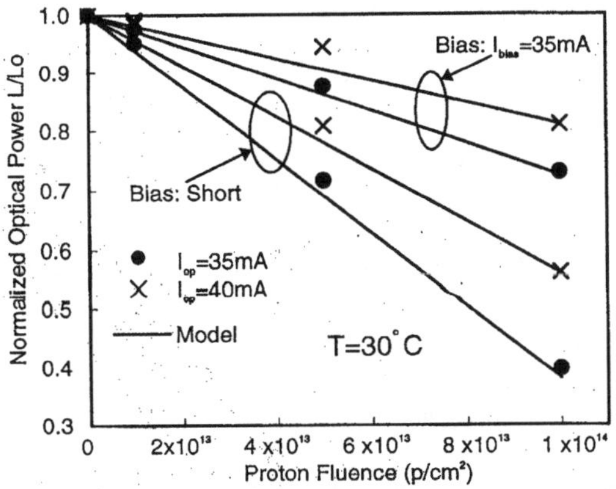

Fig. 8.43. Experimental and modeling L_Φ/L_0 vs. proton fluence for devices with different biases during irradiation and measured at different operation currents. [After Zhao et al. [8]]

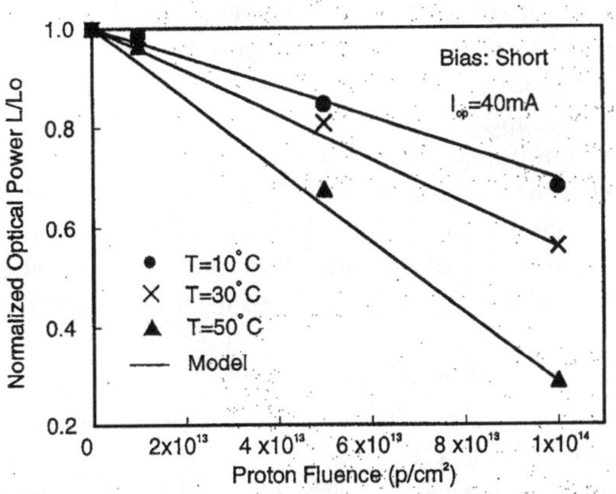

Fig. 8.44. Experimental and modeling L_Φ/L_0 vs. proton fluence for devices shorted during irradiation and measured at different operation temperatures. [After Zhao et al. [8]]

In addition, the slope efficiency of a LD may change slightly after irradiation (Fig. 8.6) [8]. This is related to optical absorption and scattering from defects induced by the irradiation. At low proton fluences ($<10^{13}$ cm^{-2}), however, η increases slightly for the device of Fig. 8.6, which is interpreted as being due to the radiation induced ordering effect, for example by relaxation of the grown-in stress in the QW [8]. At low fluences, the energy deposited by the protons can reduce the stress, which overcomes the effect of the created Frenkel defects. At higher fluences, the balance goes the other way.

It has also been observed that the lifetime damage factor is larger than the threshold current damage factor for MQW GaAs/GaAlAs laser diodes irradiated at various proton energies [82]. Such a difference is explained by the different values of the radiative lifetime when measuring the damage factors in different operation regimes. In the study, it was also difficult to find a correlation between the damage factors and the NIEL [82]. This was mainly due to the wide sample-to-sample variation (*K* values) and the nearly constant value of NIEL in the proton energy range studied.

Finally, the radiation-induced change in optical power L_0 of a LD can be expressed as follows [8]:

$$\frac{L_\Phi}{L_o} = 1 - \frac{I_{th0}}{I_{op} - I_{th0}} K_{Io} \Phi \tag{8.12}$$

In writing (8.12), it is assumed that the slope efficiency remains constant, i.e., that $\eta / \eta_o \approx 1$.

From formula (8.12) follows a few simple guidelines for the hardness of LDs [8].

(1) First, the degradation will be lower for higher L_o. This means that higher operation bias is better for the radiation tolerance. It should also be remarked here that when comparing radiation results from different devices and sources, one should take into account the operation conditions during and after irradiation. In addition, a high operation bias during exposure could produce sufficient annealing so that the correlation with NIEL may be violated.
(2) The lower I_{tho}, the less the degradation of optical power after irradiation. Thus, low I_{tho} is most suitable for radiation applications. This suggests that QW laser diodes may be more suitable for radiation applications, since they have lower I_{tho}.
(3) The higher η_o, the higher the degradation of optical power after irradiation. This suggests that LDs with lower η_o are more suitable for radiation applications, though they are not optimized lasers.
(4) When the operation temperature increases, I_{tho} increases, so that the optical power degrades more for higher operation temperatures.

A more refined numerical modeling procedure of the current threshold increase of irradiated laser diodes has been proposed by *Jolly and Vicrey* [83] and takes into account the interaction between the optical and the electrical properties of the device. In more detail, the change in the internal quantum efficiency is coupled to the additional electrical field, caused by trapped holes. This model should extend the calculations to larger (neutron) damage ranges, where the forward bias drop over the diode changes by several hundred milivolts, instead of the usual milivolts. It was also observed that there was no radiation induced shift of the spectral line width or position of the GaAlAs QW laser studied (Fig. 8.45).

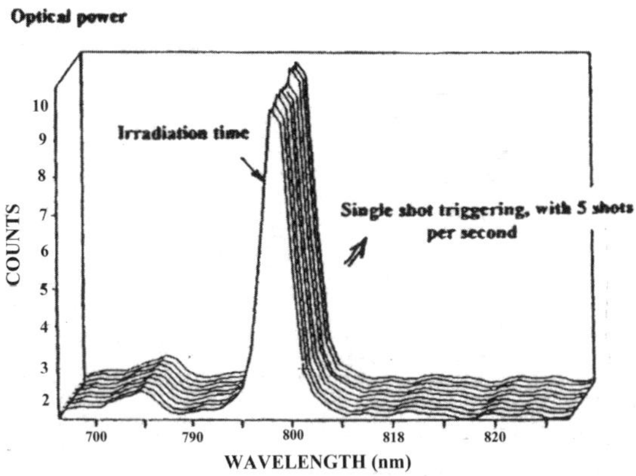

Fig. 8.45. Transient effects of pulsed 17 MeV electrons on the optical spectrum of the SDL 3230T laser bar. The laser bar is tested with I=80 A, the optical analyzer being triggered once every 5 seconds after the beginning of the electron burst. [After Jolly and Vicrey [83]]

Similar as for an LED, a LD can show bias annealing after irradiation. The annealing behavior of 780 nm MQW laser diodes after proton irradiation has been investigated in detail by *Zhao et al.* [84]. A typical response is shown in Fig. 8.46. Three stages can be identified in Fig. 8.46: during stage I, the laser diode stabilises, whereby the junction temperature increases 20-25°C, at an operation current of 35 mA. In stage II, the laser diode was irradiated by 68.5 MeV protons till a fluence of $6\text{x}10^{12}$ p/cm^2 was reached. In stage III, the device was allowed to anneal under forward bias, for 1800 s. The annealing effects are most significant during the first 100 s. It can be fitted by an exponential function:

$$L_{\Phi}(t) = L_{0} + \Delta L[1 - \exp(-t/\tau_{A})] \tag{8.13}$$

with ΔL the amplitude and τ_A the time constant of the annealing. The initial value is in the range of 50 s, followed by a slower recovery at τ_A=1000 s. Finally, the annealing proceeds even more slowly at time constants of $1.5\text{x}10^4$ to $2\text{x}10^4$ s. These three stages indicate the removal of different kinds of radiation induced defects. Simple Frenkel defects like a vacancy or a vacancy-interstitial pair will anneal easily, while the antisite defect will be very stable and hard to remove when created. The major mechanism of this forward bias annealing is not thermal dissociation and diffusion of defects, but rather related to the so-called Recombination Enhanced Defect Reaction (REDR) mechanism [84, 85]. It means that when a non-radiative electronic transition occurs, the phonon energy deposited locally can enhance the temperature at the defect and activate it. Then the defect has more energy to move and the reaction (annihilation or dissociation) rate increases. This REDR mechanism typically finds place in the active device region.

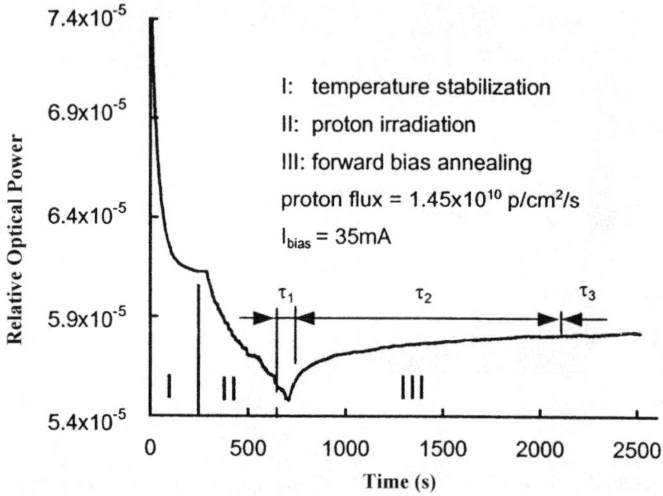

Fig. 8.46. The relative optical power vs. time, including temperature stabilization (I), proton irradiation (II), and forward-bias annealing. [After Zhao et al. [84]]

From this bias annealing, it is furthermore concluded that at the lower fluxes typical for space, the degradation of LD may be less than in a laboratory experiment [84].

The recovery of irradiated laser diodes through isochronal thermal annealing following ^{60}Co gamma irradiation has been studied by *Barnes* in the early days [1]. He found an activation energy of about 1.7 eV and an annealing stage centered around 237 °C. This corresponds approximately with the removal of the E1-E5 electron traps, associated with the V_{As}-As_i defects.

Other damage mechanisms have been reported for irradiated laser diodes. For example, facet damage (cracks) may develop during neutron exposure [6]. Also thermal effects (self-heating) play a role in the enhanced degradation [6, 9], which may be particularly important for VCSELs.

Finally, the radiation characteristics of a number of commercially available LEDs and LDs have been studied by *Lischka et al.* [86, 87], for neutron, gamma and X-ray exposure. One of their conclusions was that InGaAsP LEDs are more radiation tolerant than InGaAs, which on their turn behave better than GaAs LEDs [87].

8.4.2 Photodetectors

The degradation of irradiated InGaAs photodetectors has been studied by a number of groups [6, 10, 63, 75]. One of the primary parameters investigated is the dark reverse current, which sets the detection limit and low-frequency noise of the PD. As shown in Fig. 8.47a [10], the reverse current increases significantly with 1 MeV neutron fluence. Consequently, the photo-response will be degraded (Fig. 8.47b).

At the same time, the spectral response represented in Fig. 8.48 changes, whereby both the amplitude and the width of the spectral region reduce. The reduction in normalized response, following a 63 MeV proton exposure is given in Fig. 8.49 [7]. Upon thermal annealing in the range 100 to 300°C, a considerable recovery is found, both in the dark current (Fig. 8.50a) and in the optical response (Fig. 8.50b). A broad annealing stage in the range 400 to 500 K is observed in fair agreement with other reports [63, 88, 89]. Above 550 K a second stage, related to the annealing of more stable radiation defects is found. Such a deeper stage was also reported by *Walters et al.* [63].

The annealing rate shows an activation energy in the range 0.38 eV (dark current) to 0.41 eV (photocurrent) (Fig. 8.51), which is significantly smaller than the 1.7 eV found from irradiated GaAs laser diodes [1], or the 0.9-1.1 eV activation energies found in [89] for InGaAs photodiodes. As shown in Fig. 8.52, there is a reasonable agreement between the annealing of the E_c-0.29 eV level and the dark reverse current, for 1 MeV electron irradiated PDs [63, 88, 89]. This is supported by the activation energy derived from the Arrhenius plot of the reverse current, shown in Fig. 8.53 [88]. It is believed that the corresponding defects are V_{As} or arsenic vacancy/impurity related defect complexes [63, 88].

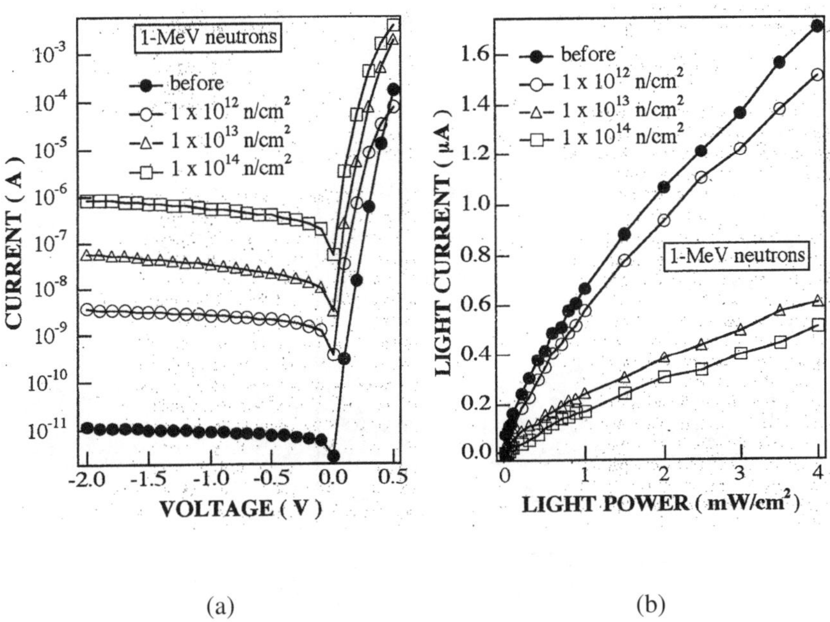

Fig. 8.47. Influence of 1 MeV neutron irradiation on the *I-V* (a) and I_L-P_L (b) curves of an InGaAs photodetector. [After Ohyama et al. [10]]

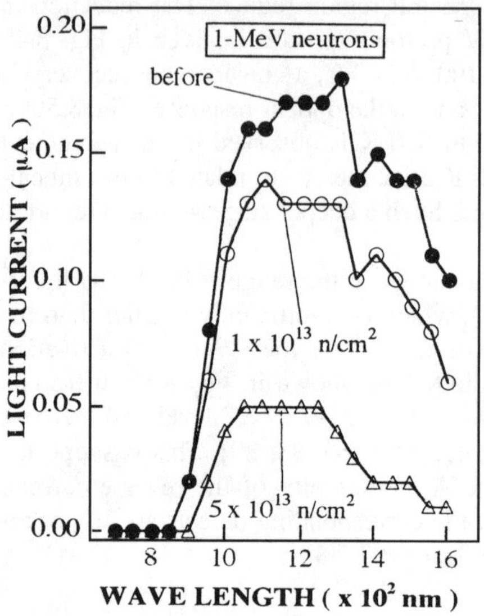

Fig. 8.48. Influence of 1 MeV neutron irradiation on I_L-λ characteristics of InGaAs photodetectors. [After Ohyama et al. [10]]

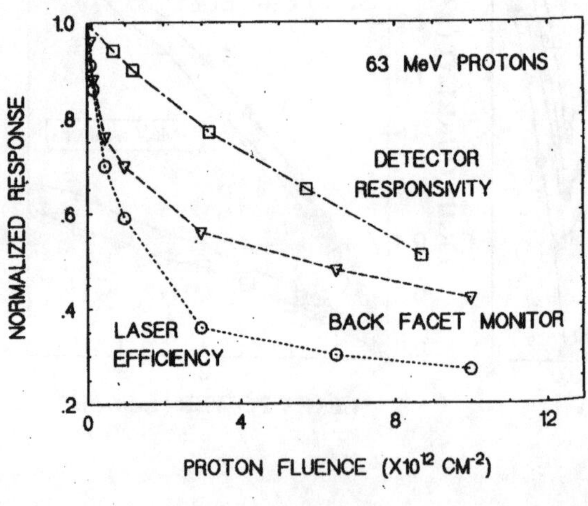

Fig. 8.49. More degradation is seen with fluence for the laser's optical and monitor outputs than for the separate InGaAs photodiode, and they vary differently with fluence. [After Marshall et al. [7]]

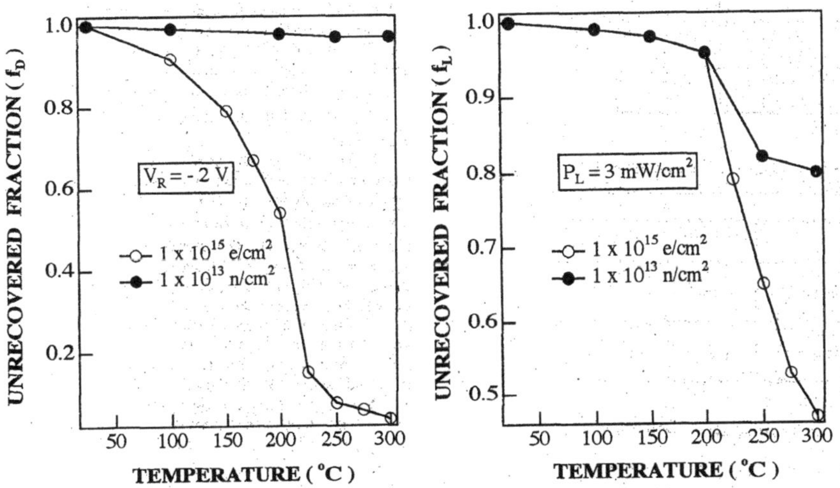

(a) (b)

Fig. 8.50. Unannealed fraction of dark current (a) and photocurrent (b) as a function of annealing temperature for a fluence of 10^{13} n/cm^2 and 10^{15} e/cm^2. [After Ohyama et al. [10]]

Ohyama and co-workers have studied in detail the radiation response of InGaAs PDs to 1 MeV neutron and electron [10, 90], 20 MeV alpha's [91] and 220 MeV C [92] exposures. Assuming that the degradation of the reverse dark and photocurrent can be modeled in the same way, i.e., by:

$$I_{D0,L0}(\Phi) = I_{D0,L0}(0) + K_{D0,L0}\Phi \tag{8.14}$$

with $I_{D0,L0}(0)$ the pre-radiation values, the damage factors $K_{D0,L0}$ for the dark (subscript D) and photocurrent (L) of Table 8.1 have been obtained. As can be seen, in first instance, there is a reasonable agreement with the calculated number of knock on atoms N_d (cm^{-3}), which scales with NIEL.

A parameter which certainly deserves further studies is the low-frequency noise of the irradiated photodectectors [75]. It has for example been found that an InGaAs multiplexed detector array on board of the SPOT 4 satellite exhibits after some time both extreme dark current values and random telegraph signals (RTSs). Extensive ground testing made clear that both phenomena are associated with the proton-irradiation damage and typically occur in high electric field regions of the structure, similar as for certain types of silicon devices (CCDs and CMOS imagers) [75].

The radiation response of IR QWIPs to high energy proton and alpha particles has been studied as well [11, 74, 93, 94]. The impact on the dark current at 77 K of 0.8 MeV protons is depicted in Fig. 8.54. A clear reduction of the dark current is observed, which may amount to two to three orders of magnitude for a fluence

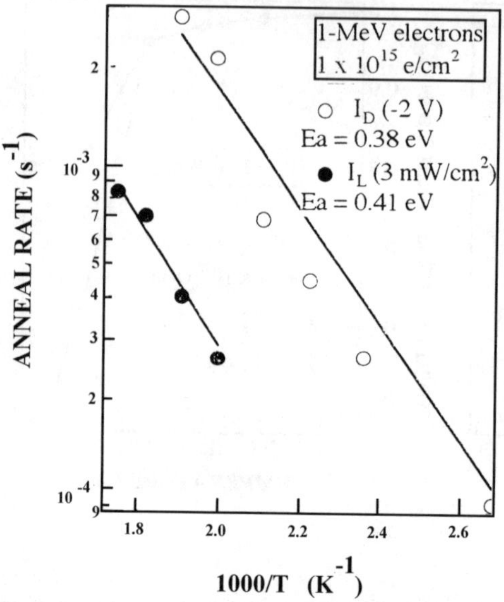

Fig. 8.51. Relationship between the annealing rate and the temperature, corresponding to 15 min isochronal annealing. [After Ohyama et al. [10]]

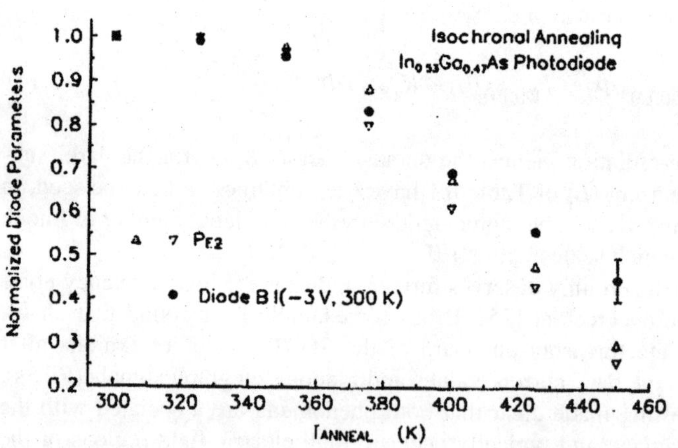

Fig. 8.52. Isochronal annealing behavior of the E2 DLTS peak height for two InGaAs photodiodes irradiated up to $6x10^{14}$ cm^{-2} 1 MeV electrons. Also shown is the annealing behavior of the reverse current of diode B as measured at -3 V and 300 K. The data are normalized to their pre-anneal values. [After Shaw et al. [88]]

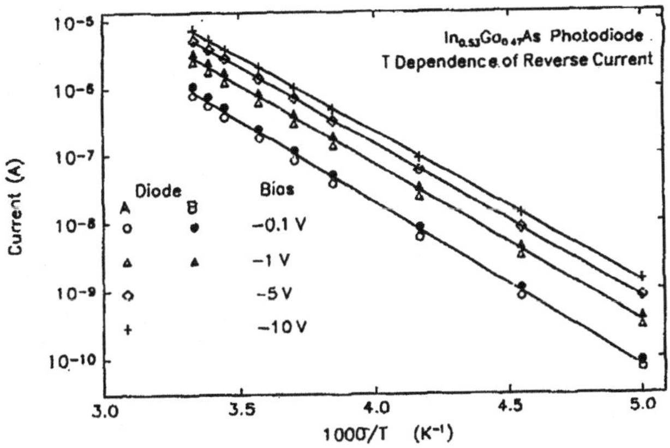

Fig. 8.53. Temperature dependence of reverse current at several voltages for diodes A and B. The solid lines are fits and E_A ranges from 0.44 to 0.4 eV as the reverse bias is increased from -0.1 to -10 V. [After Shaw et al. [88]]

Table 8.1. Dark and photocurrent damage factors defined in Eq. (8.14) and derived for irradiated InGaAs photodiodes exposed to different particles. In the 4th column, the calculated number of knock-on As atoms is indicated. No bias was applied during the irradiations

Particle	K_D (Acm^2)	K_L (Acm^2)	N_d (cm^{-3})
1 MeV e^-	$3.8x10^{-23}$	$-3.5x10^{-22}$	0.003
1 MeV n	$8.2x10^{-21}$	$-8.1x10^{-20}$	8.4
20 MeV α	$4.5x10^{-18}$	$-5.5x10^{-19}$	93[a]
220 MeV C	$1.5x10^{-18}$	$-7.0x10^{-19}$	252

K_D dark current damage factor, K_L photocurrent damage factor, N_d calculated number density of knock-on atoms.
[a]Average number of In, Al and Ga atoms.

of $2.0x10^{13}$ p/cm^2. The corresponding spectral responsivity is shown in Fig. 8.55 [95], from which a clear drop in intensity at the central wavelength is noted. The change in operability, defined as the fractional change in responsivity D^* for the same devices, under proton irradiation, is given in Fig. 8.56. Comparing the change in responsivity with the one in the dark current shows that the latter parameter is much more sensitive to irradiation. The reason is that the current across the QW is exponentially dependent on the density of electrons and, hence, the carrier removal rate in the 10^{18} cm^{-3} doped GaAs QWs, while the infrared absorption is proportional to n. The impact of the carrier removal on the current is thus much more pronounced: it generates an increase of the Schottky barrier height and a reduction of the electron mobility and lifetime. Similar results have been obtained for alpha particle and oxygen ion irradiations, whereby the degradation, as expected, increases with particle fluence and mass and with decreasing particle energy [95].

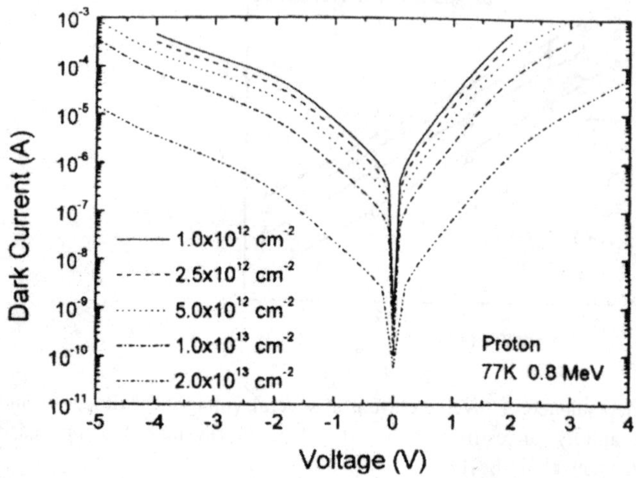

Fig. 8.54. Dark current versus bias voltage characteristics at 77 K for 0.8 MeV proton-irradiated QWIPs at different fluences. [After Li et al. [94]]

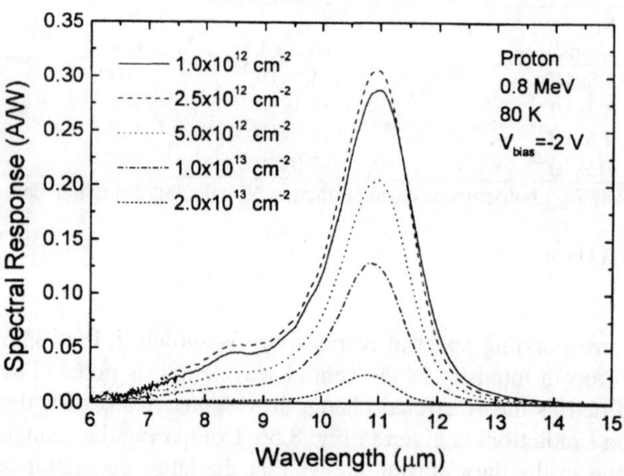

Fig. 8.55. Spectral responsivity versus wavelength characteristics at 80 K and -2 V for 0.8 MeV proton-irradiated QWIPs at different fluences. [After Li et al. [94]]

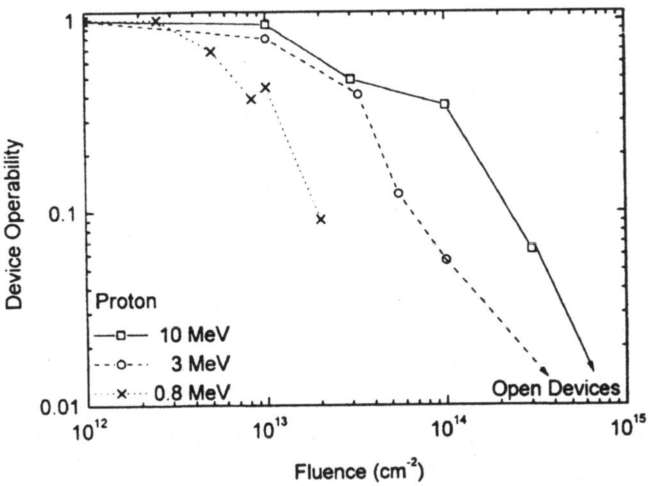

Fig. 8.56. Device operability (defined as the fractional reduction in responsivity) versus fluence for QWIPs irradiated with different high energy proton particles. [After Li et al. [94]]

8.4.3 Optocouplers

The radiation response of an optocoupler is complex, since it is composed of the degradation of the LED/LD, the PD and possibly the coupling medium and optics (lenses, fibers,...). In addition, both displacements and ionisation damage can play important roles depending of the type of light emitting and sensitive devices. Figure 8.57 illustrates the impact of the different parts of the optocoupler of Fig. 8.10b on the overall current transfer ratio [13], showing that for that particular type, displacement damage in the LED dominates the overall performance. In such a case, a silicon bipolar phototransistor is used as photosensitive device, whereby both the current gain and the photo-response degradation should be considered. It turns out that the decrease in carrier recombination lifetime is the decisive factor for the phototransistor response, while ionisation damage effects on the current gain only play a minor role in the performance degradation [13]. Using a shorter wavelength LED can substantially improve the photo-response after irradiation (e.g. 700 nm versus 850 nm), because of the smaller penetration depth of the light and the reduced sensitivity to the changes in the minority carrier diffusion length in the phototransistor [13]. Little effect of the coupling medium for the lateral structure has been found in this case.

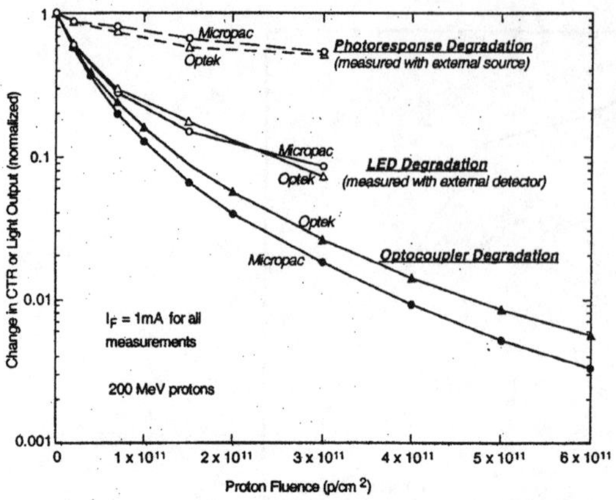

Fig. 8.57. Effects of proton degradation for photo-response, LED, and optocoupler. [After Rax et al. [13]]

The sandwich structure of Fig. 8.10a, on the other hand, hardly suffers from the LED degradation and, therefore, shows a better hardness (Fig. 8.58). This is partly related to the shorter wavelength of the LED, which reduces the sensitivity of the PD to the irradiations. It was also observed that the displacement damage produces more damage than the accompanying ionisation by the high energy protons (Fig. 8.59) [13]. In another study, it has been concluded that the CTR of optocouplers with an AlGaAs LED is limited by the light output degradation, while for GaAlP sources, the photo-response of the detector is the major mechanism [12].

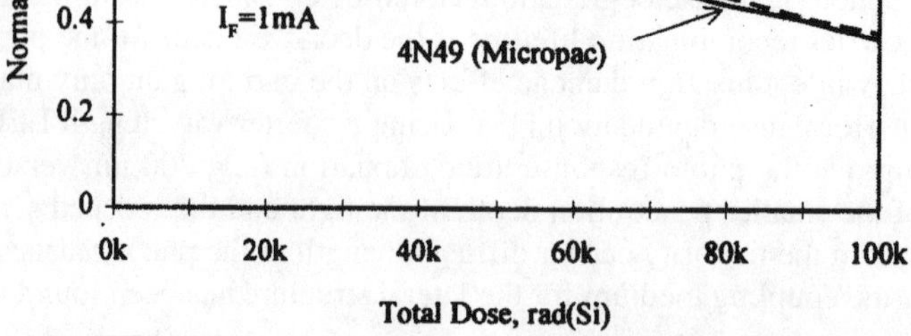

Fig. 8.58. Comparison of CTR degradation for two types of optocouplers after irradiation with cobalt-60 gamma rays. [After Rax et al. [13]]

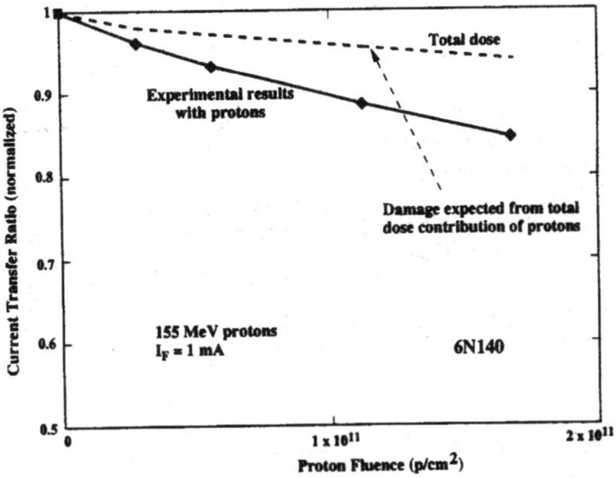

Fig. 8.59. Comparison of measured CTR degradation with calculated degradation from displacement damage of the phototransistor. The main reason for the difference is the shorter penetration depth of the 700 nm wavelength, which reduces the effect of changes in diffusion length (with displacement damage). [After Rax et al. [13]]

Another emerging issue for broad band systems is the occurrence of single event upsets and transients [12, 95]. It has been demonstrated that state-of-the art devices can be used for CERN's Large Hadron Collider (LHC), even for the parts operating at liquid argon temperatures [96]. It is furthermore shown that the radiation-induced increase of the LF noise of the LEDs and LDs is within acceptable ranges. Compared with more classical structures, QW LEDs show the largest radiation resistance (Fig. 8.60), in agreement with other reports [8].

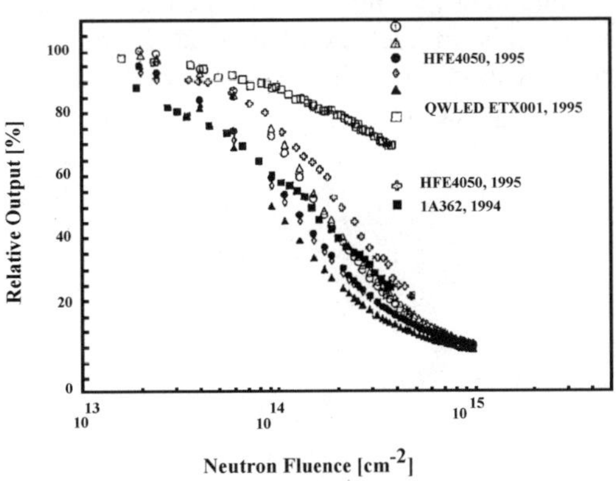

Fig. 8.60. LED output as a function of received 1 MeV neutron fluence. [After Söderqvist et al. [96]]

8.5 Conclusions

The new generation of III-V based light sources and detectors show great promise for space applications and operation in a radiation environment. However, since these technologies are under development, substantial radiation testing is necessary in order to verify their suitability for space photonics. There is also a lack of basic damage knowledge, compared with GaAs. While the latter material has been studied since the early 60ties, radiation defects in ternary compounds are still rather unexplored. Initial studies indicate that although there are definitely some links with the simple Frenkel defects in GaAs, the situation may become quite complex due to alloying. It is also clear that while for MESFET and HEMT devices carrier removal is the dominant effect, the opto-electronic components suffer mainly from a decrease of the carrier lifetime. Especially for the case of the photodetectors, LF noise studies would be welcome, in order to determine the impact on the detectivity. Additionally, they could provide fundamental information on the radiation defects. A final important observation is the fact that the operation bias of light sources during irradiation (and also during testing, although to a minor extent) dramatically changes the radiation response. Meaningful testing, therefore, should be performed as close as possible under realistic operation (biasing) conditions.

One critical issue that remains is the damage prediction of opto-electronic components. Most available data point to the fact that the NIEL is not a good metric for that purpose. This is amongst others related to the fact that the basic device parameters are no linear function with particle fluence. A conservative approach is to use the available damage factors and to linearly interpolate between proton energies. However, one should also keep in mind that for certain types of LEDs or LDs, there may be a strong sample-to-sample and batch-to-batch variation, so that a minimal radiation testing remains recommended.

Similar as for other mainstream technologies, scaling yields an enhanced hardening of opto-electronic components. It is in fact expected that future low-dimensional structures, using quantum wires or dots will show improved radiation performance, compared with QWs or heterojunctions.

9 Advanced Semiconductor Materials and Devices – Outlook

9.1 Introduction

The previous chapters have been addressing a large variety of materials and devices that are nowadays commonly used in different radiation applications. However, the radiation community also watches the trend in the microelectronics world very carefully. Therefore, new materials and device structures are already in an early phase also studied from a viewpoint of their performance in a radiation environment. This chapter aims at briefly addressing some of these materials and devices. Within the scope of the book and due to space restrictions the attention is given to potentially very promising directions with applications in the near future.

Section 9.2 briefly reviews the field of non-volatile memories. Notwithstanding the fact that these type of devices are already in use for a long time and therefore surely not new for the space community, their implementation in commercial applications has strongly increased during the last years due to extensive on-going research associated with Flash, Ferroelectric and magnetic memories. As outlined in Chap. 6, scaling of CMOS requires the application of alternative and new gate dielectrics to replace the thin SiO_2 layers.

The search for the appropriate high-*k* dielectrics surely forms also a challenge from a radiation viewpoint, as discussed in Sect. 9.3. Some materials are already around for a very long time, but only received strong interest in recent years due to their improved fabrication techniques or their potential for new applications fields that are otherwise very hard to be covered by a more classical semiconductor material. A very good example of this is SiC. Different material and device aspects are addressed in Sect. 9.4. The quality improvement of the SiC material allows the fabrication of components and devices with a good electrical performance. Although a high radiation-hardness is expected, more understanding of the fundamental radiation aspects is required. The last section gives a general conclusion and some perspectives for the future.

9.2 Non-Volatile Memories

Compared to the SRAMs and DRAMs non-volatile memories (NVMs) have the property that they do not loose their information when the power supply is switched off. With volatile memories a short power interruption can have a disastrous effect, something absolutely intolerable for space missions. Also the commercial applications of NVM devices, such as mobile phones and notebook computers, has strongly increased in recent years. These devices can be classified depending on their read-write mechanisms and the power consumption [1- 2]. Distinction is made between programmable (PROM), electrical programmable (EPROMs) using UV-rays, electrical erasable and programmable (EEPROMs) and Flash read-only memories (ROMs), and magnetic (MRAMs) and ferroelectric random access memories (FeRAMs). In 2000, NVMs were good for 23 % of the market and the share of Flash was even higher than of SRAMs. Within the Flash family, FeRAMs are gaining more and more potential [3]. Therefore this section will only review some radiation aspects of Flash and FeRAMS.

9.2.1 Flash Memories

Flash devices have been introduced in the early 80-ties, have a low access time and are mostly programmed at cell level, while blocks, sectors or pages are erased simultaneously [1]. The flash cell basically consists of a dual polysilicon gate structure, i.e., a floating (FG) bottom gate and a controlled (CG) top gate as schematically illustrated in Fig. 9.1. The threshold voltage is shifted by the injection of electrons through a dielectric layer onto or off the floating gate. The charge control on the FG involves either Fowler–Nordheim (FN) tunneling or channel hot electron (CHE) injection, and sets the threshold voltage and the "state" of the memory bit.

The flash design can be based on a NOR, NAND or split gate approach. The cell concept of the NOR and NAND flash memory is illustrated in Fig. 9.2 [4]. The NOR mostly uses CHE for writing and FN tunneling for erasing, while NAND uses tunneling for both operations. An important difference between both is the maximum operation voltage needed on chip, i.e., >12 V for NOR and > 18 V for NAND. NOR memories have a higher power consumption and are more difficult

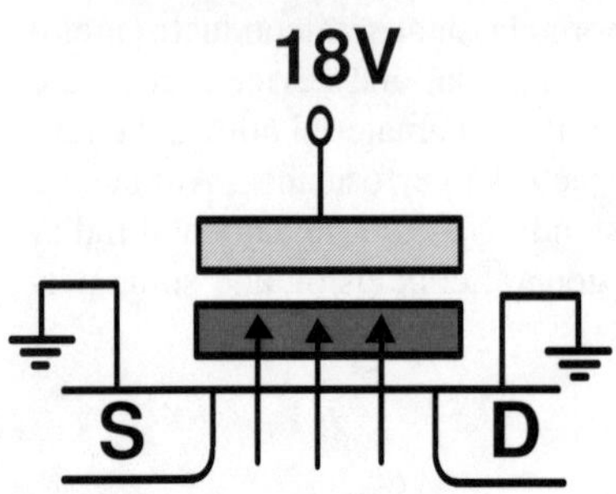

Fig. 9.1. Dual gate structure of a non-volatile memory consisting of a floating (bottom) and a control (upper) polysilicon gate

to scale to high packing densities. There is nowadays a strong breakthrough of embedded flash applications, strongly driven by the future system-on-chip (SOC) requirements. From a viewpoint of minimum feature size, flash technologies are mostly one generation behind DRAMs and key electrical parameters are the endurance and retention time [5].

Total dose effects in NVMs have been studied for more than a decade [6] and testing of Intel 64 Mb (NOR cell) and Samsumg 128 Mb (NAND cell) demonstrated that low total doses in the 2-20 krad(Si) range may already lead to severe degradation [4]. A NOR architecture cell is more radiation resistant than a NAND cell, and often the circuit failure is caused by the degradation of the internal charge pump. NVM devices are also very sensitive to single event upsets [7].

The threshold voltage shift versus fluence for a 1 MeV electron and 20 MeV proton irradiation, recorded after a read cycle is illustrated in Fig. 9.3 for a n-channel flash memory cell fabricated in a 0.8 μm CMOS technology [8]. A clear decrease with fluence is observed, while the opposite occurs for an erase cycle. For electron irradiations the damage coefficient is about three orders of magnitude lower than for protons or alphas.

A limited number of experimental data has been collected for radiation effects in flash memories irradiated at elevated temperatures [8]. For the reported studies the worst-case has been observed for room temperature irradiation, while during elevated temperature irradiations in-situ annealing of the radiation-induced charges and/or defects occur. This is illustrated in Fig. 9.4 for electron and gamma irradiations.

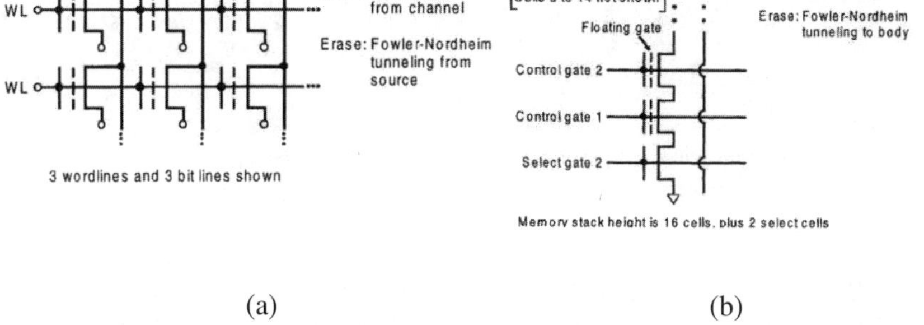

(a) (b)

Fig. 9.2. Cell architecture of a NOR (a) and a NAND (b) cell concept. The NAND memory is organised in a 16-bit stack. [After Nguyen et al. [4]]

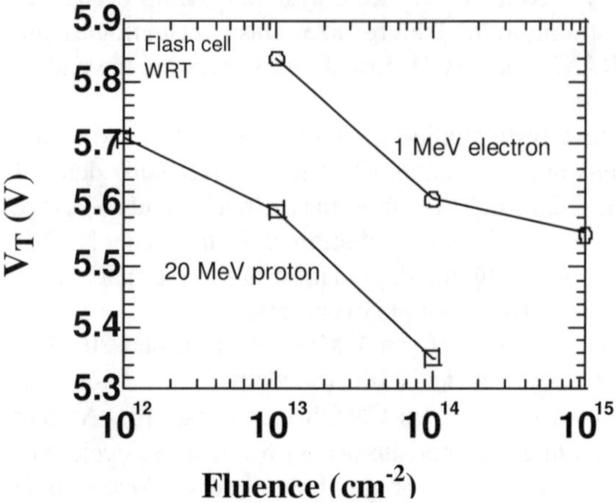

Fig. 9.3. Threshold voltage versus fluence for a 1 MeV electron and 20 MeV proton irradiation, recorded after a read cycle of a n-channel flash memory cell fabricated in a 0.8 μm CMOS technology. [After Claeys et al. [8]]

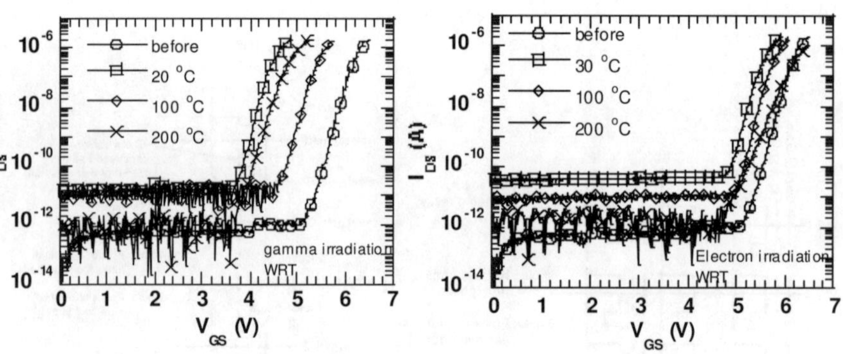

Fig. 9.4. Impact of irradiation temperature on the input characteristics (recorded after a write cycle) after 10^{15} cm^{-2} electrons and 10^8 rad(Si) gamma's, respectively [After Claeys et al. [8])]

A NVM cell is a rather complex structure as radiation-induced charges can be generated and stored at a variety of locations (see Fig. 9.1.), i.e., at the bottom SiO_2-Si interface, in the gate SiO_2, at the SiO_2 –poly 1 interface, at the poly 1-interpoly dielectric interface, in the interpoly dielectric, and at the interdielectric-poly 2 interface. In addition one has tot take into account that for these devices quite often NO or RNO dielectrics are used. The radiation performance of these dielectrics have been discussed in Sect. 6.4.2.

9.2.2 Ferroelectric Memories (FeRAMs)

The memory cell of a FeRAM is composed of a ferroelectric capacitor and a MOSFET for cell selection. During the read operation, voltage pulses with the same polarity are applied to the capacitor and the polarization reversal current of the ferroelectric film is detected. In another type of memory cell the ferroelectric capacitor is replaced with a ferroelectric-gate FET [3]. The second approach has the advantage that the read operation is non-destructive, but the structure is more difficult to fabricate as there may be an indiffusion of the ferroelectric material into the silicon substrate. FeRAMs have superior low-voltage and high-endurance operation. The most important ferroelectric materials in use are the lead zirconate titanate perovskite group (PZT family – $PbZr_xTi_{1-x}O_3$) and strontium bismuth tantalate (SBT). The performance of the layers strongly depends on the used fabrication technique (sol-gel, sputtering, pulsed laser deposition, photochemical metal-organic-deposition ...). $SrBi_2Ta_2O_9$ has successfully been used to fabricate a 64 kbit FeRAM in a 0.8 μm technology [9]. Quite often Pt top and bottom electrodes are used.

The ionising radiation response of PZT ferroelectrics has been well studied and in most cases these layers are radiation resistant to total dose levels of 1 Mrad(SiO_2) [10 - 12]. Attention has been given to the degradation of the remanent polarisation (hysteresis-loop measurement), the retained polarisation (pulse measurements), fatigue (degradation after cycling) and retention (degradation with time). However, it has been reported that radiation can lead to an asymmetry of the retention characteristics, also known as imprint, whereby repeated writing results into the formation of a preferential state in the FE film [13]. The proposed more stringent testing technique pointed out that although the older data are overestimated with 30 to 40%, FE films are still very radiation resistant.

PZT films are polycrystalline in structure, and can be considered as an insulator as the bandgap is 3.5 eV. Charge carriers in domains move and separate under the influence of a dc electric field making the effective field on micro-domains and polar micro-regions smaller than the applied field, resulting in the so-called polarisation [14]. The local polarisation in each domain produces a strong electric field introducing band bending and ionisation of traps until sufficient surface space charges have been built-up to screen the internal polarisation of the FE film. During irradiation electron-hole pairs are generated in the film, which are separated at the grain boundaries by the electric field. Within a domain the field is too weak so that the charges remain pinned at the grain surfaces and boundaries. The smaller the grain size, the larger the total surface and the more charges are trapped in the film. The PZT films also contain defects such as oxygen and lead vacancies, which because of the stress at the interface between the PZT film and the electrode are charged due to the piezoelectric effect [15]. The irradiation-induced e-h pairs can neutralise some of these charges, thereby influencing the degree of polarisation.

9.2.3 Conclusion

Flash and FeRAMs are serious candidates for being used as COTS components in future space programs. Especially for flash sufficient radiation performance data has been collected. FeRAMs are somewhat different as for the moment there is still intensive research on going to improve the properties of the FE material and the bottom and top electrodes. New materials implicitly mean additional radiation testing. However, there is strong evidence that the FE materials are sufficiently radiation resistant for future space missions.

In the field of non-volatile memories it is essential to keep in mind that research is also performed on alternative concepts such as e.g. magnetic memories (MRAMs). These devices are rather new, but their introduction for commercial applications may be faster than expected. For the moment there is a lack of information on their radiation performance. This is in contrast to other 'future' type of non-volatile memories, as for example the C-RAM, based on the use of chalcogenide alloy material such as $GeO_2Sb_2Te_5$. By applying current pulses it is possible to switch the material from a low-resistance polycrystalline state to a high-resistance amorphous phase. Initial radiation testing points out that the memory cells are at least radiation resistant up to a total dose of 1 Mrad (Si) [16]. The good radiation performance is due to the fact that in this case the switching mechanism is not based on trapping/de-trapping of charges, but rather on the structural change of the involved phase of the layer, implying that the radiation effects are mainly associated with the used CMOS technology.

9.3 High-*k* Gate Dielectrics

As outlined in Chap. 6, the scaling of the minimum feature size of CMOS technologies also requires a scaling of the thickness of the gate dielectric. However, ultra-thin SiO_2 gate dielectrics show problems with reproducibility, reliability and a reduced resistance against boron indiffusion. By increasing the dielectric constant, the equivalent oxide thickness can be increased. In Sect. 6.4.2 the impact of using nitrided and reoxidised nitrided oxides was reviewed. A further scaling necessitates the introduction of more exotic materials with a higher dielectric constant. A large variety of high-*k* materials are extensively investigated as the eventual non-availability of a replacement for SiO_2 may form a serious roadblock for the implementation of deep submicron technologies as envisaged in the International Technology Roadmap for Semiconductors (ITRS) [17].

Table 9.1 gives a summary of different high-*k* materials that are under investigation. For some of these materials very promising results has already been reported, such as e.g. for Ta_2O_5/SiO_2 [18], Hf [19], TiO_2/Si_3N_4 [20]. The main issue is that one would like to find a gate dielectric such as SiO_2, which has been around for about 5 decades. From a manufacturing viewpoint it is not practical to have to switch over to a new gate dielectric every two or three generations. A huge amount of investigations are needed before the gate dielectric and its interfaces are characterised in detail.

The fact that new gate dielectrics are being introduced may have a strong impact on the radiation hardness of the devices. Whereas the radiation performance of SiO_2 and the Si-SiO_2 interface are well understood, all the work has to be repeated for these new materials. The available information in the literature is still very limited, with the exception of SBT which was studied in view of its applicability for non-volatile memories. A comparison between the radiation hardness of 3.3 nm oxynitride and 5.4 nm Al_2O_3 films is given in Fig. 9.5 for 342 MeV Au ion irradiation with a fluence of 1.8×10^7 ions/cm^2 [21]. The Weibull distribution clearly shows that thinner oxynitride films of equivalent electrical thickness are less radiation hard.

The differences between soft (SBD) and hard breakdown (HBD) or single event gate rupture (SEGR) have been discussed in Chap. 6. For the latter the postulated model is based on a two-step nature of the irradiation-induced breakdown, whereby first a conductive pipe is created by the ionization damage and subsequently the stored energy is discharged through the pipe. The breakdown voltage scales with the square root of the physical thickness of the dielectric film [22]

$$V_{\text{HBD}} \approx \sqrt{t_{\text{physical}}} \tag{9.1}$$

This relation has been investigated for thin SiO_2 and three different alternative high-*k* dielectrics [23]. The results shown in Fig. 9.6 clearly illustrate that the power law remains valid. However, since high-*k* dielectrics have an equivalent oxide thickness (EOT) thinner than their physical thickness the practical breakdown voltage is increased compared to their equivalent SiO_2 counterparts. The 2.6 nm EOT Al_2O_3 has a higher breakdown voltage than the 2.2. nm SiO_2 film. These results are very promising for the application of these high-*k* dielectrics in space applications. In contrast to Al_2O_3 and HfO_2, Fig. 9.6 does not show an improvement for the ZrSiO film. However, one has to take into account that still a lot of work has to be done in order to improve the pre-irradiation quality of these films. In many cases the devices are suffering from a too high pre-irradiation leakage current, as was the case for the reported ZrSiO film.

Figure 9.6 also indicates the power supply voltage envisaged to be used as a function of the EOT, according to the 1997 ITRS roadmap. The post-irradiation breakdown voltages are in all cases higher than these power supply voltages, indicating that the use of commercial devices based on alternative gate dielectrics has a good potential for future space applications. As the presented data are only for heavy ion irradiations, more systematic studies with a variety of different particles are needed to obtain a complete picture of the radiation hardness performance of these dielectrics. At the end of the day this may even require the development of improved and/or new theoretical models to explain the experimental observations.

Table 9.1. Overview of high-*k* gate dielectrics under investigation

Material	Dielectric Constant	Material	Dielectric Constant
Al_2O_3	8-11.5	$NdAlO_3$	22.5
$Al_xSi_yO_z$		PrAlO3	25
(Ba, Sr) TiO_3	200-300	Si_3N_4	7
$BeAl_2O_4$	8.3-9.43	$SmAlO_3$	19
CeO_2	16.6-26	$SrTiO_3$	150-250
$CeHfO_4$	10-20	Ta_2O_5	25-45
$CoTiO_3/Si_3N_4$		Ta_2O_5-TiO_2	
$EuAlO_3$	22.5	TiO_2	86-95
HfO2	26-30	TiO_2/Si_3N_4	
Hf silicate	11	Y_2O_3	8-11.6
La_2O_3	20.8	$Y_xSi_yO_z$	
$LaAlO_3$	23.8-27	ZrO_2	22.2-28
$LaScO_3$	30	Zr-Al-O	
La_2SiO_5		Zr silicate	11-12.6
$MgAl_2O_4$	8.3-9.4	(Zr, Sn)TiO_4	40-60

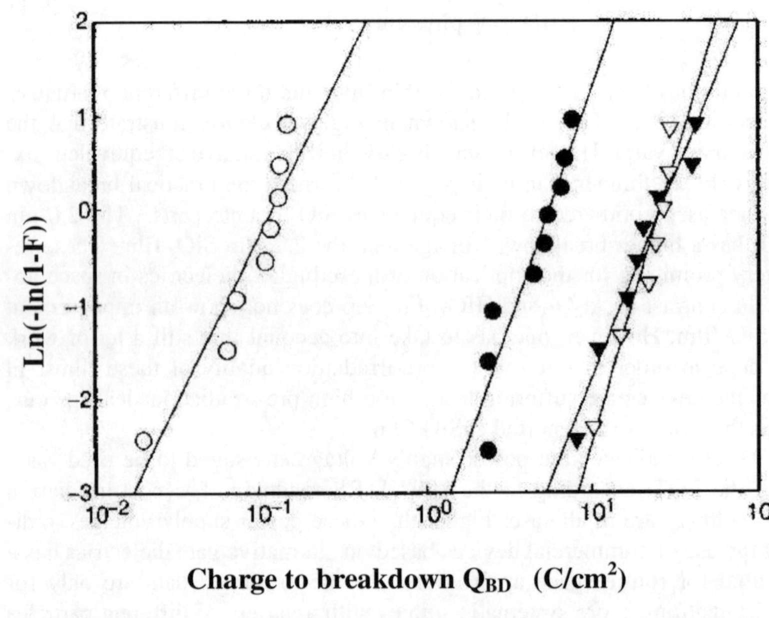

Fig. 9.5. Weibull distribution of charge to breakdown of 3.3 nm oxynitride (circles) and 5.4 nm Al_2O_3 (triangles) dielectrics. The open symbols are before and the closed ones after 342 MeV Au irradiation with a fluence of 1.8×10^7 ions/cm^2. [After Choi et al. [21]]

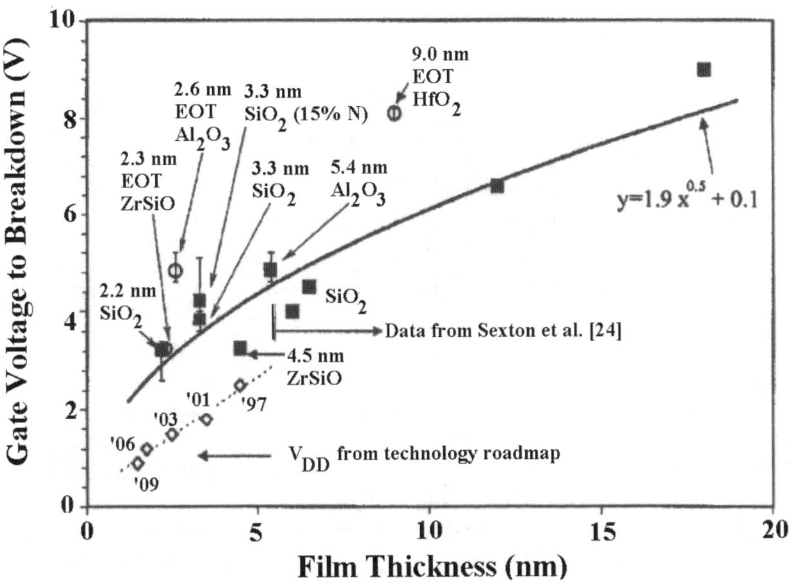

Fig. 9.6. The voltage to radiation-induced hard breakdown versus physical film thickness after a 342 MeV Au irradiation. The data from *Sexton et al.* [24] are also shown. [After Massengill et al. [23]]

9.4 Radiation Effects in SiC

In many of the early radiation effects papers, SiC was considered as a ceramic coating for nuclear fuels, or as a structural material for fusion reactors [25]. The focus in these applications was on the ability of the ceramic to withstand high temperatures (>1000°C) and high neutron fluences (>10^{20} n/cm^2) in harsh environments. When crystalline SiC became available, its potential for semiconductor devices operating at high temperatures or in intense radiation fields was soon realised [26]. Recently, strong progress has been made in the SiC crystal growth and epitaxial deposition techniques [27], so that device grade material is commercially available nowadays. In this paragraph, the fundamental radiation damage mechanisms in SiC will be described and compared with silicon and other more common semiconductor materials. In a first section, the relevant properties of SiC will be described. Next, the fundamental radiation defects will be highlighted and compared with the intrinsic point defects. In the next section, macroscopic displacement damage will be reviewed and, finally, the occurrence of ionisation damage in SiC MOSFETs and MESFETs is outlined.

9.4.1 SiC Material Properties and Analysis

SiC can crystallise in over 200 different polytypes [27], which explains the difficulties in growing high-quality substrates. Of these different polytypes only a few are of interest for device applications, namely the cubic form 3C-SiC, also termed β-SiC and the 4H- and 6H hexagonal (H) or α-SiC. In addition, some interest exists in 15R rhombohedral material. 6H-SiC is the most easily prepared and best studied crystalline material. On the other hand, 3C-SiC is the predominant type for epitaxial deposition on large-diameter silicon substrates, using low-pressure CVD [27]. The large lattice (20 %) and thermal mismatch with silicon generally results in limited material quality, characterised by a high density of threading dislocations, microtwins, stacking faults and double positioning boundaries (DPB). This implies that thicker epi layers usually have better electrical properties than thin ones, which suffer more from the defectiveness of the interface. In addition, epitaxial deposition on device quality SiC wafers is routinely applied nowadays, although the wafer diameter is limited to about 5 cm diameter. SiC substrates are available with both the n- and p-type conductivity over a wide range, while a semi-insulating SiC substrate growth technology has only recently become available [28]. Nominally undoped material is n-type due to the residual N donors. Standard dopants are Al or B in p-type and N in n-type SiC.

SiC is a so-called wide band gap material, with an E_G of ~2.3 eV for 3C-SiC, ~2.9 eV for 6H-SiC and ~3.2 eV for 4H-SiC at 300 K. From this follows that the intrinsic carrier concentration and, hence, the device leakage, at room temperature is extremely low ($<10^{-6}$ cm^{-3} at 300 K for 4H- and 6H-SiC), which makes it interesting material for high temperature operation. Furthermore, SiC has an excellent thermal conductivity and large breakdown fields, which is favorable for high-power applications. In spite of its indirect band gap, blue light emitting diodes have been achieved. However, also the ionisation energy of the typical dopant elements can be quite high [29 – 31]. This is particularly the case for Al and B, where values in the range 200 to 400 meV (6H-SiC) and 191 to 390 meV (4H-SiC) [31] are reported. In general, the dopant impurities can occupy different non-equivalent substitutional lattice sites, which are characterised by a different arrangement of the neighbouring atoms: while the nearest neighbours are the same for each Si and C atom in all polytypes, the second and third neighbour cell results in slightly altered configurations leading to crystallographically inequivalent lattice sites (3C: one cubic (k); 4H: one cubic (k) and one hexagonal (h); 6H: two cubic (k1,k2) and one hexagonal (h)) [29, 30]. As a consequence, the dopants can have two (or more) ground state energies in 6H-SiC and 4H-SiC, the lower one corresponding to the h sites [30]. The same applies for impurities or defects giving rise to deep levels in the band gap: in many cases, the deep levels observed by e.g. DLTS have a broad appearance, pointing to the fact that they are composed of two separate levels, which can not be resolved by the technique. In contrast to donors it turns out that the ionisation energy of acceptors in SiC is only weakly sensitive to a particular polytype and the different lattice sites (h or k). The fact that the ionisation energy of Al or B is quite high implies that they are not fully ionised at room temperature. Consequently, the free carrier concentration derived from e.g. *C-V* or Hall measurements in p-type SiC increases with temperature, due to the thermal ionisation.

Beside the large band gap, other beneficial electrical characteristics of SiC are the high bulk electron mobility and saturation velocity which can reach values ~1000 cm^2/s and ~$2x10^7$ cm/s in 3C-SiC at 300 K. This is because of the larger symmetry of this polytype compared with the others, leading to a lower phonon scattering. The high v_{sat} explains the interest in SiC as a substrate for high-speed circuits. Also 4H-SiC has a better μ_n than 6H-SiC [27]. Finally, SiC is highly resistant against chemical attacks (chemical inert), rendering it suitable for operation in hostile environments.

The superior radiation resistance of SiC is anticipated from the high Si-C bond energy (large band gap). This results in a high atomic displacement threshold energy E_d (see Eq. 2.8), which has been obtained by *Barry et al.* by studying the recombination lifetime damage factor of SiC LEDs versus the electron energy [32]. A value of 21.8±1.5 eV was obtained for the carbon atom. The minimum electron energy required to produce a carbon displacement was determined to be 108±7 keV; for MeV electron energies, also silicon atoms can be displaced [33]. The experimental E_d correlates well with the reciprocal lattice constant $1/a_0$, as demonstrated in Fig. 9.5 for most common semiconductors [25]. As can be derived from Fig. 9.7, E_d for SiC is significantly higher than for Si, Ge or GaAs, suggesting a higher resistance against radiation damage. This will be illustrated below.

With respect to ionisation, it should be remarked that 1Mrad(SiC) equals 1.03 Mrad(Si) [25

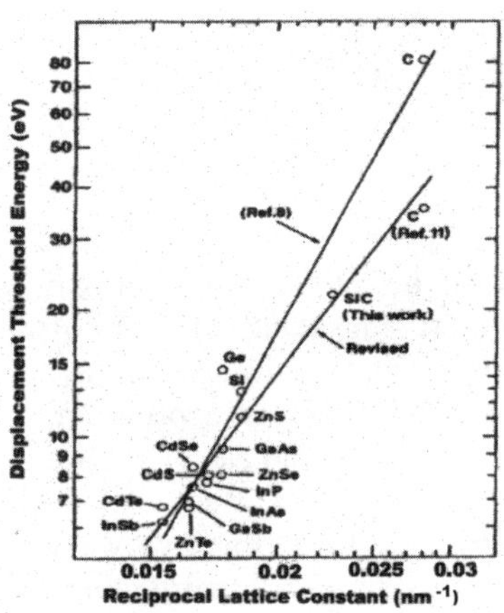

Fig. 9.7. Correlation of mean displacement threshold energy E_d with reciprocal lattice constant $1/a_0$. [After Barry et al. [32]]

For the characterisation of radiation (and other) defects in SiC, the same techniques described in Chap. 2 can be used [30, 34]. The large band gap of the material bears, however, some consequences. For DLTS, a higher temperature region has to be scanned in order to assess the levels near the mid gap (range 77 to 750 K). This enables to detect deep levels up to 1 eV from the appropriate band edge. It also implies that the middle one third of the band gap escapes from observation [30, 35]. Optical injection possibly overcomes this problem. Another issue to account for during DLTS is that heating an irradiated sample to well above room temperature can produce some annealing of the defects under investigation. It should also be remarked that most recent radiation-defect studies utilise epitaxial SiC layers grown on SiC substrates. The reason is that common SiC bulk material is rather highly doped (especially n-type), while the doping concentration is better controlled during CVD. This lower background doping density makes the material more sensitive to low defect concentrations in DLTS, as this is a relative technique (see. Eq. 2.31).

Photoluminescence has been successfully applied since the early seventies [36 - 38]. In the case of PL, one has to use excitation in the UV region, using Krypton or HeCd lasers [30]. At the same time, different detectors operating in different wavelength regimes have to be combined to cover the whole spectrum of emitted light.

For the chemical and structural identification of the radiation defects in SiC, ESR has proven very useful [39 – 42]. Additional information about the presence of open volume defects – i.e., vacancy-related defects – can be obtained by Positron Annihilation Spectroscopy (PAS) [33, 43 – 44]. In PAS, one monitors either the Doppler broadening gamma spectra resulting from electron-positron annihilation in the semiconductor or the positron lifetime is studied. Depth information of the defects can be obtained by studying the broadening parameter or the lifetime as a function of the positron energy. The presence of vacancy related defects enhances the positron lifetime and modifies the Doppler broadening parameter

9.4.2 Intrinsic and Radiation Defects in SiC

As SiC is a compound semiconductor, the intrinsic defects comprise not only vacancies and interstitials of both constituents, but also antisite defects and possibly close Frenkel pairs, i.e., I_{Si}-V_{Si} and I_C-V_C. Also the V_C-V_{Si} divacancy has to be considered in the first instance. From theoretical calculations, one expects that neither antisite (Si_C or C_{Si}) has an electrical active level in the band gap of 3C-SiC and so far, no experimental level has been assigned to it [35]. The study of radiation defects in SiC is also important for the understanding of the intrinsic grown-in defects, which may result from a non-stoichiometric composition and is, therefore, helpful in the modeling and improvement of the crystal growth. In this context, one should remark that early investigations of intrinsic and radiation defects were hindered by the poor starting quality of the material and should, therefore, be treated with care. Another important point is that one expects the simple native defects (vacancies, interstitials) to be stable at room temperature, so that they can be studied after room-temperature exposures. The fact that the different polytypes have another band gap energy and non-equivalent sites implies that the intrinsic defects are anticipated to have different parameters in α- or β-SiC.

The simplest case is in principle cubic SiC as there exists only one lattice site, so that one expects a single energy level for each defect. Combining EPR, PAS, PL and DLTS *Itoh et al.* were able to establish the level scheme of Fig. 9.8 [35]. The negatively charged silicon vacancy is thought to give rise to the T1 signal in n-type 3C-SiC and behaves, therefore, as an acceptor. A PL line at 1.913 eV was shown to correlate well with this T1 signal and with a PAS annealing stage in the range 100-200°C [35, 45]. This places the level of V_{Si}^- at about 0.5 eV above the valence band. Such a position is in line with theoretical predictions [46] and also coincides with one of the levels found in neutron irradiated 3C-SiC [47, 48]. The introduction rate for T1 after 1 MeV electron irradiation was found to be 0.017 cm^{-1} in n-type β-SiC, which is very close to the corresponding carrier removal rate of 0.014 cm^{-1} [45]. It is, therefore, concluded that V_{Si} is a dominant radiation defect in irradiated 3C-SiC. This also follows from the fact that the recovery stages of the resistivity or the carrier concentration upon annealing closely follows the annealing of the T1 centre. The negatively charged silicon vacancy not only traps electrons and, therefore, compensates the material, but acts as a carrier scattering centre, reducing the electron mobility.

Three annealing stages were found to correspond with the T1 EPR centre in n-type 3C-SiC [45]: at 150, 350 and 750°C. The highest annealing stage corresponds to a first-order reaction with an activation energy of 2.2 eV. From this, it was concluded that at this stage V_{Si}^- walks randomly and disappears at a sink, which might be related to extended defects, like dislocations. At the lower stages, it is likely that the silicon vacancy disappears through a recombination with other defects such as interstitials. In 1 MeV electron irradiated p-type material, the T1 signal is found to disappear already at 350°C annealing, which is much lower than for n-type β-SiC [45]. It is believed that this is due to the neutral charge state of the V_{Si} (Fermi level lowering) in p-SiC. In other words, the annealing (and possibly its mobility) of the silicon vacancy is charge state dependent.

Irradiation of p-type cubic SiC below ~100 K reveals another EPR signal labelled T5 [45]. It is not observed in irradiated n-type material, because of its neutral charge state there, which renders it undetectable for the technique. It has been ascribed to the carbon vacancy in a positive charge state V_C^+ [41]. Annealing of the V_C^+ centre in p-type material takes mainly place between 100 and 200°C. One assumes that the carbon vacancy moves to combine with other defects or that it recombines with interstitials, which are mobile at these temperatures. However, it has been concluded that the possibility of divacancy (V_2) formation by the combination of V_{Si} and V_C is quite low. In addition, V_C is hard to observe in PAS due to its low cross section for positron trapping when it is in its neutral charge state (n-type material) [45].

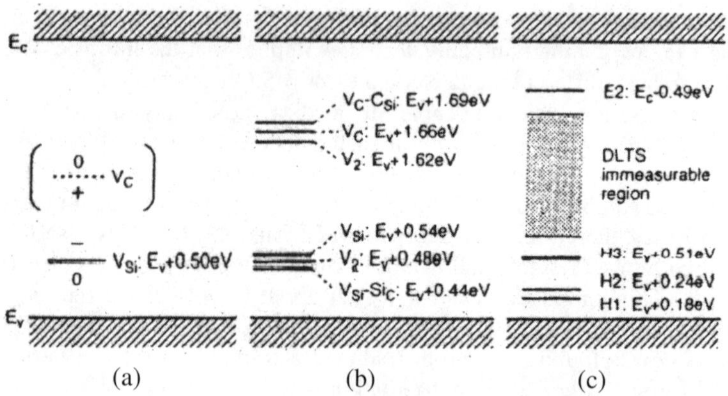

Fig. 9.8. Schematic representation of the electronic levels of point defects in 3C-SiC. a) An electronic level of V_{Si} and a possible position of a V_C level estimated from PL data; b) theoretically predicted levels of V_{Si}, V_C, V_2 and vacancy-antisite pairs (from [46]) and c) several levels obtained by DLTS for neutron irradiated 3C-SiC (from [47]). In Fig. 9.8ba, the + and – represent the charge states of V_{Si} and V_C. [After Itoh et al. [35]]

Irradiation of p-type cubic SiC below ~100 K reveals another EPR signal labelled T5 [45]. It is not observed in irradiated n-type material, because of its neutral charge state there, which renders it undetectable for the technique. It has been ascribed to the carbon vacancy in a positive charge state V_C^+ [41]. Annealing of the V_C^+ centre in p-type material takes mainly place between 100 and 200°C. One assumes that the carbon vacancy moves to combine with other defects or that it recombines with interstitials, which are mobile at these temperatures. However, it has been concluded that the possibility of divacancy (V_2) formation by the combination of V_{Si} and V_C is quite low. In addition, V_C is hard to observe in PAS due to its low cross section for positron trapping when it is in its neutral charge state (n-type material) [45].

Finally, two more EPR signals are consistently found in irradiated β-SiC, T6 and T7, which are thought to correspond to close Frenkel or vacancy-interstitial pairs [45]. They have been found to anneal around 300°C.

In comparison with β-SiC, far more DLTS studies have been devoted to irradiated hexagonal SiC: results for state-of-the-art 6H-SiC have for example been reported in Refs [29 - 32] and for 4H-SiC in Refs [52 – 54]. Recent overviews can be found in [55, 56]. It was first of all observed that most of the intrinsic defects detectable by DLTS lie in the upper one third of the band gap and show an acceptor behaviour [55]. The main results are summarised in Table 9.2 (6H-SiC) and 9.3 (4H-SiC) [56], giving the defect parameters, the stability range and a possible identification.

Table 9.2. Parameters and properties of radiation defects observed in n-type 6H-SiC [After Lebedev et al. [56]]

E_c-E_T [eV]	σ_n [cm^2]	T_A [K]	Electron irrad. Defect	Structural Defect	Possible structure
0.16-0.2	6x10^{-17}	800-950	L_1		Primary defects
0.36/0.4	2x10^{-15}	1100<...<1800	L_3/L_4	E1/E2, S	
0.5	5x10^{-15}	800-950	L_6		V_C
0.7	4x10^{-15}	1100<...<1800	L_7/L_8	Z1/Z2	V_C+V_{Si}
0.8	4x10^{-15}	1100<...<1800	L_9		
1.1-1.22	2x10^{-15}	1100<...<1800	L_{10}	R	V_C+V_{Si}

E_c-E_T activation energy, σ_n electron capture cross section, T_A annealing temperature.

Table 9.3. Parameters and properties of radiation defects observed in n-type 4H-SiC [After Lebedev et al. [56]]

E_c-E_T [eV]	σ_n [cm^2]	T_A [K]	Implantation of He$^+$	Structural Defect	Possible structure
0.18	6x10^{-15}	800-950	P1/P2		Primary defects
0.63-0.7	5x10^{-15}	1100<...<1800	Z1	Z1	V_C
0.96	5x10^{-15}	1100<...<1800	RD1/2		V_C+V_{Si}
1.0	1x10^{-16}	1100<...<1800	RD3	E_c-1.1 eV	
1.5	2x10^{-13}	1100<...<1800	RD4		

E_c-E_T activation energy, σ_n electron capture cross section, T_A annealing temperature.

According to Table 9.2, the nearest neighbour V_{Si}-V_C divacancy is a double acceptor in 6H-SiC with levels at E_c-0.7 and E_c-1.1 eV and has a high thermal stability: up to at least 1700°C in 6H-SiC and 2015°C in 4H-SiC. Associated with the Z1 or Z1/Z2 centre, observed with photoinduced electron spin resonance, are the characteristic PL lines for the D_1 defect [36, 37]. Recently, however, some doubt has been cast on the assignment of Z1/Z2 to the divacancy [51, 57 - 60], although it has been demonstrated that the Z1/Z2 centres are silicon vacancy related defects [60]. Instead, two other electron traps were assigned to V_2 [51]. A silicon vacancy involvement was also established for the E1/E2 levels [59]. The different energies of thermal ionisation for centres with supposedly the same structure in 4H and 6H-SiC, obvious from a comparison of Table 9.2 and 9.3, can be explained by different distances between the pair components [56]. It has been observed that the E_c-1.22 eV centre (R centre) has the highest introduction rate in 8 MeV proton irradiated n-type 6H-SiC [56] and that for sufficiently large fluences the resistance of the material is pinned at that level. For 4H-SiC the pinning occurs at E_c-1.25 eV for a sufficiently large Φ.

The thermal stability of vacancy related primary radiation defects has also been studied by PAS in irradiated 6H-SiC [57, 61]. Similar annealing stages as in 3C-SiC were observed for n-type material: at 100, 200, 450 and 750°C. A vacancy complex stable up to 1400°C was assigned to V_{Si}-N_C centres. The 750°C stage corresponds to the free migration of the silicon vacancies, while the 450°C stage was assigned to the annihilation of divacancies. This is at much lower temperatures than what is normally believed for V_2. V_C related annealing stages were observed at 75 and 300°C. This is a comparable stability range as in cubic SiC. The lower stage corresponds to the Frenkel pair recombination, due to the mobility of I_C, while in the range 250-450°C the carbon vacancy becomes mobile. A DLTS level at E_c-0.48-0.53 eV has been assigned to the V_C in 5 MeV electron irradiated n-type 6H-SiC [62]. Its observed annealing behaviour in the range 150-250°C and the donor character derived from the Poole-Frenkel reduction of the thermal activation energy with increasing electric field supports this identification. In addition, a dose rate dependence was observed in the production and annealing behaviour of electron irradiated 6H-SiC [61]. This is related to the ionisation-induced shift of the Fermi level during the exposure, which enhances the mobility of the I_C and I_{Si}, and, therefore, the recombination of Frenkel pairs and vacancies.

Beside particle irradiation, displacement damage can also be produced during ion implantation, which is a common local doping technique in SiC-based technologies [31]. Basically the same intrinsic radiation defects have been found after ion implantation [31, 63 - 65]. In addition, some element specific complexes can be observed. This is the case for the so-called D centre in DLTS, located at 580 to 750 meV above the valence band in 6H-SiC and 540 to 730 meV above E_v in 4H-SiC [31]. It is believed to be associated with a B_C-V complex [31]. The presence of vacancies after ion implantation leads to a so-called vacancy enhanced B diffusion. This results in an extended B tail beyond the created pn junction [64].

For certain applications, the use of semi-insulating (SI) substrates or local high-resistivity regions can be of great interest, for example for the fabrication of carrier confinement layers (isolation). As it is difficult to grow SI material, other means are looked for. This goal can be achieved by ion implantation. It has been proposed to use both high [56] and low energy [66] hydrogen implantation. Another possibility is the use of vanadium ion implantations [67].

9.4.3 Macroscopic Damage in SiC Devices

As demonstrated in Chap. 2, damage factors are very useful for the description of the radiation damage in real devices. These can be studied for different material and device parameters. The carrier removal rate K_n is a measure for the change of the free carrier concentration with fluence Φ. It has been determined in SiC for different particles and/or energies. For 1 MeV Si equivalent neutron fluences, a value of 4.5±0.5 cm^{-1} was obtained for n-channel 6H-SiC JFETs, with an initial doping density of ~$2x10^{17}$ cm^{-3} [25, 70]. This is 3 times better than for comparable silicon material, emphasising the higher radiation tolerance of SiC. In 3C-SiC epitaxial layers of n-type doping density ~10^{16} cm^{-3} a K_n~7.2 cm^{-1} was found, which is slightly lower than for silicon in the same neutron beam (7.8 cm^{-1}) [47, 48]. In a Schottky diode radiation detector on 4H-SiC, a value of 9.7±0.7 cm^{-1} was observed [69]. After 5.5 MeV alpha particle irradiations of 6H-SiC, a removal rate of $7.8x10^4$ cm^{-1} was observed, which was comparable as for InP, another highly resistant semiconductor [70].

The damage constant for the recombination lifetime K_τ has been derived for SiC LEDs at various electron energies [32]. The results are summarised in Fig. 9.9,

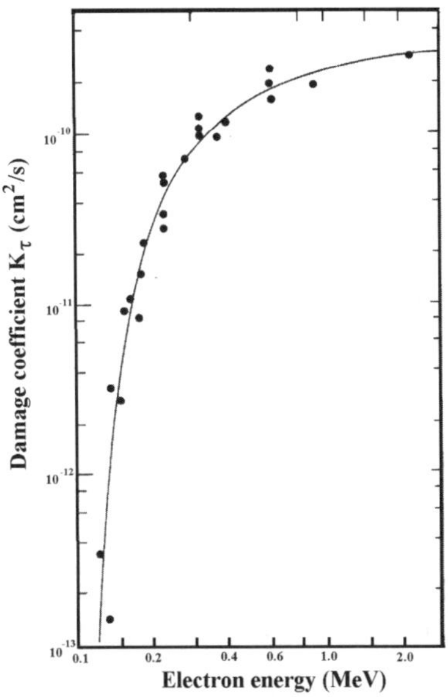

Fig. 9.9. Composite plot of K_τ versus E for several SiC LEDs. [After Barry et al. [32]]

showing a threshold electron energy for displacement damage of ~106 keV. This leads to an E_d=21.8 eV. Comparison with GaAs leads to the conclusion that K_τ for electrons with energy higher than 0.5 MeV is three orders of magnitude smaller. This illustrates once more the superior resistance of SiC against displacement damage, in most radiation environments.

9.4.4 Ionisation Damage in SiC MOSFETs and MESFETs

Gate dielectrics on SiC layers can be grown with the same oxidation processes as on silicon, making the fabrication of SiC MOSFETs rather straightforward. However, the obtained density of interface traps D_{it} is found to be significantly higher than for the Si-SiO_2 interface [71 - 73]. Typical values are in the range $2\text{x}10^{11}$ to 10^{12} eV^{-1} cm^{-1} [74] before irradiation, while corresponding oxide trap surface densities N_{ot} in the range ~10^{12} cm^{-2} have been reported. Another important remark is that due to their slow response time, near mid gap interface traps in SiC behave like slow oxide charges at room temperature [75]. This is related to the high thermal activation energy, resulting in high emission time constants. It implies that by high-frequency room-temperature *C-V* measurements one usually underestimates the true D_{it}, while D_{it} also contributes to D_{ot}.

The impact of ionisation damage on the characteristics of SiC MOS structures has been determined for the cubic [74] and the 6H polytype [76]. For 3C-SiC MOS capacitors, it was observed that for negative gate bias little effect was observed after γ-exposure. Under a positive gate bias of +3 V, it was found that the both the density of interface and oxide traps followed a $D^{2/3}$ law, with D the dose of the γ's [74]. It was concluded that this was similar as for Si MOS structures. In addition, it was shown that 3C-SiC MOSFETs are harder than their silicon counterparts. Finally, a strong dependence on the oxidation process used was reported, whereby pyrogenic oxide yielded better response than dry oxide [74].

A different behaviour has been found in 6H-SiC MOSFETs [76]. While ΔN_{ot} also follows a 2/3 power law with gamma dose, an exponent 3/2 has been reported for ΔN_{it}. This points to a different interaction mechanism between the radiation-induced charges and the precursors for the interface states. A probable origin may be the fact that the 6H-SiC structures have been fabricated on (0001) SiC substrates, while the 3H-SiC devices were fabricated on (001) silicon substrates with a β-SiC epilayer [74, 76]. It was also concluded that the 6H devices were more radiation tolerant than the 3C or the silicon MOSFETs. It should also be remarked that in the oxide layer of SiC both positive and negative charges are present [77, 78] and both types of radiation-induced charges can become trapped. The negative charges present in the non-irradiated oxide could be related to some residual C-compounds, but they can also be caused by SiOH groups after pyrogenic oxidation [78]. The fact that both negative and positive charges can be trapped during irradiation leads to a marked gate bias dependence of N_{ot} for MOS capacitors on 6H-SiC substrates [79].

It has finally been observed that hydrogen annealing at 700°C before irradiation improves the radiation response of 6H-SiC MOSFETs [77]. This is illustrated in Fig. 9.10 showing the increase in interface trap density and the normalised mobility with absorbed gamma dose. For the hydrogenated case, the 2/3 power law is retrieved again, in contrast for the non-hydrogen annealed MOSFETs. It is concluded that hydrogen annealing reduces the precursors for the radiation-induced interface traps. At the same time, a channel mobility is obtained with a much higher radiation resistance. The variation of μ_n with ΔN_{it} can be described by a scattering coefficient equal to $7.0 \pm 1.3 \times 10^{-13}$ cm^2 [77].

Beside MOSFETs, also MESFET structures can be fabricated on SiC substrates, using a metal Schottky gate on top of the active layer. Radiation studies up to 1 Grad(Si) have demonstrated the extreme radiation hardness of such devices [80].

9.4.5 Summary

Until recently, the strong potential of SiC for certain application fields could not be exploited fully due to the limited material quality. Recent progress in crystal growth and epitaxy has made available device grade substrates, suitable for the fabrication of high-performance bipolar (junction diodes), MOSFET or MESFET SiC technologies. These are expected to be extremely radiation hard, based on the high displacement damage threshold in SiC. However, there is still a lack of basic information on the dominant radiation or intrinsic defects. Defect identification is more complicated in silicon carbide due to the existence of the different polytypes and for certain of them, the number of non-equivalent lattice sites. On the other hand, the study of intrinsic defects is in principle more simple here , since single vacancies and interstitials are stable at room temperature. This means that they can be investigated after a room temperature irradiation. The available information on damage constants confirms the hardness of SiC. With respect to the SiC-SiO_2 interface properties, a large similarity with the Si-SiO_2 system has been observed, also with respect to the radiation response. One can, therefore, largely benefit from the vast amount of knowledge gained there. However, some differences exist, which are related to the presence of both types of charges (positive and negative), which may also be trapped during ionising irradiation. Another important factor is the slow response of the mid gap interface traps at room temperature, which renders the charge separation by standard measurement techniques more complicated.

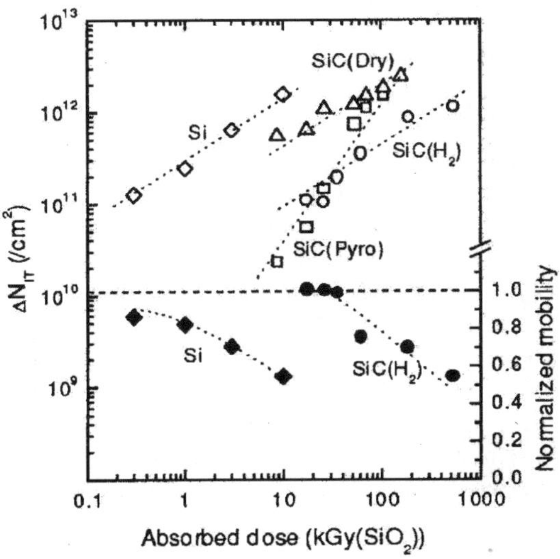

Fig. 9.10. Absorbed dose dependence of ΔN_{it} and the channel mobility for SiC(H_2) enhancement mode n-MOSFETs. The channel mobility is normalised to the pre-irradiation value. Open and closed circles represent ΔN_{it} and the normalised mobility, respectively. The reported results for ΔN_{it} for 6H-SiC MOSFETs whose gate oxide was formed using dry (open triangles) and pyrogenic oxidations (open squares) ara also shown for comparison. The reported ΔN_{it} and the normalised channel mobility for Si MOSFETs are shown as open and closed diamonds, respectively. [After Ohshima et al. [77]]

9.5 Conclusion and Outlook

The strongly growing market of non-volatile memories has been briefly reviewed. Commercial devices have already a good radiation resistance and the potential devices based on FE materials are expected to have a high radiation resistance. It is therefore anticipated that within the COTS approach more and more NVM devices will be used for space applications. Although the real winner is not known yet, it is generally accepted that further down scaling of CMOS technologies into the 100 nm and below node requires the use of new high-*k* gate dielectrics to replace SiO_2. From a radiation-hardness viewpoint this is a totally new field so that extensive radiation testing and subsequent modeling will be required. However, as new technologies are introduced every 2 to 3 years, the radiation studies will have to be performed almost in parallel with the on going development of the high-*k* dielectrics.

SiC is already a rather old material, for which only recently new application fields are gaining interest. The improved material quality leads to the fabrication of electronic components with an acceptable electric performance. Some older radiation work has to be repeated and extended in order to develop theoretical models giving a good insight into the fundamental physics involved.

Beside all the different microelectronics materials and technologies addressed in the different chapters of this book, there are several emerging technologies which have to watched very carefully as they will play an important role in the coming decade. Typical examples are micro-electromechanical systems (MEMS), polymer transistors, nano-electronics (single electron transistors, nanotubes, quantum wires…), molecular electronics etc. MEMS is nowadays finding its way for use in space applications so that reports on the radiation hardness of microsystems can already be found in the literature.

Another important trend for the near future are the system-on-chip (SOC) applications, consisting of a variety of semiconductor components and circuits integrated on a single substrate and with each their own radiation-hardness resistance. This will surely require optimised and new radiation testing methodologies.

References

Chapter 1

1. Holmes-Siedle A, Adams L (1993) Handbook of radiation effects. Oxford University Press, Oxford
2. Barth JL (1997) Modeling space radiation environments. Proc 1997 NSREC Short Course, Snowmass
3. Van Allen JA (1963) Geomagneticically trapped radiation. Space science, John Wiley & Sons, New York
4. Stassinopoulos EG, Brucker GJ, Nakamura DW, Stauffer CA, Gee GB, Barth JL (1996) Solar flare proton evaluation at geostationary orbits for engineering applications. IEEE Trans Nucl Sci 43:369-382
5. Daly EJ, Lemaire J, Heynderickx D, Rodgers DJ (1996) Problems with models of the radiation belts. IEEE Trans Nucl Sci 43:403-415
6. Dyer CS, Sims AJ, Truscott PR, Farren J, Unterwood C (1992) Radiation environments on shuttle missions using the CREA experiment. IEEE Trans Nucl Sci 39:1809-1816
7. Gussenhoven MS, Mullen EG, Brautigam DH (1996) Improved understanding of the earth's radiation belts form the CRRES satellite. IEEE Trans Nucl Sci 43:353-368
8. Huston L, Pfitzer KA (1998) A new model for the low altitude trapped proton environment. IEEE Trans Nucl Sci 4:2972-2978
9. Dyer CS, Truscott PR, Sanderson C, Watson C, Peerless CL, Knight P, Mugford R, Cousins T, Noulty R (2000) Radiation environment measurements from CREAM and CREDO during the approach of solar maximum. IEEE Trans Nucl Sci 47:2208-2217
10. Chennete DL, Chen J, Clayton E, Guzik TG, Wefel JP, Garcia-Munoz M, Lapote C, Ray KP, Mullen EG, Hardy DA (1994) The CRRESS/SPACERAD heavy ion model of the environment (CHIME) for cosmic rays and solar particle effects on electronic and biological systems in space. IEEE Trans Nucl Si 41:2332-2339
11. McKerracher PL, Kinnison JD, Maurer RH (1994) Applying new solar flare particle event models to interplanetary satellite programs. IEEE Trans Nucl Sci 41:2368-2375
12. LaBel KA, Marshall PW, Barth JL, Stassinopoulos EG, Seidleck CM, Dale CJ (1996) Commercial microelectronics technologies for applications in the satellite radiation environment. In: Proc 1996 IEEE Aerospace Applications, The IEEE, New York, pp 375-390
13. Landis GA, Appelbaum J (1990) Design considerations for Mars PV power systems. In: Proc 21st IEEE Photovoltaic Specialists Conf. The IEEE, New York, pp1623-1270
14. Appelbaum J, Segalov T, Jenkins P, Landis GA, Baraona C (1997) Verification of Mars solar radiation model based on Mars Pathfinder data. In: Proc 26st IEEE Photovoltaic Specialists Conf, The IEEE, New York, pp1039-1042

15. Holl P, Lutz G, Richter RH, Struder L, Longoni A, Sampieto M (2000) In:Proc of the 8th European symposium on semiconductor detectors. Nucl Instr & Methods in Phys Res A 439:199-685
16. Baumann RC, Smith EB (2000) Neutron-induced boron fission as a major source of soft errors in deep submicron SRAM devices. In: Proc Int Rel Phys Symp, The IEEE, New York, pp 152-157
17. May TC, Woods MH (1979) Alpha-particle-induced soft errors in dynamic memories. IEEE Trans Electron Devices 26:2-9
18. Hsieh CM, Murley PC, O'Brien RR (1991) A field funneling effect on the collection of the alpha-particle-generated carriers in silicon devices. IEEE Electron Device Lett 2:103-105
19. Hasnain Z, Ditali A (1992) Building-in reliability: Soft errors – a case study. In: Proc Int Rel Phys Symp, The IEEE, New York, pp 276-280
20. Baumann R, Hossain T, Murata S, Kitagawa H (1995) Boron compounds as a dominant source of alpha particles in semiconductor devices. In: Proc Int Rel Phys Symp, The IEEE, New York, pp 297-302
21. Normand E, Baker TJ (1993) Altitude and latitude variations in avionics SEU and atmospheric neutron fluxes. IEEE Trans Nucl Sci 40:1484-1490
22. Takia M, Kishimoto T, Ohno Y, Sayama H, Sonoda K, Satoh S, Nishimura T, Miyoshi H, Kinomura A, Horino Y, Fujii K (1996) Soft error susceptibility and immune structures in dynamic random access memories (DRAMs) investigated by nuclear microprobes. IEEE Trans Nucl Sci 43:696-704
23. Takia M, Arita Y, Abo S, Iwamatsu T, Maegawa S, Sayama H, Yamaguchi Y, Inuishi M, Nishimura T (2001) Evaluation of soft errors in DRAM and SRAM using nuclear microprobe and neutron source. In: Ryssel H, Wachutka G, Grunbacher H (eds) Proc ESSDERC 2001, Editions Frontieres, Paris, pp 17-24
24. Wolf S, Tauber RN (2000) Silicon processing for the VLSI era. Lattice press. California
25. Underwood CI, Oldfield MK (2000) Observations on the reliability of COTS-device-based solid state data recorders operating in low-earth-orbit. IEEE Trans Nucl Sci 47:647-653
26. O'Bryan MV, Label KA, Reed RA, Howard Jr JW, Ladbury RL, Barth JL, Kniffin SD, Seidleck CM, Marshall PW, Marshal CJ, Kim HS, Hawkins DK, Sanders AB, Carts MA, Forney JD, Roth DR, Kinnison JD, Nhan E, Sahu K (2000) Radiation damage and single event results for candidate spacecraft electronics. In: Proc Rad Effects Data Workshop, The IEEE, New York, pp 106-122
27. LaBel KA, Barnes CE, Marshall PW, Marshall CJ, Johnston AH, Reed RA, Barth JL, Seidleck CM, Kayali SA, O'Bryan MV (2000) A roadmap for NASA's radiation effect research in emerging microelectronics and photonics. In: Proc IEEE Aerospace Conf, The IEEE, New York, pp 535-545
28. www.estec.esa.nl/qcswww/tos_qca/
29. International Technology Roadmap for Semiconductors, 2001 Edition. www.itrs.net/2001_SIA_Roadmap/Home.html
30. Anelli G, Campbell M, Delmastro M, Faccio F, Florian S, Giraldo A, Heyne E, Jarron P, Kloukinas K, Marchioro A, Moreira P, Snoeys W (1999) Radiation-tolerant VLSI circuits in standard deep sub micron CMOS technologies for the LHC experiment: Practical design aspects. IEEE Trans Nucl Sci 46:1690-169

31. Lacoe RC, Osborn JV, Koga R, Brown S, Mayer DC (2000) Application of hardness-by-design methodologies to radiation-tolerant ASIC technologies. IEEE Trans Nucl Sci 47:2334-2341

Chapter 2

1. Van Lint VA, Flanahan TM, Leadon RE, Naber JA, Rogers VC (1980) Mechanisms of radiation effects in electronic materials. Wiley Interscience, New York
2. Ma TP, Dressendorfer PV (1989) Ionizing radiation effects in MOS devices and circuits. Wiley Interscience, New York
3. Adams L, Holmes-Siedle A (1993) Handbook of radiation effects. Oxford Scientific Publishers, Oxford
4. McLean FB, Oldham TR (1987) Basic mechanisms of radiation effects in electronic materials and devices. US Army Lab Command, HDL-TR-2129. US Army Command, Harry Diamond Laboratories, MD, USA
5. Oldham TR, McLean FB, Boesch Jr HE, McGarrity JM (1989) An overview of radiation-induced interface traps in MOS structures. Semicond Sci Technol 4:986-999
6. Heijne EHM (1985) The use of semiconductor imagers in high energy particle physics. In: Proc SPIE Solid State Imagers Appl, vol 591, pp 2-11
7. Privitera V, Coffa S, Priolo F, Rimini E (1998) Migration and interaction of point defects at room temperature in crystalline silicon. Rivista del Nuovo Cimento 21:1-52
8. Brotherton SD, Bradley P (1982) Defect production and lifetime control in electron and γ-irradiated silicon. J Appl Phys 53:5720-5732
9. Buisson J, Gaillard R, Jaureguy J-C, Poirault G (1996) Effet de déplacement induit par les protons de grande énergie dans les dispositifs électroniques semi-conducteurs. In: Proc RADECS '95. The IEEE, New York, pp 19-24
10. Dale C, Marshall P (1991) Displacement damage in Si imagers for space applications. In: Proc SPIE Charged-Coupled Devices and Solid State Optical Sensors II, vol 1447, pp 70-86
11. Meese JM (1981) A review of NTD-induced defects in silicon. In: Narayan J, Tan TY (eds) Defects in Semiconductors. North-Holland, Amsterdam, pp 225-240
12. Dreier P (1990) High resistivity silicon for detector applications. Nucl Instrum Methods A288:272-277
13. Claeys C, Simoen E, Vanhellemont J (1996) Lattice defects in high resistivity silicon. In: Claeys CL, Rai-Choudhury P, Stallhofer P, Maurits JE (eds) Proc Fourth International Symposium on High Purity Silicon. The Electrochem Soc, Pennington, NJ, vol 96-13, pp 305-324
14. Haller EE, Itoh KM, Beeman JW (1996) Neutron transmutation doped (NTD) germanium thermistors for sub-mm bolometer applications. In: Proc 30th ESLAB Symposium on Submillimetre and Far-Infrared Space Instrumentation, ESA SP-388, pp 115-118
15. Ziegler JF, Biersack JP, Littmark U (1985) The Stopping Range of Ions in Solids. Pergamon, New York
16. Linnros J, Norlin P, Hallén A (1991) Depth resolved carrier lifetime measurements of proton irradiated thyristors. In: IEDM Tech Digest. The IEEE, New York, pp 157-160

17. Mapper D, Sanderson TK, Stephen JH, Farren J, Adams L, Harboe-Sørensen R (1985) An experimental study of the effect of absorbers on the LET of the fission particles emitted by Cf-252. IEEE Trans Nucl Sci 32:4276-4281
18. Vanhellemont J, Simoen E, Claeys C, Kaniava A, Gaubas E, Bosman G, Johlander B, Adams L, Clauws P (1994) On the impact of low fluence irradiation with MeV particles on silicon diode characteristics and related material properties. IEEE Trans Nucl Sci 41:1924-1931
19. Marshall PW, Dale CJ, Summers GP, Wolicki EA, Burke EA (1989) Proton, neutron and electron-induced displacement damage in germanium. IEEE Trans Nucl Sci 36: 1882-1888
20. Goubet JJ, Stievenard D, Mathiot D, Zazoui M (1992) Electron-irradiation-induced defects in Si-Ge alloys. Phys Rev B 46:10113-10118
21. Summers GP, Burke EA, Xapsos MA, Dale CJ, Marshall PW, Petersen EL (1988) Displacement damage in GaAs structures. IEEE Trans Nucl Sci 35:1221-1226
22. Summers GP, Burke EA, Shapiro P, Messenger SR, Walters RJ (1993) Damage correlations in semiconductors exposed to gamma, electron and proton radiations. IEEE Trans Nucl Sci 40:1372-1379
23. Dale CJ, Marshall PW, Summers GP, Wolicki EA, Burke EA (1989) Displacement damage equivalent to dose in silicon devices. Appl Phys Lett 54:451-453
24. Dale CJ, Marshall PW, Burke EA, Summers GP, Bender GE (1989) The generation lifetime damage factor and its variance in silicon. IEEE Trans Nucl Sci 36:1872-1881
25. Petersen E (1981) Soft errors due to protons in the radiation belt. IEEE Trans Nucl Sci 28:3981-3986
26. McLean FB, Oldham TR (1982) Charge funneling in n- and p-type Si substrates. IEEE Trans Nucl Sci 29:2018-2020
27. McNulty PJ, Roth DR, Beauvias WJ, Abdel-Kader WG, Dinge DC (1991) Comparison of the charge collection properties of junctions and the SEU response of microelectronic circuits. Nucl Tracks Radiat Measurements 19:929-938
28. Iwata H, Ohzone T (1995) Numerical simulation of single event latchup in the temperature range 77 – 450 K. IEEE Trans Nucl Sci 42:148-154
29. Srour JR, McGarrity JM (1988) Radiation effects on microelectronics. Proc IEEE 76: 1443-1469
30. Boesch Jr HE, McGarrity JM (1976) Charge yield and dose effects in MOS capacitors at 80 K. IEEE Trans Nucl Sci 23:1520-1525
31. Boesch Jr HE, Dunn GJ (1991) Hole transport in SiO_2 and reoxidized nitrided SiO_2 gate insulators at low temperatures. IEEE Trans Nucl Sci 38:1083-1088
32. Klein RB, Saks NS, Shanfeld Z (1990) Saturation of radiation-induced threshold-voltage shifts in thin-oxide MOSFETs at 80 K. IEEE Trans Nucl Sci 37:1690-1695
33. Zupac D, Galloway KF, Schrimpf RD, Augier P (1993) Radiation-induced mobility degradation in p-channel double-diffused metal-oxide-semiconductor power transistors at 300 and 77 K. J Appl Phys 73:2910-2915
34. McWhorter PJ, Winokur PS (1986) Simple technique for separating the effects of interface traps and trapped-oxide charge in metal-oxide-semiconductor transistors. Appl Phys Lett 48:133-135
35. Witczak SC, Winokur PS, Lacoe RC, Mayer DC (2000) Charge separation technique for metal-oxide-semiconductor capacitors in the presence of hydrogen deactivated dopants. J Appl Phys 87:8206-8208

36. Groeseneken G, Maes HE, Beltran N, De Keersmaecker RF (1984) A reliable approach to charge-pumping measurements in MOS transistors. IEEE Trans Electron Devices 31:42-53
37. Fleetwood DM, Meisenheimer TL, Scofield JH (1994) 1/f noise and radiation effects in MOS devices. IEEE Trans Electron Devices 41:1953-1964
38. Chen W, Balasinski A, Ma T-P (1991) Lateral distribution of radiation-induced damage in MOSFET's. IEEE Trans Nucl Sci 38:1124-1129
39. Balasinski A, Ma T-P (1992) Ionizing radiation damage near CMOS transistor channel edges. IEEE Trans Nucl Sci 39: 1998-2003
40. Balasinski A, Ma T-P (1993) Impact of radiation-induced nonuniform damage near MOSFET junctions. IEEE Trans Nucl Sci 40:1286-1292
41. Acovic A, Dutoit M, Ilegems M (1990) Characterization of hot-electron-stressed MOSFET's by low-temperature measurements of the drain tunnel current. IEEE Trans Electron Devices 37:1467-1476
42. Rosar M, Leroy B, Schweeger G (2000) A new model for the description of gate voltage and temperature dependence of gate induced drain leakage (GIDL) in the low electric field region. IEEE Trans Electron Devices 47:154-159
43. Giebel T, Goser K (1989) Hot-carrier degradation of n-channel MOSFET's characterized by a gated-diode measurement technique. IEEE Electron Device Lett 10:76-78
44. Neugroschel A, Sah C-T, Han KM, Carroll MS, Nishida T, Kavalieros JT, Lu Y (1995) Direct-current measurements of oxide and interface traps on oxidized silicon. IEEE Trans Electron Devices 42:1657-1661
45. Speckbacher P, Berger J, Asenov A, Koch F, Weber W (1995) The "gated-diode" configuration in MOSFET's, A sensitive tool for characterizing hot-carrier degradation. IEEE Trans Electron Devices 42:1287-1295
46. Okhonin S, Hessler T, Dutoit M (1996) Comparison of gate-induced drain leakage and charge pumping measurements for determining lateral interface trap profiles in electrically stressed MOSFET's. IEEE Trans Electron Devices 43:605-611
47. Fitzgerald DJ, Grove AS (1968) Surface recombination in semiconductors. Surface Science 9:347-369
48. Becker C, Gössling C, Lichau C, Wübben T, Wüstenfeld J, Wunstorf R (2000) Gate-controlled diodes for characterization of the Si-SiO_2 interface with respect to surface effects of silicon detectors. Nucl Instrum Methods in Physics Research A 444:605-613
49. Mogro-Campero A, Chang MF, Benjamin JL (1988) Resistance changes in silicon by MeV proton implantation. J Electrochem Soc 135:172-176
50. Ntsoenzok E, Barbot JF, Desgardin P, Vernois J, Blanchard C, Isabelle DB (1994) Study of the defects induced in n-type silicon irradiated by 1-3 MeV protons. IEEE Trans Nucl Sci 41:1932-1936
51. Ntsoenzok E, Desgardin P, Saillard M, Vernois J, Barbot JF (1996) Evolution of shallow donors with proton fluence in n-type silicon. J Appl Phys 79:8274-8277
52. Yamaguchi M, Taylor SJ, Yang M-J, Matsuda S, Kawasaki O, Hisamatsu T (1996) High-energy and high-fluence proton irradiation effects in silicon solar cells. J Appl Phys 80:4916-4920
53. Matsuura H, Uchida Y, Nagai N, Hisamatsu T, Aburaya T, Matsuda S (2000) Temperature dependence of the electron concentration in type-converted silicon by $1x10^{17}$ cm^{-2} fluence irradiation of 1 MeV electrons. Appl Phys Lett 76:2092-2094
54. Keskitalo N, Hallén A (1994) Resistivity profile measurements of proton-irradiated n-type silicon. Solid-State Electron 37:55-60

55. Lemeilleur F, Glaser M, Heijne EHM, Jarron P, Occelli E (1992) Neutron-induced radiation damage in silicon detectors. IEEE Trans Nucl Sci 39:551-557
56. Bates SJ, Dezillie B, Furetta C, Glaser M, Lemeilleur F, León-Florián E (1996) Proton irradiation of various resistivity silicon detectors. IEEE Trans Nucl Sci 43:1002-1008
57. Li J (1989) Novel semiconductor substrate formed by hydrogen ion implantation into silicon. Appl Phys Lett 55:2223-2224
58. Li J (2000) The new exploration for proton-implanted silicon: the conversion of a surface-region-purification-induced p-n junction into a p-i-n electrical structure approaching silicon-on-insulator. Semicond Sci Technol 15:L6-L9
59. Wu YH, Chin A, Shih KH, Wu CC, Liao CP, Pai SC, Chi CC (2000) Fabrication of very high resistivity Si with low loss and cross talk. IEEE Electron Device Lett 21:394-396
60. Privitera V, Coffa S, Priolo F, Kyllesbech Larsen K, Mannino G (1996) Room-temperature migration and interaction of ion beam generated defects in crystalline silicon. Appl Phys Lett 68:3422-3424
61. Pease RL, Enlow EW, Dinger GL, Marshall P (1987) Comparison of proton and neutron carrier removal rates. IEEE Trans Nucl Sci 34:1140-1146
62. Gossick BR (1959) Disordered region in semiconductors bombarded by fast neutrons. J Appl Phys 30:1214-1218
63. Brudnyi VN, Gradoboev AV, Peshev VV (1999) The broad midgap deep-level transient spectroscopy band in proton (65 MeV) and fast neutron-irradiated n-GaAs. Phys Stat Sol (b) 212:229-239
64. Fourches N (1995) High defect density regions in neutron irradiated high-purity germanium: Characteristics and formation mechanisms. J Appl Phys 77:3684-3689
65. Bruzzi M, Borchi E, Baldini A (1992) Using thermally stimulated currents to visualize defect clusters in neutron-irradiated silicon. J Appl Phys 72:4007-4013
66. Watts SJ, Matheson J, Hopkins-Bond IH, Holmes-Siedle A, Mohammadzadeh A, Pace R (1996) A new model for generation-recombination in silicon depletion regions after neutron irradiation. IEEE Trans Nucl Sci 43:2587-2594
67. Gill K, Hall G, MacEvoy B (1997) Bulk damage effects in irradiated silicon detectors due to clustered divacancies. J Appl Phys 82:126-136
68. Schenk A, Krumbein U (1995) Coupled defect-level recombination: Theory and application to anomalous diode characteristics. J Appl Phys 78:3185-3192
69. Takarabe K, Landsberg PT, Liakos JK (1997) Recombination statistics involving intertrap recombination. Semicond Sci Technol 12:687-691
70. Giri PK, Mohapatra YN (2000) Capacitance transient spectroscopy models of coupled trapping kinetics among multiple defect states: Application to the study of trapping kinetics of defects in heavy-ion-damaged silicon. Phys Rev B 62: 2496-2504
71. Svensson BG, Mohadjeri B, Hallén A, Svensson JH, Corbett JW (1991) Divacancy acceptor levels in ion-irradiated silicon. Phys Rev B 43:2292-2298
72. Amekura H, Kishimoto N, Saito T (1995) Photoconductivity evolution due to carrier trapping by defects in 17 MeV-proton irradiated silicon. J Appl Phys 77:4984-4992
73. Simoen E, Vanhellemont J, Claeys C (1996) Effective generation-recombination parameters in high-energy proton irradiated silicon diodes. Appl Phys Lett 69:2858-2860
74. Markvart T (1990) Review: Radiation damage in solar cells. J Mat Sci Mat Electron 1:1-12
75. Schroder DK (1990) Semiconductor Material and Device Characterization. Wiley-Interscience, New York, Chapter 8

76. Schroder DK (1997) Carrier lifetimes in silicon. IEEE Trans Electron Devices 44:160-170
77. Bullis WM, Huff HR (1996) Interpretation of carrier recombination and diffusion length measurements in silicon. J Electrochem Soc 143:1399-1405
78. Hallén A, Keskitalo N, Masszi F, Nágl V (1996) Lifetime in proton irradiated silicon. J Appl Phys 79:3906-3914
79. Simoen E, Vanhellemont J, Claeys C, Kaniava A, Gaubas E (1996) The response of Si p-n junction diodes to proton irradiations. Semicond Sci Technol 11:1434-1442
80. Mogro-Campero A, Love RP, Chang MF, Dyer R (1986) Localized lifetime control in insulated-gate transistors by proton implantation. IEEE Trans Electron Devices 33:1667-1671
81. Hüppi MW (1990) Proton irradiation of silicon: Complete electrical characterization of the induced recombination centers. J Appl Phys 68:2702-2707
82. Hallén A, Bakowski M, Lundqvist M (1993) Multiple proton energy irradiation for improved GTO thyristors. Solid-State Electron 36:133-141
83. Simoen E, Claeys C, Ohyama H (1998) Factors determining the damage coefficients and the low-frequency noise in MeV proton-irradiated silicon diodes. IEEE Trans Nucl Sci 45:89-97
84. Watkins GD (1969) A microscopic view of radiation damage in semiconductors using EPR as a probe. IEEE Trans Nucl Sci 16:13-18
85. Lenahan PM, Conley Jr JF (1998) What can electron paramagnetic resonance tell us about the Si/SiO_2 system? J Vac Sci Technol B 16:2134-2153
86. Henderson B, Pepper M, Vranch RL (1989) Spin-dependent and localisation effects at Si/SiO_2 device interfaces. Semicond Sci Technol 4:1045-1060
87. Stathis JH (1996) Electrically detected magnetic resonance study of stress-induced leakage current in thin SiO_2. Appl Phys Lett 68:1669-1671
88. Lepine DJ (1972) Spin dependent recombination in silicon surfaces. Phys Rev B 6:436-441
89. Krick JT, Lenahan PM, Dunn GJ (1991) Direct observation of interfacial point defects generated by channel hot hole injection in n-channel metal oxide silicon field effect transistors. Appl Phys Lett 59:3437-3439
90. Jupina MA, Lenahan PM (1989) A spin dependent recombination study of radiation induced defects at and near the Si/SiO_2 interface. IEEE Trans Nucl Sci 36:1800-1807
91. Stathis JH, DiMaria DJ (1992) Identification of an interface defect generated by hot electrons in SiO_2. Appl Phys Lett 61:2887-2889
92. Laiho R, Vlasenko LS, Vlasenko MP, Kozlov VA, Kozlovski VV (1999) Electron paramagnetic resonance of radiation defects in hydrogen-implanted silicon detected by spin-dependent microwave photoconductivity. Appl Phys Lett 74:3948-3950
93. Nubile P, Bourgoin JC, Stievenard D, Deresmes D, Strobl G (1992) Defects in low-temperature electron-irradiated p-type silicon. J Appl Phys 72:2673-2679
94. Watkins GD, Troxell JR, Chatterjee AP (1979) Vacancies and interstitials in silicon. Inst Phys Conf Ser No 46:Chapter 1,16-30
95. Watkins GD (1999) Vacancies and interstitials and their interactions with other defects in silicon, In: Abe T, Bullis WM, Kobayashi S, Lin W, Wagner P (eds) Proc 3rd Int Symposium on Defects in Silicon. The Electrochem Soc, Pennington, NJ, vol 99-1, pp 38-52
96. Lenahan PM, Conley Jr JF, Wallace BD (1997) A model for hole trapping in SiO_2 films on silicon. J Appl Phys 81:6822-6824

97. Lenahan PM, Mele JJ, Conley Jr JF, Lowry RK, Woodbury D (1999) Predicting radiation response from process parameters: Verification of a physically based predictive model. IEEE Trans Nucl Sci 46:1534-1543
98. Lang DV (1974) Deep-level transient spectroscopy: A new method to characterize traps in semiconductors. J Appl Phys 45:3023-3032
99. Benton JL (1990) Characterization of defects in semiconductors by deep level transient spectroscopy. J Cryst Growth 106:116-126
100. Johnson NM, Bartelink DJ, Gold RB, Gibbons JF (1979) Constant-capacitance DLTS measurement of defect-density profiles in semiconductors. J Appl Phys 50:4828-4833
101. Watts, SJ (1998) Radiation induced defects in silicon. In: Claeys CL, Rai-Choudhury P, Watanabe M, Stallhofer P, Dawson HJ (eds) High Purity Silicon V. The Electrochem Soc, Pennington, NJ, vol 98-13, pp 355-370
102. Markvart T, Parton DP, Peters JW, Willoughby AFW (1994) DLTS of recombination centres in semiconductors. Materials Science Forum 143-147:1381-1386
103. Deixler P, Terry J, Hawkins ID, Evans-Freeman JH, Peaker AR, Rubaldo L, Maude DK, Portal J-C, Dobaczewski L, Bonde Nielsen K, Nylandsted Larsen A, Mesli A (1998) Laplace-transform deep-level transient spectroscopy studies of the G4 gold-hydrogen complex in silicon. Appl Phys Lett 73:3126-3128
104. Yao X, Mou J, Qin G (1987) Shifts and splittings of electron energy levels of A centers in silicon under uniaxial stress. Phys Rev B 35:5734-5739
105. Samara GA (1987) Pressure dependence of deep electronic levels in semiconductors: The oxygen-vacancy pair (A centre) in silicon. Phys Rev B 36:4841-4848
106. Achtziger N, Gottschalk H, Licht T, Meier J, Rüb M, Reislöhner U, Witthuhn W (1995) Recoil implantation of radioactive transition metals and their investigation in silicon by deep-level transient spectroscopy. Appl Phys Lett 66:2370-2372
107. Peaker AR, Dobaczewski L, Andersen O, Rubaldo L, Hawkins ID, Bonde Nielsen K, Evans-Freeman JH (2000) Silicon defect characterization by high resolution Laplace Deep Level Transient Spectroscopy. In: Claeys CL, Rai-Choudhury P, Watanabe M, Stallhofer P, Dawson HJ (eds) High Purity Silicon VI. The Electrochem Soc, Pennington, NJ, vol 2000-17, pp 549-560
108. Goubet JJ, Sherman Christensen J, Mejlholm P, Nylandsted Larsen A (2000) Tin-related deep levels in p- and n-type silicon In: Claeys C (ed) Proc 2nd ENDEASD Workshop, pp 137-142
109. Watkins GD, Troxell JR (1980) Negative-U properties for point defects in silicon. Phys Rev Lett 44:593-596
110. Mukashev BN, Abdullin KhA, Gorelkinskii YuV (1997) Self-interstitials in irradiated silicon. Materials Science Forum 258-263:541-546
111. Brelot A (1971) Tin as a vacancy trap in silicon at room temperature. IEEE Trans Nucl Sci 19:220-226
112. Brelot A, Charlemagne J (1971) Infrared studies of low temperature electron irradiated silicon containing germanium, oxygen and carbon. Radiation Effects 9:65-73
113. Svensson JH, Svensson BG, Monenar B (1988) Infrared absorption studies of the divacancy in silicon: New properties of the singly negative charge state. Phys Rev B 38:4192-4197
114. Ascheron C (1991) Proton beam modification of selected $A^{III}B^{V}$ compounds. Phys Stat Sol A 124:11-55
115. Davies G (1989) The optical properties of luminescence centres in silicon. Phys Reports 176:83-188

116. Thonke K, Teschner R, Sauer R (1987) New photoluminescence defect spectra in silicon irradiated at 100 K: Observation of interstitial carbon. Solid State Commun 61:241-244
117. Nakamura M, Kitamura E, Misawa Y, Suzuki T, Nagai S, Sunaga H (1994) Photoluminescence measurement of carbon in silicon crystals irradiated with high energy electrons. J Electrochem Soc 141:3576-3579
118. Nakamura M, Byrne AS, Kitamura E, Suzuki T, Nagai S, Sunaga H, Aoki Y, Naramoto H (1995) Enhanced photoluminescence detection of oxygen in silicon crystal by formation of carbon-oxygen complex through carbon implantation and electron irradiation. J Appl Phys 78:4407-4410

Chapter 3

1. Watkins GD (1999) Vacancies and interstitials and their interactions with other defects in silicon. In: Abe T, Bullis WM, Kobayashi S, Lin W, Wagner P (eds) Proc third int symposium on defects in silicon. The Electrochem Soc, Pennington, NJ, vol 99-1, pp 38-52
2. Watkins GD (2000) Intrinsic defects in silicon. Mat Sci Semicond Processing 3:227-235
3. Hazdra P, Rubeš J, Vobecky J (1999) Divacancy profiles in MeV helium irradiated silicon from reverse I-V measurement. Nucl Instrum Meth Phys Research B 159:207-217
4. Simoen E, Claeys C, Gaubas E, Ohyama H (2000) Impact of the divacancy (?) on the generation-recombination properties of 10 MeV proton irradiated Float-Zone silicon diodes. Nucl Instrum Meth Phys Research A 439:310-318
5. Lee Y-H, Corbett JW (1976) EPR studies of defects in electron-irradiated silicon: A triplet state of vacancy-oxygen complexes. Phys Rev B 13:2653-2666
6. Trauwaert M-A, Vanhellemont J, Maes HE, Van Bavel A-M, Langouche G, Clauws P (1995) Low-temperature anneal of the divacancy in p-type silicon: A transformation from V_2 to V_xO_y complexes? Appl Phys Lett 66:3057-3058
7. Privitera V, Coffa S, Priolo F, Rimini E (1998) Migration and interaction of point defects at room temperature in crystalline silicon. Rivista Nuovo Cimento 21:1-52
8. Kimerling LC, Asom MT, Benton JL, Drevinsky PJ, Caefer CE (1989) Interstitial defect reactions in silicon. Mat Science Forum 38-41:141-150
9. Trauwaert M-A, Vanhellemont J, Maes HE, Van Bavel A-M, Langouche G, Stesmans A, Clauws P (1996) Influence of oxygen and carbon on the generation and annihilation of radiation defects in silicon. Mat Science Engineering B36:196-199
10. Davies G (1989) The optical properties of luminescence in silicon. Phys Reports 176:83-188
11. Lee Y-H, Gerasimenko NN, Corbett JW (1976) EPR study of neutron-irradiated silicon: A positive charge state of the <100> split di-interstitial. Phys Rev B 14:4506-4520
12. Lee YH (1998) Silicon di-interstitial in ion-implanted silicon. Appl Phys Lett 73:1119-1121
13. Lefèvre H (1980) Trap-centers of self-interstitials in silicon. Appl Phys 22:15-22
14. Lefèvre H (1982) Annealing behavior of trap-centers in silicon containing A-swirl defects. Appl Phys A 29:105-111

15. Watts SJ, Matheson J, Hopkins-Bond IH, Holmes-Siedle A, Mohammadzadeh A, Pace R (1996) A new model for generation-recombination in silicon depletion regions after neutron irradiation. IEEE Trans Nucl Sci 43:2587-2594
16. Watts SJ (1998) Radiation induced defects in silicon. In: Claeys CL, Rai-Choudhury P, Watanabe M, Stallhofer P, Dawson HJ (eds) High purity silicon V. The Electrochem Soc, Pennington, NJ, vol 98-13, pp 355-370
17. Stolk PA, Gossmann H-J, Eaglesham DJ, Jacobson DC, Poate JM, Luftman HS (1995) Trap-limited interstitial diffusion and enhanced boron clustering in silicon. Appl Phys Lett 66:568-570
18. Benton JL, Libertino S, Kringhøj P, Eaglesham DJ, Poate JM (1997) Evolution from point to extended defects in ion implanted silicon. J Appl Phys 82:120-125
19. Giles MD (1991) Transient phosphorus diffusion below the amorphization threshold. J Electrochem Soc 138:1160-1165
20. Hobler G, Pelaz L, Rafferty CS (2000) Dose, energy, and ion species dependence of the effective plus factor for transient enhanced diffusion. J Electrochem Soc 147:3494-3501
21. Benton JL, Libertino S, Eaglesham DJ, Coffa S (1998) Defect evolution in ion implanted silicon. In: Claeys CL, Rai-Choudhury P, Watanabe M, Stallhofer P, Dawson HJ (eds) High purity silicon V. The Electrochem Soc, Pennington, NJ, vol 98-13, pp 328-340
22. Benton JL, Halliburton K, Libertino S, Eaglesham DJ, Coffa S (1998) Electrical signatures and thermal stability of interstitial clusters in ion implanted silicon. J Appl Phys 84:4749-4756
23. Claverie A, Colombeau B, Ben Assayag G, Bonafos C, Cristiano F, Omri M, de Mauduit B (2000) Thermal evolution of extended defects in implanted Si: impact on dopant diffusion. Mat Sci Semicond Processing 3:269-277
24. Cowern NEB, Mannino G, Stolk PA, Roozeboom F, Huizing HGA, van Berkum JGM, Cristiano F, Claverie A, Jaraíz M (1999) Energetics of self-interstitial clusters in Si. Phys Rev Lett 82:4460-4463
25. Cowern NEB, Mannino G, Stolk PA, Roozeboom F, Huizing HGA, van Berkum JGM, Cristiano F, Claverie A, Jaraíz M (1999) Cluster ripening and transient enhanced diffusion in silicon. Mat Sci Semicond Processing 2:369-376
26. Schiettekatte F, Roorda S, Poirier R, Fortin MO, Chazal S, Héliou R (2000) Direct evidence for 8-interstitial-controlled nucleation of extended defects in c-Si. Appl Phys Lett 77:4322-4324
27. Davies G, Lightowlers EC, Ciechanowska ZE (1987) The 1018 meV (W or I_1) vibronic band in silicon. J Phys C: Solid State Phys 20:191-205
28. Schultz PJ, Thompson TD, Elliman RG (1992) Activation energy for the photoluminescence W center in silicon. Appl Phys Lett 60:59-61
29. Nakamura M, Nagai S, Aoki Y, Naramoto H (1998) Oxygen participation in the formation of the photoluminescence W center and the center's origin in ion-implanted crystals. Appl Phys Lett 72:1347-1349
30. Estreicher SK, Weber J, Derecskei-Kovacs A, Marynick DS (1997) Noble-gas-related defects in Si and the origin of the 1018 meV photoluminescence line. Phys Rev B 55:5037-5044
31. Giri PK, Coffa S, Rimini E (2001) Evidence for small interstitial clusters as the origin of photoluminescence W band in ion-implanted silicon. Appl Phys Lett 78:291-293

32. Schmidt DC, Svensson BG, Seibt M, Jagadish C, Davies G (2000) Photoluminescence, deep level transient spectroscopy and transmission electron microscopy measurements on MeV self-ion implanted and annealed n-type silicon. J Appl Phys 88:2309-2317
33. Hallén A, Keskitalo N, Masszi F, Nágl V (1996) Lifetime in proton irradiated silicon. J Appl Phys 79:3906-3914
34. Bleichner H, Jonsson P, Keskitalo N, Nordlander E (1996) Temperature and injection dependence of the Shockley-Read-Hall lifetime in electron irradiated n-type silicon. J Appl Phys 79:9142-9148
35. Keskitalo N, Jonsson P, Nordgren K, Bleichner H, Nordlander E (1998) Temperature and injection dependence of the Shockley-Read-Hall lifetime in electron-irradiated p-type silicon. J Appl Phys 83:4206-4212
36. Evwaraye AO, Baliga BJ (1977) The dominant recombination centers in electron-irradiated semiconductor devices. J Electrochem Soc 124:913-916
37. Kaniava A, Vanhellemont J, Simoen E, Claeys C, Gaubas E (1996) Characterisation of high-energy proton irradiation induced recombination centers in silicon. Solid State Phenomena 47-48:371-376
38. Baliga BJ, Evwaraye AO (1983) Correlation of lifetime with recombination centers in electron-irradiated p-type silicon. J Electrochem Soc 130:1916-1918
39. Hüppi MW (1990) Proton irradiation of silicon: Complete electrical characterization of the induced recombination centers. J Appl Phys 68:2702-2707
40. Kuchinskii PV, Lomako VM (1986) The effect of thermal and radiation defects on the recombination properties of the base region of diffused silicon p-n structures. Solid-State Electron 29:1041-1051
41. Germaná RS, Campisano SU (1992) Room-temperature recombination of point defects produced in silicon p-n junctions by light ion irradiation. Appl Phys Lett 60:1726-1728
42. Simoen E, Vanhellemont J, Claeys C (1996) Effective generation-recombination parameters in high-energy proton irradiated silicon diodes. Appl Phys Lett 69:2858-2860
43. Iles PA (2001) Evolution of space solar cells. Solar Energy Mat & Solar Cells 68:1-13
44. Gill K, Hall G, MacEvoy B (1997) Bulk damage effects in irradiated silicon detectors due to clustered divacancies. J Appl Phys 82:126-136
45. MacEvoy BC, Hall G (2000) Defect kinetics in novel detector materials. Mat Sci Semicond Processing 3:243-249
46. Feick H, Fretwurst E, Moll M, Lindström G (1997) Correlation of radiation damage effects in high resistivity silicon detectors with results from deep level spectroscopy. IEEE Trans Nucl Sci 44:825-833
47. Svensson BG, Mohadjeri B, Hallén A, Svensson JH, Corbett JW (1991) Divacancy acceptor levels in ion-irradiated silicon. Phys Rev B 43:2292-2298
48. Moll M, Feick H, Fretwurst E, Lindström G, Schütze C (1997) Comparison of defects produced by fast neutrons and ^{60}Co-gammas in high-resistivity silicon detectors using deep-level transient spectroscopy. Nucl Instrum Methods Phys Research A 388:335-339
49. Hardalov ChM, Stefanov KD, Sueva D (1995) On the applicability of deep-level transient spectroscopy for the investigation of deep centers in silicon created by fast neutron irradiation. Appl Phys A 61:107-109
50. Stefanov KD, Hardalov ChM, Sueva D (1997) Investigation of radiation damage in silicon produced by fast neutron irradiation with lifetime measurements and deep level transient spectroscopy. Phys Stat Sol (a) 163:27-32

51. Passeri D, Bilei GM, Ciampolini P (1999) A comprehensive analysis of low-resistivity silicon radiation detectors. IEEE Trans Nucl Sci 46:260-265
52. Passeri D, Ciampolini P, Bilei GM, Moscatelli F (2001) Comprehensive modeling of bulk-damage effects in silicon radiation detectors. In: Proc RADECS 2000, pp. 188-193
53. Osborne MD, Hobson PR, Watts SJ (2000) Numerical simulation of neutron radiation effects in avalanche photodiodes. IEEE Trans Nucl Sci 47:529-535
54. Khan A, Yamaguchi M, Kaneiwa M, Saga T, Abe T, Annzawa O, Matsuda S (2000) Influence of the dopant species on radiation-induced defects in Si single crystals. J Appl Phys 87:8389-8392
55. Dezillie B, Li Z, Eremin V, Bruzzi M, Pirollo S, Pandey SU, Li CJ (1999) Improved neutron radiation hardness for Si detectors: Application of low resistivity starting material and/or manipulation of N_{eff} by selective filling of radiation-induced traps at low temperatures. IEEE Trans Nucl Sci 46:221-227
56. Lindström G, Fretwurst E, Feick H, Moll M, Vasilescu A, Watts SJ (2000) Radiation hard silicon detectors – Developments by the RD48 (ROSE) collaboration. In: Claeys C (ed) Proc 2nd ENDEASD Workshop, pp 225-243
57. Ruzin A, Casse G, Glaser M, Lemeilleur F, Matheson J, Watts S, Zanet A (2000) Radiation effects in silicon detectors processed on carbon and oxygen-rich substrates. Mat Sci Semicond Processing 3:257-261
58. Simoen E, Claeys C, Ohyama H (1998) Factors determining the damage coefficients and the low-frequency noise in MeV proton-irradiated silicon diodes. IEEE Trans Nucl Sci 45:89-97
59. Brelot A (1971) Tin as a vacancy trap in silicon at room temperature. IEEE Trans Nucl Sci 19:220-226
60. Brelot A, Charlemagne J (1971) Infrared studies of low temperature electron irradiated silicon containing germanium, oxygen and carbon. Radiation Effects 9:65-73
61. Watkins GD (1975) Defects in irradiated silicon: EPR of the tin-vacancy pair. Phys Rev B 12:4383-4390
62. Simoen E, Claeys C, Neimash VB, Kraitchinskii A, Kras'ko M, Puzenko O, Blondeel A, Clauws P (2000) Deep levels in high-energy proton-irradiated tin-doped n-type Czochralski silicon. Appl Phys Lett 76:2838-2840
63. Nylandsted Larsen A, Goubet JJ, Mejlholm P, Sherman Christensen J, Fanciulli M, Gunnlaugsson HP, Weyer G, Wulff Petersen J, Resende A, Kaukonen M, Jones R, Öberg S, Briddon PR, Svensson BG, Lindström JL, Dannefaer S (2000) Tin-vacancy acceptor levels in electron-irradiated n-type silicon. Phys Rev B 62:4535-4544
64. Goubet JJ, Sherman Christensen J, Mejlholm P, Nylandsted Larsen A (2000) Tin-related deep levels in p- and n-type silicon In: Claeys C (ed) Proc 2nd ENDEASD Workshop, pp 137-142
65. Da Viá C (2000) Heavily irradiated Si detectors operated at cryogenic temperatures: the Lazarus effect. In: Claeys C (ed) Proc 2nd ENDEASD Workshop, pp 110-116
66. Fage-Pedersen J, Nylandsted Larsen A, Mesli A (2000) Irradiation-induced defects in Ge studied by transient spectroscopies. Phys Rev B 62:10116-10125
67. Wolf J, Katterloher R, Lemke D, Grözinger U, Hermans L, Frenzl O, Engemann D, Beeman J, Fabbricotti M (1996) Far-infrared stressed Ge:Ga array for FIRST. In: Proc 30th ESLAB symposium on submillimetre and far-infrared space instrumentation, ESA SP-388, pp 25-28

68. Patrashin M, Fouks B, Grözinger U, Lemke D, Wolf J (1997) Residual conductivity of stressed Ge:Ga photoconductors after low-dose gamma irradiation. J Appl Phys 82:1450-1453
69. Beeman JW, Hansen WL, Haller EE (1996) Extrinsic germanium photoconductors for far-IR astronomy: Research results and works in progress. In: Proc 30th ESLAB symposium on submillimetre and far-infrared space instrumentation, ESA SP-388, pp. 21-24
70. Haller EE, Itoh KM, Beeman JW (1996) Neutron transmutation doped (NTD) germanium thermistors for sub-mm bolometer applications. In: Proc 30th ESLAB symposium on submillimetre and far-infrared space instrumentation, ESA SP-388, pp 115-118
71. Simoen E, Clauws P, Broeckx J, Vennik J, Van Sande M, De Laet L (1982) Correlation between DLTS-measurements and the performance of high purity germanium detectors. IEEE Trans Nucl Sci 29:789-792
72. Blondeel A, Clauws P (1999) Photoinduced current transient spectroscopy of deep defects in n-type ultrapure germanium. J Appl Phys 86:940-945
73. Chung MA, Meier DL, Szedon JR, Bartko J (1988) Electron radiation and annealing of MOCVD GaAs and GaAs/Ge solar cells. In: Proc 20th photovoltaic specialists conference. The IEEE, New York, pp 924-929
74. Anspaugh BE (1991) Proton and electron damage coefficients for GaAs/Ge solar cells. In: Proc 22nd photovoltaic specialists conference. The IEEE, New York, pp 1593-1598
75. Gobeli GW (1958) Alpha-particle irradiation of Ge at 4.2 K. Phys Rev 112:732-739
76. MacKay JW, Klontz EE (1959) Low-temperature annealing studies in Ge. J Appl Phys 30:1269-1274
77. Callcot TA, MacKay JW (1967) Irradiation damage in n-type germanium at 4.2 K. Phys Rev 161:698-710
78. Bourgoin J, Mollot F (1971) Behaviour of primary defects in electron-irradiated germanium. Phys Stat Sol B 43:343-355
79. Ehrhart P, Zillgen H (1999) Vacancies and interstitial atoms in e^--irradiated germanium. J Appl Phys 85:3503-3511
80. Fukuoka N, Saito H (1974) The defects produced by electron irradiation and annealed at about 360K in n-type germanium. Jpn J Appl Phys 13:1524-1532
81. Fukuoka N, Saito H (1976) Radiation induced defects in germanium single crystals of high purity. Jpn J Appl Phys 15:237-242
82. Emtsev VV, Mashovets TV, Miknovich VV (1992) Frenkel pairs in germanium and silicon. Sov Phys Semicond 26:12-25
83. Mooney PM, Poulin F, Bourgoin JC (1983) Annealing of electron-induced defects in n-type germanium. Phys Rev B 28:3372-3377
84. Fourches N, Huck A, Walter G, Bourgoin JC (1989) Fast neutron irradiation induced defects in high purity germanium. Materials Science Forum 38-41:1233-1238
85. Fourches N, Walter G, Bourgoin JC (1991) Neutron-induced defects in high-purity germanium. J Appl Phys 69:2033-2043
86. Swanson ML (1960) Low-temperature neutron-irradiation damage and its recovery in high-purity germanium. Can J Phys 44:2181-2199
87. Haesslein H, Sielemann R, Zistl C (1998) Vacancies and self-interstitials in germanium observed by perturbed angular correlation spectroscopy. Phys Rev Lett 80:2626-2629
88. Poulin F, Bourgoin JC (1982) Characteristics of the electron traps produced by electron irradiation in n-type germanium. Phys Rev B 26:6788-6794

89. Ito K, Corbett JW (1983) Interaction of point defects with hydrogen in germanium. Jpn J Appl Phys 22:L724-L726
90. Fukuoka N, Saito H (1983) The hole trapping defects in irradiated germanium as studied by DLTS. Physica 116B:343-348
91. Li SS, Choi CG, Loo RY (1985) Studies of radiation-induced defects in one-MeV electron and low energy proton irradiated germanium and $Al_xGa_{1-x}As$ p-n junction solar cells. In: Proc 18th photovoltaic specialists conference. The IEEE, New York, pp 640-645
92. Fukuoka N, Honda M, Nishioka Y, Atobe K, Matsukawa T (1995) Property of radiation-induced defects in germanium single crystals. Jpn J Appl Phys 34:3204-3208
93. Nagesh V, Farmer JW (1988) Study of irradiation-induced defects in germanium. J Appl Phys 63:1549-1553
94. Marie P, Levalois M, Bogdanski P (1993) Irradiation-induced defects in n-type germanium. J Appl Phys 74:868-871
95. Marie P, Levalois M, Paumier E (1996) Swift-heavy-ion induced damage in germanium: An evaluation of defect introduction rates. J Appl Phys 79:7555-7562
96. Colder A, Levalois M, Marie P (2000) Study of electron, proton and swift heavy ion irradiation of n-type germanium using deep level transient spectroscopy. J Appl Phys 88:3082-3084
97. Fukuoka N, Saito H (1982) Defect states in n-type germanium irradiated with 1.5 MeV electrons. Jpn J Appl Phys 21:930-935
98. Fourches N (1995) High defect density regions in neutron irradiated high-purity germanium: Characteristics and formation mechanisms. J Appl Phys 77:3684-3689
99. Patrashin M, Shibai H, Okuda H, Hiromoto N, Fujiwara M, Fouks B (1996) Suppression of the radiation-induced effects in Ge:Ga detectors. In: Proc 30th ESLAB symposium on submillimetre and far-infrared space instrumentation, ESA SP-388, pp 29-32
100. Lang DV, People R, Bean JC, Sergent AM (1985) Measurement of the band gap of Ge_xSi_{1-x}/Si strained-layer heterostructures. Appl Phys Lett 47:1333-1335
101. Maiti CK, Bera LK, Chattopadhyay S (1998) Strained-Si heterostructure field effect transistors. Semicond Sci Technol 13:1225-1246
102. Honda T, Suezawa M, Sumino K (1996) Growth and characterization of bulk Si-Ge single crystals. Jpn J Appl Phys 35:5980-5985
103. Jain SC, Hayes W (1991) Structure, properties and applications of Ge_xSi_{1-x} strained layers and superlattices. Semicond Sci Technol 6:547-576
104. Osten HJ (1998) Band-gap changes and band offsets for ternary $Si_{1-x-y}Ge_xC_y$ alloys on Si(001). J Appl Phys 84:2716-2721
105. Franz M, Pressel K, Gaworzewski P (1998) Alloy effects in boron doped Si-rich SiGe bulk crystals. J Appl Phys 84:709-712
106. Wolf M, Brendel R, Werner JH, Queisser HJ (1998) Solar cell efficiency and carrier multiplication in $Si_{1-x}Ge_x$ alloys. J Appl Phys 83:4213-4221
107. Said K, Poortmans J, Libezny M, Caymax M, Nijs J, Mertens R, Vinckier C, Vyncke D, Seifert W, Kittler M, Silier I, Gutjahr A, Konuma M (1997) Low-temperature passivation for SiGe-alloy solar cells. In: Proc 14th European photovoltaic solar energy conference, pp 986-992
108. LeGoues FL, Meyerson BS, Morar JF, Kirchner PD (1992) Mechanism and conditions for anomalous strain relaxation in graded thin films and superlattices. J Appl Phys 71:4230-4243

109. Cressler JD (1995) Status and trends in the cryogenic operation of SiGe bipolar technology. In: Claeys CL, Raider SI, Kirschman R, Brown WD (eds) Proc symposium on low temperature electronics and high temperature superconductivity. The Electrochem Soc, Pennington, NJ, proc vol 95-9, pp 159-177
110. Nayak DK, Woo JCS, Park JS, Wang KL, MacWilliams KP (1991) Enhancement-mode quantum-well Ge_xSi_{1-x} PMOS. IEEE Electron Device Lett 12:154-156
111. O'Neill AG, Antoniadis DA (1996) Deep submicron CMOS based on silicon germanium technology. IEEE Trans Electron Devices 43:911-918
112. Welser J, Hoyt JL, Gibbons JF (1994) Electron mobility enhancement in strained-Si n-type metal-oxide-semiconductor field-effect transistor. IEEE Electron Device Lett 15:100-102
113. De Meyer K, Biesemans S, Collaert N, Kubicek S, Verheyen P (1998) New architectures for deep submicron MOSFETs. In: Touboul A, Danto Y, Klein JP, Grünbacher H (eds) Proc ESSDERC '98. Editions Frontières, Paris, France, pp 63-68
114. Karunasiri RPG, Wang KL (1991) Quantum devices using SiGe/Si heterostructures. J Vac Sci Technol B 9:2064-2071
115. Nur O, Willander M, Turan R, Sardela Jr MR, Hansson GV (1996) Metal-semiconductor junctions on p-type strained $Si_{1-x}Ge_x$ layers. Appl Phys Lett 68:1084-1086
116. Dentel D, Kubler L, Bischoff JL, Chattopadhyay S, Bera LK, Ray SK, Maiti CK (1998) Molecular beam epitaxial growth of strained $Si_{1-x}Ge_x$ layers on graded $Si_{1-y}Ge_y$ for Pt silicide Schottky diodes. Semicond Sci Technol 13:214-219
117. Jung TG, Chang CY, Liu CS, Chang TC, Lin HC, Tsai WC, Huang GW, Chen LP (1994) Characterization of the Si/SiGe heterojunction diode grown by ultrahigh vacuum chemical vapor deposition. Appl Phys Lett 76:4921-4923
118. Vonsovici A, Vescan L, Apetz R, Koster A, Schmidt K (1998) Room temperature photocurrent spectroscopy of SiGe/Si p-i-n photodiodes grown by selective epitaxy. IEEE Trans Electron Devices 45:538-542
119. Chang S-J, Nayak DK, Shiraki Y (1998) 1.54 μm electroluminescence from erbium-doped SiGe light emitting diodes. J Appl Phys 83:1426-1428
120. Budtz-Jørgensen CV, Kringhøj P, Nylansted Larsen A, Abrosimov NV (1998) Deep-level transient spectroscopy of the Ge-vacancy pair in Ge-doped n-type silicon. Phys Rev B 58:1110-1113
121. Kawasuso A, Okada S, Yonenaga I, Honda T, Suezawa M (1997) Positron annihilation study of electron-irradiated silicon-germanium bulk alloys. Materials Science Forum 258-263:127-132
122. Kawasuso A, Okada S, Suezawa M, Honda T, Yonenaga I (1997) Positron annihilation in electron-irradiated Si_xGe_{1-x} bulk crystal. J Appl Phys 81:2916-2918
123. Schmalz K, Emtsev VV (1994) Radiation-induced defects in Czochralski-grown silicon doped with germanium. Appl Phys Lett 65:1575-1577
124. Mesli A, Nylandsted Larsen A (1999) Vacancy in relaxed p-type $Si_{1-x}Ge_x$ alloys: Evidence for strong disorder induced relaxation. Phys Rev Lett 83:148-151
125. Goubet JJ, Stievenard D, Mathiot D, Zazoui M (1992) Electron-irradiation-induced defects in Si-Ge alloys. Phys Rev B 46:10113-10118
126. Goubet JJ, Stievenard D (1995) Annealing study of electron irradiation-induced defects in SiGe alloys. Appl Phys Lett 66:1409-1411
127. Kringhøj P, Nylandsted Larsen A (1995) Irradiation-induced defect states in epitaxial n-type $Si_{1-x}Ge_x$ alloy layers. Phys Rev B 52:16333-16336

128. Nylandsted Larsen A (1997) Defects in SiGe. Materials Science Forum 258-263:83-90
129. Monakhov EV, Nylandsted Larsen A, Kringhøj P (1997) Electronic defect levels in relaxed, epitaxial p-type $Si_{1-x}Ge_x$ layers produced by MeV proton irradiation. J Appl Phys 81:1180-1183
130. Drevinsky PJ, Caefer CE, Tobin SP, Mikkelsen Jr JC, Kimerling LC (1988)Influence of oxygen and boron on defect production in irradiated silicon. Mat Res Soc Symp Proc 104:167-172
131. Goodman SA, Auret FD, Nauka K, Malherbe JB (1997) Defect characterization of n-type $Si_{1-x}Ge_x$ after 1.0 keV helium-ion etching. J Electron Mater 26:463-469
132. Goodman SA, Auret FD, Mamor M, Deenapanray PNK, Meyer WE (1997) Electronic properties of defects introduced in n- and p-type $Si_{1-x}Ge_x$ during ion etching. Materials Science Forum 258-263:133-138
133. Vanhellemont J, Trauwaert M-A, Poortmans J, Caymax M, Clauws P (1992) 1 MeV electron irradiation induced degradation of boron doped strained $Si_{1-x}Ge_x$ layers. Thin Solid Films 222:166-172
134. Vanhellemont J, Trauwaert M-A, Poortmans J, Caymax M, Clauws P (1993) Fast degradation of boron-doped strained $Si_{1-x}Ge_x$ layers by 1-MeV electron irradiation. Appl Phys Lett 62:309-311
135. Ohyama H, Vanhellemont J, Sunaga H, Poortmans J, Caymax M, Clauws P (1994) On the degradation of 1-MeV electron irradiated $Si_{1-x}Ge_x$ diodes. IEEE Trans Nucl Sci 41:487-494
136. Ohyama H, Vanhellemont J, Takami Y, Hayama K, Sunaga H, Poortmans J, Caymax M, Clauws P (1994) Germanium content dependence of radiation damage in strained $Si_{1-x}Ge_x$ epitaxial devices. IEEE Trans Nucl Sci 41:2437-2442
137. Ohyama H, Vanhellemont J, Takami Y, Hayama K, Sunaga H, Poortmans J, Caymax M (1995) Degradation of $Si_{1-x}Ge_x$ epitaxial heterojunction bipolar transistors by 1-MeV fast neutrons, IEEE Trans Nucl Sci 42:1550-1557
138. Ohyama H, Vanhellemont J, Takami Y, Hayama K, Sunaga H, Poortmans J, Caymax M (1996) Radiation source dependence of degradation and recovery of irradiated $Si_{1-x}Ge_x$ epitaxial devices. In: Proc RADECS '95. The IEEE, New York, pp 66-71
139. Ohyama H, Hayama K, Vanhellemont J, Poortmans J, Caymax M, Takami Y, Sunaga H, Nashiyama I, Uwatoko Y (1996) Degradation of $Si_{1-x}Ge_x$ epitaxial devices by proton irradiation. Appl Phys Lett 69:2429-2431
140. Ohyama H, Vanhellemont J, Takami Y, Hayama K, Sunaga H, Nashiyama I, Uwatoko Y, Poortmans J, Caymax M (1996) Degradation and recovery of proton irradiated $Si_{1-x}Ge_x$ epitaxial devices. IEEE Trans Nucl Sci 43:3089-3096
141. Ohyama H, Vanhellemont J, Takami Y, Sunaga H, Nashiyama I, Uwatoko Y, Poortmans J, Caymax M (1997) Degradation of SiGe devices by proton irradiation. Radiat Phys Chem 50:341-346
142. Ohyama H, Simoen E, Claeys C, Vanhellemont J, Takami Y, Sunaga H, Poortmans J, Caymax M (1997) Lattice defects in $Si_{1-x}Ge_x$ devices by proton irradiation and their effect on device performance. Solid State Phenomena 57-58:239-244
143. Ohyama H, Simoen E, Claeys C, Takami Y, Hayama K, Hakata T, Kobayashi K, Sunaga H, Poortmans J, Caymax M (1998) The impact of the Ge content on the characteristics of strained $Si_{1-x}Ge_x$ epitaxial diodes before and after degradation by high energy particles. In: Touboul A, Danto Y, Klein JP, Grünbacher H (eds), Proc ESSDERC '98. Les Editions Frontières, Paris, France, pp 548-551

144. Ohyama H, Simoen E, Claeys C, Takami Y, Hayama K, Hakata T, Sunaga H, Poortmans J, Caymax M (1998) Impact of high energy particle irradiation on the electrical performance of $Si_{1-x}Ge_x$ epitaxial diodes. In: Proc 2nd international conference on materials for microelectronics, The Institute of Materials, London, UK, pp 11-18
145. Wang KL, Lee YH, Corbett JW (1978) Defect distribution near the surface of electron-irradiated silicon. Appl Phys Lett 33:547-548
146. Mamor M, Auret FD, Goodman SA, Myburg G (1998) Electrical characterization of defects introduced in p-$Si_{1-x}Ge_x$ during electron-beam deposition of Sc Schottky barrier diodes. Appl Phys Lett 72:1069-1071
147. Mamor M, Auret FD, Goodman SA, Myburg G, Deenapanray PNK, Meyer WE (1997) Electrical characterization of electron beam induced defects in epitaxially grown $Si_{1-x}Ge_x$. Materials Science Forum 258-263:115-120
148. Lie DYC (1998) Doping and processing epitaxial $Si_{1-x}Ge_x$ films on Si(100) by ion implantation for Si-based heterojunction devices applications. J Electron Mater 27:377-401
149. Haynes TE, Holland OW (1992) Damage accumulation during ion implantation of unstrained $Si_{1-x}Ge_x$ alloy layers. Appl Phys Lett 61:61-63
150. Holland OW, Haynes TE (1992) Damage saturation during high-energy ion implantation of $Si_{1-x}Ge_x$. Appl Phys Lett 61:3148-3150
151. Nylandsted Larsen A, O'Raifeartaigh C, Barklie RC, Holm B, Priolo F, Franzo G, Lulli G, Bianconi M, Nipoti R, Lindner JKN, Mesli A, Grob JJ, Cristiano F, Hemment PLF (1997) MeV ion implantation induced damage in relaxed $Si_{1-x}Ge_x$. J Appl Phys 81:2208-2218
152. Lindner JKN (1996) Radiation damage of 2 MeV Si ions in $Si_{0.75}Ge_{0.25}$: optical measurements and damage modelling. Nucl Instrum Methods Phys Research B 112:316-320
153. Willander M, Shen G-D, Xu D-X, Ni W-X (1988) Current transport in strained n-$Si_{1-x}Ge_x$/p-Si heterojunction diodes. J Appl Phys 63:5036-5039
154. Xu DX, Shen GD, Willander M, Knall J, Hasan M-A, Hansson GV (1990) The influence of defects on device performance of MBE-grown Si homojunction and strained $Si_{1-x}Ge_x$/Si heterostructures. J Electron Materials 19:1033-1041
155. Park JS, Lin TL, Jones EW, Gunapala SD, Soli GA, Wilson BA (1993) Proton irradiation effects on strained $Si_{1-x}Ge_x$/Si heterostructures. Appl Phys Lett 63:3497-3499

Chapter 4

1. Farmer JW, Look DC (1979) Type conversion in electron-irradiated GaAs. J Appl Phys 50:2970-2972
2. Sellami L, Aubin M, Aktik C, Carlone C, Houdayer A, Hinrichsen P (1999) Photoluminescence study of gallium arsenide irradiated with 15 MeV alpha particles. IEEE Trans Nucl Sci 46:1603-1607
3. Pons D, Bourgoin J (1981) Anisotropic-defect introduction in GaAs by electron irradiation. Phys Rev Lett 47:1293-1296
4. Bourgoin JC, von Bardeleben HJ, Stiévenard D (1988) Native defects in gallium arsenide. J Appl Phys 64:R65-R91
5. Guillot G (1988) EL2-related studies in irradiated and implanted GaAs. Revue Phys Appl 23:833-846

6. Goswami NK, Newman RC, Whitehouse JE (1981) The observation of high concentrations of arsenic anti-site defects in electron irradiated n-type GaAs by X-band EPR. Solid State Commun 40:473-477
7. Tajima M (1987) Radiative recombination mechanism of EL2 level in GaAs. Jpn J Appl Phys 26:L885-L888
8. Kuriyama K, Yokoyama K, Satoh A (1991) Thermally stimulated current in neutron-transmutation-doped semi-insulating GaAs. Appl Phys Lett 59:1326-1328
9. Kuriyama K, Tomizawa K, Koga K, Hayashi N, Watanabe H, Ikeda Y, Maekawa H (1993) Thermally stimulated current in fast neutron irradiated semi-insulating GaAs: Ga antisite related new trap. Appl Phys Lett 63:1966-1968
10. McGregor DS, Rojeski RA, Knoll GF, Terry Jr FL, East J, Eisen Y (1994) Evidence for field enhanced electron capture by EL2 centers in semi-insulating GaAs and the effect on GaAs radiation detectors. J Appl Phys 75:7910-7915
11. Bourgoin JC, Neffati T (1999) The energy level of the EL2 defect in GaAs. Solid-State Electron 43:153-158
12. Rezazadeh AA, Palmer DW (1985) An electron-trapping defect level associated with the 235 K annealing stage in electron-irradiated n-GaAs. J Phys C: Solid State Phys 18:43-54
13. Siyanbola WO, Palmer DW (1991) Electronic energy levels of defects that anneal in the 280-K stage in irradiated n-type gallium arsenide. Phys Rev Lett 66:56-59
14. Jorio A, Wang A, Parenteau M, Carlone C, Rowell NL, Khanna SM (1994) Optical identification of the gallium vacancy in neutron-irradiated gallium arsenide. Phys Rev B 50:1557-1566
15. Bingwen L, Yuanxi Z, Binglin Z, Milnes AG (1987) Minority carrier diffusion lengths in bulk n-type GaAs. J Electron Mat 16:177-180
16. Fang Z-Q, Schlesinger TE, Milnes AG (1987) Evidence for EL6 (E_c-0.35 eV) acting as the dominant recombination center in n-type horizontal Bridgman GaAs. J Appl Phys 61:5047-5050
17. Nissen MK, Steiner T, Beckett DJS, Thewalt MLW (1990) Photoluminescence transitions of the deep EL2 defect in gallium arsenide. Phys Rev Lett 65:2282-2285
18. Nissen MK, Villemaire A, Thewalt MLW (1991) Photoluminescence studies of the EL2 defect in gallium arsenide under external perturbations. Phys Rev Lett 67:112-115
19. von Bardeleben HJ, Stiévenard D, Deresmes D, Huber A, Bourgoin JC (1986) Identification of a defect in a semiconductor: EL2 in GaAs. Phys Rev B 34:7192-7202
20. Meyer BK, Hofmann DM, Niklas JR, Spaeth J-M (1987) Arsenic antisite defect As_{Ga} and EL2 in GaAs. Phys Rev B 36:1332-1335
21. Manasreh MO, McDonald PF, Kivlighn SA, Minton JT, Covinton BC (1988) Electron paramagnetic resonance of isolated As_{Ga}^+ antisite defect in neutron-transmutation doped semi-insulating GaAs. Solid State Commun 65:1267-1269
22. Manasreh MO, Fischer DW (1989) Quenching and recovery characteristics of the EL2 defect in GaAs under monochromatic-light illumination. Phys Rev B 40:11756-11763
23. Saarinen K, Kuisma S, Hautojärvi P, Corbel C, LeBerre C (1994) Metastable vacancy in the EL2 defect in GaAs studied by positron-annihilation spectroscopies. Phys Rev B 49:8005-8016
24. Thommen K (1970) Recovery of low temperature electron irradiation-induced damage in n-type GaAs. Rad Effects 2:201-210

25. Lang DV, Kimerling LC, Leung SY (1976) Recombination-enhanced annealing of the E1 and E2 defect levels in 1-MeV-electron-irradiated n-GaAs. J Appl Phys 47:3587-3591
26. Pons D, Mooney PM, Bourgoin JC (1980) Energy dependence of deep level introduction in electron irradiated GaAs. J Appl Phys 51:2038-2042
27. Pons D, Mircea A, Bourgoin J (1980) An annealing study of electron irradiation-induced defects in GaAs. J Appl Phys 51:4150-4157
28. Martin GM, Makram-Ebeid S (1983) Manifestations of deep level point defects in GaAs. Physica 116B:371-383
29. Stievenard D, Boddaert X, Bourgoin JC (1986) Irradiation-induced defects in p-GaAs. Phys Rev B 34:4048-4058
30. Stiévenard D, Boddaert X, Bourgoin JC, von Bardeleben HJ (1990) Behavior of electron-irradiation-induced defects in GaAs. Phys Rev B 41:5271-5279
31. Stiévenard D, Bourgoin JC (1986) Defect-enhanced annealing by carrier recombination in GaAs. Phys Rev B 33:8410-8415
32. Look DC, Sizelove JR (1987) Defect production in electron-irradiated, n-type GaAs. J Appl Phys 62:3660-3664
33. Ziebro B, Hemsky JW, Look DC (1992) Defect models in electron-irradiated n-type GaAs. J Appl Phys 72:78-81
34. Siyanbola WO, Palmer DW (1990) Low temperature annealing of deep electron traps produced by proton irradiation of n-GaAs. Semicond Sci Technol 5:7-15
35. Hautojärvi P, Moser P, Stucky M, Corbel C, Plazaola F (1986) Detection of Ga vacancies in electron irradiated GaAs by positrons. Appl Phys Lett 48:809-810
36. Corbel C, Pierre F, Hautojärvi P, Saarinen K, Moser P (1990) Native and irradiation-induced monovacancies in n-type and semi-insulating GaAs. Phys Rev B 41:10632-10641
37. Jia YQ, von Bardeleben HJ, Stiévenard D, Delerue C (1992) Electron-paramagnetic-resonance observation of gallium vacancy in electron-irradiated p-type GaAs. Phys Rev B 45:1645-1649
38. Goodman SA, Auret FD (1993) Deep level transient spectroscopy characterization of defects introduced in n-GaAs after alpha irradiation at 15 K. Jpn J Appl Phys 32:L1120-L1122
39. Siyanbola WO, Palmer DW (1990) Optical-MCTS of proton-irradiated n-type gallium arsenide. Solid State Commun 74:209-213
40. Khanna SM, Jorio A, Carlone C, Parenteau M, Houdayer A, Gerdes Jr JW (1995) Particle dependence of the gallium vacancy production in irradiated n-type gallium arsenide. IEEE Trans Nucl Sci 42:2095-2103
41. Jorio A, Parenteau M, Aubin M, Carlone C, Khanna SM, Gerdes Jr JW (1994) A mobility study of the radiation induced order effect in gallium arsenide. IEEE Trans Nucl Sci 41:1937-1944
42. Khanna SM, Houdayer A, Jorio A, Carlone C, Parenteau M, Gerdes Jr JW (1996) Nuclear radiation displacement damage prediction in gallium arsenide through low temperature photoluminescence measurements. IEEE Trans Nucl Sci 43:2601-2608
43. Hashizume T, Nagabuchi H (1989) Characterisation of deep electron states in LEC grown GaAs material. Semicond Sci Technol 4:427-434
44. Hashizume T, Hasegawa H (1990) Variation of deep electron traps created by γ irradiation of GaAs. J Appl Phys 68:4598-4603

45. Lai ST, Nener BD, Faraone L, Nassibian AG, Hotchkis MAC (1993) Characterization of deep-level defects in GaAs irradiated by 1 MeV electrons. J Appl Phys 73:640-647
46. Lai ST, Nener BD (1994) Annealing behavior of deep-level defects in 1 MeV electron irradiated GaAs. J Appl Phys 75:2354-2357
47. Lai ST, Alexiev D, Nener BD (1995) Comparison between deep level defecs in GaAs induced by gamma, 1 MeV electron, and neutron irradiation. J Appl Phys 78:3686-3690
48. Ikoma T, Michizuki Y (1985) Identification of the "EL2 Family" midgap levels in GaAs. Jpn J Appl Phys 24:L935-L937
49. Hausmann H, Ehrhart P (1995) Ga-vacancies and As_{Ga}-antisites in electron irradiated GaAs. Mat Sci Forum 196-201:1255-1260
50. Yousefi GH, Webb JB, Rousina R, Khanna SM (1995) Electron irradiation induced defects and Schottky diode characteristics for metalorganic vapor phase epitaxy and molecular beam epitaxial n-GaAs. J Electron Mat 24:15-20
51. Stiévenard D, Bourgoin JC (1986) Impurity defect interactions in GaAs. J Appl Phys 59:743-747
52. Khanna SM, Rejeb C, Jorio A, Parenteau M, Carlone C, Gerdes Jr JW (1993) Electron and neutron radiation-induced order effect in gallium arsenide. IEEE Trans Nucl Sci 40:1350-1359
53. Jorio A, Rejeb C, Parenteau M, Carlone C, Khanna SM (1993) Radiation induced carrier enhancements and intrinsic defect formation in n-GaAs. J Appl Phys 74:2310-2317
54. Makram-Ebeid S, Boher P (1988) Defect pairs and clusters related to the EL2 centre in GaAs. Revue Phys Appl 23:847-862
55. Brudnyi VN, Gradoboev AV, Peshev VV (1999) The broad midgap deep-level transient spectroscopy band in proton (65 MeV) and fast neutron-irradiated n-GaAs. Phys Stat Sol (b) 212:229-239
56. Barnes CE, Zipperian TE, Dawson LR (1985) Neutron-induced trapping levels in aluminum gallium arsenide. J Electron Mater 14:95-118
57. Tan HH, Williams JS, Jagadish C (1995) Characterization of deep levels and carrier compensation created by proton irradiation in undoped GaAs. J Appl Phys 78:1481-1487
58. Jorio A, Carlone C, Parenteau M, Aktik C, Rowell NL (1996) Formation of EL2, As_{Ga} and U band in irradiated GaAs: Effects of annealing. J Appl Phys 80:1364-1369
59. Lai ST, Nener BD, Alexiev D, Faraone L, Ku TC, Dytlewski N (1995) Comparison of neutron and electron irradiation on the EL2 defect in GaAs. J Appl Phys 77:2985-2988
60. Webb JB, Yousefi GH, Khanna SM (1993) Structure of the U band in neutron irradiated MOVPE grown n-GaAs epilayers. Electron Lett 29:1994-1996
61. Goltzene A, Meyer B, Schwab C (1989) Persistent photoquenching and anion antisite defects in neutron-irradiated GaAs. Appl Phys Lett 54:907-909
62. Greene PD (1979) Transmutation doping of GaAs by thermal neutrons. Solid State Commun 32:325-326
63. Low TS, Kim MH, Lee B, Skromme BJ, Lepkowski TR, Stillman GE (1985) Neutron transmutation doping of high purity GaAs. J Electron Materials 14:477-511
64. Satoh M, Yokoyama K, Kuriyama K (1990) The role of Ga antisite defect in the activation process of transmuted impurities in neutron-transmutation-doped semi-insulating GaAs. J Appl Phys 68:363-366
65. Satoh M, Kuriyama K (1989) Quenching phenomenon of hopping conduction in neutron-transmutation-doped semi-insulating GaAs. Phys Rev B 40:3473-3475

66. Kato Y, Shimada T, Shiraki Y, Komatsubara KF (1974) Electrical conductivity of disordered layers in GaAs crystal produced by ion implantation. J Appl Phys 45:1044-1049
67. Satoh M, Kuriyama K, Yagahi M, Iwamura K, Kim C, Kawakubo T, Yoneda K, Kimura I (1987) Influence of fast neutrons on electrical properties in neutron transmutation doped GaAs: New annealing stage. Appl Phys Lett 50:580-582
68. Manasreh MO, Fischer DW (1989) Infrared absorption properties of the EL2 and the isolated As_{Ga} defects in neutron-transmutation-doped GaAs: Generation of an EL2-like defect. Phys Rev B 39:3239-3249
69. Manasreh MO, Mudare SM (1989) Photoluminescence bands of deep centres in neutron-transmutation-doped GaAs. Semicond Sci Technol 4:435-438
70. Satoh M, Kuriyama K, Makita Y (1989) Photoluminescence study of the annealing behavior of transmuted impurities in neutron-transmutation-doped semi-insulating GaAs. J Appl Phys 65:2248-2253
71. Kuriyama K, Yokoyama K, Tomizawa K (1991) Annealing behavior of Ga and Ge antisite defects in neutron-transmutation-doped semi-insulating GaAs. J Appl Phys 70:7315-7317
72. Zakharenkov LF, Kozlovskii VV, Shustrov BA (1990) Transmutation doping of indium phosphide and gallium arsenide due to protons and α-particles. Phys Stat Sol (a) 117:85-90
73. Carlone C, Parenteau M, Houdayer A, Hinrichsen P, Vincent J (1997) Photoluminescence study of gallium vacancy defects in gallium arsenide irradiated by relativistic protons. IEEE Trans Nucl Sci 44:1856-1861
74. Samitier J, Morante JR, Giraudet L, Gourrier S (1986) Optical behavior of the U band in relation to EL2 and EL6 levels in boron-implanted GaAs. Appl Phys Lett 48:1138-1140
75. Morante JR, Pérez-Rodríguez A, Samitier J, Romano-Rodríguez A (1991) On the artificial creation of the EL2 center by means of boron implantation in gallium arsenide. J Appl Phys 70:4202-4210
76. Yokota K, Kuchii H, Nakamura K, Sakaguchi M, Takano H, Ando Y (2000) EL2, EL3, and EL6 defects in GaAs highly implanted with sulfur. J Appl Phys 88:5017-5021
77. Allsopp DWE, Peaker AR (1986) Damage related deep electron levels in ion implanted GaAs. Solid-State Electron 29:467-470
78. Eden RC, Welch BM, Zucca R (1978) Planar GaAs IC technology: Applications for digital LSI. IEEE J Solid-State Circ 13:419-425
79. D'Avanzo DC (1982) Proton isolation for GaAs integrated circuits. IEEE Trans Electron Devices 29:1051-1059
80. Short KT, Pearton SJ (1988) Implant isolation of GaAs. J Electrochem Soc 135:2835-2840
81. De Souza JP, Danilov I, Boudinov H (1997) Thermal stability of the electrical isolation in n-type GaAs: Effects of damage and carrier concentrations. Nucl Instrum Methods in Phys Research B 122:51-54
82. Eisen FH, Bachem K, Klausman E, Koehler K, Haddad R (1992) Ion irradiation damage in n-type GaAs in comparison with its electron irradiation damage. J Appl Phys 72:5593-5601
83. Arpatzanis N, Vlastou R, Konstantinidis G, Assmann W, Papastamatiou M, Gazis E, Papaioannou GJ (1998) Ion irradiation induced defects in epitaxial GaAs layers. Solid-State Electron 42:277-282

84. Boudinov H, Kucheyev SO, Williams JS, Jagadish C, Li G (2001) Electrical isolation of GaN by MeV ion irradiation. Appl Phys Lett 78:943-945
85. de Souza JP, Danilov I, Boudinov H (1998) Electrical isolation of n-type GaAs layers by proton bombardment: Effects of the irradiation temperature. J Appl Phys 84:4757-4760
86. de Souza JP, Danilov I, Boudinov H (1997) Thermal stability of the electrical isolation in n-type gallium arsenide layers irradiated with H, He, and B ions. J Appl Phys 81:650-655
87. de Souza JP, Danilov I, Boudinov H (1996) Electrical isolation in GaAs by light ion irradiation: The role of antisite defects. Appl Phys Lett 68:535-537
88. Pearton SJ, Hobson WS, Von Neida AE, Haegel NM, Jones KS, Morris N, Sealy BJ (1990) Implant activation and redistribution in $Al_xGa_{1-x}As$. J Appl Phys 67:2396-2409
89. Comedi D, Zhao J, Jankowska K, Thompson DA, Simmons J (1993) High-resistivity regions in n-type InGaAsP produced by $^4He^+$ ion bombardment at 80 and 300 K. Appl Phys Lett 63:2126-2128
90. Polyakov AY, Chelniy AA, Govorkov AV, Smirnov NB, Milnes AG, Pearton SJ, Wilson RG, Balmashnov AA, Aluev AN, Markov AV (1995) Effects of proton implantation and hydrogen plasma passivation on electrical properties of InGaAlP and InGaP. Solid-State Electron 38:1131-1135
91. Danilov I, Pataro LL, de Castro MPP, Frateschi NC (2000) Electrical isolation and transparency in ion-irradiated p-InGaP/GaAs/InGaAs structures. J Appl Phys 88:7354-7356
92. Pearton SJ, Kuo JM, Ren F, Katz A, Perley AP (1991) Ion implantation doping and isolation of $In_{0.5}Ga_{0.5}P$. Appl Phys Lett 59:1467-1469
93. Binari SC, Dietrich HB, Kelner G, Rowland LB, Doverspike K, Wickenden DK (1995) H, He, and N implant isolation of n-type GaN. J Appl Phys 78:3008-3011
94. Pearton SJ, Wilson RG, Zavada JM, Han J, Shul RJ (1998) Thermal stability of ^2H-implanted n- and p-type GaN. Appl Phys Lett 73:1877-1879
95. Uzan-Saguy C, Salzman J, Kalish R, Richter V, Tish U, Zamir S, Prawer S (1999) Electrical isolation of GaN by ion implantation damage: Experiment and model. Appl Phys Lett 74:2441-2443
96. Cao XA, Pearton SJ, Dang GT, Zhang AP, Ren F, Wilson RG, Van Hove JM (2000) Creation of high resistivity GaN by implantation of Ti, O, Fe, or Cr. J Appl Phys 87:1091-1095
97. Pearton SJ, Abernathy CR, Panish MB, Hamm RA, Lunardi LM (1989) Implant-induced high-resistivity regions in InP and InGaAs. J Appl Phys 66:656-662
98. Sargunas V, Thompson DA, Simmons JG (1995) High resistivity in n-type InP by He^+ bombardment at 300 and 60 K. Solid-State Electron 38:75-81
99. Ghandhi SK, Kwan P, Bhat KN, Borrego JM (1982) Ion beam damage effects during the low energy cleaning of GaAs. IEEE Electron Device Lett 3:48-50
100. Pang SW, Geis MW, Efremow NN, Lincoln GA (1985) Effects of ion species and adsorbed gas on dry etching induced damage in GaAs. J Vac Sci Technol B 3:398-401
101. Neffati T, Lu GN, Barret C (1988) Electrical characterization of Schottky diodes on very low energy ion-etched GaAs surfaces. Solid-State Electron 31:1335-1342
102. Yuba Y, Ishida T, Gamo K, Namba S (1988) Characterization of ion beam etching induced defects in GaAs. J Vac Sci Technol B 6:253-256
103. Meyer E, Heymann G (1989) Defect generation by Schottky contacts on n-GaAs. J Vac Sci Technol B 7:491-496

104. He L, Anderson WA (1992) A study of defects induced in GaAs by plasma etching. Solid-State Electron 35:151-156
105. Auret FD, Goodman SA, Myburg G, Meyer WE (1992) Electrical characteristics of Ar-ion sputter induced defects in epitaxially grown n-GaAs. J Vac Sci Technol B 10:2366-2370
106. Auret FD, Goodman SA (1996) Electronic properties of defects introduced during low-energy He-ion bombardment of epitaxially grown n-GaAs. Appl Phys Lett 68:3275-3277
107. Christou A, Anderson Jr WT, Day HM (1985) Trap distribution in gold-refractory/GaAs Schottky barriers. Solid-State Electron 28:329-338
108. Behle AF, Zuleeg R (1972) Fast neutron tolerance of GaAs JFET's operating in the hot electron range. IEEE Trans Electron Devices 19:993-995
109. Knudson AR, Campbell AB, Stapor WJ, Shapiro P, Mueller GP, Zuleeg R (1985) Radiation damage effects of electrons and H, He, O, Cl and Cu ions on GaAs JFETS. IEEE Trans Nucl Sci 32:4388-4392
110. Campbell AB, Knudson AR, Stapor WJ, Summers G, Xapsos MA, Jessee M, Palmer T, Zuleeg R, Dale CJ (1986) Particle damage effects in GaAs JFET test structures. IEEE Trans Nucl Sci 33:1435-1441
111. Burke EA, Dale CJ, Campbell AB, Summers GP, Palmer T, Zuleeg R (1987) Energy dependence of proton-induced displacement damage in gallium arsenide. IEEE Trans Nucl Sci 34:1220-1226
112. Summers GP, Burke EA, Xapsos MA, Dale CJ, Marshall PW, Petersen EL (1988) Displacement damage in GaAs structures. IEEE Trans Nucl Sci 35:1221-1226
113. Williams JG, Patel JU, Ougouag AM, Yang S-Y (1991) Carrier removal and changes in electrical properties of neutron irradiated GaAs. J Appl Phys 70:4931-4937
114. Nishigushi M, Hashinaga T, Nishizawa H, Hayashi H, Okazaki N, Kitagawa M, Fujino T (1990) Radiation tolerant GaAs MESFET with a highly-doped thin active layer grown by OMVPE. IEEE Trans Nucl Sci 37:2071-2075
115. Aukerman LW, Graft RD (1962) Annealing of electron-irradiated GaAs. Phys Rev 127:1576-1583
116. Parenteau M, Carlone C, Khanna SM (1992) Damage coefficient associated with free exciton lifetime in GaAs irradiated with neutrons and electrons. J Appl Phys 71:3747-3753
117. Ghamlouch H, Aubin M, Carlone C, Khanna SM (1993) Electron irradiation of GaAs: Improvement of transport properties and observation of DX-like centers at ambient pressure. J Appl Phys 74:4357-4362
118. Messenger SR, Burke EA, Summers GP, Xapsos MA, Walters RJ, Jackson EM, Weaver BD (1999) Nonionizing energy loss (NIEL) for heavy ions. IEEE Trans Nucl Sci 46:1595-1602
119. Barry AL, Maxseiner R, Wojcik R, Briere MA, Braunig D (1990) An improved displacement damage monitor. IEEE Trans Nucl Sci 37:1726-1731
120. Moss SC, Halle LF, Marvin DC (1995) Effects of electron beam irradiation on transient photoluminescence measurements of GaAs and AlGaAs double heterostructures. IEEE Trans Nucl Sci 42:2058-2065
121. Parenteau M, Carlone C, Morris D, Khanna SM (1997) Time-resolved spectroscopy of irradiated n-GaAs. IEEE Trans Nucl Sci 44:1849-1855
122. Yamaguchi M, Ando K (1988) Mechanism for radiation resistance of InP solar cells. J Appl Phys 63:5555-5562

123. Loo RY, Kamath GS, Li SS (1990) Radiation damage and annealing in GaAs solar cells. IEEE Trans Electron Devices 37:485-497
124. Johnson MB, McGill TC, Paulter NG (1989) Carrier lifetimes in ion-damaged GaAs. Appl Phys Lett 54:2424-2426
125. Lambsdorff M, Kuhl J, Rosenzweig J, Axmann A, Schneider J (1991) Subpicosecond carrier lifetimes in radiation-damaged GaAs. Appl Phys Lett 58:1881-1883
126. Jagadish C, Tan HH, Krotkus A, Marcinkevicius S, Korona KP, Kaminska M (1996) Ultrafast carrier trapping in high energy ion implanted gallium arsenide. Appl Phys Lett 68:2225-2227
127. Carlone C, Parenteau M, Houdayer A, Hinrichsen P, Vincent J (1997) Photoluminescence study of gallium vacancy defects in gallium arsenide irradiated by relativistic protons. IEEE Trans Nucl Sci 44:1856-1861
128. Carlone C, Parenteau M, Khanna SM (1998) Gigaelectron-volt heavy ion irradiation of gallium arsenide. J Appl Phys 83:5164-5170
129. Chen C-L, Wise KD (1982) Gate formation in GaAs MESFET's using ion-beam etching technology. IEEE Trans Electron Devices 29:1522-1529
130. Ejimanya JI (1986) Current-voltage characteristics of proton-bombarded Au-GaAs contacts. Solid-State Electron 29:841-844
131. Myburg G, Auret D (1992) Influence of the electron beam evaporation rate of Pt and the semiconductor carrier density on the characteristics of Pt/n-GaAs Schottky contacts. J Appl Phys 71:6172-6176
132. Thurzo I, Hrubčín L, Bartoš J, Pinčik E (1993) Current-voltage characteristics and charge DLTS spectra of proton-bombarded Schottky diodes on semi-insulating GaAs. Nucl Instr Meth in Phys Research B 83:145-152
133. Arulkumaran S, Arokiaraj J, Dharmarasu N, Kumar J, Magudapathy P, Nair KGM (1997) Current-voltage characteristics of low energy proton and alpha particle irradiated Au and Ag/n-GaAs Schottky barrier diodes. Solid-State Electron 41:802-805
134. Jayavel P, Asokan K, Kanjilal D, Kumar J (2000) Investigations on the annealing behavior of high-energy carbon irradiated Au/n-GaAs Schottky barrier diodes. Mat Sci Semicond Processing 3:195-199
135. McGregor DS, Knoll GF, Eisen Y, Brake R (1992) Development of bulk GaAs room temperature radiation detectors. IEEE Trans Nucl Sci 39:1226-1236
136. Nava F, Alietti M, Canali C, Cavallini A, Chiossi C, del Papa C, Re V, Lanzieri C (1996) Performances of SI GaAs detectors fabricated with implanted ohmic contacts. IEEE Trans Nucl Sci 43:1130-1136
137. Bates R, Da Via C, O'Shea V, Pickford A, Raine C, Smith K (1997) Radiation-induced damage in GaAs particle detectors. IEEE Trans Nucl Sci 44:1705-1707
138. Ferrini R, Galli M, Guizzetti G, Patrini M, Nava F, Canali C, Vanni P (1997) Optical evaluation of the ionized EL2 fraction in proton (24 GeV) irradiated semi-insulating GaAs. Appl Phys Lett 71:3084-3086
139. Xianbi X, Wenhui D, Xiulan C, Xianbo L (1998) Electron irradiation and thermal annealing effects on GaAs solar cells. Solar Energy Materials & Solar Cells 55:313-322
140. Summers GP, Messenger SR, Burke EA, Xapsos MA, Walters RJ (1997) Low energy proton-induced displacement damage in shielded GaAs solar cells in space. Appl Phys Lett 71:832-834
141. Li SS, Wang W-L, Lai P-W, Loo RY, Kamath GS, Knechtli RC (1980) Deep-level defects, recombination mechanisms, and their correlation to the performance of low-

energy proton-irradiated AlGaAs-GaAs solar cells. IEEE Trans Electron Devices 27:857-864
142. Li SS, Loo RY (1991) Deep-level defects and numerical simulation of radiation damage in GaAs solar cells. Solar Cells 31:349-377
143. Debney BT (1979) A theoretical evaluation and optimization of the radiation resistance of gallium arsenide solar-cell structures. J Appl Phys 50:7210-7219
144. Wilson JW, Walker GH, Outlaw RA (1984) Proton damage in GaAs solar cells. IEEE Trans Electron Devices 31:421-422
145. Yeh CS, Li SS, Loo RY (1987) Computer simulations of the proton irradiated (AlGa)As-GaAs solar cells. J Appl Phys 62:4611-4616
146. Moreno EG, Alcubilla R, Prat L, Castañer L (1988) Radiation damage evaluation on AlGaAs/GaAs solar cells. IEEE Trans Nucl Sci 35:1067-1071
147. Augustine G, Rohatgi A, Jokerst NM (1992) Base doping optimization for radiation-hard Si, GaAs, and InP solar cells. IEEE Trans Electron Devices 39:2395-2400

Chapter 5

1. Schrimpf RD (1996) Recent advances in understanding total-dose effects in bipolar transistors. IEEE Trans Nucl Sci 43:787-796
2. Johnston AH, Plaag RE (1987) Models for total dose degradation of linear integrated circuits. IEEE Trans Nucl Sci 34:1474-1480
3. Summers GP, Burke EA, Dale CJ, Wolicki EA, Marshall PW, Gehlhausen MA (1987) Correlation of particle-induced displacement damage in silicon. IEEE Trans Nucl Sci 34:1134-1139
4. Dale CJ, Marshall PW, Summers GP, Wolicki EA (1989) Displacement damage equivalent to dose in silicon devices. Appl Phys Lett 54:451-453
5. Dale C, Marshall P (1991) Displacement damage in Si imagers for space applications. In: Proc SPIE Charged-Coupled Devices and Solid State Optical Sensors II, vol 1447, pp 70-86
6. Enlow EW, Pease RL, Combs WE, Schrimpf RD, Nowlin RN (1991) Response of advanced bipolar processes to ionizing radiation. IEEE Trans Nucl Sci 38:1342-1351
7. Nowlin RN, Enlow EW, Schrimpf RD, Combs WE (1992) Trends in the total-dose response of modern bipolar transistors. IEEE Trans Nucl Sci 39:2026-2035
8. Nowlin RN, Fleetwood DM, Schrimpf RD, Pease RL, Combs WE (1993) Hardness-assurance and testing issues for bipolar/BiCMOS devices. IEEE Trans Nucl Sci 40:1686-1693
9. Wei A, Kosier SL, Schrimpf RD, Combs WE, DeLaus M (1995) Excess collector current due to an oxide-trapped-charge-induced emitter in irradiated NPN BJTs. IEEE Trans Electron Devices 42:923-926
10. Kosier SL, Combs WE, Wei A, Schrimpf RD, Fleetwood DM, DeLaus M, Pease RL (1994) Bounding total-dose response of modern bipolar transistors. IEEE Trans Nucl Sci 41:1864-1870
11. Wei A, Kosier SL, Schrimpf RD, Fleetwood DM, Combs WE (1994) Dose-rate effects on radiation-induced bipolar junction transistor gain degradation. Appl Phys Lett 65:1918-1920

12. Nowlin RN, Fleetwood DM, Schrimpf RD (1994) Saturation of the dose-rate response of bipolar transistors below 10 rad(SiO_2)/s: Implications for hardness assurance. IEEE Trans Nucl Sci 41:2637-2641
13. Shaneyfelt MR, Schwank JR, Witczak SC, Fleetwood DM, Pease RL, Winokur PS, Riewe LC, Hash GL (2000) Thermal-stress effects and enhanced low dose rate sensitivity in linear bipolar ICs. IEEE Trans Nucl Sci 47:2539-2545
14. Pease RL, Cohn LM, Fleetwood DM, Gehlhausen MA, Turflinger TL, Brown DB, Johnston AH (1997) A proposed hardness assurance test methodology for bipolar linear circuits and devices in a space ionizing radiation environment. IEEE Trans Nucl Sci 44:1981-1988
15. Fleetwood DM, Kosier SL, Nowlin RN, Schrimpf RD, Reber Jr RA, DeLaus M, Winokur PS, Wei A, Combs WE, Pease RL (1994) Physical mechanisms contributing to enhanced bipolar gain degradation at low dose rates. IEEE Trans Nucl Sci 41:1871-1883
16. Fleetwood DM, Riewe LC, Schwank JR, Witczak SC, Schrimpf RD (1996) Radiation effects at low electric fields in thermal, SIMOX, and bipolar-base oxides. IEEE Trans Nucl Sci 43:2537-2546
17. Freitag RK, Brown DB (1998) Study of low dose rate radiation effects on commercial linear bipolar ICs. IEEE Trans Nucl Sci 45:2649-2658
18. Belyakov VV, Pershenkov VS, Shalnov AV, Shvetzov-Shilovsky IN (1995) Use of MOS structures for the investigation of low-dose-rate effects in bipolar transistors. IEEE Trans Nucl Sci 42:1660-1666
19. Pershenkov VS, Maslov VB, Cherepko SV, Shvetzov-Shilovsky IN, Belyakov VV, Sogoyan AV, Rusanovsky VI, Ulimov VN, Emelianov VV, Nasibullin VS (1997) The effect of emitter junction bias on the low dose-rate radiation response of bipolar devices. IEEE Trans Nucl Sci 44:1840-1848
20. Witczak SC, Galloway KF, Schrimpf RD, Suehle JS (1995) Relaxation of Si-SiO_2 interfacial stress in bipolar screen oxides due to ionizing radiation. IEEE Trans Nucl Sci 42:1689-1697
21. Kosier SL, Schrimpf RD, Nowlin RN, Fleetwood DM, DeLaus M, Pease RL, Combs WE, Wei A, Chai F (1993) Charge separation for bipolar transistors. IEEE Trans Nucl Sci 40:1276-1285
22. Graves RJ, Schmidt DM, Kosier SL, Wei A, Schrimpf RD, Galloway KF (1994) Visualization of ionizing-radiation and hot-carrier stress response of polysilicon emitter BJTs. In: IEDM Tech Dig, The IEEE, New York, pp 233-236
23. Kosier SL, Wei A, Schrimpf RD, Fleetwood DM, DeLaus MD, Pease RL, Combs WE (1995) Physically based comparison of hot-carrier-induced and ionizing-radiation-induced degradation in BJTs. IEEE Trans Electron Devices 42:436-443
24. Roldán JM, Niu G, Ansley WE, Cressler JD, Clark SD, Ahlgren DC (1998) An investigation of the spatial location of proton-induced traps in SiGe HBTs. IEEE Trans Nucl Sci 45:2424-2429
25. Jenkins KA, Cressler JD (1988) Electron beam damage of advanced silicon bipolar transistors and circuits. In: IEDM Tech Dig, The IEEE, New York, pp 30-33
26. Jenkins KA, Cressler JD, Warnock JD (1991) Use of electron-beam irradiation to study performance degradation of bipolar transistors after reverse-bias stress. In: IEDM Tech Dig, The IEEE, New York, pp 873-876
27. Pease RL, Kosier SL, Schrimpf RD, Combs WE, Davey M, DeLaus M, Fleetwood DM (1994) Comparison of hot-carrier and radiation induced increases in base current in bipolar transistors. IEEE Trans Nucl Sci 41:2567-2573

28. Johnston AH, Lee CI, Rax BG (1996) Enhanced damage in bipolar devices at low dose rates: Effects at very low dose rates. IEEE Trans Nucl Sci 43:3049-3059
29. Jenkins KA (1989) Frequency response of bipolar junction transistors after electron-beam irradiation. IEEE Trans Electron Devices 36:1722-1724
30. Jenkins KA, Cressler JD (1991) Electron-beam damage of self-aligned silicon bipolar transistors and circuits. IEEE Trans Electron Devices 38:1450-1457
31. Simoen E, Decoutere S, Merron B, Deferm L, Claeys C, Berger G, Ryckewaert G, Ohyama H, Sunaga H (1998) High-energy particle irradiation effects in 0.5 μm BiCMOS polysilicon emitter bipolar junction transistors. In: Barbottin G, Dressendorfer P (eds) Proc RADECS '97, The IEEE, New York, pp 102-107
32. Hiemstra DM (1999) Dose rate dependence of the current noise performance of an ultra-low noise precision bipolar operational amplifier. IEEE Trans Nucl Sci 46:1674-1679
33. Witczak SC, Lacoe RC, Mayer DC, Fleetwood DM, Schrimpf RD, Galloway KF (1998) Space charge limited degradation of bipolar oxides at low electric fields. IEEE Trans Electron Devices 45:2339-2351
34. Witczak SC, Lacoe RC, Shaneyfelt MR, Mayer DC, Schwank JR, Winokur PS (2000) Implications of radiation-induced dopant deactivation for npn bipolar junction transistors. IEEE Trans Nucl Sci 47:2281-2288
35. Witczak SC, Winokur PS, Lacoe RC, Mayer DC (2000) Charge separation technique for metal-oxide-silicon capacitors in the presence of hydrogen deactivated dopants. J Appl Phys 87:8206-8208
36. Beaucour J, Carrière T, Gach A, Laxague D, Poirot P (1994) Total dose effects on negative voltage regulator. IEEE Trans Nucl Sci 41:2420-2426
37. Johnston AH, Swift GM, Rax BG (1994) Total dose effects in conventional bipolar transistors and linear integrated circuits. IEEE Trans Nucl Sci 41:2427-2436
38. McClure S, Pease RL, Will W, Perry G (1994) Dependence of total dose response of bipolar linear microcircuits on applied dose rate. IEEE Trans Nucl Sci 41:2544-2549
39. Johnston AH, Rax BG, Lee CI (1995) Enhanced damage in linear bipolar integrated circuits at low dose rate. IEEE Trans Nucl Sci 42:1650-1659
40. Barnaby H, Tausch HJ, Turfler R, Cole P, Baker P, Pease RL (1996) Analysis of bipolar linear circuit response mechanisms for high and low dose rate total dose irradiations. IEEE Trans Nucl Sci 43:3040-3048
41. Schmidt DM, Fleetwood DM, Schrimpf RD, Pease RL, Graves RJ, Johnson GH, Galloway KF, Combs WE (1995) Comparison of ionizing-radiation-induced gain degradation in lateral, substrate, and vertical PNP BJTs. IEEE Trans Nucl Sci 41:1541-1549
42. Schrimpf RD, Graves RJ, Schmidt DM, Fleetwood DM, Pease RL, Combs WE, DeLaus M (1995) Hardness-assurance issues for lateral PNP bipolar junction transistors. IEEE Trans Nucl Sci 42:1641-1649
43. Schmidt DM, Wu A, Schrimpf RD, Fleetwood DM, Pease RL (1996) Modeling ionizing radiation induced gain degradation of the lateral PNP bipolar junction transistor. IEEE Trans Nucl Sci 43:3032-3039
44. Wu A, Schrimpf RD, Barnaby HJ, Fleetwood DM, Pease RL, Kosier SL (1997) Radiation-induced gain degradation in lateral PNP BJTs with lightly and heavily doped emitters. IEEE Trans Nucl Sci 44:1914-1921
45. Barnaby HJ, Cirba C, Schrimpf RD, Kosier S, Fouillat P, Montagner X (1998) Minimizing gain degradation in lateral pnp bipolar junction transistors using gate control. IEEE Trans Nucl Sci 46:1652-1659

46. Babcock JA, Cressler JD, Vempati LS, Clark SD, Jaeger RC, Harame DL (1995) Ionizing radiation tolerance and low-frequency noise degradation in UHV/CVD SiGe HBTs. IEEE Electron Device Lett 16:351-353
47. Babcock JA, Cressler JD, Vempati LS, Clark SD, Jaeger RC, Harame DL (1995) Ionizing radiation tolerance of high-performance SiGe HBTs grown by UHV/CVD. IEEE Trans Nucl Sci 42:1558-1566
48. Banerjee G, Niu G, Cressler JD, Clark SD, Palmer MJ, Ahlgren DC (1999) Anomalous dose rate effects in gamma irradiated SiGe heterojunction bipolar transistors. IEEE Trans Nucl Sci 46:1620-1626
49. Roldán JM, Ansley WE, Cressler JD, Clark SD, Nguyen-Ngoc D (1997) Neutron radiation tolerance of advanced UHV/CVD SiGe HBT BiCMOS technology. IEEE Trans Nucl Sci 44:1965-1973
50. Cressler JD, Hamilton MC, Mullinax GS, Li Y, Niu G, Marshall CJ, Marshall PW, Kim HS, Palmer MJ, Joseph AJ, Freeman G (2000) The effects of proton irradiation on the lateral and vertical scaling of UHV/CVD SiGe HBT BiCMOS technology. IEEE Trans Nucl Sci 47:2515-2520
51. Ohyama H, Vanhellemont J, Takami Y, Hayama K, Sunaga H, Poortmans J, Caymax M (1995) Degradation of $Si_{1-x}Ge_x$ epitaxial heterojunction bipolar transistors by 1-MeV fast neutrons. IEEE Trans Nucl Sci 42:1550-1557
52. Zhang S, Niu G, Cressler JD, Clark SD, Ahlgren DC (1999) The effects of proton irradiation on the RF performance of SiGe HBTs. IEEE Trans Nucl Sci 46:1716-1721
53. Zhang S, Niu G, Cressler JD, Mathew SJ, Gogineni U, Clark SD, Zampardi P, Pierson RL (2000) A comparison of the effects of gamma irradiation on SiGe HBT and GaAs HBT technologies. IEEE Trans Nucl Sci 47:2521-2527
54. Niu G, Juraver JB, Borgarino M, Jin Z, Cressler JD, Plana R, Llopis O, Mathew S, Zhang S, Clark S, Joseph AJ (2001) Impact of gamma irradiation on the RF phase noise capability of UHV/CVD SiGe HBTs. Solid-State Electron 45:107-112

Chapter 6

1. Walters M, Reisman A (1991) Radiation-induced neutral electron trap generation in electrically biased insulated gate field effect transistor gate insulators. J Electrochem Soc 138:2756-2762
2. Ceschia M, Paccagnella A, Cester A, Scarpa A, Ghidini G (1998) Radiation induced leakage current and stress induced leakage current in ultra-thin gate oxides. IEEE Trans Nucl Sci 45:2375-2382
3. Saks NS, Ancona MG, Modolo JA (1986) Generation of interface states by ionizing radiation in very thin oxides. IEEE Trans Nucl Sci 33:1185-1190
4. Ma TP, Dressendorfer PV (1989) Ionizing radiation effects in MOS devices and circuits. Wiley, New York.
5. Oldham TR (1999) Ionizing radiation effects in MOS oxides. In: Sah CT (ed), World Scientific.
6. Share S, Martin RA (1975) Effects of ionizing radiation on short-channel thin-oxide (200-Å) MOSFETs. IEEE Trans Electron Devices 22:619-620
7. Huang JST, Schrankler JW (1985) Flat-band voltage dependence on channel length in short-channel threshold model. IEEE Trans Electron Devices 32:1001-1002

8. Schrankler JW, Reich RK, Holt MS, Ju DH, Huang JST, Kirchner GD, Hughes HL (1985) CMOS scaling implications for total dose radiation. IEEE Trans Nucl Sci 32: 3988-3990
9. Chen JY, Patterson DO, Martin R (1981) Radiation hardness on submicron NMOS. IEEE Trans Nucl Sci 28:4314-4317
10. Bhattacharya PK, Reisman A (1990) A study of X-ray damage effects on the short channel behavior of IGFETs. J Electron Materials 19:727-732
11. Scarpulla J, Amram AL, Gin VW, Morse TC, Nakamura KT (1992) Gate size dependence of the radiation-produced changes in threshold voltage, mobility, and interface state density in bulk CMOS. IEEE Trans Nucl Sci 39:1990-1997
12. Shaneyfelt MR, Fleetwood DM, Winokur PS, Schwank JR, Meisenheimer TL (1993) Effects of device scaling and geometry on MOS radiation hardness assurance. IEEE Trans Nucl Sci 40:1678-1685
13. Djezzar B, Smatti A, Amrouche A, Kechouane M (2000) Channel-length impact on radiation-induced threshold-voltage shift in n-MOSFET's devices at low gamma rays radiation doses. IEEE Trans Nucl Sci 47:1872-1878
14. Paccagnella A, Ceschia M, Verzellesi G, Dalla Betta GF, Bellutti P, Fuochi PG, Soncini G (1996) Forward and reverse characteristics of irradiated MOSFETs. IEEE Trans Nucl Sci 43:797-804
15. Vasudevan V, Vasi J (1994) A two-dimensional numerical simulation of oxide charge buildup in MOS transistors due to radiation. IEEE Trans Electron Devices 41:383-390
16. Fleetwood DM (1992) Border traps in MOS devices. IEEE Trans Nucl Sci 39:269-271
17. Peckerar MC, Brown DB, Lin HC, Ma DI (1983) Modeling total dose effects in narrow-channel devices. IEEE Trans Electron Devices 30:1159-1164
18. Chen W, Balasinski A, Ma T-P (1991) Lateral distribution of radiation-induced damage in MOSFETs. IEEE Trans Nucl Sci 38:1124-1129
19. Balasinski A, Ma T-P (1992) Ionizing radiation damage near CMOS transistor channel edges. IEEE Trans Nucl Sci 39:1998-2003
20. Balasinski A, Ma T-P (1993) Impact of radiation-induced nonuniform damage near MOSFET junctions. IEEE Trans Nucl Sci 40:1286-1292
21. Pantelakis DC, Hemmenway DF, van Vonno N, Sanders TJ (1991) Freeze-out characteristics of radiation hardened n^+ polysilicon gate CMOS transistors. IEEE Trans Nucl Sci 38:1289-1296
22. Acovic A, Hsu CC-H, Hsia L-C, Balasinski A, Ma T-P (1992) Effects of X-ray irradiation on GIDL in MOSFETs. IEEE Electron Device Lett 13:189-191
23. Das NC, Nathan V, Tallon R, Maier RJ (1992) Radiation effects on gate induced drain leakage current in metal oxide semiconductor transistors. J Appl Phys 72:4958-4962
24. Gdula RA (1979) The effects of processing on radiation damage in SiO_2. IEEE Trans Electron Devices 26:644-647
25. Hsu CC-H, Wang LK, Wordeman MR, Ning TH (1989) Hot-electron-induced instability in 0.5-μm p-channel MOSFETs patterned using synchrotron X-ray lithography. IEEE Electron Device Lett 10:327-329
26. Miller PA, Fleetwood DM, Schubert WK (1991) Damage due to electron, ion, and x-ray lithography. J Appl Phys 69:488-494
27. Lelis AJ, Oldham TR (1993) Reliability effects of X-ray lithography exposures on submicron-channel MOSFETs. IEEE Trans Nucl Sci 40:1367-1371
28. Yue J, Lo E, Flanery M (1997) Total dose radiation response of plasma-damaged NMOS devices. IEEE Electron Device Lett 18:532-534

29. Aubuchon KG (1971) Radiation hardening of PMOS devices by optimization of thermal SiO_2 gate insulator. IEEE Trans Nucl Sci 18:117-125
30. Dawes, Jr WR, Derbenwick GF, Gregory BL (1976) Process technology for radiation-hardened CMOS integrated circuits. IEEE J Solid State Circuits SC11:459-465
31. Winokur PS, Erret EB, Fleetwood DM, Dressendorfer PV, Turpin DC (1985) Optimizing and controlling the radiation hardness of Si-gate CMOS process. IEEE Trans Nucl Sci 32:3954-3960
32. Schwank JR, Fleetwood DM (1988) Effect of post-oxidation anneal temperature on radiation-induced charge trapping in metal-oxide-semiconductor devices. Appl Phys Lett 53:770-772
33. Seager CH, Schubert WK (1988) Hole trapping in oxides grown by rapid thermal processing. J Appl Phys 63:2869-2871
34. Hofmann K, Young DR, Rubloff GW (1987) Hole trapping in SiO_2 films annealed in low-pressure oxygen atmosphere. J Appl Phys 62:925-930
35. Schubert WK (1991) Comparison of the effects of post-oxidation anneals on the initial properties and the radiation response of rapid thermally processed oxides. J Appl Phys 69:3159-3166
36. Flament O, Leray JL, Martin F, Orsier E, Pelloie JL, Truche R, Devine RAB (1995) Effect of rapid thermal annealing on radiation hardening of MOS devices. IEEE Trans Nucl Sci 42:1667-1673
37. Yoshii I, Hama K, Maeguchi K (1989) Radiation effects on p^+ poly gate MOS structures with thin oxides. IEEE Trans Nucl Sci 36:2124-2130
38. Chang-Liao K-S, Chuang C-C (1994) Improvement of radiation hardness in poly-Si gate MOS capacitor by use of amorphous-Si. Electron Lett 30:1540-1542
39. Kasama K, Toyokawa F, Tsukiji M, Sakamoto M, Kobayashi K (1986) Mechanical stress dependence of radiation effects in MOS structures. IEEE Trans Nucl Sci 33: 1210-1215
40. Kasama K, Tsujiki M, Kobayashi K (1987) Correlation between mechanical stress and hydrogen-related effects on radiation-induced damage in MOS structures. IEEE Trans Nucl Sci 34:1202-1207
41. Wang Y, Nishioka Y, Ma TP, Barker RC (1988) Radiation and hot-electron effects on SiO_2/Si interfaces with oxides grown in O_2 containing small amounts of trichloroethane. Appl Phys Lett 52:573-575
42. Nishioka Y, Ohyu K, Ohji Y, Ma T-P (1989) Channel length and width dependence of hot-carrier hardness in fluorinated MOSFETs. IEEE Electron Device Lett 10:540-542
43. Nishioka Y, Ohyu K, Ohji Y, Natsuaki N, Mukai K, Ma TP (1989) The effect of fluorine implantation on the interface radiation hardness of Si-gate metal-oxide-semiconductor transistors. J Appl Phys 66:3909-3912
44. Naiman ML, Terry FL, Burns JA, Raffel JI, Aucoin R (1980) Properties of thin oxynitride gate dielectrics produced by thermal nitridation of silicon dioxide. IEDM Tech Dig, The IEEE, New York, pp 562-564
45. Terry Jr FL, Aucoin RJ, Naiman ML, Senturia SD (1983) Radiation effects in nitrided oxides. IEEE Electron Device Lett 4:191-193
46. Pancholy RK, Erdmann FM (1983) Radiation effects on oxynitride gate dielectrics. IEEE Trans Nucl Sci 30:4141-4145
47. Ruggles GA, Monkowski JR (1986) An investigation of fixed charge buildup in nitrided oxides. J Electrochem Soc 133:787-793

48. Sundaresan R, Matloubian MM, Bailey WE (1986) Rapid-thermal nitridation of SiO_2 for radiation-hardened MOS gate dielectrics. IEEE Trans Nucl Sci 33:1223-1227
49. Dunn GJ (1991) Effect of synchrotron X-ray radiation on the channel hot-carrier reliability of reoxidized nitrided silicon dioxide. IEEE Electron Device Lett 12:8-9
50. Dunn GJ, Wyatt PW (1989) Reoxidized nitrided oxide for radiation-hardened MOS devices. IEEE Trans Nucl Sci 36:2161-2168
51. Lo GQ, Shih DK, Ting W, Kwong DL (1989) Effects of post-nitridation anneals on radiation hardness in rapid thermal nitrided gate oxides. Appl Phys Lett 55:2405-2407
52. Severi M, Dori L, Impronta M, Guerri S (1989) Process dependence of hole trapping in thin nitrided SiO_2 films. IEEE Trans Electron Devices 36:2447-2451
53. Dunn GJ, Jayaraman R, Yang W, Sodini CG (1988) Radiation effects in low-pressure reoxidised nitrided oxide gate dielectrics. Appl Phys Lett 52:1713-1715
54. Wu Y-L, Hwu J-G (1993) Improvement in radiation hardness of reoxidised nitrided oxide (RNO) in the absence of post-oxidation anneal. IEEE Electron Device Lett 14:1-3
55. Lu W-S, Lin K-C, Hwu J-G (1993) Process dependency of radiation hardness of rapid thermal reoxidized nitrided gate oxides. IEEE Trans Electron Devices 40:1597-1602
56. Neumeier K, Bruemmer HP (1996) A systematic investigation of the influence of processing parameters on the radiation hardness of reoxidized nitrided oxides. In: Proc RADECS '95, The IEEE, New York, pp 131-136
57. Lu W-S, Chan C-H, Hwu J-G (1995) A systematic study of the inital electrical and radiation hardness properties of reoxidized nitrided oxides by rapid thermal processing. IEEE Trans Nucl Sci 42:167-173
58. Dunn GJ (1989) Hole trapping in reoxidized nitrided silicon dioxide. J Appl Phys 65: 4879-4883
59. Krisch KS, Gross BJ, Sodini CG (1991) Positive-charge trapping in nitrided oxide and reoxidized nitrided oxide. J Appl Phys 70:2185-2194
60. Mallik A, Vasi J, Chandorkar AN (1993) The nature of the hole traps in reoxidized nitrided oxide gate dielectrics. J Appl Phys 74:2665-2668
61. Mallik A, Chandorkar AN, Vasi J (1995) Capture cross-section of hole traps in reoxidized nitrided oxide measured by irradiation. Solid-State Electron 38:1851-1853
62. Conley Jr JF, Lenahan PM, Lelis AJ, Oldham TR (1995) Electron spin resonance evidence that E'_γ centers can behave as switching oxide traps. IEEE Trans Nucl Sci 42:1744-1749
63. Dunn GJ (1988) Generation of interface states in nitrided oxide gate dielectrics by ionizing radiation and Fowler-Nordheim stressing. Appl Phys Lett 53:1650-1651
64. Ramgopal Rao V, Vasi J (1992) Radiation-induced interface-state generation in reoxidized nitrided SiO_2. J Appl Phys 71:1029-1031
65. Bhat N, Vasi J (1992) Interface-state generation under radiation and high-field stressing in reoxidized nitrided oxide MOS capacitors. IEEE Trans Nucl Sci 39:2230-2235
66. Cartier E, Buchanan DA, Dunn GJ (1994) Atomic hydrogen-induced interface degradation of reoxidized-nitrided silicon dioxide on silicon. Appl Phys Lett 64:901-903
67. Yount JT, Lenahan PM, Wyatt PW (1995) The effects of thermal nitridation and reoxidation on the interfacial stress and structure of silicon dioxide gate dielectrics. J Appl Phys 77:699-705
68. Mallik A, Vasi J, Chandorkar AN (1993) Electron trapping during irradiation in reoxidized nitrided oxide. IEEE Trans Nucl Sci 40:1380-1387
69. Krantz RJ, Scarpulla J, Cable JS (1991) Total dose-induced charge buildup in nitrided-oxide MOS devices. IEEE Trans Nucl Sci 38:1746-1753

70. de Castro AJ, Fernández M, Sacedón JL (1993) Effects of thermal nitridation on the radiation hardness of the SiO_2/Si interface. J Appl Phys 73:7465-7470
71. Ramgopal Rao V, Sharma DK, Vasi J (1996) Neutral electron trap generation under irradiation in reoxidized nitrided gate dielectrics. IEEE Trans Electron Devices 43:1467-1470
72. Dunn GJ, Gross BJ, Sodini CG (1992) Radiation-induced increase in the inversion layer mobility of reoxidized nitrided oxide MOSFETs. IEEE Trans Electron Devices 39:677-684
73. Mallik A, Vasi J, Chandorkar AN (1993) A study of radiation effects on reoxidized nitrided oxide MOSFETs, including effects on mobility. Solid-State Electron 36:1359-1361
74. Boesch Jr HE, Dunn GJ (1991) Hole transport in SiO_2 and reoxidized nitrided SiO_2 gate insulators at low temperature. IEEE Trans Nucl Sci 38:1083-1088
75. Yount JT, Lenahan PM, Dunn GJ (1992) Electron spin resonance study of radiation-induced point defects in nitrided and reoxidized nitrided oxides. IEEE Trans Nucl Sci 39:2211-2219
76. Hwang H, Ting W, Maiti B, Kwong D-L, Lee J (1990) Electrical characteristics of ultrathin oxynitride gate dielectric prepared by rapid thermal oxidation of Si in N_2O. Appl Phys Lett 57:1010-1011
77. Lo GQ, Joshi AB, Kwong DL (1993) Radiation hardness of MOSFETs with N_2O-nitrided gate oxides. IEEE Trans Electron Devices 40:1565-1567
78. Wu Y-L, Kuo K-M, Hwu J-G (1994) Reduction of radiation-induced degradation in n-channel metal-oxide-semiconductor field-effect transistors (MOSFETs) with gate oxides prepared by repeated rapid thermal N_2O annealing. Jpn J Appl Phys 33:L916-L917
79. Wu Y-L, Hwu J-G (1994) Improvement in radiation hardness of gate oxides in metal-oxide semiconductor devices by repeated rapid thermal oxidations in N_2O. Appl Phys Lett 64:3136-3138
80. Saks NS, Simons M, Fleetwood DM, Yount JT, Lenahan PM, Klein RB (1994) Radiation effects in oxynitrides grown in N_2O. IEEE Trans Nucl Sci 41:1854-1863
81. Fleetwood DM, Saks NS (1996) Oxide, interface, and border traps in thermal, N_2O, and N_2O-nitrided oxides. J Appl Phys 79:1583-1594
82. Anthony CJ, Uren MJ, Nayar V (1996) Radiation hardness of N_2O grown oxynitrides assessed using the conductance technique. Appl Phys Lett 69:2104-2106
83. Scarpa A, Paccagnella A, Montera F, Ghibaudo G, Pananakakis G, Ghidini G, Fuochi PG (1997) Ionizing radiation induced leakage current on ultra-thin gate oxides. IEEE Trans Nucl Sci 44:1818-1825
84. Candelori A, Paccagnella A, Cammarata M, Ghidini G, Fuochi PG (1998) Fowler-Nordheim characteristics of electron irradiated MOS capacitors. IEEE Trans Nucl Sci 45:2383-2390
85. Houssa M, De Gendt S, de Bokx P, Mertens PW, Heyns MM (1999) Effect of x-ray irradiation on the electrical characteristics of ultra-thin gate oxides. Semicond Sci Technol 14:741-746
86. Ling CH, Ang CH, Ang DS (2000) Characterization of leakage current in thin gate oxide subjected to 10 keV X-ray irradiation. IEEE Trans Electron Devices 47:650-652
87. Cho BJ, Chong PF, Chor EF, Joo MS, Yeo IS (2000) Electron-beam irradiation-induced gate oxide degradation. J Appl Phys 88:6731-6735

88. Goguenheim D, Bravaix A, Moragues JM, Lambert P, Boivin P (2000) Comparison of oxide leakage currents induced by ion implantation and high field electric stress. Microelectron Reliab 40:751-754
89. Ceschia M, Paccagnella A, Sandrin S, Ghidini G, Wyss J, Lavale M, Flament O (2000) Low field leakage current and soft breakdown in ultra-thin gate oxides after heavy ions, electrons or X-ray irradiation. IEEE Trans Nucl Sci 47:566-573
90. Candelori A, Ceschia M, Paccagnella A, Wyss J, Bisello D, Ghidini G (2000) Thin oxide degradation after high energy ion irradiation. In: Proc RADECS Workshop 2000, pp 66-74
91. Ceschia M, Paccagnella A, Turrini M, Candelori A, Ghidini G, Wyss J (2000) Heavy ion irradiation of thin gate oxides. IEEE Trans Nucl Sci 47:2648-2655
92. Riess P, Ceschia M, Paccagnella A, Ghibaudo G, Pananakakis G (2000) Comparison of the electrical and thermal stability of stress- or radiation-induced leakage current in thin oxides. Appl Phys Lett 76:1158-1160
93. Cho BJ, Kim SJ, Ling CH, Joo M-S, Yeo I-S (2000) A comparison between leakage currents in thin gate oxides subjected to X-ray radiation and electrical stress degradation. Solid-State Electron 44:1289-1292
94. Ang CH, Ling CH, Cheng ZY, Kim SJ, Cho BJ (2000) A comparative study of radiation- and stress-induced leakage currents in thin gate oxides. Semicond Sci Technol 15:961-964
95. Ang C-H, Ling C-H, Cho B-J, Kim S-J, Cheng Z-Y (2000) Radiation and electrical stress-induced hole trap-assisted tunneling currents in ultrathin gate oxides. Solid-State Electron 44:2001-2007
96. Ceschia M, Paccagnella A, Scarpa A, Cester A, Ghidini G (1999) Total dose dependence of radiation-induced leakage current in ultra-thin gate oxides. Microelectron Reliab 39:221-226
97. Ang C-H, Ling C-H, Cheng Z-Y, Kim S-J, Cho B-J (2000) Reduction of radiation-induced leakage currents in thin oxides by application of a low post-irradiation gate bias. Jpn J Appl Phys 39:L757-L759
98. Ang C-H, Ling C-H, Cheng Z-Y, Kim S-J, Cho B-J (2000) Bias and thermal annealings of radiation-induced leakage currents in thin-gate oxides. IEEE Trans Nucl Sci 47:2758-2764
99. Larcher L, Paccagnella A, Ceschia M, Ghidini G (1999) A model of radiation induced leakage current (RILC) in ultra-thin gate oxides. IEEE Trans Nucl Sci 46:1553-1561
100. Miranda E, Suñé J, Rodríguez R, Nafría M, Aymerich X (1999) A function-fit model for the soft breakdown failure mode. IEEE Electron Device Lett 20:265-267
101. Sexton FW, Fleetwood DM, Shaneyfelt MR, Dodd PE, Hash GL (1997) Single event gate rupture in thin gate oxides. IEEE Trans Nucl Sci 44:2345-2352
102. Sexton FW, Fleetwood DW, Shaneyfelt MR, Dodd PE, Hash GL, Schanwald LP, Loemker RA, Krisch KS, Green ML, Weir BE, Silverman PJ (1998) Precursor ion damage and angular dependence of single event gate rupture in thin oxides. IEEE Trans Nucl Sci 45:2509-2518
103. Johnston AH, Swift GM, Miyahira T, Edmonds LD (1998) Breakdown of gate oxides during irradiation with heavy ions. IEEE Trans Nucl Sci 45:2500-2508
104. Wheatley CF, Titus JL, Burton DI (1994) Single-event gate rupture in vertical power MOSFETs: An original empirical expression. IEEE Trans Nucl Sci 41:2152-2159

105. Mouret I, Allenspach M, Titus JL, Wheatley CF, LaBel KA, Calvet M-C, Schrimpf RD, Galloway KF (1996) Measurement of a cross-section for single-event gate rupture in power MOSFETs. IEEE Electron Device Lett 17:163-165
106. Kim S-H, Lee H-J, Han C-H, Lee K, Choi S-S, Jeon Y-J, Di Fabrizio E, Gentili M (1995) The effects of X-ray irradiation-induced damage on reliability in MOS structures. Solid-State Electron 38:95-99
107. Houssa M, De Gendt S, de Bokx P, Mertens PW, Heyns MM (1999) X-ray irradiation effect on the reliability of ultra-thin gate oxides and oxynitrides. Microelectron Engineering 48:43-46
108. Wilk GD, Wallace RM, Anthony JM (2001) High-κ gate dielectrics: Current status and materials properties considerations. J Appl Phys 89:5243-5275
109. Brisset C, Ferlet-Cavrois V, Flament O, Musseau O, Leray JL, Pelloie JL, Escoffier R, Michez A, Cirba C, Bordure G (1996) Two-dimensional simulation of total dose effects on NMOSFET with lateral parasitic transistor. IEEE Trans Nucl Sci 43:2651-2658
110. Flament O, Autran JL, Paillet P, Roche P, Faynot O, Truche R (1996) Charge pumping analysis of radiation effects in LOCOS parasitic transistors. IEEE Trans Nucl Sci 44:1930-1938
111. Nishioka Y, Itoga T, Ohyu K, Kato M, Ma T-P (1990) Radiation effects on fluorinated field oxides and associated devices. IEEE Trans Nucl Sci 37:2026-2032
112. Shaneyfelt MR, Dodd PE, Draper BL, Flores RS (1998) Challenges in hardening technologies using shallow-trench isolation. IEEE Trans Nucl Sci 45:2584-2591
113. Youk GU, Khare PS, Schrimpf RD, Massengill, LW, Galloway KF (1999) Radiation-enhanced short channel effects due to multi-dimensional influence from charge at trench isolation oxides. IEEE Trans Nucl Sci 46:1830-1835
114. Brady FT, Maimon JD, Hurt MJ (1999) A scaleable, radiation hardened shallow trench isolation. IEEE Trans Nucl Sci 46:1836-1840
115. Manasevit HM, Simpson WI (1964) Single crystal SOS substrates. J Appl Phys 35:1349-1351
116. Manesevit HM (1974) A survey of the heteroepitaxial growth of semiconductor films on insulating substrates. J Cryst Growth 22:125-148
117. Colinge JP (1997) Silicon-on-Insulator Technology: Materials to VLSI. Kluwer, Boston
118. Sadana DK (2001) SOI for CMOS logic and memory applications. In: Claeys CL, Gonzalez F, Murato J, Saraswat K (eds) ULSI Process Integration II, The Electrochem Soc Proc Ser PV 2001-2:474-488
119. Maurits JA (1978) SOS wafers – Some comparisons to silicon wafers. IEEE Trans Electron Devices 25:859-863
120. Hsu ST (1978) Electron mobility in SOS films. IEEE Trans Electron Devices 25:913-916
121. Lau SS, Matteson S, Mayer JW, Revesz P, Gyulai J, Roth J, Sigmon TW, Cass T (1979) Improvement of crystalline quality of epitaxial Si layers by ion implantation techniques. Appl Phys Lett 34:76-78
122. Yoshii T, Taguchi S, Inoue T, Tango H (1982) Improvement of SOS device performance by solid phase epitaxy. Jpn J Appl Phys 21:175-179
123. Garcia GA, Reedy RE, Burgener ML (1988) High-quality CMOS in thin (100 nm) silicon on sapphire. IEEE Electron Device Lett 9:32-34

124. Dumin DJ, Dabral S, Freytag MH, Robertson PJ, Carver GP, Novotny DB (1989) High mobility CMOS transistors fabricated in very thin SOS films. IEEE Trans Electron Devices 36:596-598
125. Kjar RA, Peel J (1974) Radiation induced leakage current in n-channel SOS transistors. IEEE Trans Nucl Sci 21:208-210
126. Neamen D, Shedd W, Buchanan B (1974) Radiation induced charge trapping at the silicon sapphire substrate interface. IEEE Trans Nucl Sci 21:211-216
127. Buchanan BL, Neamen DA, Shedd WM (1978) SOS Device radiation effects and hardening. IEEE Trans Electron Devices 25:959-970
128. Halle LF, Zietlow CT, Barnes CE (1988) JFE/SOS Devices – Part II: Gamma-radiation-induced effects. IEEE Trans Electron Devices 35:359-364
129. Neamen D, Buchanan B (1976) Silicon-sapphire interface charge trapping effects on sapphire type and epi growth conditions. IEEE Trans Nucl Sci 23:1590-1593
130. Repace JL, Goodman AM (1978) The effect of processing variations on interfacial and radiation induced charge in silicon-on-sapphire capacitors. IEEE Trans Electron Devices 25:978-982
131. Johnston RA, de la Houssaye PR, Chang CE, Chen PF, Wood ME, Garcia GA, Lagnado I, Asbeck PM (1998) Advanced thin film silicon-on-sapphire technology: Microwave circuit applications. IEEE Trans Electron Devices 45:1047-1053
132. Lyons G (1998) Commercial SOS technology for radiation-tolerant space applications. IEEE Radiation Effects Data Workshop 96-99
133. Mathew SJ, Niu G, Clark SD, Cressler JD, Palmer MJ, Dubbelday WB (1999) radiation-induced back-channel leakage in SiGe CMOS on silicon-on-sapphire (SOS) technology. IEEE Trans Nucl Sci 46:1848-1853
134. Lagnado I, de la Houssaye PR, Koester SJ, Hammond R, Chu JO, Ott JA, Mooney PM, Perraud L, Jenkins KA (2001) Silicon-on-sapphire technology: Quo vadis II. A competitive alternative for rf systems. In: Cristoloveanu S, Hemment PLF, Izumi K, Celler GK, Assaderaghi F, Kim YW (eds) Proc 9th Int Symp on Silicon-on-Insulator Technology and Devices. The Electrochem Soc Proc Ser PV 2001-3:265-270
135. Kamins TI, Lee KF, Gibbons JF (1980) Crystal structure and thermal oxidation of laser-recrystallised polycrystalline silicon. Appl Phys Lett 36:550-553
136. Im JS, Chen CK, Thompson CV, Geis MW, Tomita H (1987) Liquid-solid interface morphologies and defect structures in zone-melting recrystallised SOI. In: Sturm JC, Chen CK, Pfeiffer L, Hemment PLF (eds) Proc Mat Res Soc 107: 169-174
137. Watanabe M, Tooi A (1966) Formation of SiO_2 films by oxygen-ion bombardment. J Appl Phys 5:737-738
138. Izumi K, Doken M, Ariyoshi H (1978) CMOS devices fabricated on buried SiO_2 layers by oxygen implantation into silicon. Electron Lett 14:593-594
139. Bruel M, Margail J, Jaussaud C, Auberton-Herve AJ, Stoemenos J (1988) Silicon-on-Insulator by oxygen implantation: An advanced technology. Microelectr Eng 8:149-161
140. Stoemenos J, Aspar B, Margail J (1994) Mechanisms of SIMOX synthesis and related microstructural properties. In: Christoloveanu S (ed) Proc 6th Int Symp on Silicon-on-Insulator Technology and Devices. The Electrochem Soc Proc Ser PV 94-11:16-27
141. Maszara WP, Goetz G, Caviglia A, McKitterick JB (1988) Bonding of silicon wafers for silicon-on-insulator. J Appl Phys 64:4943-4950
142. Mitai K, Goesele U (1992) Wafer bonding technology for silicon-on-insulator applications: A review. J Electronic Mat 21:669-676

143. Annamalai NK, Bockman JF, McGruer NE, Chapski J (1990) A comparison of buried oxide characteristics of single and multiple implant SIMOX and bond and etch back wafers. IEEE Trans Nucl Sci 37:2001-2007
144. Bruel M (1995) Silicon on insulator material technology. Electronics Lett 31:1201-1202
145. Auberthon-Hervé AJ, Barge T, Metral F, Bruel M, Aspar B, Moriceau H (1998) Silicon on insulator wafers using the smart cut technology. In: Huff HR, Gösele U, Tsuya H (eds) Semiconductor Silicon 1998. The Electrochem Soc Proc Ser PV 98-1:1341-1360
146. Tong QY, Gösele U (1996) A model of low-temperature wafer bonding and its applications. J Electrochem Soc 143:1773-1779
147. Moriceau H, Fournel F, Rayssac O, Soubie A, Bataille B, Aspar B, Rousseau K, Magnea N, Maleville C, Barge T (2001) Overview of some recent advances in wafer bonding technologies. The Electrochem Soc Extended Abstr PV 2001-2:1450
148. Guttierrez-D EA, Claeys C, Simoen E (2001) Silicon devices and circuits. In: Guttierrez-D EA, Deen MJ, Claeys CL (eds) Low Temperature Electronics – Physics, Devices, Circuits and Applications, Academic Press, San Diego, Chap 2
149. Rotondaro ALP, Magnusson UK, Claeys C, Flandre D, Terao A, Colinge JP (1993) Evidence for different conduction mechanisms in accumulation-mode p-channel SOI MOSFETs at room and liquid helium temperature. IEEE Trans Electron Devices 40:727-733
150. Flandre D, Terrao A, Francis P, Gentinne B, Colinge JP (1993) Demonstration of the potential of accumulation-mode MOS transistors on SOI substrates for high-temperature operation (150-300ºC). IEEE Electron Device Lett 14:10-13
151. Balestra F, Cristoloveanu S, Nemachir M, Brini J, Elewa T (1987) Double gate silicon-on-insulator transistors with volume inversion. IEEE Electron Device Lett 8:410-412
152. Colinge JP, Gao MH, Romano-Rodriguez A, Maes H, Claeys C (1990) Silicon-on-insulator gate-all-around SOI transistors. In: IEDM Tech Digest. The IEEE, New York, pp 595-598
153. Schwank JR (1997) Advantages and limitations of silicon-on-insulator technology in radiation environments. Microelectron Eng 36:335-342
154. Fleetwood DM, Tsao SS, Winokur PS (1988) Total dose hardness assurance issues for SOI MOSFETs. IEEE Trans Nucl Sci 35:1361-1367
155. Lawrence RK, Hughes HL, Stahlbush (1990) Radiation sensitivity of buried oxides. J Electron Mat 19:665-670
156. Mrstik BJ, McMarr PJ, Lawrence RK (1994) Relationship between radiation response and density of buried oxide in separation-by-implantation-of-oxygen material. Appl Phys Lett 65:2993-2995
157. Vanheusden K, Schwank JR, Warren WL, Fleetwood DM, Devine RAB (1997) Radiation-induced H^+ trapping in buried SiO_2. Microelectron Eng 36:241-244
158. Allen LP, Alles ML, Dolan RP, Hughes HL, McMarr P (1997) Thin BOX SIMOX silicon-on-insulator substrates for radiation tolerant advanced electronics. Microelectron Eng 37:383-386
159. Mrstik BJ, Afanas'ev VV, Stesmans A, McMarr PJ, Lawrence RK (1999) Relationship between oxide density and charge trapping in SiO_2 films. J Appl Phys 85:6577-6588
160. MacFarlane PJ, Stahlbush RE (2000) Near interface H^+ trapping and its influence on H^+ transport in hydrogenated unibond buried oxides. Appl Phys Lett 77:3081-3083
161. Schwank JR, Shaneyfelt MR, Dodd PE, Ferlet-Cavrois V, Loemker RA, Winokur PS, Fleetwood DM, Paillet P, Leray JL, Draper BL, Witczak SC, Riewe LC (2000) Corre-

lation between Co-60 and X-ray radiation-induced charge buildup in silicon-on-insulator buried oxides. IEEE Trans Nucl Sci 47:2175-2182
162. Lawrence RK, Colinge JP, Hughes HL (1991) Radiation effects in gate-all-around structures. In: Proc Int. SOI Conf. The IEEE, New York, pp. 80-81
163. Colinge JP, Terao A (1993) Effects of total-dose irradiation on gate-all-around (GAA) devices. IEEE Trans Nucl Sci 40:78-82
164. Francis P, Flandre D, Colinge JP (1995) Theoretical consideration for SRAM total-dose hardening. IEEE Trans Nucl Sci 42:82-91
165. Simoen E, Claeys C, Coenen S, Decreton M (1995) DC and low frequency noise characteristics of γ-irradiated gate-all-around silicon-on-insulator MOS transistors. Solid State Electron 38:1-8
166. Vandooren A, Colinge JP (1999) Drain conductance degradation in gate-all-around inversion-mode nMOSFETs with total dose. In: Hemment P (ed) Proc 9th Int Symp on Silicon-on-Insulator Technology and Devices. The Electrochem Soc Proc Ser PV 99-3:323-328
167. Claeys C, Simoen E (1994) The perspectives of silicon-on-insulator technologies for cryogenic applications. J Electrochem Soc 141:2533-2532
168. Guttierrez-D, Deen MJ, Claeys CL (2001) Low Temperature Electronics: Physics, Devices, Circuits and Applications. Academic Press
169. Claeys C, Simoen E (1999) Is there a future for cryogenic SOI. In: Hemment P (ed) Proc 9th Int Symp on Silicon-on-Insulator Technology and Devices. The Electrochem Soc Proc Ser PV 99-3:36-50
170. Saks N, Klein RB, Griscom DL (1988) Formation of interface traps in MOSFETs during annealing following low temperature irradiation. IEEE Trans Nucl Sci 35:1234-1240
171. Saks NS, Klein RB, Yoon S, Griscom DL (1991) Formation of interface traps in metal-oxide-semiconductor devices during isochronal annealing after irradiation at 78 K. J Appl Phys 70:7434-7442
172. Jenkins WC, Liu ST (2000) Total dose performance at 77 K of a radiation hard 0.35 μm CMOS SOI technology. IEEE Trans Nucl Sci 47:2204-2207
173. Fourches N, Abbon Ph, Chipaux R, Delagnes E, Orsier E, Pailler P, du Port de Pontcharra J, Rouger M, Sueur M, Truche R (1997) Thick film SOI technology: characteristics of devices and performance of circuits for high-energy physics at cryogenic temperatures; effect of ionizing radiation. Nucl Instr Meth A401:229-237

Chapter 7

1. Anderson WT, Binari SC (1983) Radiation effects in GaAs devices and ICs. In: IEEE Int Reliability Proc, The IEEE, New York, pp 316-319
2. Bourgoin JC, von Bardeleben JC, Stiévenard D (1988) Native defects in gallium arsenide. J Appl Phys 64:R65-R91
3. Guillot G (1988) EL2-related studies in irradiated and implanted GaAs. Revue Phys Appl 23:833-846
4. Makram-Ebeid S, Boher P (1988) Defect pairs and clusters related to the EL2 centre in GaAs. Revue Phys Appl 23:847-862

5. Bourgoin JC, Lannoo M (1988) A critical look at EL2 models. Revue Phys Appl 23: 863-869
6. Brunkov PN, Kalinovsky VS, Nikitin VG, Sobolev MM (1992) Generation of the EL2 defect in n-GaAs irradiated by high energy protons. Semicond Sci Technol 7:1237-1240
7. Hiemstra DM (1995) Dose rate and total dose dependence of the 1/f noise performance of a GaAs operational amplifier during irradiation. IEEE Trans Nucl Sci 42:1615-1621
8. Mooney PM (1990) Deep donor levels (DX centers) in III-V semiconductors. J Appl Phys 67:R1-R26
9. Papaioannou GJ, Papastamatiou MJ, Christou A (1995) He ion radiation effects in high electron mobility transistors. J Appl Phys 78:3066-3076
10. Nishiguchi M, Hashinaga T, Nishizawa H, Hayashi H, Okazaki N, Kitagawa M, Fujino T (1990) Radiation tolerant GaAs MESFET with a highly-doped thin active layer grown by OMVPE. IEEE Trans Nucl Sci 37:2071-2075
11. Janousek BK, Yamada WE, Krantz RJ, Bloss WL (1988) Neutron radiation effects in GaAs ion-implanted metal-semiconductor field-effect transistors. J Appl Phys 63: 1678-1686
12. Bloss WL, Yamada WE, Young AM, Janousek BK (1988) Characteristics of GaAs MESFET inverters exposed to high energy neutrons. IEEE Trans Nucl Sci 35:1074-1079
13. Su C, Rohdin H, Stolte C (1983) 1/f noise in GaAs MESFETs. In: IEDM Tech Dig, The IEEE, New York, pp 601-604
14. Subramanian S, Sarkar A, Ungier L, Goodnick SM (1997) Integrity of III-V heterojunction interfaces under gamma irradiation. IEEE Trans Nucl Sci 44:1862-1869
15. Ohyama H, Simoen E, Kuroda S, Claeys C, Takami Y, Hakata T, Sunaga H (1998) Impact of high energy particles on InGaP/InGaAs pseudomorphic HEMTs. IEEE Trans Nucl Sci 45:2861-2866
16. Anderson WT, Simons M, Tseng WF, Herb JA, Bandy S (1987) Transient radiation effects in AlGaAs/GaAs MODFETs. IEEE Trans Nucl Sci 34:1669-1675
17. Krantz RJ, Bloss WL, O'Loughlin MJ (1988) High energy neutron irradiation effects in GaAs modulation-doped field effect transistors (MODFETs): Threshold voltage. IEEE Trans Nucl Sci 35:1438-1443
18. Zuleeg R (1989) Radiation effects in GaAs FET devices. Proc IEEE 77:389-407
19. Anderson WT, Simons M, Christou A, Beall J (1985) GaAs MMIC technology radiation effects. IEEE Trans Nucl Sci 32:4040-4045
20. Meulenberg A, Hung H-LA, Peterson KL, Anderson WT (1988) Total dose and transient radiation effects on GaAs MMICs. IEEE Trans Nucl Sci 35:2125-2132
21. Galashan AF, Bland SW (1990) Neutron and proton irradiation of shallow channel GaAs direct-coupled field-effect transistor logic devices and circuits. J Appl Phys 67:173-179
22. Derewonko H, Bosella A, Pataut G, Perié D, Pinsard JL, Sentubery C, Verbeck C, Bressy P, Augier, P (1996) Evaluation of GaAs low noise and power MMIC technologies to neutron, ionizing dose and dose rate effects. IEEE Trans Nucl Sci 43:837-844
23. Listvan MA, Vold PJ, Arch DK (1987) Ionizing radiation hardness of GaAs technologies. IEEE Trans Nucl Sci 34:1664-1668
24. Chang JY, Badawi MH, DeCicco A (1989) Neutron/gamma induced damage mechanisms and synergistic effects in GaAs MESFETs. IEEE Trans Nucl Sci 36:2068-2075

25. Campbell AB, Knudson AR, Stapor WJ, Summers G, Xapsos MA, Jessee M, Palmer T, Zuleeg R, Dale CJ (1986) Particle damage effects in GaAs test structures. IEEE Trans Nucl Sci 33:1435-1441
26. Behle AF, Zuleeg R (1972) Fast neutron tolerance of GaAs JFETs operating in the hot electron regime. IEEE Trans Electron Devices 19:993-995
27. Tan HH, Williams JS, Jagadish C (1995) Characterization of deep levels and carrier compensation created by proton irradiation in undoped GaAs. J Appl Phys 78:1481-1487
28. Khanna SM, Rejeb C, Jorio A, Parenteau M, Carlone C, Gerdes Jr JW (1993) Electron and neutron radiation-induced order effect in gallium arsenide. IEEE Trans Nucl Sci 40:1350-1359
29. Jorio A, Parenteau M, Aubin M, Carlone C, Khanna SM, Gerdes Jr JW (1994) A mobility study of the radiation induced order effect in gallium arsenide. IEEE Trans Nucl Sci 41:1937-1944
30. Khanna SM, Jorio A, Carlone C, Parenteau M, Houdayer A, Gerdes Jr JW (1995) Particle dependence of the gallium vacancy production in irradiated n-type gallium arsenide. IEEE Trans Nucl Sci 42:2095-2103
31. Citterio M, Rescia S, Radeka V (1995) Radiation effects at cryogenic temperatures in Si-JFET, GaAs MESFET, and MOSFET devices. IEEE Trans Nucl Sci 42:2266-2270
32. Hiemstra DM, Kizeevi AA, Hou LZ, Salama CAT (1998) Dose rate and total dose dependence of low frequency noise performance, I-V curves and sidegating for GaAs MESFETs. IEEE Trans Nucl Sci 45:2616-2623
33. Goronkin H, Birrittella MS, Seelbach WC, Vaitkus RL (1982) Backgating and light sensitivity in ion-implanted GaAs integrated circuits. IEEE Trans Electron Devices 29:845-850
34. Subramanian S, Bhattacharya PK, Staker KJ, Ghosh CL, Badawi MH (1985) Geometrical and light-induced effects on back-gating in ion-implanted GaAs MESFET's. IEEE Trans Electron Devices 32:28-33
35. Carruthers TF, Anderson WT, Weller JF (1985) Optically induced backgating transients in GaAs FET's. IEEE Electron Device Lett 6:580-582
36. Shaw GJ, Xapsos MA, Weaver BD, Summers GP (1993) Low temperature proton irradiation of GaAs MESFETs. IEEE Trans Nucl Sci 40:1300-1306
37. Siyanbola WO, Palmer DW (1990) Low temperature annealing of deep electron traps produced by proton irradiation of n-GaAs. Semicond Sci Technol 5:7-15
38. Siyanbola WO, Palmer DW (1991) Electronic energy levels of defects that anneal in the 280-K stage in irradiated n-type gallium arsenide. Phys Rev Lett 66:56-59
39. Meneghesso G, Paccagnella A, Camin DV, Fedyakin N, Pessina G, Canali C (1997) Study of neutron damage in GaAs MESFETs. IEEE Trans Nucl Sci 44:840-846
40. Anderson WT, Gerdes J, Roussos JA (1993) Temperature dependent GaAs MMIC radiation effects. IEEE Trans Nucl Sci 40:1735-1739
41. O'Loughlin MJ (1988) Radiation effects in high electron mobility transistors: Total dose gamma irradiation. IEEE Trans Nucl Sci 34:1808-1811
42. Arpatzanis N, Papastamatiou M, Papaioannou GJ, Hatzopoulos Z, Konstandinides G (1995) The gamma ray radiation effects in high-electron-mobility transistors. Semicond Sci Technol 10:1445-1451
43. Papastamatiou M, Arpatzanis N, Papaioannou GJ, Papastergiou C, Christou A (1997) Neutron radiation effects in high electron mobility transistors. IEEE Trans Electron Devices 44:364-372

44. Dupont-Nivet E, Pasquali M (1992) Neutron effects on HEMT devices. In: Proc RADECS '91, The IEEE, New York, pp 189-193
45. Ziegler JF, Biersack JP, Littmark P (1987) The Stopping and Range of Ions in Solids. Pergamon, New York, p 120
46. Janousek BK, Krantz RJ, Bloss WL, Yamada WE, Brown S, Remke RL, Witmer S (1989) Characteristics of GaAs heterojunction FETs (HFETs) and source follower FET logic (SFFL) inverters exposed to high energy neutrons. IEEE Trans Nucl Sci 36:2223-2228
47. Hakata T, Ohyama H, Kuroda S, Simoen E, Claeys C, Kudou T, Kobayashi K, Nakabayashi M, Yoneoka M, Takami Y, Sunaga H, Miyahara K (1999) Impact of induced lattice defects on performance degradation of AlGaAs/GaAs p-HEMTs. Physica B 273-274:1034-1036
48. Ohyama H, Simoen E, Kuroda S, Claeys C, Takami Y, Hakata T, Hayama K, Tajiri T, Nakabayashi M, Kobayashi K, Yoneoka M, Sunaga H (2000) Radiation damage in AlGaAs/GaAs pseudomorphic HEMTs. In: Proc of RADECS '99, The IEEE, New York, pp 295-298
49. Ohyama H, Simoen E, Kuroda S, Claeys C, Takami Y, Hakata T, Sunaga H (1999) Radiation induced lattice defects in InGaP/InGaAs p-HEMTs and their effect on device performance. Solid State Phenom 69-70:563-568
50. Cai SJ, Tang YS, Li R, Wei YY, Wong L, Chen YL, Wang KL, Chen M, Zhao YF, Schrimpf RD, Keay JC, Galloway KF (2000) Annealing behavior of a proton irradiated $Al_xGa_{1-x}N$/GaN high electron mobility transistor grown by MBE. IEEE Trans Electron Devices 47:304-307
51. Tanaka N, Kawanishi H, Ishikawa T (1993) Photoluminescence study of electron-beam damage in GaAs/AlGaAs quantum-well structures. Jpn J Appl Phys 32:540-543
52. Fink T, Smith DD, Braddock WD (1990) Electron-beam-induced damage study in GaAs-AlGaAs heterostructures as determined by magnetotransport characterization. IEEE Trans Electron Devices 37:1422-1425
53. Kanayama T, Takeuchi Y, Sugiyama Y, Tacano M (1992) In situ Hall measurement of two-dimensional electron gas at $Al_{0.3}Ga_{0.7}Al$/GaAs interface irradiated with 10-keV Ar ions. Appl Phys Lett 61:1402-1404
54. Wada T, Kanayama T, Ichimura S, Sugiyama Y, Misawa S, Komuro M (1993) Effects of electron irradiation on two-dimensional electron gas in AlGaAs/GaAs heterostructure. Jpn J Appl Phys 32:6262-6267
55. Wada T, Kanayama T, Tada T, Wang X-L, Sugiyama Y, Komuro M (1995) Damages in AlGaAs/GaAs heterostructures induced by keV-electron-beam irradiation. Materials Science Forum 196-201:1455-1460
56. Anderson WT, Knudson AR, Meulenberg A, Hung H-L, Roussos JA, Kiriakidis G (1990) Heavy ion total fluence effects in GaAs devices. IEEE Trans Nucl Sci 37:2065-2070
57. Ohyama H, Yajima K, Simoen E, Katoh T, Claeys C, Takami Y, Kobayashi K, Yoneoka M, Nakabayashi M, Hakata T, Takizawa H (2000) Impact of 20-MeV α-ray irradiation on the V-band performance of AlGaAs pseudomorphic HEMTs. IEEE Trans Nucl Sci 47:2546-2550

Chapter 8

1. Barnes CE (1970) Effects of Co^{60} gamma irradiation on epitaxial GaAs laser diodes. Phys Rev B 1:4735-4747
2. Barnes CE (1979) A comparison of gamma-irradiation-induced degradation in amphoterically Si-doped GaAs LED's and Zn-diffused GaAs LED's. IEEE Trans Electron Devices 26:739-745
3. Rose BH, Barnes CE (1982) Proton damage effects on light emitting diodes. J Appl Phys 53:1772-1780
4. Dimiduk KC, Ness CQ, Foley JK (1985) Electron irradiation of GaAsP LEDs. IEEE Trans Nucl Sci 32:4010-4015
5. Evans BD, Hager HE, Hughlock BW (1993) 5.5-MeV proton irradiation of a strained quamtum-well laser diode and a multiple quantum-well broad-band LED. IEEE Trans Nucl Sci 40:1645-1654
6. Carson RF, Chow WW (1989) Neutron effects in high-power GaAs laser diodes. IEEE Trans Nucl Sci 36:2076-2082
7. Marshall PW, Dale CJ, Burke EA (1992) Space radiation effects on optoelectronic materials and components for a 1300 nm fiber optic data bus. IEEE Trans Nucl Sci 39: 1982-1989
8. Zhao YF, Patwary AR, Schrimpf RD, Neifeld MA, Galloway KF (1997) 200 MeV proton damage effects on multi-quantum well laser diodes. IEEE Trans Nucl Sci 44: 1898-1905
9. Paxton AH, Carson RF, Schöne H, Taylor EW, Choquette KD, Hou HQ, Lear KL, Warren ME (1997) Damage from proton irradiation of vertical-cavity surface-emitting lasers. IEEE Trans Nucl Sci 44:1893-1897
10. Ohyama H, Vanhellemont J, Takami Y, Hayama K, Kudou T, Hakata T, Kohiki S, Sunaga H (1996) Degradation and recovery of $In_{0.53}Ga_{0.47}As$ photodiodes by 1-MeV fast neutrons. IEEE Trans Nucl Sci 43:3019-3026
11. Khanna SM, Liu HC, Wilson PH, Li L, Buchanan M (1996) High energy proton and alpha radiation effects on GaAs/AlGaAs quantum well infrared photodetectors. IEEE Trans Nucl Sci 43:3012-3018
12. Reed RA, Marshall PW, Johnston AH, Barth JL, Marshall CJ, LaBel KA, D'Ordine M, Kim HS, Carts MA (1998) Emerging optocoupler issues with energetic particle-induced transients and permanent radiation damage. IEEE Trans Nucl Sci 45:2833-2841
13. Rax BG, Lee CI, Johnston AH, Barnes CE (1996) Total dose and proton damage in optocouplers. IEEE Trans Nucl Sci 43:3167-3173
14. Burkig VC, McNichols JL, Ginell WS (1969) Infrared absorption in neutron-irradiated GaAs. J Appl Phys 40:3268-3273
15. Mudron J, Müllerová J, Dubecky F (1998) Optical properties of semi-insulating GaAs irradiated by neutrons. Solid-State Electron 42:243-246
16. Carlone C, Bernier G, Tannous E, Khanna SM, Anderson WT, Gerdes JW (1990) The photoluminescent spectrum of neutron irradiated GaAs. IEEE Trans Nucl Sci 37:1718-1725
17. Dyment JC, North JC, D'Asaro LA (1973) Optical and electrical properties of proton-bombarded p-type GaAs. J Appl Phys 44:207-213

18. Khanna SM, Carlone C, Hallé S, Parenteau M, Béliveau A, Aktik C, Gerdes Jr JW (1991) The photoconductivity spectrum of electron and neutron irradiated n lightly doped GaAs. IEEE Trans Nucl Sci 38:1145-1152
19. Ascheron C (1991) Proton beam modifications of selected $A^{III}B^{V}$ compounds. Phys Stat Sol (a) 124:11-55
20. Manasreh MO, von Bardeleben HJ, Mousalitin AM, Khokhlov DR (1999) Electron irradiation effects on the intersubband transitions in InGaAs/AlGaAs multiple quantum wells. J Appl Phys 85:630-632
21. Manasreh MO, Ballet P, Smathers JB, Salamo GJ, Jagadish C (1999) Proton irradiation effects on the intersubband transition in GaAs/AlGaAs multiple quantum wells with bulk or superlattice barriers. Appl Phys Lett 75:525-527
22. Hegeler F, Manasreh MO, Morath C, Ballet P, Yang H, Salamo GJ, Tan HH, Jagadish C (2000) Thermal annealing recovery of intersubband transitions in proton-irradiated GaAs/AlGaAs multiple quantum wells. Appl Phys Lett 77:2867-2869
23. Berhane Y, Manasreh MO, Weaver BD (2001) He^{+}-ion irradiation effect on intersubband transitions in GaAs/AlGaAs multiple quantum wells. J Appl Phys 89:3517-3519
24. Ressel P, Strusny H, Gramlich S, Zeimer U, Sebastian J, Vogel K (1993) Optimised proton implantation step for vertical-cavity surface-emitting lasers. Electron Lett 29:918-919
25. Kundrotas J, Dargys A, Valusis G, Asmontas S, Granja C, Pospisil S, Köhler K (2001) Changes of MQW photoluminescence under alpha particle irradiation. Superlatt and Microstruct 29:281-285
26. Gramlich S, Nebauer E, Sebastian J, Beister G (2001) Damage profile of He implantation in AlGaAs laser diode material detected by photoluminescence. Electron Lett 37:463-464
27. Kalish R, Kramer L-Y, Law K-K, Merz JL, Feldman LC, Jacobson DC, Weir BE (1992) Local intermixing of GaAs/GaAlAs quantum structures by individual ion implant tracks. Appl Phys Lett 61:2589-2591
28. Leon R, Swift GM, Magness B, Taylor WA, Tang YS, Wang KL, Dowd P, Zhang YH (2000) Changes in luminescence emission induced by proton irradiation: InGaAs/GaAs quantum wells and quantum dots. Appl Phys Lett 76:2074-2076
29. Rao MV, Hong W-P, Caneau C, Chang G-K, Papanicolaou N, Dietrich HB (1991) $In_{0.53}Ga_{0.47}As$ metal-semiconductor-metal photodetector using proton bombarded p-type material. J Appl Phys 70:3943-3945
30. Marsh JH (1993) Quantum well intermixing. Semicond Sci Technol 8:1136-1155
31. Kash K, Tell B, Grabbe P, Dobisz EA, Craighead HG, Tamargo MC (1988) Aluminum ion-implantation enhanced intermixing of GaAs-AlGaAs quantum-well structures. J Appl Phys 63:190-194
32. Hirayama Y (1989) Mechanism of Ga implantation-induced intermixing of GaAs-AlGaAs material. Jpn J Appl Phys 28:L162-L165
33. Charbonneau S, Poole PJ, Piva PG, Aers GC, Koteles ES, Fallahi M, He J-J, McCaffrey JP, Buchanan M, Dion M, Goldberg RD, Mitchell IV (1995) Quantum-well intermixing for optoelectronic integration using high energy ion implantation. J Appl Phys 78:3697-3705
34. Cibert J, Petroff PM, Werder DJ, Pearton SJ, Gossard AC, English JH (1986) Kinetics of implantation enhanced interdiffusion of Ga and Al at $GaAs\text{-}Ga_xAl_{1-x}As$ interfaces. Appl Phys Lett 49:223-225

35. Elman B, Koteles ES, Melman P, Armiento A (1989) GaAs/AlGaAs quantum-well intermixing using shallow ion implantation and rapid thermal annealing. J Appl Phys 66:2104-2107
36. Piva PG, Poole PJ, Buchanan M, Champion G, Templeton I, Aers GC, Williams R, Wasilewski ZR, Koteles ES, Charbonneau S (1994) Enhanced compositional disordering of quantum wells in GaAs/AlGaAs and InGaAs/GaAs using focused Ga^+ ion beams. Appl Phys Lett 65:621-623
37. Vieu C, Schneider M, Planel R, Launois H, Descouts B, Gao Y (1991) Mixing of GaAs/(Ga,Al)As interfaces by Ga^+ implantation. J Appl Phys 70:1433-1443
38. Cibert J, Petroff PM, Dolan GJ, Pearton SJ, Gossard AC, English JH (1986) Optically detected carrier confinement to one and zero dimension in GaAs quantum well wires and boxes. Appl Phys Lett 49:1275-1277
39. Hiramoto T, Hirakawa K, Iye Y, Ikoma T (1987) One-dimensional GaAs wires fabricated by focused ion beam implantation. Appl Phys Lett 51:1620-1622
40. Hirayama Y, Tarucha S, Suzuki Y, Okamoto H (1988) Fabrication of a GaAs quantum-well-wire structure by Ga focused-ion-beam implantation and its optical properties. Phys Rev B 37:2774-2777
41. Laruelle F, Hu P, Simes R, Kubena R, Robinson W, Merz J, Petroff PM (1989) Implantation enhanced interdiffusion in GaAs/GaAlAs quantum structures. J Vac Sci Technol B 7:2034-2038
42. Petroff PM, Li YZ, Xu Z, Beinstingl W, Sasa S, Ensslin K (1991) Nanostructures processing by focused ion beam implantation. J Vac Sci Technol B 9:3074-3078
43. Allard LB, Aers GC, Charbonneau S, Jackman TE, Williams RL, Templeton IM, Buchanan M, Stevanovic D, Almeida FJD (1992) Fabrication of nanostructures in strained InGaAs/GaAs quantum wells by focused-ion-beam implantation. J Appl Phys 72:422-428
44. Poole PJ, Piva PG, Buchanan M, Aers GC, Roth AP, Dion M, Wasilewski ZR, Koteles ES, Charbonneau S, Beauvais J (1994) The enhancement of quantum well intermixing through repeated ion implantation. Semicond Sci Technol 9:2134-2137
45. Tan HH, Williams JS, Jagadish C, Burke PT, Gal M (1996) Large energy shifts in GaAs-AlGaAs quantum wells by proton irradiation-induced intermixing. Appl Phys Lett 68:2401-2403
46. Tan HH, Jagadish C (1997) Wavelength shifting in GaAs quantum well lasers by proton irradiation. Appl Phys Lett 71:2680-2682
47. Dao LV, Johnston MB, Gal M, Fu L, Tan HH, Jagadish C (1998) Improved carrier collection in intermixed InGaAs/GaAs quantum wells. Appl Phys Lett 73:3408-3410
48. Fu L, Tan HH, Johnston MB, Gal M, Jagadish C (1999) Proton irradiation-induced intermixing in InGaAs/(Al)GaAs quantum wells and quantum-well lasers. J Appl Phys 85:6786-6789
49. Piva PG, Charbonneau S, Mitchell IV, Goldberg RD (1996) Effect of implantation dose on photoluminescence decay times in intermixed GaAs/AlGaAs quantum wells. Appl Phys Lett 68:2252-2254
50. Johnston MB, Gal M, Li N, Chen Z, Liu X, Li N, Lu W, Shen SC, Fu L, Tan HH, Jagadish C (1999) Interdiffused quantum-well infared photodetectors for color sensitive arrays. Appl Phys Lett 75:923-925
51. Kahen KB, Rajeswaran G (1989) Study of the interdiffusion of GaAs-AlGaAs interfaces during rapid thermal annealing of ion-implanted structures. J Appl Phys 66:545-551

52. Leon R, Williams DRM, Krueger J, Weber ER, Melloch MR (1997) Diffusivity transients and radiative recombination in intermixed $In_{0.5}Ga_{0.5}As/GaAs$ quantum structures. Phys Rev B 56:R4336-R4339
53. Yousefi GH, Webb JB, Rousina R, Khanna SM (1995) Electron irradiation induced defects and Schottky diode characteristics for Metalorganic Vapor Phase Epitaxy and Molecular Beam Epitaxial n-GaAs. J Electron Materials 24:15-20
54. Zaidi MA, Maaref H, Zazoui M, Bourgoin JC (1993) Defects in electron-irradiated GaAlAs alloys. J Appl Phys 74:284-290
55. Chaabane H, Bourgoin JC (1994) Irradiation effect in electron transport through GaAlAs barriers. Appl Phys Lett 64:1006-1008
56. Lang DV, Hartman RL, Schumaker NE (1976) Capacitance spectroscopy studies of degraded $Al_xGa_{1-x}As$ DH stripe-geometry lasers. J Appl Phys 47:4986-4992
57. Lang DV, Logan RA, Kimerling LC (1977) Identification of the defect state associated with a gallium vacancy in GaAs and $Al_xGa_{1-x}As$. Phys Rev B 15:4874-4882
58. Papastamatiou M, Arpatzanis N, Papaioannou GJ, Papastergiou C, Christou A (1997) Neutron radiation effects in high electron mobility transistors. IEEE Trans Electron Devices 44:364-372
59. Irvine AC, Palmer DW (1994) Demonstration of gallium-defect annealing at 280 K in irradiated GaAs and $Al_xGa_{1-x}As$. Phys Rev B 49:5695-5698
60. Barnes CE, Zipperian TE, Dawson LR (1985) Neutron-induced trapping levels in Aluminum Gallium Arsenide. J Electron Mater 14:95-118
61. Muñoz E, García F, Jimenez B, Calleja E, Gomez A, Alcober V (1985) EL2-related defects in neutron irradiated $GaAs_{1-x}P_x$ alloys. Appl Phys Lett 47:798-800
62. Garcia F, Muñoz E, Calleja E, Alcober V (1986) Damage constant and deep-level transient spectroscopy in neutron irradiated GaAsP alloys. J Electron Mater 15:133-139
63. Walters RJ, Shaw GJ, Summers GP, Burke EA, Messenger SR (1992) Radiation effects in $Ga_{0.47}In_{0.53}As$ devices. IEEE Trans Nucl Sci 39:2257-2264
64. Kudou T, Ohyama H, Simoen E, Claeys C, Vanhellemont J, Sigaki K, Takami Y, Fujii A (1999) Effect of irradiation on InGaAs photo devices. J Radioanalyt Nucl Chem 239:361-364
65. Ohyama H, Kudou T, Simoen E, Claeys C, Takami Y, Sunaga H (1999) Radiation damage of $In_{0.53}Ga_{0.47}As$ photodiodes by high energy particles. J Mater Sci: Mat in Electron 10:403-405
66. Onoda S, Hirao T, Laird JS, Kudou T, Ohyama H, Mori H, Okamoto T, Itoh H (2000) Gamma and electron radiation degradation of $In_{0.53}Ga_{0.47}As$ p-i-n photodiodes. In: Proc RADECS 2000 Workshop, Louvain-la-Neuve, Belgium, pp 105-109
67. Ohyama H, Simoen E, Claeys C, Hakata T, Kudou T, Yoneoka M, Kobayashi K, Nakabayashi M, Takami Y, Sunaga H (1999) Radiation-induced lattice defects in InGaAsP laser diodes and their effects on device performance. Physica B 273-274:1031-1033
68. Zaidi MA, Maaref H, Zazoui M, Bourgoin JC (1994) Defects in electron irradiated GaP and GaInP. Mater Sci Forum 143-147:295-298
69. Khan A, Yamaguchi M, Bourgoin JC, de Angelis N, Takamoto T (2000) Room-temperature minority-carrier injection-enhanced recovery of radiation-induced defects in p-InGaP and solar cells. Appl Phys Lett 76:2559-2561
70. Khan A, Yamaguchi M, Bourgoin JC, Ando K, Takamoto T (2001) Recombination enhanced defect reactions in 1 MeV electron irradiated p InGaP. J Appl Phys 89:4263-4268

71. Barry AL, Houdayer AJ, Hinrichsen PF, Letourneau WG, Vincent J (1995) The energy dependence of lifetime damage constants in GaAs LEDs for 1 - 500 MeV protons. IEEE Trans Nucl Sci 42:2104-2107
72. Summers GP, Burke EA, Xapsos MA, Dale CJ, Marshall PW, Petersen EL (1988) Displacement damage in GaAs structures. IEEE Trans Nucl Sci 35:1221-1226
73. Reed RA, Marshall PW, Marshall CJ, Ladbury RL, Kim HS, Nguyen LX, Barth JL, LaBel KA (2000) Energy dependence of proton damage in AlGaAs light-emitting diodes. IEEE Trans Nucl Sci 47:2492-2499
74. Khanna SM, Estan D, Liu HC, Gao M, Buchanan M, SpringThorpe AJ (2000) 1-15 MeV proton and alpha particle radiation effects on GaAs quantum well light emitting diodes. IEEE Trans Nucl Sci 47:2508-2514
75. Barde S, Ecoffet R, Costeraste J, Meygret A, Hugon X (2000) Displacement damage effects in InGaAs detectors: Experimental results and semi-empirical model prediction. IEEE Trans Nucl Sci 47:2466-2472
76. Barry AL, Wojcik R, MacDiarmid AL (1989) Response of GaAs displacement damage monitors to protons, electrons and gamma irradiation. IEEE Trans Nucl Sci 36: 2400-2404
77. Moss SC, Halle LF, Marvin DC (1995) Effects of electron beam irradiation on transient photoluminescence measurements of GaAs and AlGaAs double heterostructures. IEEE Trans Nucl Sci 42:2058-2065
78. Johnston AH, Rax BG, Selva LE, Barnes CE (1999) Proton degradation of light-emitting diodes. IEEE Trans Nucl Sci 46:1781-1789
79. Johnston AH, Miyahara TF (2000) Characterization of proton damage in light-emitting diodes. IEEE Trans Nucl Sci 47:2500-2507
80. Johnston AH (2000) Proton displacement damage in light-emitting and laser diodes. In: Proc RADECS 2000 Workshop, Louvain-la-Neuve, Belgium, pp 75-81
81. Li X, Gu SQ, Reuter EE, Verdeyen JT, Bishop SG, Coleman JJ (1996) Effect of e-beam irradiation on a p-n junction GaN light emitting diode. J Appl Phys 80:2687-2690
82. Lee SC, Zhao YF, Schrimpf RD, Neifeld MA, Galloway KF (1999) Comparison of lifetime and threshold current damage factors for multi-quantum-well (MQW) GaAs/GaAlAs laser diodes irradiated at different proton energies. IEEE Trans Nucl Sci 46:1797-1803
83. Jolly A, Vicrey J (1994) Modelling threshold shift of power laser diodes under neutronic and photonic irradiation. In: Proc RADECS '93, The IEEE, New York, pp 232-238
84. Zhao YF, Schrimpf RD, Patwary AR, Neifeld MA, Al-Johani AW, Weller RA, Galloway KF (1998) Annealing effects on multi-quantum well laser diodes after proton irradiation. IEEE Trans Nucl Sci 45:2826-2832
85. Kimerling LC (1978) Recombination enhanced defect reactions. Solid-State Electron 21:1391-1401
86. Lischka H, Henschel H, Lennartz W, Schmidt HU (1992) Radiation sensitivity of light emitting diodes (LED), Lased Diodes (LD) and Photodiodes (PD). In: Proc RADECS 1991, The IEEE, New York, pp 423-427
87. Lischka H, Henschel H, Köhn O, Lennartz W, Schmidt HU (1994) Radiation effects in light emitting diodes, laser diodes, photodiodes and optocouplers. In: Proc RADECS 1993, The IEEE, New York, pp 226-231
88. Shaw GJ, Messenger SR, Walters RJ, Summers GP (1993) Radiation-induced reverse dark currents in $In_{0.53}Ga_{0.47}As$ photodiodes. J Appl Phys 73:7244-7249

89. Shaw GJ, Walters RJ, Messenger SR, Summers GP (1993) Time dependence of radiation-induced generation currents in irradiated InGaAs photodiodes. J Appl Phys 74:1629-1635
90. Ohyama H, Vanhellemont J, Takami Y, Hayama K, Kudou T, Kohiki S, Sunaga H, Hakata T (1996) Degradation of InGaAs pin photodiodes by neutron irradiation. Semicond Sci Technol 11:1461-1463
91. Ohyama H, Simoen E, Claeys C, Takami Y, Kudou T, Sunaga H (1998) Radiation source dependence of degradation and recovery of irradiated $In_{0.53}Ga_{0.47}As$ PIN photodiodes. In: Barbottin G, Dressendorfer P (eds) Proc RADECS '97, The IEEE, New York, pp 108-113
92. Kudou T, Ohyama H, Simoen E, Claeys C, Takami Y, Shigaki K, Fujii A, Sunaga H (1998) Radiation damage of InGaAs photodiodes by high energy particles. Mat Res Soc Proc 487:471-476
93. Liu HC, Wilson PH, Buchanan M, Khanna SM (1996) Nuclear radiation effects on GaAs/AlGaAs quantum well infrared photodetectors. In: Proc SPIE 2746:134-141
94. Li L, Liu HC, Wilson PH, Buchanan M, Khanna SM (1997) Influence of high energy particle radiation on GaAs/AlGaAs quantum well infrared photodetectors. Semicond Sci Technol 12:947-952
95. Hogsed MR, Paxton AH, Sanchez A, Marquez R, Joshi AM, Walsh DS, Schöne H (2000) Radiation-induced transient effects in InGaAs photodiodes. In: Proc RADECS 2000 Workshop, Louvain-la-Neuve, Belgium, pp 88-92
96. Söderqvist J, Eek LO, Collot J, Coulon JP, Dinkespiler B, Hostachy JY, Jevaud M, Lund-Jensen B, Merkel B, Monnier E, Olivetto C, de Saintignon P, Tisserant S (1997) Radiation hardness evaluation of an analogue optical link for operation at cryogenic temperatures. IEEE Trans Nucl Sci 44:861-865

Chapter 9

1. Groeseneken G, Maes HE, Van Houdt J, Witters S (1997) Basic of non-volatile semiconductor memory devices. In: Brown WB, Brewer JE (eds) Non-volatile semiconductor memory technology, a comprehensive guide to understanding and using NVM devices, The IEEE Press, pp 1-88
2. Baldi L, Maurelli A (1999) Embedded non volatile memories in deep-submicron CMOS. In: Maes HE, Mertens RP, Declerck G, Grünbacher H (eds), Proc ESSDERC 99, Editions Frontières, Neuilly France, pp 127-134
3. Ishiwara H (2001) Current status and prospects of FET-type ferroelectric memories. J Semicond Techn Sci 1:1-14
4. Nguygen DN, Guertin SM, Swift GM, Johnston AH (1999) Radiation effects in flash memories. IEEE Trans Nucl Sci 46:1744-1756
5. De Santi G, Clementi C, Rebora A (1999) Process integration evolution in manufacturing techniques for flash memories. In: Claeys C, Iwai H, Bronner G, Fair R (eds) Proc ULSI Process Integration, The Electrochem Soc, Pennington PV 99-18:93-106
6. Kynett VN, Baker A, Fandrich ML, Hoekstra GP, Jungroth O, Kreifels JA, Wells S, Winston MD (1988) An in-system reprogrammable 32kx8 CMOS flash memory. IEEE J Solid State Circ 23:1157-1162

7. Schwartz HR, Nichols DK, Johnston AH (1997) Single-event upset in flash memories. IEEE Nucl Sci 44:2315-2324
8. Claeys C, Ohyama H, Simoen E, Nakabayashi N, Kobayashi K (2002) Radiation damage in flash memory cells. Nucl Instr Method B (in press)
9. Yang B, Lee SS, Kang YM, Noh KH, Hong SK, Oh SH, Kang EY, Lee SW, Kim JG, Shu CW, Seong JY, Lee CG, Kang NS, Park YJ (2001) Integrated process and reliabolity for $SrBi_2Ta_2O_9$ based ferroelectric memories. J Semicond Techn Sci 1:141-157
10. Benedetto JM, Moore RA, McLean FB, Brody PS, Dey SK (1990) The effect of ionising radiation on sol-gel ferroelectric PZT capacitors. IEEE Trans Nucl Sci 37:1713-1717
11. Schwank JR, Nasby RF, Miller SL, Rodgers MS, Dressendorfer PV (1990). IEEE Trans Nucl Sci 37:1703-1712
12. Moore RA, Benedetto JM, Rod BJ (1993) Total dose effects on ferroelectric PZT capacitors used as non-volatile storage element. IEEE Trans Nucl Sci 40:1591-1596
13. Moore RA, Benedetto JM (1995) Ionizing radiation-induced asymmetries of the retention characteristics of ferroelectric thin films. IEEE Trans Nucl Sci 42:1575-1584
14. Cheng Z, Zhang L, Yao X (1992) Effect of space charge on micro-macro domain transition of PZT. IEEE Trans Electr Insul 27:773-776
15. Gao J, Zheng L, Huang B, Song Z, Yang L, Fan Y, Zhu S, Line C (1999) Total dose radiation effects on Pt/PZT/Pt ferroelectric capacitors fabricated by LP method. Semicon Sci Technol 14:836-839
16. Bernacki S, Hunt K, Tyson S, Hudgens S, Pashmakov B, Czubatyj W (2000) Total dose radiation response of high temperature imprint characteristics of chalcogenide based RAM resistor elements. IEEE Trans Nucl Sci 47: 2528-2533
17. International Technology Roadmap for Semiconductors, 2001 Edition. www.itrs.net/2001_SIA_Roadmap/Home.htm
18. Degraeve R, Kaczer B, Houssa M, Groeseneken G, Heyns M, Jeons JS, Halliyal A (1999) Analysis of high voltage TDDB measurements on Ta_2O_5/SiO_2 stack. In: IEDM Tech Digest. The IEEE, New York, pp 327-330
19. Lee BH, Kang L, Qi WJ, Nieh R, Jeon Y, Onishi K, Lee JC (1999) Ultrathin hafnium oxide with low leakage and excellent reliability for alternative gate dielectric applications. In: IEDM Tech Digest. The IEEE, New York, pp 133-136
20. Guo X, Wang X, Luo Z, Ma TP, Tamagawa T (1999) High quality ultra-thin (1.5 nm) TiO_2/Si_3N_4 gate dielectric for deep sub-micron CMOS technology. In: IEDM Tech Digest. The IEEE, New York, pp 137-140
21. Choi BK, Fleetwood DM, Massengill LW, Schrimpf RD, Galloway KF, Shaneyfelt MR, Meisenheimer TL, Dodd PE, Schwank JR, Lee YM, Johnson RS, Lucovsky G (2002) Reliability degradation of ultra-thin oxynitride and Al_2O_3 gate dielectric films owing to heavy-ion irradiation. Electron Lett 38:157-158
22. Alam MA, Bude J, Weir B, Silverman P, Ghetti A, Monroe D, Cheung KP, Moccio S (1999) An anode hole injection percolation model for oxide breakdown-The doom's day scenario revisited. In: IEDM Tech Digest. The IEEE, New York, pp 715-718
23. Massengill LW, Choi BK, Fleetwood DM, Schrimpf, RD, Galloway KF, Shaneyfelt MR, Meisenheimer TL, Dodd PE, Schwank JR, Lee YM, Johnson RS, Lucovsky G (2001) Heavy-ion-induced breakdown in ultra-thin gate oxides and high-k dielectrics. IEEE Trans Nucl Sci 48:1904-1912
24. Sexton FW, Fleetwood DM, Shaneyfelt MR, Dodd PE, Hash GL (1997) Single event gate rupture in thin oxides. IEEE Trans Nucl Sci 44:2345-2352

25. McGarrity JM, McLean FB, DeLancey WM, Palmour J, Carter C, Edmond J, Oakley RE (1992) Silicon carbide JFET radiation response. IEEE Trans Nucl Sci 39:1974-1981
26. Babcock R (1965) Radiation damage in SiC. IEEE Trans Nucl Sci 12:43-47
27. MorkoÇ H, Strite S, Gao GB, Lin ME, Sverdlov B, Burns M (1994) Large-band-gap SiC, III-V nitride, and II-VI ZnSe-based semiconductor device technologies. J Appl Phys 76:1363-1398
28. McD Hobgood H, Glass RC, Augustine G, Hopkins RH, Jenny J, Skowronski M, Mitchel WC, Roth M (1995) Semi-insulating 6H-SiC grown by physical vapor transport. Appl Phys Lett 66:1364-1366
29. Ikeda M, Matsunami H, Tanaka T (1980) Site effect on the impurity levels in 4H, 6H, and 15R SiC. Phys Rev B 22:2842-2854
30. Pensl G, Choyke WJ (1993) Electrical and optical characterization of SiC. Physica B 185:264-283
31. Troffer T, Schadt M, Frank T, Itoh H, Pensl G, Heindl J, Strunk HP, Maier M (1997) Doping of SiC by implantation of boron and aluminum. Phys Stat Sol (a) 162:277-298
32. Barry AL, Lehmann B, Fritsch D, Bräunig D (1991) Energy dependence of electron damage and displacement threshold energy in 6H silicon carbide. IEEE Trans Nucl Sci 38:1111-1115
33. Rempel AA, Schaefer H-E (1995) Irradiation-induced atomic defects in SiC studied by positron annihilation. Appl Phys A 61:51-53
34. Janzén E, Henry A, Bergman JP, Ellison A, Magnusson B (2001) Material characterization need for SiC-based devices. Materials Sci in Semicond Process 4:181-186
35. Itoh H, Kawasuso A, Ohshima T, Yoshikawa M, Nashiyama I, Tanigawa S, Misawa S, Okumura H, Yoshida S (1997) Intrinsic defects in cubic silicon carbide. Phys Stat Sol (a) 162:173-198
36. Choyke WJ, Patrick L (1971) Photoluminescence of radiation defects in cubic SiC: Localized modes and Jahn-Teller effect. Phys Rev B 4:1843-1847
37. Patrick L, Choyke WJ (1972) Photoluminescence of radiation defects in ion-implanted 6H SiC. Phys Rev B 5:3253-3259
38. Freitas Jr JA, Bishop SG, Edmond JA, Ryu J, Davis RF (1987) Photoluminescence spectroscopy of ion-implanted 3C-SiC grown by chemical vapor deposition. J Appl Phys 61:2011-2016
39. de S Balona LA, Loubser JHN (1970) Esr in irradiated silicon carbide. J Phys C: Solid St Phys 3:2344-2351
40. Itoh H, Yoshikawa M, Nashiyama I, Misawa S, Okumura H, Yoshida S (1990) Radiation induced defects in CVD-grown 3C-SiC. IEEE Trans Nucl Sci 37:1732-1738
41. Itoh H, Yoshikawa M, Nashiyama I, Misawa S, Okumura H, Yoshida S (1992) Electron Spin Resonance study of defects in CVD-grown 3C-SiC irradiated with 2 MeV protons. J Electron Materials 21:707-710
42. Il'in VA, Ballandovich VA (1993) EPR and DLTS of point defects in silicon carbide crystals. Diff and Defect Data 103-105:633-644
43. Dannefaer S, Craigen D, Kerr D (1995) Carbon and silicon vacancies in electron-irradiated 6H-SiC. Phys Rev B 51:1928-1930
44. Brauer G, Anwand W, Coleman PG, Knights AP, Plazaola F, Pacaud Y, Skorupa W, Störmer J, Willutzki P (1996) Positron studies of defects in ion-implanted SiC. Phys Rev B 54:3084-3092

45. Itoh H, Yoshikawa M, Nashiyama I, Okumura H, Misawa S, Yoshida S (1995) Photoluminescence of radiation induced defects in 3C-SiC epitaxially grown on Si. J Appl Phys 77:837-842
46. Talwar DN, Feng ZC (1991) Tight-binding description for the bound electronic states of isolated single and paired native defects in β-SiC. Phys Rev B 44:3191-3198
47. Nagesh V, Farmer JW, Davis RF, Kong HS (1990) Defects in cubic SiC on Si. Rad Eff and Defects in Solids 112:77-84
48. Nagesh V, Farmer JW, Davis RF, Kong HS (1987) Defects in neutron irradiated SiC. Appl Phys Lett 50:1138-1140
49. Hemmingsson C, Son NT, Kordina O, Janzén E, Lindström JL (1998) Capture cross sections of electron irradiation induced defects in 6H-SiC. J Appl Phys 84:704-708
50. Gong M, Fung S, Beling CD, You Z (1999) A deep level transient spectroscopy study of electron irradiation induced deep levels in p-type 6H-SiC. J Appl Phys 85:7120-7122
51. Gong M, Fung S, Beling CD, You Z (1999) Electron-irradiation-induced deep levels in n-type 6H-SiC. J Appl Phys 85:7604-7608
52. Doyle JP, Aboelfotoh MO, Linnarsson MK, Svensson BG, Schöner A, Nordell N, Harris C, Lindström JL, Janzén E, Hemmingsson C (1996) Mat Res Soc Proc Vol 423:519-524
53. Hemmingsson C, Son NT, Kordina O, Bergman JP, Janzén E, Lindström JL, Savage S, Nordell N (1997) Deep level defects in electron-irradiated 4H SiC epitaxial layers. J Appl Phys 81:6155-6159
54. Doyle JP, Linnarsson MK, Pellegrino P, Keskitalo N, Svensson BG, Schöner A, Nordell N, Lindström JL (1998) Electrically active point defects in n-type 4H-SiC. J Appl Phys 84:1354-1357
55. Dalibor T, Pensl G, Matsunami H, Kimoto T, Choyke WJ, Schöner A, Nordell N (1997) Deep defect centers in silicon carbide monitored with deep level transient spectroscopy. Phys Stat Sol (a) 162:199-225
56. Lebedev AA, Veinger AI, Davydov DV, Kozlovski VV, Savkina NS, Strel'chuk AM (2000) Doping of n-type 6H-SiC and 4H-SiC with defects created with a proton beam. J Appl Phys 88:6265-6271
57. Kawasuso A, Itoh H, Okada S, Okumura H (1996) Annealing processes of vacancy-type defects in electron-irradiated and as-grown 6H-SiC studied by positron lifetime spectroscopy. J Appl Phys 80:5639-5645
58. Storasta L, Carlsson FHC, Sridhara SG, Bergman JP, Henry A, Egilsson T, Hallén A, Janzén E (2001) Pseudodonor nature of the D_1 defect in 4H-SiC. Appl Phys Lett 78:46-48
59. Kawasuso A, Redmann F, Krause-Rehberg R, Frank T, Weidner M, Pensl G, Sperr P, Itoh H (2001) Vacancies and deep levels in electron-irradiated 6H SiC epilayers studied by positron annihilation and deep level transient spectroscopy. J Appl Phys 90:3377-3382
60. Kawasuso A, Redmann F, Krause-Rehberg R, Weidner M, Frank T, Pensl G, Sperr P, Triftshäuser W, Itoh H (2001) Annealing behavior of vacancies and $Z_{1/2}$ levels in electron-irradiated 4H-SiC studied by positron annihilation and deep-level transient spectroscopy. Appl Phys Lett 79:3950-3952
61. Kawasuso A, Itoh H, Ohshima T, Abe K, Okada S (1997) Vacancy production by 3 MeV electron irradiation in 6H-SiC studied by positron lifetime spectroscopy. J Appl Phys 82:3232-3238

62. Ballandovich VS (1999) Deep-level transient spectroscopy of radiation-induced levels in 6H-SiC. Semiconductors 33:1188-1192
63. Gong M, Reddy CV, Beling CD, Fung S, Brauer G, Wirth H, Skorupa W (1998) Deep level traps in the extended tail region of boron-implanted n-type 6H-SiC. Appl Phys Lett 72:2739-2741
64. Fung S, Gong M, Beling CD, Brauer G, Wirth H, Skorupa W (1998) Aluminum-implantation-induced deep levels in n-type 6H-SiC. J Appl Phys 84:1152-1154
65. Raynaud C, Ghaffour K, Ortolland S, Locatelli M-L, Souifi K, Guillot G, Chante J-P (1998) Electrical characterization of silicon carbide n^+pp^+ diodes with an N-implanted n^+ emitter. J Appl Phys 84:3073-3077
66. Nadella RK, Capano MA (1997) High-resistance layers in n-type 4H-silicon carbide by hydrogen ion implantation. Appl Phys Lett 70:886-888
67. Kimoto T, Nakajima T, Matsunami H, Nakata T, Inoue M (1996) Formation of semi-insulating 6H-SiC layers by vanadium ion implantations. Appl Phys Lett 69:1113-1115
68. McLean FB, McGarrity JM, Scozzie CJ, Tipton CW, DeLancey WM (1994) Analysis of neutron damage in high-temperature silicon carbide JFETs. IEEE Trans Nucl Sci 41:1884-1894
69. Seshadri S, Dulloo AR, Ruddy FH, Seidel JG, Rowland LB (1999) Demonstration of an SiC neutron detector for high-radiation environments. IEEE Trans Nucl Sci 46:567-571
70. Rybicki GC (1995) Deep level defects in alpha particle irradiated 6H silicon carbide. J Appl Phys 78:2996-3000
71. Ouisse T, Bécourt N, Jaussaud C, Templier F (1994) Low-frequency, high-temperature conductance and capacitance measurements on metal-oxide-silicon carbide capacitors. J Appl Phys 75:604-607
72. Brown DM, Ghezzo M, Kretchmer J, Downey E, Pimbley J, Palmour J (1994) SiC MOS interface characteristics. IEEE Trans Electron Devices 41:618-620
73. Di Ventra M (2001) Can we make the SiC-SiO_2 interface as good as the Si-SiO_2 interface? Appl Phys Lett 79:2402-2404
74. Yoshikawa M, Itoh H, Morita Y, Nashiyama I, Misawa S, Okumura H, Yoshida S (1991) Effects of gamma-ray irradiation on cubic silicon carbide metal-oxide-semiconductor structure. J Appl Phys 70:1309-1312
75. Bano E, Ouisse T, Di Cioccio L, Karmann S (1994) Surface potential fluctuations in metal-oxide-semiconductor capacitors fabricated on different silicon carbide polytypes. Appl Phys Lett 65:2723-2724
76. Ohshima T, Yoshikawa M, Itoh H, Aoki Y, Nashiyama I (1998) Generation of interface traps and oxide-trapped charge in 6H-SiC metal-oxide-semiconductor transistors by gamma-ray irradiation. Jpn J Appl Phys 37:L1002-L1004
77. Ohshima T, Itoh H, Yoshikawa M (2001) Effect of gamma-ray irradiation on the characteristics of 6H silicon carbide metal-oxide-semiconductor field effect transistor with hydrogen-annealed gate oxide. J Appl Phys 90:3038-3041
78. Yoshikawa M, Saitoh K, Ohshima T, Itoh H, Nashiyama I, Yoshida S, Okumura H, Takahashi Y, Ohnishi K (1996) Depth profile of trapped charges in oxide layer of 6H-SiC metal-oxide-semiconductor structures. J Appl Phys 80:282-287
79. Yoshikawa M, Saitoh K, Ohshima T, Itoh H, Nashiyama I, Takahashi Y, Ohnishi K, Okumura H, Yoshida S (1998) Generation mechanisms of trapped charges in oxide layers of 6H-SiC MOS structures irradiated with gamma-rays. Mat Sci Forum 264-268:1017-1020

80. Brisset C, Noblanc O, Picard C, Joffre F, Brylinski C (2000) 4H-SiC MESFETs behavior after high dose irradiation. IEEE Trans Nucl Sci 47:598-603

Springer Series in

MATERIALS SCIENCE

Editors: R. Hull R. M. Osgood, Jr. J. Parisi

1 **Chemical Processing with Lasers***
By D. Bäuerle

2 **Laser-Beam Interactions with Materials**
Physical Principles and Applications
By M. von Allmen and A. Blatter
2nd Edition

3 **Laser Processing of Thin Films and Microstructures**
Oxidation, Deposition and Etching of Insulators
By. I. W. Boyd

4 **Microclusters**
Editors: S. Sugano, Y. Nishina, and S. Ohnishi

5 **Graphite Fibers and Filaments**
By M. S. Dresselhaus, G. Dresselhaus, K. Sugihara, I. L. Spain, and H. A. Goldberg

6 **Elemental and Molecular Clusters**
Editors: G. Benedek, T. P. Martin, and G. Pacchioni

7 **Molecular Beam Epitaxy**
Fundamentals and Current Status
By M. A. Herman and H. Sitter 2nd Edition

8 **Physical Chemistry of, in and on Silicon**
By G. F. Cerofolini and L. Meda

9 **Tritium and Helium-3 in Metals**
By R. Lässer

10 **Computer Simulation of Ion-Solid Interactions**
By W. Eckstein

11 **Mechanisms of High Temperature Superconductivity**
Editors: H. Kamimura and A. Oshiyama

12 **Dislocation Dynamics and Plasticity**
By T. Suzuki, S. Takeuchi, and H. Yoshinaga

13 **Semiconductor Silicon**
Materials Science and Technology
Editors: G. Harbeke and M. J. Schulz

14 **Graphite Intercalation Compounds I**
Structure and Dynamics
Editors: H. Zabel and S. A. Solin

15 **Crystal Chemistry of High-T_c Superconducting Copper Oxides**
By B. Raveau, C. Michel, M. Hervieu, and D. Groult

16 **Hydrogen in Semiconductors**
By S. J. Pearton, M. Stavola, and J. W. Corbett

17 **Ordering at Surfaces and Interfaces**
Editors: A. Yoshimori, T. Shinjo, and H. Watanabe

18 **Graphite Intercalation Compounds II**
Editors: S. A. Solin and H. Zabel

19 **Laser-Assisted Microtechnology**
By S. M. Metev and V. P. Veiko
2nd Edition

20 **Microcluster Physics**
By S. Sugano and H. Koizumi
2nd Edition

21 **The Metal-Hydrogen System**
By Y. Fukai

22 **Ion Implantation in Diamond, Graphite and Related Materials**
By M. S. Dresselhaus and R. Kalish

23 **The Real Structure of High-T_c Superconductors**
Editor: V. Sh. Shekhtman

24 **Metal Impurities in Silicon-Device Fabrication**
By K. Graff 2nd Edition

25 **Optical Properties of Metal Clusters**
By U. Kreibig and M. Vollmer

26 **Gas Source Molecular Beam Epitaxy**
Growth and Properties of Phosphorus Containing III–V Heterostructures
By M. B. Panish and H. Temkin

27 **Physics of New Materials**
Editor: F. E. Fujita 2nd Edition

28 **Laser Ablation**
Principles and Applications
Editor: J. C. Miller

Springer Series in

MATERIALS SCIENCE

Editors: R. Hull R. M. Osgood, Jr. J. Parisi

29 **Elements of Rapid Solidification**
Fundamentals and Applications
Editor: M. A. Otooni

30 **Process Technology for Semiconductor Lasers**
Crystal Growth and Microprocesses
By K. Iga and S. Kinoshita

31 **Nanostructures and Quantum Effects**
By H. Sakaki and H. Noge

32 **Nitride Semiconductors and Devices**
By H. Morkoç

33 **Supercarbon**
Synthesis, Properties and Applications
Editors: S. Yoshimura and R. P. H. Chang

34 **Computational Materials Design**
Editor: T. Saito

35 **Macromolecular Science and Engineering**
New Aspects
Editor: Y. Tanabe

36 **Ceramics**
Mechanical Properties, Failure Behaviour, Materials Selection
By D. Munz and T. Fett

37 **Technology and Applications of Amorphous Silicon**
Editor: R. A. Street

38 **Fullerene Polymers and Fullerene Polymer Composites**
Editors: P. C. Eklund and A. M. Rao

39 **Semiconducting Silicides**
Editor: V. E. Borisenko

40 **Reference Materials in Analytical Chemistry**
A Guide for Selection and Use
Editor: A. Zschunke

41 **Organic Electronic Materials**
Conjugated Polymers and Low Molecular Weight Organic Solids
Editors: R. Farchioni and G. Grosso

42 **Raman Scattering in Materials Science**
Editors: W. H. Weber and R. Merlin

43 **The Atomistic Nature of Crystal Growth**
By B. Mutaftschiev

44 **Thermodynamic Basis of Crystal Growth**
P–T–X Phase Equilibrium and Non-Stoichiometry
By J. Greenberg

45 **Thermoelectrics**
Basic Principles and New Materials Developments
By G. S. Nolas, J. Sharp, and H. J. Goldsmid

46 **Fundamental Aspects of Silicon Oxidation**
Editor: Y. J. Chabal

47 **Disorder and Order in Strongly Nonstoichiometric Compounds**
Transition Metal Carbides, Nitrides and Oxides
By A. I. Gusev, A. A. Rempel, and A. J. Magerl

48 **The Glass Transition**
Relaxation Dynamics in Liquids and Disordered Materials
By E. Donth

49 **Alkali Halides**
A Handbook of Physical Properties
By D. B. Sirdeshmukh, L. Sirdeshmukh, and K. G. Subhadra

50 **High-Resolution Imaging and Spectrometry of Materials**
Editors: F. Ernst and M. Rühle

Printing (Computer to Film): Saladruck Berlin
Binding: Stürtz AG, Würzburg